03
CON

Applied Bayesian Modelling

WILEY SERIES IN PROBABILITY AND STATISTICS

Established by WALTER A. SHEWHART and SAMUEL S. WILKS

A complete list of the titles in this series appears at the end of this volume.

Applied Bayesian Modelling

PETER CONGDON
Queen Mary, University of London, UK

WILEY

Telephone (+44) 1243 779777
Email (for orders and customer service enquiries): cs-books@wiley.co.uk

Visit our Home Page on www.wileyeurope.com or www.wiley.com

Other Wiley Editorial Offices

John Wiley & Sons Inc., 111 River Street,
Hoboken, NJ 07030, USA

Jossey-Bass, 989 Market Street,
San Francisco, CA 94103-1741, USA

Wiley-VCH Verlag GmbH, Boschstr.
12, D-69469 Weinheim, Germany

John Wiley & Sons Australia Ltd, 33 Park Road, Milton,
Queensland 4064, Australia

John Wiley & Sons (Asia) Pte Ltd, 2 Clementi Loop #02-01,
Jin Xing Distripark, Singapore 129809

John Wiley & Sons Canada Ltd, 22 Worcester Road,
Etobicoke, Ontario, Canada M9W 1L1

Wiley also publishes its books in a variety of electronic formats. Some content that appears in print may not be available in electronic books.

Library of Congress Cataloging-in-Publication Data

Congdon, Peter.
Applied Bayesian modelling / Peter Congdon.
p. cm. – (Wiley series in probability and statistics)
Includes bibliographical references and index.
ISBN 0-471-48695-7 (cloth : alk. paper)
1. Bayesian statistical decision theory. 2. Mathematical statistics. I. Title. II. Series.

QA279.5 .C649 2003
519.542–dc21 2002035732

British Library Cataloguing in Publication Data
A catalogue record for this book is available from the British Library

ISBN 0 471 48695 7

Typeset in 10/12 pt Times by Kolam Information Services, Pvt. Ltd., Pondicherry, India
Printed and bound in Great Britain by Biddles Ltd, Guildford, Surrey.

This book is printed on acid-free paper responsibly manufactured from sustainable forestry in which at least two trees are planted for each one used for paper production.

Contents

Preface

This book follows *Bayesian Statistical Modelling* (Wiley, 2001) in seeking to make the Bayesian approach to data analysis and modelling accessible to a wide range of researchers, students and others involved in applied statistical analysis. Bayesian statistical analysis as implemented by sampling based estimation methods has facilitated the analysis of complex multi-faceted problems which are often difficult to tackle using 'classical' likelihood based methods.

The preferred tool in this book, as in *Bayesian Statistical Modelling*, is the package WINBUGS; this package enables a simplified and flexible approach to modelling in which specification of the full conditional densities is not necessary and so small changes in program code can achieve a wide variation in modelling options (so, *inter alia*, facilitating sensitivity analysis to likelihood and prior assumptions). As Meyer and Yu in the *Econometrics Journal* (2000, pp. 198–215) state, "any modifications of a model including changes of priors and sampling error distributions are readily realised with only minor changes of the code." Other sophisticated Bayesian software for MCMC modelling has been developed in packages such as S-Plus, Minitab and Matlab, but is likely to require major reprogramming to reflect changes in model assumptions; so my own preference remains WINBUGS, despite its possible slower performance and convergence than tailored made programs.

There is greater emphasis in the current book on detailed modelling questions such as model checking and model choice, and the specification of the defining components (in terms of priors and likelihoods) of model variants. While much analytical thought has been put into how to choose between two models, say M_1 and M_2, the process underlying the specification of the components of each model is subject, especially in more complex problems, to a range of choices. Despite an intention to highlight these questions of model specification and discrimination, there remains considerable scope for the reader to assess sensitivity to alternative priors, and other model components. My intention is not to provide fully self-contained analyses with no issues still to resolve.

The reader will notice many of the usual 'specimen' data sets (the Scottish lip cancer and the ship damage data come to mind), as well as some more unfamiliar and larger data sets. Despite recent advantages in computing power and speed which allow estimation via repeated sampling to become a serious option, a full MCMC analysis of a large data set, with parallel chains to ensure sample space coverage and enable convergence to be monitored, is still a time-consuming affair.

Some fairly standard divisions between topics (e.g. time series vs panel data analysis) have been followed, but there is also an interdisciplinary emphasis which means that structural equation techniques (traditionally the domain of psychometrics and educational statistics) receive a chapter, as do the techniques of epidemiology. I seek to review the main modelling questions and cover recent developments without necessarily going into the full range of questions in specifying conditional densities or MCMC sampling options (one of the benefits of WINBUGS means that this is a possible strategy). I recognise the ambitiousness of such a broad treatment, which the more cautious might not attempt. I am pleased to receive comments (nice and possibly not so nice) on the success of this venture, as well as any detailed questions about programs or results via e-mail at p.congdon@qmul.ac.uk. The WINBUGS programs that support the examples in the book are made available at ftp://ftp.wiley.co.uk/pub/books/congdon.

Peter Congdon

CHAPTER 1

The Basis for, and Advantages of, Bayesian Model Estimation via Repeated Sampling

1.1 INTRODUCTION

Bayesian analysis of data in the health, social and physical sciences has been greatly facilitated in the last decade by advances in computing power and improved scope for estimation via iterative sampling methods. Yet the Bayesian perspective, which stresses the accumulation of knowledge about parameters in a synthesis of prior knowledge with the data at hand, has a longer history. Bayesian methods in econometrics, including applications to linear regression, serial correlation in time series, and simultaneous equations, have been developed since the 1960s with the seminal work of Box and Tiao (1973) and Zellner (1971). Early Bayesian applications in physics are exemplified by the work of Jaynes (e.g. Jaynes, 1976) and are discussed, along with recent applications, by D'Agostini (1999). Rao (1975) in the context of smoothing exchangeable parameters and Berry (1980) in relation to clinical trials exemplify Bayes reasoning in biostatistics and biometrics, and it is here that many recent advances have occurred.

Among the benefits of the Bayesian approach and of recent sampling methods of Bayesian estimation (Gelfand and Smith, 1990) are a more natural interpretation of parameter intervals, whether called credible or confidence intervals, and the ease with which the true parameter density (possibly skew or even multi-modal) may be obtained. By contrast, maximum likelihood estimates rely on Normality approximations based on large sample asymptotics. The flexibility of Bayesian sampling estimation extends to derived or 'structural' parameters[1] combining model parameters and possibly data, and with substantive meaning in application areas (Jackman, 2000), which under classical methods might require the delta technique.

New estimation methods also assist in the application of Bayesian random effects models for pooling strength across sets of related units; these have played a major role in applications such as analysing spatial disease patterns, small domain estimation for survey outcomes (Ghosh and Rao, 1994), and meta-analysis across several studies (Smith *et al.*, 1995). Unlike classical techniques, the Bayesian method allows model comparison across non-nested alternatives, and again the recent sampling estimation

[1] See, for instance, Example 2.8 on geriatric patient length of stay.

Applied Bayesian Modelling P. Congdon
 ISBN: 0-471-48695-7

developments have facilitated new methods of model choice (e.g. Gelfand and Ghosh, 1998; Chib, 1995). The MCMC methodology may be used to augment the data and this provides an analogue to the classical EM method – examples of such data augmentation are latent continuous data underlying binary outcomes (Albert and Chib, 1993) and the multinomial group membership indicators (equalling 1 if subject i belongs to group j) that underlie parametric mixtures. In fact, a sampling-based analysis may be made easier by introducing this extra data – an example is the item analysis model involving 'guessing parameters' (Sahu, 2001).

1.1.1 Priors for parameters

In classical inference the sample data y are taken as random while population parameters θ, of dimension p, are taken as fixed. In Bayesian analysis, parameters themselves follow a probability distribution, knowledge about which (before considering the data at hand) is summarised in a prior distribution $\pi(\theta)$. In many situations, it might be beneficial to include in this prior density the available cumulative evidence about a parameter from previous scientific studies (e.g. an odds ratio relating the effect of smoking over five cigarettes daily through pregnancy on infant birthweight below 2500 g). This might be obtained by a formal or informal meta-analysis of existing studies. A range of other methods exist to determine or elicit subjective priors (Berger, 1985, Chapter 3; O'Hagan, 1994, Chapter 6). For example, the histogram method divides the range of θ into a set of intervals (or 'bins') and uses the subjective probability of θ lying in each interval; from this set of probabilities, $\pi(\theta)$ may then be represented as a discrete prior or converted to a smooth density. Another technique uses prior estimates of moments, for instance in a Normal $N(m, V)$ density[2] with prior estimates m and V of the mean and variance.

Often, a prior amounts to a form of modelling assumption or hypothesis about the nature of parameters, for example, in random effects models. Thus, small area death rate models may include spatially correlated random effects, exchangeable random effects with no spatial pattern, or both. A prior specifying the errors as spatially correlated is likely to be a working model assumption, rather than a true cumulation of knowledge.

In many situations, existing knowledge may be difficult to summarise or elicit in the form of an 'informative prior' and to reflect such essentially prior ignorance, resort is made to non-informative priors. Examples are flat priors (e.g. that a parameter is uniformly distributed between $-\infty$ and $+\infty$) and Jeffreys prior

$$\pi(\theta) \propto \det\{I(\theta)\}^{0.5}$$

where $I(\theta)$ is the expected information[3] matrix. It is possible that a prior is improper (doesn't integrate to 1 over its range). Such priors may add to identifiability problems (Gelfand and Sahu, 1999), and so many studies prefer to adopt minimally informative priors which are 'just proper'. This strategy is considered below in terms of possible prior densities to adopt for the variance or its inverse. An example for a parameter

[2] In fact, when θ is univariate over the entire real line then the Normal density is the maximum entropy prior according to Jaynes (1968); the Normal density has maximum entropy among the class of densities identified by a summary consisting of mean and variance.

[3] If $\ell(\theta) = \log(L(\theta))$ then $I(\theta) = -E\left\{\frac{\delta^2\ell(\theta)}{\delta\ell(\theta_i)\delta\ell(\theta_j)}\right\}$

distributed over all real values might be a Normal with mean zero and large variance. To adequately reflect prior ignorance while avoiding impropriety, Spiegelhalter *et al.* (1996) suggesting a prior standard deviation at least an order of magnitude greater than the posterior standard deviation.

1.1.2 Posterior density vs. likelihood

In classical approaches such as maximum likelihood, inference is based on the likelihood of the data alone. In Bayesian models, the likelihood of the observed data y given parameters θ, denoted $f(y|\theta)$ or equivalently $L(\theta|y)$, is used to modify the prior beliefs $\pi(\theta)$, with the updated knowledge summarised in a posterior density, $\pi(\theta|y)$. The relationship between these densities follows from standard probability equations. Thus

$$f(y, \theta) = f(y|\theta)\pi(\theta) = \pi(\theta|y)m(y)$$

and therefore the posterior density can be written

$$\pi(\theta|y) = f(y|\theta)\pi(\theta)/m(y)$$

The denominator $m(y)$ is known as the marginal likelihood of the data and found by integrating (or 'marginalising') the likelihood over the prior densities

$$m(y) = \int f(y|\theta)\pi(\theta)d\theta$$

This quantity plays a central role in some approaches to Bayesian model choice, but for the present purpose can be seen as a proportionality factor, so that

$$\pi(\theta|y) \propto f(y|\theta)\pi(\theta) \tag{1.1}$$

Thus, updated beliefs are a function of prior knowledge and the sample data evidence. From the Bayesian perspective the likelihood is viewed as a function of θ given fixed data y, and so elements in the likelihood that are not functions of θ become part of the proportionality in Equation (1.1).

1.1.3 Predictions

The principle of updating extends to future values or predictions of 'new data'. Before the study a prediction would be based on random draws from the prior density of parameters and is likely to have little precision. Part of the goal of the a new study is to use the data as a basis for making improved predictions 'out of sample'. Thus, in a meta-analysis of mortality odds ratios (for a new as against conventional therapy) it may be useful to assess the likely odds ratio z in a hypothetical future study on the basis of the observed study findings. Such a prediction is based is based on the likelihood of z averaged over the posterior density based on y:

$$f(z|y) = \int f(z|\theta)\pi(\theta|y)d\theta$$

where the likelihood of z, namely $f(z|\theta)$ usually takes the same form as adopted for the observations themselves.

One may also take predictive samples order to assess the model performance. A particular instance of this, useful in model assessment (see Chapters 2 and 3), is in cross-validation based on omitting a single case. Data for case i is observed, but a prediction of y_i is nevertheless made on the basis of the remaining data $y_{[i]} = \{y_1, y_2, .. y_{i-1}, y_{i+1}, .. y_n\}$. Thus in a regression example with covariates x_i, the prediction z_i would be made based on a model fitted to $y_{[i]}$; a typical example might be a time series model for $t = 1, .. n$, including covariates that are functions of time, where the model is fitted only up to $i = n - 1$ (the likelihood is defined only for $i = 1, .. n - 1$), and the prediction for $i = n$ is based on the updated time functions. The success of a model is then based on the match between the replicate and actual data.

One may also derive

$$f(y_i|y_{[i]}) = \int f(y_i|\theta)\pi(\theta|y_{[i]})d\theta$$

namely the probability of y_i given a model based on the data excluding it (Gelfand *et al.*, 1992). This is known as the Conditional Predictive Ordinate (CPO) and has a role in model diagnostics (see Section 1.5). For example, a set of count data (without covariates) could be modelled as Poisson (with case i excluded) leading to a mean $\theta_{[i]}$. The Poisson probability of case i could then be evaluated in terms of that parameter.

This type of approach (n-fold cross-validation) may be computationally expensive except in small samples. Another option is for a large dataset to be randomly divided into a small number k of groups; then cross-validation may be applied to each partition of the data, with $k - 1$ groups as 'training' sample and the remaining group as the validation sample (Alqalaff and Gustafson, 2001). For large datasets, one might take 50% of the data as the training sample and the remainder as the validation sample (i.e. $k = 2$).

One may also sample new or replicate data based on a model fitted to all observed cases. For instance, in a regression application with predictors x_i for case i, a prediction z_i would make use of the estimated regression parameters β and the predictors as they are incorporated in the regression means, for example $\mu_i = x_i\beta$ for a linear regression These predictions may be used in model choice criteria such as those of Gelfand and Ghosh (1998) and the expected predictive deviance of Carlin and Louis (1996).

1.1.4 Sampling parameters

To update knowledge about the parameters requires that one can sample from the posterior density. From the viewpoint of sampling from the density of a particular parameter θ_k, it follows from Equation (1.1) that aspects of the likelihood which are not functions of θ may be omitted. Thus, consider a binomial example with r successes from n trials, and with unknown parameter p representing the binomial probability, with a beta prior $B(a, b)$, where the beta density is

$$\frac{\Gamma(a + b)}{\Gamma(a)\Gamma(b)} p^{a-1}(1 - p)^{b-1}$$

The likelihood is then, viewed as a function of p, proportional to a beta density, namely

$$f(p) \propto p^r(1 - p)^{n-r}$$

and the posterior density for p is then a beta density with parameters $r + a$ and $n + b - r$:

$$p \sim B(r + a, n + b - r) \tag{1.2}$$

Therefore, the parameter's posterior density may be obtained by sampling from the relevant beta density, as discussed below. Incidentally, this example shows how the prior may in effect be seen to provide a prior sample, here of size $a + b - 2$, the size of which increases with the confidence attached to the prior belief. For instance, if $a = b = 2$, then the prior is equivalent to a prior sample of 1 success and 1 failure.

In Equation (1.2), a simple analytic result provides a method for sampling of the unknown parameter. This is an example where the prior and the likelihood are conjugate since both the prior and posterior density are of the same type. In more general situations, with many parameters in θ and with possibly non-conjugate priors, the goal is to summarise the marginal posterior of a particular parameter θ_k given the data. This involves integrating out all the parameters but this one

$$P(\theta_k|y) = \int P(\theta_1, \ldots, \theta_{k-1}, \theta_{k+1}, \ldots \theta_p|y)d\theta_1 \ldots d\theta_{k-1}d\theta_{k+1} \ldots d\theta_p$$

Such integrations in the past involved demanding methods such as numerical quadrature.

Monte Carlo Markov Chain (MCMC) methods, by contrast, use various techniques which ultimately amount to simulating repeatedly from the joint posterior of all the parameters

$$P(\theta_1, \theta_2, \ldots \theta_p|y)$$

without undertaking such integrations. However, inferences about the form of the parameter densities are complicated by the fact that the samples are correlated. Suppose S samples are taken from the joint posterior via MCMC sampling, then marginal posteriors for, say, θ_k may be estimated by averaging over the S samples $\theta_{k1}, \theta_{k2} \ldots . \theta_{kS}$. For example, the mean of the posterior density may be taken as the average of the samples, and the quantiles of the posterior density are given by the relevant points from the ranked sample values.

1.2 GIBBS SAMPLING

One MCMC algorithm is known as Gibbs sampling[4], and involves successive sampling from the complete conditional densities

$$P(\theta_k|y, \theta_1, \ldots \theta_{k-1}, \theta_{k+1}, \ldots . \theta_p)$$

which condition on both the data and the other parameters. Such successive samples may involve simple sampling from standard densities (gamma, Normal, Student t, etc.) or sampling from non-standard densities. If the full conditionals are non-standard but of a certain mathematical form (log-concave), then adaptive rejection sampling (Gilks and Wild, 1992) may be used within the Gibbs sampling for those parameters. In other cases, alternative schemes based on the Metropolis–Hastings algorithm, may be used to sample from non-standard densities (Morgan, 2000). The program WINBUGS may be applied with some or all parameters sampled from formally coded conditional densities;

[4] This is the default algorithm in BUGS.

however, provided with prior and likelihood WINBUGS will infer the correct conditional densities using directed acyclic graphs[5].

In some instances, the full conditionals may be converted to simpler forms by introducing latent data w_i, either continuous or discrete (this is known as 'data augmentation'). An example is the approach of Albert and Chib (1993) to the probit model for binary data, where continuous latent variables w_i underlie the observed binary outcome y_i. Thus the formulation

$$w_i = \beta x_i + u_i \text{ with } u_i \sim \mathrm{N}(0, 1)$$
$$y_i = I(w_i > 0)$$

is equivalent to the probit model[6]. Latent data are also useful for simplifying survival models where the missing failure times of censored cases are latent variables (see Example 1.2 and Chapter 9), and in discrete mixture regressions, where the latent categorical variable for each case is the group indicator specifying to which that case belongs.

1.2.1 Multiparameter model for Poisson data

As an example of a multi-parameter problem, consider Poisson data y_i with means λ_i, which are themselves drawn from a higher stage density. This is an example of a mixture of densities which might be used if the data were overdispersed in relation to Poisson assumptions. For instance, if the λ_i are gamma then the y_i follow a marginal density which is negative binomial. Suppose the λ_i are drawn from a Gamma density with parameters α and β, which are themselves unknown parameters (known as hyperparameters). So

$$y_i \sim \mathrm{Poi}(\lambda_i)$$
$$f(\lambda_i|\alpha, \beta) = \lambda_i^{\alpha-1} e^{-\beta\lambda_i} \beta^\alpha / \Gamma(\alpha)$$

Suppose the prior densities assumed for α and β are, respectively, an exponential[7] with parameter a and a gamma with parameters $\{b, c\}$, so that

[5] Estimation via BUGS involves checking the syntax of the program code (which is enclosed in a model file), reading in the data, and then compiling. Each statement involves either a relation ~ (meaning distributed as) which corresponds to solid arrows in a directed acyclic graph, or a deterministic relation <- which corresponds to a hollow arrow in the DAG. Model checking, data input and compilation involve the model menu in WINBUGS- though models may also be constructed directly by graphical means. The number of chains (if in excess of one) needs to be specified before compilation. If the compilation is successful the initial parameter value file or files ('inits files') are read in. If, say, three parallel chains are being run three inits files are needed. Syntax checking involves highlighting the entire model code, or just the first few letters of the word model, and then choosing the sequence model/specification/check model. To load a data file either the whole file is highlighted or just the first few letters of the word 'list'. For ascii data files the first few letters of the first vector name need to be highlighted. Several separate data files may be read in if needed. After compilation the inits file (or files) need not necessarily contain initial values for all the parameters and some may be randomly generated from the priors using 'gen inits'. Sometimes doing this may produce aberrant values which lead to numerical overflow, and generating inits is generally excluded for precision parameters. An expert system chooses the sampling method, opting for standard Gibbs sampling if conjugacy is identified, and for adaptive rejection sampling (Gilks and Wild, 1992) for non-conjugate problems with log-concave sampling densities. For non-conjugate problems without log-concavity, Metropolis–Hastings updating is used, either slice sampling (Neal, 1997) or adaptive sampling (Gilks *et al.*, 1998). To monitor parameters (i.e. obtain estimates from averaging over sampled values) go inference/samples and enter the relevant parameter name. For parameters which would require extensive storage to be monitored fully an abbreviated summary (for say the model means of all observations in large samples, as required for subsequent calculation of model fit formulas) is obtained by inference/summary and then entering the relevant parameter name.

[6] $I(u)$ is 1 if u holds and zero otherwise.

[7] The exponential density with parameter θ is equivalent to the gamma density $G(\theta, 1)$.

$$\alpha \sim \mathrm{E}(a)$$

$$\beta \sim \mathrm{G}(b,\ c)$$

where *a, b* and *c* are taken as constants with known values (or briefly 'taken as known'). Then the posterior density of $\theta = (\lambda_1 .. \lambda_n, \alpha, \beta)$ is

$$f(\lambda_1, \ldots . \lambda_n, \alpha, \beta | y_1, \ldots y_n) \propto$$

$$e^{-a\alpha}\beta^{b-1}e^{-c\beta}\ [\beta^\alpha/\Gamma(\alpha)]^n \prod_{i=1}^{n} e^{-\lambda_i}\lambda_i^{yi} \prod_{i=1}^{n} \lambda_i^{\alpha-1}e^{-\beta\lambda_i} \tag{1.3}$$

If elements of this density which do not involve the λ_i are regarded as constants, it can be seen that the conditional density of the λ_i is a gamma with parameters $y_i + \alpha$ and $\beta + 1$. Similarly, disregarding elements not functions of β, the conditional density of β is gamma with parameters $b + n\alpha$ and $c + \Sigma\lambda_i$. The full conditional density of α is

$$f(\alpha | y, \beta, \underset{\sim}{\lambda}) \propto e^{-a\alpha}[\beta^\alpha/\Gamma(\alpha)]^n \left(\prod_{i=1}^{n} \lambda_i\right)^{\alpha-1}$$

This density is non-standard but log-concave (see George *et al.*, 1993). Adaptive rejection sampling might then be used, and this is the default in BUGS, for example. Another option is to establish a grid of probability values according to possible values $\alpha_j (j = 1, \ldots J)$ of α; this is described as 'griddy Gibbs' by Tanner (1993). At each iteration the densities at each value of α are calculated, namely

$$G_j = e^{-a\alpha_j}[\beta^{\alpha_j}/\Gamma(\alpha_j)]^n \left(\prod_{i=1}^{n} \lambda_i\right)^{\alpha_j-1}$$

and then scaled to sum to 1, with the choice among possible values α_j decided by a categorical indicator. In practice, a preliminary run might be used to ascertain the support for α, namely the range of values across which its density is significant, and so define a reasonable grid $\alpha_j,\ j = 1, .. J$.

If the Poisson counts (e.g. deaths, component failures) are based on different exposures E_i (populations, operating time), then

$$y_i \sim \mathrm{Poi}(E_i\lambda_i)$$

and the posterior density in Equation (1.3) is revised to

$$f(\lambda_1, \ldots . \lambda_n, \alpha, \beta | y_1, \ldots y_n) \propto$$

$$e^{-a\alpha}\beta^{b-1}e^{-c\beta}[\beta^\alpha/\Gamma(\alpha)]^n \prod_{i=1}^{n} e^{-E_i\lambda_i}\lambda_i^{yi} \prod_{i=1}^{n} \lambda_i^{\alpha-1}e^{-\beta\lambda_i}$$

(Note that E_i raised to the power y_i drops out as a constant.) Then the conditional density of the λ_i is a gamma with parameters $\alpha + y_i$ and $\beta + E_i$. The conditional densities of α and β are as above.

Example 1.1 Consider the power pumps failure data of Gaver and O'Muircheartaigh (1987), where failures y_i of the *i*th pump, operating for time E_i are Poisson with

$$y_i \sim \mathrm{Poi}(\kappa_i)$$

where $\kappa_i = E_i\lambda_i$. The data are as follows:

Pump	E_i	Y_i
1	94.5	5
2	15.7	1
3	62.9	5
4	126	14
5	5.24	3
6	31.4	19
7	1.05	1
8	1.05	1
9	2.1	4
10	10.5	22

The BUGS coding for direct sampling from the full conditionals including the grid prior on α is as follows:

```
  model {for (i in 1 : n) { lambda[i] ~ dgamma(A[i],B[i])
              A[i] <- alpha+y[i];  B[i] <- beta+E[i]
              log.lam[i] <- log(lambda[i])}
# product lambda[1]*lambda[2] . . . *lambda[N] obtained by exponentiating sum
# of log lambda's
              Prod.lam <- exp(sum(log.lam[]))
              for (j in 1:50) {alph[j] <- j*0.05;
# grid based on conditional posterior of alpha; prior of alpha an E(1)
G[j] <- exp(-alph[j])*pow(pow(beta,alph[j])/exp(loggam(alph[j])),N)
* pow(Prod.lam,alph[j]-1)
# scale grid probabilities to sum to 1
              G.Pr[j] <- G[j]/sum(G[])}
# sample from grid (discrete prior)
              k.alph ~ dcat(G.Pr[])
              alpha <- alph[k.alph]
              C <- n*alpha+0.1
              D <- 1+sum(lambda[])
              beta ~ dgamma(C,D)}
```

The grid for α ranges between 0.05 and 2.5. This coding adopts the priors used by George *et al.* (1993), namely, an exponential density with known parameter 1 for α, and a Gamma(0.1, 1) prior for β. The inits file just contains initial values for α and β, while those of λ_i may be generated using 'gen inits'. Then the posterior means and standard deviations of α and β, from a single long chain of 50 000 iterations with 5000 burn in, are 0.70 (0.27) and 0.94 (0.54).

However, this coding may be avoided by specifying just the priors and likelihood, as follows:

```
  model {for (i in 1 : n) {
        lambda[i] ~ dgamma(alpha, beta)
        kappa[i] <- lambda[i]* E[i]
```

```
y[i] ~ dpois(kappa[i])}
alpha ~ dexp(1)
beta ~ dgamma(0.1, 1.0)}
```

1.2.2 Survival data with latent observations

As a second example, consider survival data assumed to follow a Normal density – note that usually survival data are non-Normal. Survival data, and more generally event history data, provide the most familiar examples of censoring. This occurs if at the termination of observation, certain subjects are (right) censored in that they have yet to undergo the event (e.g. a clinical end-point), and their unknown duration or survival time t is therefore not observed. Instead, the observation is the time t^* at which observation of the process ceased, and for censored cases, it must be the case that $t \geq t^*$. The unknown survival times for censored subjects provide additional unknowns (as augmented or latent data) to be estimated.

For the Normal density, the unknown distributional parameters are the mean and variance $\{\mu, \sigma^2\}$. In Bayesian modelling there are potential simplifications in considering the specification of the prior, and updating to the posterior, in terms of the inverse of the variance, or precision, $\tau = \sigma^{-2}$. Since both the variance and precision are necessarily positive, an appropriate prior density is constrained to positive values. Though improper reference priors for the variance or precision are often used, consider prior densities $P(\tau)$, which are proper in the sense that the integral over possible values is defined. These include the uniform density over a finite range, such as

$$\tau \sim \mathrm{U}(0, 1000)$$

or a gamma density which allows for various types of skewness. This has the form

$$\tau \sim \mathrm{G}(f, g)$$

so that

$$P(\tau) \propto \tau^{f-1} \exp(-g\tau) \tag{1.4}$$

where f and g are taken as known constants, and where the prior mean of τ is then f/g with variance f/g^2. For instance, taking $f = 1$, $g = 0.001$ gives a prior on τ which still integrates to 1 (is proper) but is quite diffuse in the sense of not favouring any value. A similar diffuse prior takes $f = g = 0.001$ or some other common small value[8]. Substituting $f = 1$ and $g = 0.001$ in Equation (1.4) shows that for these values of f and g the prior in (1.4) is approximately (but not quite)

$$P(\tau) \propto 1 \tag{1.5}$$

Setting $g = 0$ is an example of an improper prior, since then $P(\tau) \propto 1$, and

$$\int P(\tau) d\tau = \infty$$

So taking $f = 1$, $g = 0.001$ in Equation (1.4) represents a 'just proper' prior.

In fact, improper priors are not necessarily inadmissible for drawing valid inferences providing the posterior density, given by the product of prior and likelihood, as in

[8] In this case, the prior on τ is approximately $P(\tau) \propto 1/\tau$.

Equation (1.1), remains proper (Fraser *et al.*, 1997). Certain improper priors may qualify as reference priors, in that they provide minimal information (for example, that the variance or precision is positive), still lead to proper posterior densities, and also have valuable analytic properties, such as invariance under transformation. An example is the Jeffreys prior for $\sigma = \tau^{-0.5}$, namely

$$P(\sigma) = 1/\sigma$$

In BUGS, priors in this form may be implemented over a finite range using a discrete grid method and then scaling the probabilities to sum to 1. This preserves the shape implications of the prior, though obviously they are no longer improper.

1.2.3 Natural conjugate prior

In a model with constant mean μ over all cases, a joint prior for $\{\mu, \tau\}$, known as the 'natural conjugate' prior, may be specified for Normally or Student t distributed data which assumes a Gamma form for τ, and a conditional prior distribution for μ given τ which is Normal. Thus, the prior takes the form

$$P(\mu, \tau) = P(\tau)P(\mu|\tau)$$

One way to specify the prior for the precision τ is in terms of a prior 'guess' at the variance V_0 and a prior sample size ν (possibly non-integer) which represents the strength of belief (usually slight) in this guess. Typical values are $\nu = 2$ or lower. Then the prior for τ takes the form

$$\tau \sim G(\nu/2, \nu V_0/2) \quad (1.6)$$

and taking $V_0 = 0.001$ and $\nu = 2$, gives a 'just proper' prior

$$\tau \sim G(1, 0.001)$$

as discussed earlier. Given the values of τ drawn from this prior, the prior for μ takes the form

$$\mu \sim N(M_0, (n_0\tau)^{-1})$$

where M_0 is a prior guess at the unknown mean of the Normal density. Since higher values of the precision $n_0\tau$ mean lower variances, it can be seen that higher values of n_0 imply greater confidence in this guess. Usually, n_0 is taken small (1 or less, as for ν). So the entire prior has the form

$$P(\mu, \tau) \propto \tau^{0.5} \exp\{-0.5n_0\tau(\mu - M_0)^2\} \exp(-0.5\nu V_0\tau)\tau^{0.5\nu-1}$$

1.2.4 Posterior density with Normal survival data

In the survival example, suppose initially there is only one group of survival times, and that all times are known (i.e. there is no censoring). Let the observed mean survival time be

$$M = n^{-1}\sum_i t_i$$

and observed variance be

$$V = \Sigma(t_i - M)^2/(n - 1)$$

Then one may show that the posterior density of $\{\mu, \tau\}$ given data $\{t_1, t_2, \ldots. t_n\}$ is proportional to the product of

1. A Normal density for μ with precision $n_1\tau$, where $n_1 = n_0 + n$, and with mean M_1, which is a weighted average of data and prior means, namely $M_1 = w_0 M_0 + w_1 M$, with weights $w_0 = n_0/n_1$ and $w_1 = n/n_1$; and
2. A Gamma density for τ of the form in Equation (1.6) which has 'sample size' $\nu_1 = \nu + n$ and variance

$$V_1 = (\nu + n)^{-1}[V_0\nu + n_0 M_0^2 + (n - 1)V + nM^2 - n_1 M_1^2]$$

Thus

$$P(\tau,\ \mu|y) \propto \tau^{0.5} \exp[-0.5\tau n_1(\mu - M_1)^2]\ \tau^{0.5\nu_1 - 1} \exp(-0.5\tau\nu_1 V_1)$$

The Gibbs sampling approach considers the distributions for τ and μ conditional on the data and the just sampled value of the other. The full conditional for μ (regarding τ as a constant) can be seen to be a Normal with mean M_1 and precision τn_1. Then just having drawn μ at iteration t, the next iteration samples from the full conditional density for τ which is a gamma density with shape (first parameter) $0.5\nu_1 + 0.5$ and scale $0.5[\nu_1 V_1 + n_1(\mu - M_1)^2]$.

If some event times were in fact censored when observation ceased, then these are extra parameters drawn from the Normal density with mean μ and precision τ subject to being at least t^*. That is, constrained sampling from the Normal above (with mean $\mu^{(t)}$ and precision $\tau^{(t)}$ at the tth iteration) is used, disregarding sampled values which are lower than t^*. The subsequently updated values of M and V include these imputations as well as the uncensored t_i.

It can be seen that even for a relatively standard problem, namely updating the parameters of a Normal density, the direct coding in terms of full conditional densities becomes quite complex. The advantage with BUGS is that it is only necessary to specify the priors

$$\tau \sim \mathrm{G}(\nu/2,\ V_0\nu/2)$$

$$\mu \sim \mathrm{N}(M_0, (n_0\tau)^{-1})$$

and the form of the likelihood for the data, namely

$$t_i \sim \mathrm{N}(\mu, \tau^{-1}) \qquad (t \text{ uncensored})$$

$$t_i \sim \mathrm{N}(\mu, \tau^{-1})\ I(t_i^*,) \qquad (\text{censoring at } t^*)$$

and the full conditionals are inferred. The $I(a$, b) symbol denotes a range within which sampling is confined. For uncensored data the t_i are observed, but for the censored data the observations are t_i^* and the true t_i are latent (or 'augmented', or 'missing') data.

Example 1.2 Leukaemia remission times Consider the frequently analysed data of Gehan (1965) on leukaemia remission times under two therapies, the latter denoted $Tr[\,]$ in the code below with $Tr[i] = 1$ for the new treatment. Here delayed remission (longer survival) indicates a better clinical outcome. There is extensive censoring of times under the new therapy, with censored times coded as NA, and sampled to have minimum defined by the censored remission time.

Assume independent Normal densities differing in mean and variance according to treatment, and priors

$$\mu_j \sim N(0, \tau_j^{-1}) \text{ with } n_0 = 1$$

$$\tau_j \sim G(1,\ 0.001)$$

in treatment groups j. Then the code for BUGS (which parameterises the Normal with the inverse variance) and with the data from Gehan, is as follows:

```
model { for (i in 1:42) { t[i] ~ dnorm(mu[Tr[i]], tau[Tr[i]]) I(min[i],)}
          for (j in 1:2) {mu[j] ~ dnorm(0,tau[j])
# inverse variances
                              tau[j] ~ dgamma(1,0.001)}}
```

with data file

list(t=c(NA,6,6,6,7,NA, NA,10,NA,13,16,NA,NA,NA,22,23,NA,NA,NA, NA,NA,1,1, 2,2,3,4,4,5,5,8,8,8,8,11,11,12,12,15,17,22,23), min=c(6,0,0,0,0,9,10,0,11,0,0,17,19,20,0,0,25,32,32,34,35,0,0, 0,0,0,0,0,0,0,0,0,0,0,0,0,0,0,0,0,0,0),Tr=c(1,1,1,1,1,1,1,1,1,1,1,1,1,1,1,1,1,1, 1,1,1,2,2,2,2,2,2,2,2,2,2,2,2,2,2,2,2,2,2,2,2,2))

The inits file

list(mu=c(10,10),tau=c(1,1))

specifies the initial Normal parameters only, and the missing survival times (the remaining unknowns) may be sampled from the prior (using 'gen inits'). The average remission times under new and old therapies are 24.8 and 8.3 months. An example of a skew posterior density is provided by the imputed survival times for censored subjects. While an unspecified upper sampling limit for censored times is one option, there may be subject matter considerations ruling out unusually high values (e.g. survival exceeding five years). Since setting $n_0 = 1$ (or any other value) may appear arbitrary, one may also assume independent priors for μ_j and τ_j (Gelman *et al.*, 1995, Chapter 3), such as $\mu_j \sim N(0, K_j)$, with K_j typically large, and $\tau_j \sim G(1, 0.001)$ as before. With $K_j = 1000$, the average remission times become 25.4 and 8.6.

1.3 SIMULATING RANDOM VARIABLES FROM STANDARD DENSITIES

Parameter estimation by MCMC methods and other sampling-based techniques requires simulated values of random variables from a range of densities. As pointed out by Morgan (2000), sampling from the uniform density $U(0, 1)$ is the building block for sampling the more complex densities; in BUGS 0, 1 this involves the code

$$U \sim \text{dunif}(0,\ 1)$$

Thus, the Normal univariate density[9] is characterised by a mean μ and variance ϕ, with

[9] BUGS parameterises the Normal in terms of the inverse variance, so priors are specified on $P = \phi^{-1}$ and μ, and samples of ϕ may be obtained by specifying $\phi = P^{-1}$. With typical priors on μ and P, this involves the coding

$$X \sim \mathrm{N}(\mu, \phi)$$

A sample from Normal density with mean 0 and variance 1 may be obtained by considering two independent draws U_1 and U_2 from a $U(0, 1)$ density. Then with $\pi \approx 3.1416$, the pair

$$Z_1 = [-2\ \ln(U_1)]^{0.5} \sin(2\pi U_2)$$

$$Z_2 = [-2\ \ln(U_1)]^{0.5} \cos(2\pi U_2)$$

are independent draws from an N(0,1) density. Then using either of these draws (say $Z = Z_1$), a sample from N(μ, ϕ) is obtained via

$$X = \mu + Z\sqrt{\phi}$$

An approximately Normal N(0, 1) variable may also be obtained using central limit theorem ideas: take n draws $U_1, U_2, .. U_n$ from a $U(0, 1)$ then

$$X = \left(\sum_i U_i - 0.5n\right)\left(\frac{12}{n}\right)^{0.5}$$

is approximately N(0, 1) for large n. In fact $n = 12$ is often large enough and simplifies the form of X.

1.3.1 Binomial and negative binomial

Another simple application of sampling from the uniform U(0, 1) is if a sample of an outcome Y_i (either 0 or 1) from a Bernoulli density with probability ρ is required. Thus, if U_i is a sample from U(0, 1) and $U_i \leq \rho$ then Y_i is taken as 1, whereas $U_i > \rho$ leads to $Y_i = 0$. So the unit interval is in effect split into sections of length ρ and $1 - \rho$. This principle can be extending to simulating 'success' counts r from a binomial with n subjects at risk of an event with probability ρ. The sampling from $U(0, 1)$ is repeated n times and the number of times for which $U_i \leq \rho$ is the simulated success count.

Similarly, consider the negative binomial density, with

$$\Pr(x) = \binom{x-1}{r-1} p^r (1-p)^{x-r} \quad x = r, r+1, r+2, ..$$

In this case a sequence $U_1, U_2, \ldots$ may be drawn from the U(0,1) density until r of them are less than or equal to p, with x given by the number of draws U_i needed to reach this threshold.

1.3.2 Inversion method

A further fundamental building block based on the uniform density follows from the fact that if U_i is a draw from $U(0, 1)$ then

$$X_i = -1/\mu \ln(U_i)$$

is a draw from an exponential[10] with mean μ. The exponential density is defined by

---

```
phi <- 1/P
X ~ dnorm(mu,P)
mu ~ dnorm(0,0.001)
P ~ dgamma(0.001,0.001).
```

[10] In BUGS the appropriate code is x ~dexp(mu).

$$f(\mathrm{x}) = \mu \exp(-\mu x)$$

with mean $1/\mu$ and variance $1/\mu^2$, and is often a baseline model for waiting or inter-event times.

This way of sampling the exponential is an example of the inversion method for simulation of a continuous variable with distribution function F, the inverse of which is readily available. If $u \sim U(0, 1)$ then

$$\Pr[F^{-1}(u) \leq x] = \Pr[u \leq F(x)] = F(x)$$

and the quantities $x = F^{-1}(u)$ are then draws from a random variable with cumulative density $F(x)$. The same principle may be used to obtain draws from a logistic distribution $x \sim \text{Logistic}(\mu, \tau)$, a heavy tailed density (as compared to the Normal) with cdf

$$F(x) = 1/\{1 + \mathrm{e}^{-\tau(x-\mu)}\}$$

and pdf

$$f(x) = \tau \mathrm{e}^{-\tau(x-\mu)}/[1 + \mathrm{e}^{-\tau(x-\mu)}]^2$$

This distribution has mean μ, variance $\pi^2/(3\tau^2)$, and a draw may be obtained by the transformation

$$x = \log_e(U/[1 - U])/\tau - \mu.$$

The Pareto, with density

$$f(x) = ab^a/[x^{a+1}] \quad x \geq b > 0$$

may be obtained as

$$x = b/(1 - U)^{1/a}$$

or equivalently,

$$x = b/U^{1/a}.$$

1.3.3 Further uses of exponential samples

Simulating a draw x from a Poisson with mean μ can be achieved by sampling $U_i \sim U(0, 1)$ and taking x as the maximum n for which the cumulative sum of $L_i = -\ln(U_i)$,

$$S_i = L_1 + L_2 + \ldots L_i$$

remains below μ. From above, the L_i are exponential with rate 1, and so viewed as inter-event times of a Poisson process with rate 1, $N = N(\mu)$ equals the number of events which have occurred by time μ. Equivalently, x is given by n, where $n + 1$ draws from an exponential density with parameter μ are required for the sum of the draws to first exceed 1.

The Weibull density is a generalisation of the exponential also useful in event history analysis. Thus, if $t \sim \text{Weib}(\alpha, \lambda)$, then

$$f(t) = \alpha\lambda t^{\alpha-1} \exp(-\lambda t^{\alpha}), \qquad t > 0$$

If x is exponential with rate λ, then $t = x^{1/\alpha}$ is Weib(α, λ). Thus in BUGS the codings

```
t[i] ~ dweib(alpha,lambda)
```

and

```
x[i] ~ dexp(lambda)
t[i] <- pow(x[i],1/alpha)
```

generate the same density.

1.3.4 Gamma, chi-square and beta densities

The gamma density is central to the modelling of variances in Bayesian analysis, and as a prior for the Poisson mean. It has the form

$$f(x) = [\beta^\alpha/\Gamma(\alpha)]x^{\alpha-1}\exp(-\beta x), \quad x > 0$$

with mean α/β and variance α/β^2. Several schemes are available for generating a gamma variate. For $\alpha = K$ an integer, drawing a sample $x_1, x_2, .. x_K$ from an exponential with mean β and then taking the sum $y = \Sigma_K^{i=1} x_i$ generates a draw from a Gamma density. Note also that if $x \sim G(\alpha, 1)$, then $y = x/\beta$ is a $G(\alpha, \beta)$ variable.

Since a $G(\alpha, \beta)$ density is often used as a prior for a precision parameter, it is also worth noting that it follows that the variance then follows an inverse gamma density, with the same parameters. The inverse gamma has the form

$$f(x) = [\beta^\alpha/\Gamma(\alpha)]x^{-\alpha-1}\exp(-\beta/x), \quad x > 0$$

with mean $\beta/(\alpha - 1)$ and variance $\beta^2/[(\alpha - 1)^2(\alpha - 2)]$. One possibility (for x approximately Normal) for setting a prior on the variance is to take the prior mean of σ^2 to be the square of one sixth of the anticipated range of x (since the range is approximately 6σ for a Normal variable). Then for $\alpha = 2$ (or just exceeding 2 to ensure finite variance), one might set $\beta = (\text{range}/6)^2$.

From the gamma density may be derived a number of other densities, and hence ways of sampling from them. The chi-square is also used as a prior for the variance, and is the same as a gamma density with $\alpha = \nu/2$, $\beta = 0.5$. Its expectation is then ν, usually interpreted as a degrees of freedom parameter. The density (1.6) above is sometimes known as a scaled chi-square. The chi-square may also be obtained for ν an integer, by taking ν draws $x_1, x_2, \ldots x_\nu$ from an N(0,1) density and taking the sum of $x_1^2, x_2^2, .. x_\nu^2$. This sum is chi-square with ν degrees of freedom.

The beta density is used as a prior for the probability p in the binomial density, and can accommodate various degrees of left and right skewness. It has the form

$$f(p) = \Gamma(\alpha + \beta)/[\Gamma(\alpha)\Gamma(\beta)]\, p^{\alpha-1}(1 - p)^{\beta-1} \quad \alpha, \beta > 0;\ 0 < p < 1$$

with mean $\alpha/(\alpha + \beta)$. Setting $\alpha = \beta$ implies a symmetrical density with mean 0.5, whereas $\alpha > \beta$ implies positive skewness and $\alpha < \beta$ implies negative skewness. The total $\alpha + \beta - 2$ defines a prior sample size as in Equation (1.2).

If y and x are gamma densities with equal scale parameters (say $\omega = 1$), and if $y \sim G(\alpha, \omega)$ and $x \sim G(\beta, \omega)$, then

$$x = y/(y + x)$$

is a $B(\alpha, \beta)$ density. The beta has mean $\alpha/(\alpha + \beta)$ and variance $\alpha\beta/[(\alpha + \beta)^2 (\alpha + \beta + 1)]$.

1.3.5 Univariate and Multivariate *t*

For continuous data, the Student *t* density is a heavy tailed alternative to the Normal, though still symmetric, and is more robust to outlier points. The heaviness of the tails is governed by an additional degrees of freedom parameter ν as compared to the Normal density. It has the form

$$f(x) = \Gamma(0.5\nu + 0.5)/[\Gamma(0.5\nu)(\sigma^2\nu\pi)^{0.5}]\ [1 + (x - \mu)^2/(\nu\sigma^2)]^{-0.5(\nu+1)}$$

with mean μ and variance $\nu\sigma^2/(\nu - 2)$. If z is a draw from a standard Normal, N(0, 1), and y is a draw from the Gamma G(0.5ν, 0.5) density, then $x = \mu + \sigma z\sqrt{\nu}/\sqrt{y}$ is a draw form a Student $t_\nu(\mu, \sigma^2)$. Equivalently, let y be a draw from a Gamma density, G(0.5ν,0.5ν), then the Student *t* is obtained by sampling from N(μ, σ^2/y). The latter scheme is the best form for generating the scale mixture version of the Student *t* density (see Chapter 2).

A similar relationship holds between the multivariate Normal and multivariate *t* densities. Let x be a d-dimensional continuous outcome. Suppose x is multivariate Normal with mean $\mu = (\mu_1, \mu_2, ..\, \mu_d)$ and $d \times d$ dispersion matrix V. This is denoted $x \sim MVN(\mu, V)$ or $x \sim N_d(\mu, V)$, and

$$f(x) = (2\pi)^{-d/2}|V|^{-0.5}\exp[-0.5(X - \mu)V^{-1}(X - \mu)]$$

Sampling from this density involves the Cholesky decomposition[11] of V, namely $V = AA^T$, where A is also $d \times d$. Then if $z_1, z_2, ..\, z_d$ are independent univariate draws from a standard Normal,

$$x = \mu + Az$$

is a draw from the multivariate Normal. The multivariate Student *t* with ν degrees of freedom, mean $\mu = (\mu_1, \mu_2, ..\, \mu_d)$ and dispersion matrix V is defined by

$$f(x) = K|V|^{-0.5}\{1 + (1/\nu)(x - \mu)V^{-1}(x - \mu)\}^{-0.5(\nu+d)}$$

where K is a constant ensuring the integrated density sums to unity. This density is useful for multivariate data with outliers or other sources of heavy tails, and may be sampled from by taking a single draw Y from a Gamma density, $\lambda \sim G(0.5\nu, 0.5\nu)$ and then sampling the vector

$$x \sim N_d(\mu,\ V/\lambda)$$

The Wishart density, the multivariate generalisation[12] of the gamma or of the chi-square density, is the most common prior structure assumed for the inverse of the dispersion

[11] This matrix may be obtained (following an initialisation of A) as:
for $i = 1$ to d
for $j = 1$ to $i - 1$

$$A_{ij} = \left(V_{ij} - \sum_{k=1}^{j-1} A_{ik}A_{jk}\right)/A_{jj}$$

$$A_{ji} = 0$$

$$A_{ii} = \left(V_{ii} - \sum_{k=1}^{i-1} A_{ik}^2\right)^{0.5}$$

[12] Different parameterisations are possible. The form in WINBUGS generalises the chi-square.

matrix V, namely the precision matrix $T = V^{-1}$. One form for this density, for a degrees of freedom $\nu \geq d$, and a scale matrix S

$$T \propto |S|^{\nu/2}|T|^{0.5(\nu-d-1)} \exp(-0.5\text{tr}[ST])$$

The matrix S/ν is a prior guess at the dispersion matrix, since $E(T) = \nu S^{-1}$.

1.3.6 Densities relevant to multinomial data

The multivariate generalisation of the Bernoulli and binomial densities allows for a choice among $C > 2$ categories, with probabilities $\pi_1, \pi_2, ..\pi_C$ summing to 1. In BUGS the multivariate generalisation of the Bernoulli may be sampled from in two ways:

$$Y[i] \sim \text{dcat}(\text{pi}[1{:}\, C])$$

which generates a choice j between 1 and C, or

$$Z[i] \sim \text{dmulti}(\text{pi}[1{:}\, C],\ 1)$$

This generates a choice indicator, $Z_{ij} = 1$ if the jth category is chosen, and $Z_{ij} = 0$ otherwise.

For example, the code

```
{for (i in 1:100) {Y[i] ~ dcat(pi[1:3])}}
```

with data in the list file

```
list( pi=c(0.8,0.1,0.1)}
```

would on average generate 80 one's, 10 two's and 10 three's. The coding

```
{for (i in 1:100) {Y[i,1:3] ~ dmulti(pi[1:3],1)}}
```

with data as above would generate a 100×3 matrix, with each row containing a one and two zeroes, and the first column of each row being 1 for 8 out of 10 times on average.

A commonly used prior for the probability vector $\prod = (\pi_1, ..\pi_C)$ is provided by the Dirichlet density. This is a multivariate generalisation of the beta density, as can be seen from its density

$$f(\pi_1, .., \pi_C) = \Gamma(\alpha_1 + \alpha_2 + \ldots + \alpha_C)/[\Gamma(\alpha_1)\Gamma(\alpha_2)\ldots\Gamma(\alpha_C)]$$
$$\pi_1^{\alpha_1-1}\pi_2^{\alpha_2-1}\ldots\pi_C^{\alpha_C-1}$$

where the parameters $\alpha_1, \alpha_2, ..\alpha_C$ are positive. The Dirichlet may be drawn from directly in WINBUGS and a common default option sets $\alpha_1 = \alpha_2 = \ldots \alpha_C = 1$. However, an alternative way of generation is sometimes useful. Thus, if $Z_1, Z_2, ..Z_C$ are gamma densities with equal scale parameters (say $\omega = 1$), and if

$$Z_1 \sim G(\alpha_1, \omega),\ Z_2 \sim G(\alpha_2, \omega), ..Z_C \sim G(\alpha_C, \omega)$$

then the quantities

$$Z_j = \alpha_j / \sum_k \alpha_k \quad j = 1, ..C$$

are draws from the Dirichlet with prior weights vector $(\alpha_1, ..\alpha_C)$.

1.4 MONITORING MCMC CHAINS AND ASSESSING CONVERGENCE

An important practical issue involves assessment of convergence of the sampling process used to estimate parameters, or more precisely update their densities. In contrast to convergence of optimising algorithms (maximum likelihood or minimum least squares, say), convergence here is used in the sense of convergence to a density rather than single point. The limiting or equilibrium distribution $P(\theta|Y)$ is known as the target density. The sample space is then the multidimensional density in p-space; for instance, if $p = 2$ this density may be approximately an ellipse in shape.

The above two worked examples involved single chains, but it is preferable in achieving convergence to use two or more parallel chains[13] to ensure a complete coverage of this sample space, and lessen the chance that the sampling will become trapped in a relatively small region. Single long runs may, however, often be adequate for relatively straightforward problems, or as a preliminary to obtain inputs to multiple chains.

A run with multiple chains requires overdispersed starting values, and these might be obtained from a preliminary single chain run; for example, one might take the 1st and 99th percentiles of parameters from a trial run as initial values in a two chain run (Bray, 2002), or the posterior means from a trial run combined with null starting values. Another option might combine parameters obtained as a random draw[14] from a trial run with null parameters. Null starting values might be zeroes for regression parameters, one for precisions, and identity matrices for precision matrices. Note that not all parameters need necessarily be initialised, and parameters may instead be initialised by generating[15] from their priors.

A technique often useful to aid convergence, is the over-relaxation method of Neal (1998). This involves generates multiple samples of each parameter at the next iteration and then choosing the one that is least correlated with the current value, so potentially reducing the tendency for sampling to become trapped in a highly correlated random walk.

1.4.1 Convergence diagnostics

Convergence for multiple chains may be assessed using the Gelman-Rubin scale reduction factors, which are included in WINBUGS, whereas single chain diagnostics require use of the CODA or BOA packages[16] in Splus or R. The scale reduction factors compare variation in the sampled parameter values within and between chains. If parameter samples are taken from a complex or poorly identified model then a wide divergence in the sample paths between different chains will be apparent (e.g. Gelman, 1996, Figure 8.1) and variability of sampled parameter values between chains will considerably exceed the variability within any one chain. Therefore, define

$$V_j = \sum_{t=s}^{T+s} \left(\theta_j^{(t)} - \bar{\theta}_j\right)^2 / (T-1)$$

[13] In WINBUGS this involves having separate inits files for each chain and changing the number of chains from the default value of 1 before compiling.

[14] For example, by using the state space command in WINBUGS.

[15] This involves 'gen ints' in WINBUGS.

[16] Details of these options and relevant internet sites are available on the main BUGS site.

as the variability of the samples $\theta_j^{(t)}$ within the jth chain ($j = 1, \ldots J$). This is assessed over T iterations after a burn in of s iterations. An overall estimate of variability within chains is the average V_W of the V_j. Let the average of the chain means $\bar{\theta}_j$ be denoted $\bar{\theta}_\bullet$. Then the between chain variance is

$$V_B = \frac{T}{J-1}\sum_{j=1}^{J}(\bar{\theta}_j - \bar{\theta}_\bullet)^2$$

The Scale Reduction Factor (SRF) compares a pooled estimator of var(θ), given by

$$V_P = V_B/T + TV_W/(T-1)$$

with the within sample estimate V_W. Specifically, the SRF is

$$(V_P/V_W)^{0.5}$$

and values of the SRF, or 'Gelman-Rubin statistic', under 1.2 indicate approximate convergence.

The analysis of sampled values from a single MCMC chain or parallel chains may be seen as an application of time series methods (see Chapter 5) in regard to problems such as assessing stationarity in an autocorrelated sequence. Thus, the autocorrelation at lags 1, 2, and so on, may be assessed from the original series of sampled values $\theta^{(t)}$, $\theta^{(t+1)}$, $\theta^{(t+2)\ldots}$, or from more widely spaced sub-samples K steps apart $\theta^{(t)}$, $\theta^{(t+K)}$, $\theta^{(t+2K)}$. Geweke (1992) developed a t-test applicable to assessing convergence in runs of sampled parameter values, both in single and multiple chain situations. Let $\bar{\theta}_a$ be the posterior mean of scalar parameter θ from the first n_a iterations in a chain (after burn-in), and $\bar{\theta}_b$ be the mean from the last n_b draws. If there is a substantial run of intervening iterations, then the two samples should be independent. Let V_a and V_b be the variances of these averages[17]. Then the statistic

$$Z = (\bar{\theta}_a - \bar{\theta}_b)/(V_a + V_b)^{0.5}$$

should be approximately N(0, 1). This test may be obtained in CODA or the BOA package.

1.4.2 Model identifiability

Problems of convergence of MCMC sampling procedures may reflect problems in model identifiability due to over-fitting or redundant parameters. Use of diffuse priors increases the chances of a poorly identified model, especially in complex hierarchical models (Gelfand and Sahu, 1999), and elicitation of more informative priors may assist identification and convergence. Slow convergence will show in poor 'mixing' with high autocorrelation in the successive sampled values of parameters, apparent graphically in trace plots that wander rather than rapidly fluctuating around a stable mean.

[17] If by chance the successive samples $\theta_a^{(t)}, t = 1, \ldots n_a$ and $\theta_b^{(t)}, t = 1, \ldots n_b$ were independent, then V_a and V_b would be obtained as the population variance of the $\theta^{(t)}$, namely $V(\theta)$, divided by n_a and n_b. In practice, dependence in the sampled values is likely, and V_a and V_b must be estimated by allowing for the autocorrelation. Thus

$$V_a = (1/n_a)\left[\gamma_0 + \sum_{j=1}^{n_a-1}\gamma_j\left(\frac{n_a - j}{n_a}\right)\right]$$

where γ_j is the autocovariance at lag j. In practice, only a few lags may be needed.

Conversely, running multiple chains often assists in diagnosing poor identifiability of models. Examples might include random effects in nested models, for instance

$$y_{ij} = \mu + \eta_i + u_{ij} \quad i = 1,..n; j = 1,..m \tag{1.7}$$

where $\eta_i \sim N(0, \sigma^2_\eta)$, $u_{ij} \sim N(0, \sigma^2_u)$. Poor mixing may occur because the mean of the η_i and the global mean μ are confounded: a constant may be added to the η_i and subtracted from μ without altering the likelihood (Gilks and Roberts, 1996). Vines, Gilks and Wild (1996) suggest the transformation (or reparameterisation) in Equation (1.7),

$$\nu = \mu + \bar{\eta}; \; \alpha_i = \eta_i - \bar{\eta}$$

leading to the model

$$\begin{aligned} y_{ij} &= \nu + \alpha_i + u_{ij} \\ \alpha_1 &\sim N(0, (m-1)\sigma^2_\eta/m) \\ \alpha_j &\sim N\left(-\sum_{k=1}^{j-1}\alpha_k, \frac{m-j}{m-j+1}\sigma^2_\eta\right) \\ \alpha_m &= -\sum_{k=1}^{m-1}\alpha_k \end{aligned}$$

More complex examples occur in a spatial disease model with unstructured and spatially structured errors (Gelfand *et al.*, 1998), sometimes known as a spatial convolution model and considered in Example 1.3 below, and in a particular kind of multiple random effects model, the age-period-cohort model (Knorr-Held and Rainer, 2001). Identifiability issues also occur in discrete mixture regressions (Chapter 3) and structural equation models (Chapter 8) due to label switching during the MCMC sampling. Such instances of non-identifiability will show as essentially nonconvergent parameter series between chains, whereas simple constraints on parameters will typically achieve identifiability. For example, if a structural equation model involved a latent construct such as alienation and loadings on this construct were not suitably constrained, then one chain might fluctuate around a loading of −0.8 on social integration (the obverse of alienation) and another chain fluctuate around a loading of 0.8 on alienation.

Correlation between parameters within the parameter set $\underset{\sim}{\theta} = (\theta_1, \theta_2, \ldots \theta_p)$, such as between θ_1 and θ_2, also tends to delay convergence and to increase the dependence between successive iterations. Re-parameterisation to reduce correlation – such as centring predictor variables in regression – may improve convergence (Gelfand *et al.*, 1995; Zuur *et al.*, 2002). In nonlinear regressions, a log transform of a parameter may be better identified than its original form (see Chapter 10 for examples in dose-response modelling).

1.5 MODEL ASSESSMENT AND SENSITIVITY

Having achieved convergence with a suitably identified model a number of processes may be required to firmly establish the models credibility. These include model choice (or possibly model averaging), model checks (e.g. with regard to possible outliers) and, in a Bayesian analysis, an assessment of the relation of posterior inferences to prior assumptions. For example, with small samples of data or with models where the random effects are to some extent identified by the prior on them, there is likely to be sensitivity in posterior estimates and inferences to the prior assumed for parameters. There may also be sensitivity if an informative prior based on accumulated knowledge is adopted.

1.5.1 Sensitivity on priors

One strategy is to consider a limited range of alternative priors and assess changes in inferences; this is known as 'informal' sensitivity analysis (Gustafson, 1996). One might also consider more formal approaches to robustness based perhaps on non-parametric priors (such as the Dirichlet process prior) or on mixture ('contamination') priors. For instance, one might assume a two group mixture with larger probability $1-p$ on the 'main' prior $\pi_1(\theta)$, and a smaller probability such as $p = 0.2$ on a contaminating density $\pi_2(\theta)$, which may be any density (Gustafson, 1996; Berger, 1990). One might consider the contaminating prior to be a flat reference prior, or one allowing for shifts in the main prior's assumed parameter values (Berger, 1990). For instance, if $\pi_1(\theta)$ is N(0, 1), one might take $\pi_2(\theta) \sim N(m_2, v_2)$, where higher stage priors set $m_2 \sim U(-0.5, 0.5)$ and $v_2 \sim U(0.7, 1.3)$.

In large datasets, regression parameters may be robust to changes in prior unless priors are heavily informative. However, robustness may depend on the type of parameter and variance parameters in random effects models may be more problematic, especially in hierarchical models, where different types of random effect coexist in a model (Daniels, 1999; Gelfand *et al.*, 1998). While a strategy of adopting just proper priors on variances (or precisions) is often advocated in terms of letting the data speak for themselves (e.g. gamma(a, a) priors on precisions with $a = 0.001$ or $a = 0.0001$), this may cause slow convergence and relatively weak identifiability, and there may be sensitivity in inferences between analyses using different supposedly vague priors (Kelsall and Wakefield, 1999). One might introduce stronger priors favouring particular values more than others (e.g. a gamma(5, 1) prior on a precision), or even data based priors loosely based on the observed variability. Mollié (1996) suggests such a strategy for the spatial convolution model. Alternatively the model might specify that random effects and/or their variances interact with each other; this is a form of extra information.

1.5.2 Model choice and model checks

Additional forms of model assessment common to both classical and Bayesian methods involve measuring the overall fit of the model to the dataset as a basis for model choice, and assessing the impact of particular observations on model estimates and/or fit measures. Model choice is considered in Chapter 2 and certain further aspects which are particularly relevant in regression modelling are discussed in Chapter 3. While marginal likelihood, and the Bayes factor based on comparing such likelihoods, defines the canonical model choice, in practice (e.g. for complex random effects models or models with diffuse priors) this method may be relatively difficult to implement. Relatively tractable approaches based on the marginal likelihood principle include those of Newton and Raftery (1994) based on the harmonic average of likelihoods, the importance sampling method of Gelfand and Dey (1994), as exemplified by Lenk and Desarbo (2000), and the method of Chib (1995) based on the marginal likelihood identity (Equation (2.4) in Chapter 2).

Methods such as cross-validation by single case omission lead to a form of pseudo Bayes factor based on multiplying the CPO for model 1 over all cases and comparing the result with the same quantity under model 2 (Gelfand, 1996, p. 150). This approach when based on actual omission of each case in turn may (with current computing technology) be only practical with relatively small samples. Other sorts of partitioning of the data into training samples and hold-out (or validation) samples may be applied, and are less computationally intensive.

In subsequent chapters, the main methods of model choice are (a) those based on predictive criteria, comparing model predictions z with actual observations[18], as advocated by Gelfand and Ghosh (1998) and others, and (b) modifications of classical deviance tests to reflect the effective model dimension, as in the DIC criterion discussed in Chapter 2 (Spiegelhalter *et al.*, 2002). These are admittedly not formal Bayesian choice criteria, but are relatively easy to apply over a wide range of models including non-conjugate and heavily parameterised models.

The marginal likelihood approach leads to posterior probabilities or weights on different models, which in turn are the basis for parameter estimates derived by model averaging (Wasserman, 2000). Model averaging has particular relevance for regression models, especially for smaller datasets where competing specifications provide closely comparable explanations for the data, and so there is a basis for weighted averages of parameters over different models; in larger datasets by contrast, most model choice diagnostics tend to overwhelmingly support one model. A form of model averaging also occurs under predictor selection methods, such as those of George and McCulloch (1993) and Kuo and Mallick (1998), as discussed in Chapter 3.

1.5.3 Outlier and influence checks

Outlier and influence analysis in Bayesian modelling may draw in a straightforward fashion from classical methods. Thus in a linear regression model with Normal errors

$$y_i = \beta_1 + \beta_2 x_{1i} + \ldots \beta_{p+1} x_{pi} + e_i$$

the posterior mean of $\hat{e}_i = y_i - \hat{\beta} x_i$ compared to its posterior standard deviation provides an indication of outlier status (Pettitt and Smith, 1985; Chaloner, 1998) – see Example 3.12. In frequentist applications of this regression model, the influence of a particular case is apparent in the ratio of $\text{Var}(\hat{e}_i) = \sigma^2(1 - v_i)$ to the overall residual variance σ^2, where $v_i = x_i'[X'X]^{-1}x_i$, with X the $n \times (p+1)$ covariate matrix for all cases; a similar procedure may be used in Bayesian analysis.

Alternatively, the CPO predictive quantity $f(y_i|y_{[-i]})$ may be used as an outlier diagnostic and as the basis for influence measures. Weiss and Cho (1998) consider possible divergence criteria in terms of the ratios $a_i = [\text{CPO}_i / f(y_i|\theta)]$, such as the L_1 norm, with the influence of case i on the totality of model parameters then repre-

[18] A simple approach to predictive fit generalises the method of Laud and Ibrahim (1995) – see Example 3.2 – and is mentioned by Gelfand and Ghosh (1998), Sahu *et al.* (1997) and Ibrahim *et al.* (2001). Let y_i be the observed data, ϕ be the parameters, and z_i be 'new' data sampled from $f(z|\phi)$. Suppose v_i and ς_i are the posterior mean and variance of z_i, then one possible criterion for any $w > 0$ is

$$C = \sum_{i=1}^{n} \varsigma_i + [w/(w+1)] \sum_{i=1}^{n} (v_i - y_i)^2$$

Typical values of w at which to compare models might be $w = 1$, $w = 10$ and $w = 100{,}000$. Larger values of w put more stress on the match between v_i and y_i and so downweight precision of predictions. Gelfand and Ghosh (1998) develop deviance-based criteria specific for non-Normal outcomes (see Chapter 3), though these assume no missingness on the response.

sented by $d(a_i) = 0.5|a_i - 1|$ – see Example 1.4. Specific models, such as those introducing latent data, lead to particular types of Bayesian residual (Jackman, 2000). Thus, in a binary probit or logit model, underlying the observed binary y are latent continuous variables z, confined to negative or positive values according as y is 0 or 1. The estimated residual is then $z - \hat{\beta}x_i$ analogously to a Normal errors model.

Example 1.3 Lung cancer in London small areas As an example of the possible influence of prior specification on regression coefficients and random effects, consider a small area health outcome: female lung cancer deaths y_i in the three year period 1990–92 in 758 London small areas[19] (electoral wards). If we focus first on regression effects, there is overwhelming accumulated evidence that ill health and mortality (especially lung cancer deaths) are higher in more deprived, lower income areas. Having allowed for the impact of age differences via indirect standardisation (to provide expected deaths E_i) variations in this type of mortality are expected to be positively related to a deprivation score x_i, which is in standard form (zero mean, variance 1). The following model is assumed

$$y_i \sim \text{Poi}(\mu_i)$$
$$\mu_i = E_i\rho_i$$
$$\log(\rho_i) = \beta_1 + \beta_2 x_i$$

The only parameters, β_1 and β_2, are assigned diffuse but proper N(0,1000) priors. Since the sum of observed and expected deaths is the same and x is standardised, one might expect β_1 to be near zero. Two sets initial values of adopted $\beta = (0, 0)$ and $\beta = (0, 0.2)$ with the latter the mean of a trial (single chain) run. A two chain run then shows early convergence via Gelman-Rubin criteria (at under 250 iterations) and from iterations 250–2500 pooled over the chains a 95% credible interval for β_2 of (0.18,0.24) is obtained.

However, there may well be information which would provide more informative priors. Relative risks ρ_i between areas for major causes of death (from chronic disease) reflect, albeit imperfectly, gradients in risk for individuals over attributes such as income, occupation, health behaviours, household tenure, ethnicity, etc. These gradients typically show at most five fold variation between social categories except perhaps for risk behaviours directly implicated in causing a disease. Though area contrasts may also be related to environmental influences (usually less strongly) accumulated evidence, including evidence for London wards, suggests that extreme relative contrasts in standard mortality ratios ($100 \times \rho_i$) between areas are unlikely to exceed 10 or 20 (i.e. SMRs ranging from 30 to 300, or 20 to 400 at the outside). Simulating with the known covariate x_i and expectancies E_i it is possible to obtain or 'elicit' priors consistent with these prior beliefs. For instance one might consider taking a N(0,1) prior on β_1 and a N(0.5,1) prior on β_2. The latter favours positive values, but still has a large part of its density over negative values.

Values of y_i are simulated (see Model 2 in Program 1.3) with these priors; note that initial values are by definition generated from the priors, and since this is pure simulation there is no notion of convergence. Because relative risks tend to be skewed, the median relative risks (i.e. y_i/E_i) from a run of 1000 iterations are considered as

[19] The first is the City of London (1 ward), then wards are alphabetic within boroughs arranged alphabetically (Barking, Barnet, ...,Westminster). All wards have five near neighbours as defined by the nearest wards in terms of crow-fly distance.

summaries of contrasts between areas under the above priors. The extreme relative risks are found to be 0 and 6 (SMRs of 0 and 600) and the 2.5% and 97.5% percentiles of relative risk are 0.37 and 2.99. So this informative prior specification appears broadly in line with accumulated evidence.

One might then see how far inference about β_2 is affected by adopting the N(0.5, 1) prior instead of the N(0, 1000) diffuse prior[20] when the observations are restored. In fact, the 95% credible interval from a two chain run (with initial values as before and run length of 2500 iterations) is found to be the same as under the diffuse prior.

A different example of sensitivity analysis involves using a contamination prior on β_2. Thus, suppose $\pi_1(\beta_2)$ is N(0.5, 1) as above, but that for $\pi_2(\beta_2)$ a Student t with 2 degrees of freedom but same mean zero and variance is adopted, and $p = 0.1$. Again, the same credible interval for β_2 is obtained as before (Model 3 in Program 1.3). One might take the contaminating prior to be completely flat (dflat( ) in BUGS), and this is suggested as an exercise. In the current example, inferences on β_2 appear robust here to alternative priors, and this is frequently the case with regression parameters in large samples – though with small datasets there may well be sensitivity.

An example where sensitivity in inferences concerning random effects may occur is when the goal in a small area mortality analysis is not the analysis of regressor effects but the smoothing of unreliable rates based on small event counts or populations at risk (Manton *et al.*, 1987). Such smoothing or 'pooling strength' uses random effects over a set of areas to smooth the rate for any one area towards the average implied under the density of the effects. Two types of random effect have been suggested, one known as unstructured or 'white noise' variation, whereby smoothing is towards a global average, and spatially structured variation whereby smoothing is towards the average in the 'neighbourhood' of adjacent wards. Then the total area effect α_i consists of an unstructured or 'pure heterogeneity' effect υ_i and a spatial effect ϕ_i. While the data holds information about which type of effect is more predominant, the prior on the variances σ^2_υ and σ^2_ϕ may also be important in identifying the relative roles of the two error components.

A popular prior used for specifying spatial effects, the CAR(1) prior of Besag *et al.* (1991), introduces an extra identifiability issue in that specifies differences in risk between areas i and j, $\phi_i - \phi_j$, but not the average level (i.e. the location) of the spatial risk (see Chapter 7). This prior can be specified in a conditional form, in which

$$\phi_i \sim \mathrm{N}\Big(\sum_{j\in A_i} e_j,\ \sigma^2_\phi/M_i\Big)$$

where M_i is the number of areas adjacent to area i, and $j \in A_i$ denotes that set of areas. To resolve the identifiability problem one may centre the sampled ϕ_i at each MCMC iteration and so provide a location, i.e. actually use in the model to predict $\log(\rho_i)$ the shifted effects $\phi'_i = \phi_i - \bar{\phi}$. In fact, following Sun *et al.* (1999), identifiability can also be gained by introducing a correlation parameter γ

$$\phi_i \sim \mathrm{N}\Big(\gamma\sum_{j\in A_i} \phi_j,\ \sigma^2_\phi/M_i\Big) \qquad (1.8)$$

which is here taken to have prior $\gamma \sim U(0, 1)$. Issues still remain in specifying priors on σ^2_υ and σ^2_ϕ (or their inverses) and in identifying both these variances and the separate risks υ_i and ϕ_i in each area in the model

[20] The prior on the intercept is changed to N(0, 1) also.

$$\log(\rho_i) = \beta_1 + \upsilon_i + \phi_i$$

where the prior for ϕ_i is taken to be as in Equation (1.8) and where $\upsilon_i \sim N(0, \sigma_\upsilon^2)$. A 'diffuse prior' strategy might be to adopt gamma priors $G(a_1, a_2)$ on the precisions $1/\sigma_\upsilon^2$ and $1/\sigma_\phi^2$, where $a_1 = a_2 = a$ and a is a small constant such as $a = 0.001$, but possible problems in doing this are noted above. One might, however, set priors on a_1 and a_2 themselves rather than presetting them (Daniels and Kass, 1999), somewhat analogous to contamination priors in allowing for higher level uncertainty.

Identifiability might also be improved by instead linking the specification of υ_i and ϕ_i in some way (see Model 4 in Program 1.3). For example, one might adopt a bivariate prior on these random effects as in Langford *et al.* (1998) and discussed in Chapter 7. Or one might still keep υ_i and ϕ_i as univariate errors, but recognise that the variances are interdependent, for instance taking $\sigma_\upsilon^2 = c\sigma_\phi^2$ so that one variance is conditional on the other and a pre-selected value of c. Bernardinelli *et al.* (1995) recommend $c = 0.7$. A prior on c might also be used, e.g. a gamma prior with mean 0.7. One might alternatively take a bivariate prior (e.g. bivariate Normal) on $\log(\sigma_\upsilon^2)$ and $\log(\sigma_\phi^2)$. Daniels (1999) suggests uniform priors of the ratio of one variance to the sum of the variances, for instance a U(0, 1) prior on $\sigma_\upsilon^2/[\sigma_\upsilon^2 + \sigma_\phi^2]$, though the usual application of this approach is to other forms of hierarchical model.

Here we first consider independent G(0.5, 0.0005) priors on $1/\sigma_\upsilon^2$ and $1/\sigma_\phi^2$ in a two chain run. One set of initial values is provided by 'default' values, and the other by setting the model's central parameters to their mean values under an initial single chain run. The problems possible with independent diffuse priors show in the relatively slow convergence of σ_ϕ; not until 4500 iterations does the Gelman-Rubin statistic fall below 1.1. As an example of inferences on relative mortality risks, the posterior mean for the first area, where there are three deaths and 2.7 expected (a crude relative risk of 1.11), is 1.28, with 95% interval from 0.96 to 1.71. The risk for this area is smoothed upwards to the average of its five neighbours, all of which have relatively high mortality. This estimate is obtained from iterations 4500–9000 of the two chain run. The standard deviations σ_υ and σ_ϕ of the random effects have posterior medians 0.041 and 0.24.

In a second analysis the variances[21] are interrelated with $\sigma_\upsilon^2 = c\sigma_\phi^2$ and c taken as G(0.7, 1). This is relatively informative prior structure, and reflects the expectation that any small area health outcome will probably show both types of variability. Further, the prior on $1/\sigma_\phi^2$ allows for uncertainty in the parameters, i.e. instead of a default prior such as $1/\sigma_\phi^2 \sim G(1, 0.001)$, it is assumed that

$$1/\sigma_\phi^2 \sim G(a_1, a_2)$$

with

$$a_1 \sim \text{Exp}(1)$$

$$a_2 \sim G(1, 0.001)$$

The priors for a_1 and a_2 reflect the option sometimes used for a diffuse prior on precisions such as $1/\sigma_\phi^2$, namely a G(1, v) prior on precisions (with v preset at a small constant, such as $v = 0.001$).

This model achieves convergence in a two chain run of 10 000 iterations at around 3000 iterations, and yields a median for c of 0.12, and for σ_υ and σ_ϕ of 0.084 and 0.24. The posterior medians of a_1 and a_2 are 1.7 and 0.14. Despite the greater element of pure

[21] In BUGS this inter-relationship involves precisions.

heterogeneity the inference on the first relative risk is little affected, with mean 1.27 and 95% credible interval (0.91, 1.72).

So some sensitivity is apparent regarding variances of random effects in this example despite the relatively large sample, though substantive inferences may be more robust. A suggested exercise is to experiment with other priors allowing interdependent variances or errors, e.g. a U(0, 1) prior on $\sigma_\upsilon^2/[\sigma_\upsilon^2 + \sigma_\phi^2]$. A further exercise might involve summarising sensitivity on the inferences about relative risk, e.g. how many of the 758 mean relative risks shift upward or downward by more than 2.5%, and how many by more than 5%, in moving from one random effects prior to another.

Example 1.4 Gessel score To illustrate possible outlier analysis, we follow Pettitt and Smith (1985) and Weiss and Cho (1998), and consider data for $n = 21$ children on Gessel adaptive score (y) in relation to age at first word (x in months). Adopting a Normal errors model with parameters $\beta = (\beta_1, \beta_2)$, estimates of the CPO_i may be obtained by single case omission, but an approximation based on a single posterior sample avoids this. Thus for T samples (Weiss, 1994),

$$\text{CPO}_i^{-1} \approx T^{-1} \sum_{t=1}^{T} [f(y_i|\beta^{(t)}, x_i)]^{-1}$$

or the harmonic mean of the likelihoods of case i. Here an initial run is used to estimate the CPOs in this way, and a subsequent run produces influence diagnostics, as in Weiss and Cho (1998). It is apparent (Table 1.1) that child 19 is both a possible outlier and influential on the model parameters, but child 18 is influential without being an outlier.

Table 1.1 Diagnostics for Gessel score

Child	CPO	Influence (Kullback K1)	Influence (L1 norm)	Influence (chi square)
1	0.035	0.014	0.066	0.029
2	0.021	0.093	0.161	0.249
3	0.011	0.119	0.182	0.341
4	0.025	0.032	0.096	0.071
5	0.024	0.022	0.081	0.047
6	0.035	0.015	0.068	0.031
7	0.034	0.015	0.068	0.031
8	0.035	0.014	0.067	0.029
9	0.034	0.016	0.071	0.035
10	0.029	0.023	0.083	0.051
11	0.019	0.067	0.137	0.166
12	0.033	0.017	0.072	0.035
13	0.011	0.119	0.182	0.341
14	0.015	0.066	0.137	0.163
15	0.032	0.015	0.068	0.032
16	0.035	0.014	0.067	0.030
17	0.025	0.022	0.081	0.047
18	0.015	1.052	0.387	75.0
19	0.000138	2.025	0.641	56.2
20	0.019	0.042	0.111	0.098
21	0.035	0.014	0.067	0.030

As Pettitt and Smith (1985) note, this is because child 18 is outlying in the covariate space, with age at first word (x) much later than other children, whereas child 19 is outlying in the response (y) space.

1.6 REVIEW

The above worked examples are inevitably selective, but start to illustrate some of the potentials of Bayesian methods but also some of the pitfalls in terms of the need for 'cautious inference'. The following chapters consider similar modelling questions to those introduced here, and include a range of worked examples. The extent of possible model checking in these examples is effectively unlimited, and a Bayesian approach raises additional questions such as sensitivity of inferences to assumed priors.

The development in each chapter draws on contemporary discussion in the statistical literature, and is not confined to reviewing Bayesian work. However, the worked examples seek to illustrate Bayesian modelling procedures, and to avoid unduly lengthy discussion of each, the treatments will leave scope for further analysis by the reader employing different likelihoods, prior assumptions, initial values, etc.

Chapter 2 considers the potential for pooling information across similar units (hospitals, geographic areas, etc.) to make more precise statements about parameters in each unit. This is sometimes known as 'hierarchical modelling', because higher level priors are specified on the parameters of the population of units. Chapter 3 considers model choice and checking in linear and general linear regressions. Chapter 4 extends regression to clustered data, where regression parameters may vary randomly over the classifiers (e.g. schools) by which the lowest observation level (e.g. pupils) are classified.

Chapters 5 and 6 consider time series and panel models, respectively. Bayesian specifications may be relevant to assessing some of the standard assumptions of time series models (e.g. stationarity in ARIMA models), give a Bayesian interpretation to models commonly fitted by maximum likelihood such as the basic structural model of Harvey (1989), and facilitate analysis in more complex problems, for example, shifts in means and/or variances of series. Chapter 6 considers Bayesian treatments of the growth curve model for continuous outcomes, as well as models for longitudinal discrete outcomes, and panel data subject to attrition. Chapter 7 considers observations correlated over space rather than through time, and models for discrete and continuous outcomes, including instances where regression effects may vary through space, and where spatially correlated outcomes are considered through time.

An alternative to expressing correlation through multivariate models is to introduce latent traits or classes to model the interdependence. Chapter 8 considers a variety of what may be termed structural equation models, the unity of which with the main body of statistical models is now being recognised (Bollen, 2001).

The final two chapters consider techniques frequently applied in biostatistics and epidemiology, but certainly not limited to those application areas. Chapter 9 considers Bayesian perspectives on survival analysis and chapter 10 considers ways of using data to develop support for causal mechanisms, as in meta-analysis and dose-response modelling.

REFERENCES

Albert, J. and Chib, S (1993) Bayesian analysis of binary and polychotomous response data. *J. Am. Stat. Assoc.* **88**, 669–679.

Alqallaf, F. and Gustafson, P. (2001) On cross-validation of Bayesian models. *Can. J. Stat.* **29**, 333–340.

Berger, J. (1985) *Statistical Decision Theory and Bayesian Analysis*. New York: Springer-Verlag.

Berger, J. (1990) Robust Bayesian analysis: Sensitivity to the prior. *J. Stat. Plann. Inference* **25**(3), 303–328.

Bernardinelli, L., Clayton, D., Pascutto, C., Montomoli, C., Ghislandi, M. and Songini, M. (1995) Bayesian-analysis of space-time variation in disease risk. *Statistics Medicine* **14**, 2433–2443.

Berry, D. (1980) Statistical inference and the design of clinical trials. *Biomedicine* **32**(1), 4–7.

Bollen, K. (2002) Latent variables in psychology and the social sciences. *Ann. Rev. Psychol.* **53**, 605–634.

Box, G. and Tiao, G. (1973) *Bayesian Inference in Statistical Analysis*. Addison-Wesley.

Bray, I. (2002) Application of Markov chain Monte Carlo methods to projecting cancer incidence and mortality. *J. Roy. Statistics Soc. Series C*, **51**, 151–164.

Carlin, B. and Louis, T. (1996) *Bayes and Empirical Bayes Methods for Data Analysis*. Monographs on Statistics and Applied Probability. 69. London: Chapman & Hall.

Chib, S. (1995) Marginal likelihood from the Gibbs output. *J. Am. Stat. Assoc.* **90**, 1313–1321.

D'Agostini, G. (1999) Bayesian Reasoning in High Energy Physics: Principles and Applications. *CERN Yellow Report 99–03*, Geneva.

Daniels, M. (1999) A prior for the variance in hierarchical models. *Can. J. Stat.* **27**(3), 567–578.

Daniels, M. and Kass, R. (1999) Nonconjugate Bayesian estimation of covariance matrices and its use in hierarchical models. *J. Am. Stat. Assoc.* **94**, 1254–1263.

Fraser, D. McDunnough, P. and Taback, N. (1997) Improper priors, posterior asymptotic Normality, and conditional inference. In: Johnson, N. L. *et al.*, (eds.) *Advances in the Theory and Practice of Statistics*. New York: Wiley, pp. 563–569.

Gaver, D. P. and O'Muircheartaigh, I. G. (1987) Robust empirical Bayes analyses of event rates. *Technometrics*, **29**, 1–15.

Gehan, E. (1965) A generalized Wilcoxon test for comparing arbitrarily singly-censored samples. *Biometrika* **52**, 203–223.

Gelfand, A., Dey, D. and Chang, H. (1992) Model determination using predictive distributions with implementation via sampling-based methods. In: Bernardo, J. M., Berger, J. O., Dawid, A. P. and Smith, A. F. M. (eds.) *Bayesian Statistics 4*, Oxford University Press, pp. 147–168.

Gelfand, A. (1996) Model determination using sampling-based methods. In: Gilks, W., Richardson, S. and Spiegelhalter, D. (eds.) *Markov Chain Monte Carlo in Practice* London: Chapman: & Hall, pp. 145–161.

Gelfand, A. and Dey, D. (1994) Bayesian model choice: Asymptotics and exact calculations. *J. Roy. Stat. Soc., Series B* **56**(3), 501–514.

Gelfand, A. and Ghosh, S (1998) Model choice: A minimum posterior predictive loss approach. *Biometrika* **85**(1), 1–11.

Gelfand, A. and Smith, A. (1990) Sampling-based approaches to calculating marginal densities. *J. Am. Stat. Assoc.* **85**, 398–409.

Gelfand, A., Sahu, S. and Carlin, B. (1995) Efficient parameterizations for normal linear mixed models. *Biometrika* **82**, 479–488.

Gelfand, A., Ghosh, S., Knight, J. and Sirmans, C. (1998) Spatio-temporal modeling of residential sales markets. *J. Business & Economic Stat.* **16**, 312–321.

Gelfand, A. and Sahu, S. (1999) Identifiability, improper priors, and Gibbs sampling for generalized linear models. *J. Am. Stat. Assoc.* **94**, 247–253.

Gelman, A., Carlin, J. B. Stern, H. S. and Rubin, D. B. (1995) *Bayesian Data Analysis, 1st ed.* Chapman and Hall Texts in Statistical Science Series. London: Chapman & Hall.

Gelman, A. (1996) Inference and monitoring convergence. In: Gilks, W., Richardson, S. and Spiegelhalter, D. (eds.). *Practical Markov Chain Monte Carlo*, London: Chapman & Hall, pp. 131–143.

George, E., Makov, U. and Smith, A. (1993) Conjugate likelihood distributions. *Scand. J. Stat.* **20**(2), 147–156.

Geweke, J. (1992). Evaluating the accuracy of sampling-based approaches to calculating posterior moments. In: Bernardo, J. M., Berger, J. O., Dawid, A. P. and Smith, A. F. M. (eds.), *Bayesian Statistics 4*. Oxford: Clarendon Press.

Ghosh, M. and Rao, J. (1994) Small area estimation: an appraisal. *Stat. Sci.* **9**, 55–76.

Gilks, W. R. and Wild, P. (1992) Adaptive rejection sampling for Gibbs sampling. *Appl. Stat.* **41**, 337–348.

Gilks, W. and Roberts, C. (1996) Strategies for improving MCMC. In: Gilks, W., Richardson, S. and Spiegelhalter, D. (eds.), *Practical Markov Chain Monte Carlo*. London: Chapman & Hall, pp. 89–114.

Gilks, W. R., Roberts, G. O. and Sahu, S. K. (1998) Adaptive Markov chain Monte Carlo through regeneration. *J. Am. Stat. Assoc.* **93**, 1045–1054.

Gustafson, P. (1996) Robustness considerations in Bayesian analysis. *Stat. Meth. in Medical Res.* **5**, 357–373.

Harvey, A. (1993) *Time Series Models, 2nd ed.* Hemel Hempstead: Harvester-Wheatsheaf.

Jackman, S. (2000) Estimation and inference are 'missing data' problems: unifying social science statistics via Bayesian simulation. *Political Analysis*, **8**(4), 307–322.

Jaynes, E. (1968) Prior probabilities. *IEEE Trans. Syst., Sci. Cybernetics* **SSC-4**, 227–241.

Jaynes, E. (1976) Confidence intervals vs Bayesian intervals. In: Harper, W. and Hooker, C. (eds.), *Foundations of Probability Theory, Statistical Inference, and Statistical Theories of Science*. Dordrecht: Reidel.

Kelsall, J. E. and Wakefield, J. C. (1999) Discussion on Bayesian models for spatially correlated disease and exposure data (by N. G. Best *et al.*). In: Bernardo, J. *et al.* (eds.), *Bayesian Statistics 6: Proceedings of the Sixth Valencia International Meeting*. Oxford: Clarendon Press.

Knorr-Held, L. and Rainer, E. (2001) Prognosis of lung cancer mortality in West Germany: a case study in Bayesian prediction. *Biostatistics* **2**, 109–129.

Langford, I., Leyland, A., Rasbash, J. and Goldstein, H. (1999) Multilevel modelling of the geographical distributions of diseases. *J. Roy. Stat. Soc., C*, **48**, 253–268.

Lenk, P. and Desarbo, W. (2000) Bayesian inference for finite mixtures of generalized linear models with random effects. *Psychometrika* **65**(1), 93–119.

Manton, K., Woodbury, M., Stallard, E., Riggan, W., Creason, J. and Pellom, A. (1989) Empirical Bayes procedures for stabilizing maps of US cancer mortality rates. *J. Am. Stat. Assoc.* **84**, 637–650.

Mollié, A. (1996) Bayesian mapping of disease. In: Gilks, W., Richardson, S. and Spieglehalter, D. (eds.), *Markov Chain Monte Carlo in Practice*. London: Chapman & Hall, pp. 359–380.

Morgan, B. (2000) *Applied Stochastic Modelling*. London: Arnold.

Neal, R. (1997) Markov chain Monte Carlo methods based on 'slicing' the density function. *Technical Report No.9722*, Department of Statistics, University of Toronto.

Neal, R. (1998) Suppressing random walks in Markov chain Monte Carlo using ordered over-relaxation. In: Jordan, M. (ed.), *Learning in Graphical Models*, Dordrecht: Kluwer Academic, pp. 205–225.

Newton, D. and Raftery, J. (1994) Approximate Bayesian inference by the weighted bootstrap. *J. Roy. Stat. Soc. Series B*, **56**, 3–48.

O'Hagan, A. (1994) *Bayesian Inference, Kendalls Advanced Theory of Statistics*. London: Arnold.

Rao, C. (1975) Simultaneous estimation of parameters in different linear models and applications to biometric problems. *Biometrics* **31**(2), 545–549.

Sahu, S. (2001) *Bayesian estimation and model choice in item response models*. Faculty of Mathematical Studies, University of Southampton.

Smith, T., Spiegelhalter, D. and Thomas, A. (1995) Bayesian approaches to random-effects meta-analysis: a comparative study. *Stat. in Medicine* **14**, 2685–2699.

Spiegelhalter, D, Best, N, Carlin, B and van der Linde, A (2002) Bayesian measures of model complexity and fit, J. Royal Statistical Society, 64B, 1–34.

Spiegelhalter, D., Best, N., Gilks, W. and Inskip, H. (1996) Hepatitis B: a case study of Bayesian methods. In: Gilks, W., Richardson, S. and Spieglehalter, D. (eds.), *Markov Chain Monte Carlo in Practice*. London: Chapman & Hall, pp. 21–43.

Sun, D., Tsutakawa, R. and Speckman, P. (1999) Posterior distribution of hierarchical models using CAR(1) distributions. *Biometrika* **86**, 341–350.

Tanner, M. (1993) *Tools for Statistical Inference. Methods for the Exploration of Posterior Distributions and Likelihood Functions, 2nd ed.* Berlin: Springer-Verlag.

Vines, S., Gilks, W. and Wild, P. (1996) Fitting Bayesian multiple random effects models. *Stat. Comput.* **6**, 337–346.

Wasserman, L. (2000) Bayesian model selection and model averaging. *J. Math. Psychol.* **44**, 92–107.

Weiss, R. (1994) Pediatric pain, predictive inference, and sensitivity analysis. *Evaluation Rev.*, **18**, 651–677.

Weiss, R. and Cho, M. (1998) Bayesian marginal influence assessment. *J. Stat. Planning & Inference* **71**, 163–177.

Zellner, A. (1971) *An Introduction to Bayesian Inference in Econometrics*. New York: Wiley.

Zuur, G., Gartwaite, P. and Fryer, R. (2002) Practical use of MCMC methods: lessons from a case study. *Biometrical J.* **44**, 433–455.

CHAPTER 2

Hierarchical Mixture Models

2.1 INTRODUCTION: SMOOTHING TO THE POPULATION

A relatively simple Bayesian problem, but one which has motivated much research, is that of ensemble estimation, namely estimating the parameters of a common distribution thought to underlay a collection of outcomes for similar types of units. Among possible examples are medical, sports, or educational: death rates for geographical areas, batting averages for baseball players, Caesarian rates in maternity units, and exam success rates for schools. Given the parameters of the common density, one seeks to make conditional estimates of the true outcome rate in each unit of observation. Because of this conditioning on the higher stage densities, such estimation for sets of similar units is also known as 'hierarchical modelling' (Kass and Steffey, 1989; Lee, 1997, Chapter 8). For instance, in the first stage of the Poisson-gamma model considered below, the observed counts are conditionally independent given the unknown means that are taken to have generated them. At the second stage, these means are themselves determined by the gamma density parameters, while the density for the gamma parameters forms the third stage.

These procedures, whether from a full or empirical Bayes perspective, usually result in a smoothing of estimates for each unit towards the average outcome rate, and have generally been shown to have greater precision and better out of sample predictive performance. Specifically, Rao (1975) shows that with respect to a quadratic loss function, empirical Bayes estimators outperform classical estimators in problems of simultaneous inference regarding a set of related parameters. These procedures may, however, imply a risk of bias as against unadjusted maximum likelihood estimates – this dilemma is known as the bias-variance trade-off.

Such procedures for 'pooling strength' rest on implicit assumptions: that the units are exchangeable (similar enough to justify an assumption of a common density), and that the smoothing model chosen is an appropriate one. It may be that units are better considered exchangeable within sub-groups of the data (e.g. outcomes for randomised trials in one sub-group vs. outcomes for observational studies in another). Model choice is an additional uncertainty (e.g. does one take parametric or non-parametric approach to smoothing, and if a non-parametric discrete mixture, how many components?).

Therefore, this chapter includes some guidelines as to model comparison and choice, which will be applicable to this and later chapters. There are no set 'gold standard' model choice criteria, though some arguably come closer to embodying true Bayesian

Applied Bayesian Modelling P. Congdon
 ISBN: 0-471-48695-7

principles than others. Often one may compare 'classical' fit measures such as deviance or the Akaike Information Criterion (Bozdogan, 2000), either averages over an MCMC chain (e.g. averages of deviances $D^{(t)}$ attaching to parameters $\theta^{(t)}$ at each iteration), or in terms of the deviance at the posterior mean. These lead to a preliminary sifting of models and more comprehensive model assessment, and selection is reserved to a final stage of the analysis involving a few closely competing models.

2.2 GENERAL ISSUES OF MODEL ASSESSMENT: MARGINAL LIKELIHOOD AND OTHER APPROACHES

There is usually uncertainty about appropriate error structures and predictor variables to include in models. Adding more parameters may improve fit, but maybe at the expense of identifiability and generalisability. Model selection criteria assess whether improvements in fit measures such as likelihoods, deviances or error sum of squares justify the inclusion of extra parameters in a model. Classical and Bayesian model choice methods may both involve comparison either of measures of fit to the current data or cross validatory fit to out of sample data. For example, the deviance statistics of general linear models (with Poisson, normal, binomial or other exponential family outcomes) follow standard densities for comparisons of models nested within one another, at least approximately in large samples (McCullagh and Nelder, 1989). Penalised measures of fit (Bozdogan, 2000; Aikake, 1973) may be used, involving an adjustment to the model log-likelihood or deviance to reflect the number of parameters in the model.

Thus, suppose L denotes the likelihood and D the deviance of a model involving p parameters. The deviance may be simply defined as minus twice the log likelihood, $D = -2\log L$, or as a scaled deviance:

$$D' = -2\log(L/L_s),$$

where L_s is the saturated likelihood obtained by an exact fit of predicted to observed data. Then to allow for the number of parameters (or 'dimension' of the model), one may use criteria such as the Akaike Information Criterion (or AIC), expressed either as[1]

$$D + 2p$$

or

$$D' + 2p$$

So when the AIC is used to compare models, an increase in likelihood and reduction in deviance is offset by a greater penalty for more complex models.

Another criterion used generally as a penalised fit measure, though also justified as an asymptotic approximation to the Bayesian posterior probability of a model, is the Schwarz Information Criterion (Schwarz, 1978). This is also often called the Bayes Information Criterion. Depending on the simplifying assumptions made, it may take different forms, but the most common version is, for sample of size n,

$$\text{BIC} = D + p\log_e(n)$$

[1] So a model is selected if it has lowest AIC. Sometimes the AIC is obtained as $L - p$ with model selection based on *maximising* the AIC.

Under this criterion models with lower BIC are chosen, and larger models (with more parameters) are more heavily penalised than under the AIC. The BIC approximation for model j is derived by considering the posterior probability for the model M_j as in Equation (2.1) below, and by expanding minus twice the log of that quantity around the maximum likelihood estimate (or maybe some other central estimate).

In Bayesian modelling, prior information is introduced on the parameters, and the fit of the model to the data at hand and the resulting posterior parameter estimates are constrained to some degree by adherence also to this prior 'data'. One option is to simply compare averages of standard fit measures such as the deviance or BIC over an MCMC run, e.g. consider model choice in terms of a model which has minimum average AIC or BIC. Approaches similar in some ways to classical model validation procedures are often required because the canonical Bayesian model choice methods (via Bayes factors) are infeasible or difficult to apply in complex models or large samples (Gelfand and Ghosh, 1998; Carlin and Louis, 2000, p. 220). The Bayes factor may be sensitive to the information contained in diffuse priors, and is not defined for improper priors.

Monitoring fit measures such as the deviance over an MCMC run has utility if one seeks penalised fit measures taking account of model dimension. A complication is that the number of parameters in complex random effects models is not actually defined. Here work by Spiegelhalter *et al.* (2002) may be used to estimate the effective number of parameters, denoted p_e. Specifically, for data y and parameters θ, p_e is approximated by the difference between the expected deviance $E(D|y, \theta)$, as measured by the posterior mean of sampled deviances $D^{(t)} = D(\theta^{(t)})$ at iterations $t = 1, .., T$ in a long MCMC run, and the deviance $D(\bar{\theta}|y)$, evaluated at the posterior mean $\bar{\theta}$ of the parameters. Then one may define a penalised fit measure analogous to the Akaike information criterion as

$$D(\bar{\theta}|y) + 2p_e$$

and this has been termed the Deviance Information Criterion. Alternatively a modified Bayesian Information Criterion BIC $= D(\bar{\theta} \mid y) + p_e \log(n)$ may be used, as this takes account of both sample size and complexity (Upton, 1991; Raftery, 1995). Note that p_e might also be obtained by comparing an average likelihood with the likelihood at the posterior mean and then multiplying by 2. Related work on effective parameters when the average likelihoods of two models are compared appears in Aitkin (1991).

The Bayesian approach to model choice and its implementation via MCMC sampling methods has benefits in comparisons of non-nested models – for instance, in comparing two nonlinear regressions or comparing a beta-binomial model as against a discrete mixture of binomials (Morgan, 2000). A well known problem in classical statistics is in likelihood comparisons of discrete mixture models involving different numbers of components, and here the process involved in Bayesian model choice is simpler.

2.2.1 Bayes model selection using marginal likelihoods

The formal Bayesian model assessment scheme involves marginal likelihoods, and while it follows a theoretically clear procedure may in practice be difficult to implement. Suppose K models, denoted M_k, $k = 1, .. K$, have prior probabilities $\phi_k = P(M_k)$ assigned to them of being true, with $\Sigma_{k=1,K}\phi_k = 1$. Let θ_k be the parameter set in model k, with prior $\pi(\theta_k)$. Then the posterior probabilities attaching to each model after observing data y are

$$P(M_k|y) = P(M_k)\int f(y|\theta_k)\pi(\theta_k)d\theta_k \Big/ \sum_{j=1}^{K}\{P(M_j)\int f(y|\theta_j)\pi(\theta_j)d\theta_j\} \tag{2.1}$$

where $f(y|\theta_k) = L(\theta_k|y)$ is the likelihood of the data under model k. The integrals in both the denominator and numerator of Equation (2.1) are known as prior predictive densities or marginal likelihoods (Gelfand and Dey, 1994). They give the probability of the data conditional on a model as

$$P(y|M_k) = m_k(y) = \int f(y|\theta_k)\pi(\theta_k)d\theta_k \tag{2.2}$$

The marginal density also occurs in Bayes Formula for updating the parameters θ_k of model k, namely

$$\pi(\theta_k|y) = f(y|\theta_k)\pi(\theta_k)/m_k(y) \tag{2.3}$$

where $\pi(\theta_k|y)$ denotes the posterior density of the parameters. This is also expressible as the 'marginal likelihood identity' (Chib, 1995; Besag, 1989):

$$m_k(y) = f(y|\theta_k)\pi(\theta_k)/\pi(\theta_k|y) \tag{2.4}$$

Model assessment can often be reduced to a sequential set of choices between two competing models – though an increased emphasis is now being placed on averaging inferences over models. It is in such comparisons that marginal likelihoods play a role. The formal method for comparing two competing models in a Bayesian framework involves deriving posterior odds after estimating the models separately. For equal prior odds on two models M_1 and M_2, with parameters θ_1 and θ_2 of dimension p_1 and p_2, this is equivalent to examining the Bayes factor on model 2 versus model 1. The Bayes factor is obtained as the ratio of marginal likelihoods $m_1(y)$ and $m_2(y)$, such that

$$\begin{array}{ccccc} \dfrac{P(M_1|y)}{P(M_2|y)} & = & \dfrac{P(y|M_1)}{P(y|M_2)} & & \dfrac{P(M_1)}{P(M_2)} \\ \text{PosteriorOdds} & & \text{Bayesfactor} & & \text{PriorOdds} \end{array} \tag{2.5}$$

$$(= [m_1(y)/m_2(y)]\ [\phi_1/\phi_2])$$

The integral in Equation (2.2) can in principle be evaluated by sampling from the prior and calculating the resulting likelihood, and is sometimes available analytically. However, more complex methods are usually needed, and in highly parameterised or non-conjugate models a fully satisfactory procedure has yet to be developed. Several approximations have been suggested, some of which are described below. Another issue concerns Bayes factor stability when flat or just proper non-informative priors are used on parameters. It can be demonstrated that such priors lead (when models are nested within each other) to simple models being preferred over more complex models – this is Lindley's paradox (Lindley, 1957), with more recent discussions in Gelfand and Dey (1994) and DeSantis and Spezzaferri (1997). By contrast, likelihood ratios used in classical testing tend to favour more complex models by default (Gelfand and Dey, 1994). Even under proper priors, with sufficiently large sample sizes the Bayes factor tends to attach too little weight to the correct model and too much to a less complex or null model. Hence, some advocate a less formal view to Bayesian model selection based on predictive criteria other than the Bayes factor (see Section 2.2.4). These may lead to model checks analogous to classical p tests or to pseudo-Bayes factors of various kinds.

2.2.2 Obtaining marginal likelihoods in practice

MCMC simulation methods are typically applied to deriving posterior densities $f(\theta|y)$ or sampling predictions y_{new} in models considered singly. However, they have extended to include parameter estimation and model choice in the joint parameter and model space $\{\theta_k, M_k\}$ for $k = 1, .., K$ (Carlin and Chib, 1995). Thus, at iteration t there might be a switch between models (e.g. from M_j to M_k) and updating only on the parameters in model k. For equal prior model probabilities, the best model is the one chosen most frequently, and the posterior odds follow from Equation (2.5). The reversible jump algorithm of Green (1995) also provides a joint space estimation method.

However, following a number of studies such as Chib (1995), Lenk and Desarbo (2000) and Gelfand and Dey (1994), the marginal likelihood of a single model may be approximated from the output of MCMC chains. The most simple apparent estimator of the marginal likelihood would apply the usual Monte Carlo methods for estimating integrals in Equation (2.2). Thus for each of a large number of draws, $t = 1, .., T$ from the prior density of θ, one may evaluate the likelihood $L^{(t)} = L(\theta^{(t)}|y)$ at each draw, and calculate the average. Subject to possible numerical problems, this may be feasible with a moderately informative prior, but would require a considerable number of draws (T perhaps in the millions).

Since Equation (2.4) is true for any point, this suggests another estimator for $m(y)$ based on an approximation for the posterior density $\hat{\pi}(\theta|y)$, perhaps at a high density point such as the mean $\bar{\theta}$. So taking logs throughout,

$$\log(m(y)) \approx \log(f(y|\bar{\theta}) + \log \pi(\bar{\theta}) - \log \hat{\pi}(\bar{\theta}|y) \tag{2.6}$$

Alternatively, following DiCiccio *et al.* (1997), Gelfand and Dey (1994, p. 511), and others, importance sampling may be used. In general, the integral of a function $h(u)$ may be written as

$$H = \int h(u)du = \int \{h(u)/g(u)\} \mathrm{g(u)du}$$

where $g(u)$ is the importance function. Suppose $u^{(1)}, u^{(2)}, .., u^{(T)}$ are a series of draws from this function g which approximates h, whereas h itself which is difficult to sample from. An estimate of H is then

$$T^{-1} \sum_{t=1}^{T} h(u^{(t)})/g(u^{(t)})$$

As a particular example, the marginal likelihood might be expressed as

$$m(y) = \int f(y|\theta)\ \pi(\theta)\ d\theta = \int [f(y|\theta)\pi(\theta)/g(\theta)]\, g(\theta)d\theta$$

where g is a normalised importance function for $f(y|\theta)\pi(\theta)$. The sampling estimate of is then

$$\hat{m}(y) = T^{-1} \sum_{t=1}^{T} L(\theta^{(t)})\pi(\theta^{(t)})/g(\theta^{(t)})$$

where $\theta^{(1)}, \theta^{(2)}, .., \theta^{(T)}$ are draws from the importance function g. In practice, only an unnormalised density g^* may be known, and the normalisation constant is estimated as $T^{-1}\Sigma_{t=1}^{T}\ \pi(\theta^{(t)})/g^*(\theta^{(t)})$, with corresponding sampling estimate

$$\hat{m}(y) = \sum_{t=1}^{T} L(\theta^{(t)})w(\theta^{(t)}) \sum_{t=1}^{T} w(\theta^{(t)}) \tag{2.7}$$

where $w(\theta^{(t)}) = \pi(\theta^{(t)})/g^*(\theta^{(t)})$. Following Geweke (1989), it is desirable that the tails of the importance function g decay slower than those of the posterior density that the importance function is approximating. So if the posterior density is multivariate Normal (for analytic reasons or by inspection of MCMC samples), then a multivariate Student t with low degrees of freedom is most appropriate as an importance density.

A special case occurs if $g^* = L\pi$, leading to cancellation in Equation (2.7) and to the harmonic mean of the likelihoods as an estimator for $m(y)$, namely

$$\hat{m}(y) = T/[\sum_t \{1/L^{(t)}\}] \tag{2.8}$$

For small samples this estimator may, however, be subject to instability (Chib, 1995). For an illustration of this criterion in disease mapping, see Hsiao *et al.* (2000).

Another estimator for the marginal likelihood based on importance sampling ideas is obtainable from the relation[2]

$$[m(y)]^{-1} = \int \frac{g(\theta)}{L(\theta|y)\pi(\theta)} \pi(\theta|y)\, d\theta$$

so that

$$\begin{aligned} m(y) &= \int L(\theta|y)\pi(\theta)d\theta \\ &= 1/E[g(\theta)/\{L(\theta|y)\,\pi(\theta)\}] \end{aligned}$$

where the latter expectation is with respect to the posterior distribution of θ.
The marginal likelihood may then be approximated by

$$\hat{m}(y) = 1/[T^{-1} \textstyle\sum_t g^{(t)}/\{L^{(t)}\pi^{(t)}\}] = \mathrm{T}/[\textstyle\sum_t \mathrm{g}^{(t)}/\{\mathrm{L}^{(t)}\ \pi^{(t)}\}] \tag{2.9}$$

Evidence on the best form of $g()$ to use in Equation (2.9) is still under debate, but it is generally recommended to be a function (or product of separate functions) that approximates $\pi(\theta|y)$. So in fact two phases of sampling are typically involved: an initial MCMC analysis to provide approximations g to $f(\theta|y)$ or its components; and a second run recording $g^{(t)}$, $L^{(t)}$ and $\pi^{(t)}$ at iterations $t = 1, \ldots, T$, namely the values of the importance density, the likelihood and the prior as evaluated at the sampled values $\theta^{(t)}$, which are either from the posterior (after convergence), or from g itself. The importance density and prior value calculations, $g^{(t)}$ and $\pi^{(t)}$, may well involve a product over relevant components for individual parameters.

[2] For a normalised density $1 = \int g(\theta)d\theta = \int g(\theta)[m(y)\pi(\theta|y)/\{L(\theta|y)\pi(\theta)\}]d\theta$, where the term enclosed in square brackets follows from Equation (2.3).

For numeric reasons (i.e. underflow of likelihoods $L^{(t)}$ in larger samples), it may be more feasible to obtain estimates of $\log[\hat{m}(y)]$ in Equation (2.9), and then take exponentials to provide a Bayes factor. This involves monitoring

$$\delta^{(t)} = \log[g^{(t)}/\{L^{(t)}\pi^{(t)}] = \log(g^{(t)}) - [\log(L^{(t)}) + \log(\pi^{(t)})]$$

for T iterations. Then a spreadsheet[3] might be used to obtain

$$\Delta^{(t)} = \exp[\delta^{(t)}] \tag{2.10}$$

and then *minus* the log of the average of the $\Delta^{(t)}$ calculated, so that

$$\log[\hat{m}(y)] = -\log(\bar{\Delta})$$

If exponentiation in Equation (2.10) leads to numeric overflow, a suitable constant (such as the average of the $\delta^{(t)}$ can be subtracted from the $\delta^{(t)}$ before they are exponentiated, and then also subtracted from $-\log(\bar{\Delta})$.

2.2.3 Approximating the posterior

In Equations (2.6) and (2.9), an estimate of the marginal likelihood involves a function g that approximates the posterior $\pi(\theta|y)$ using MCMC output. One possible approximation entails taking moment estimates of the joint posterior density of all parameters, or a product of moment estimate approximations of posterior densities of individual parameters or subsets of parameters. Suppose θ is of dimension q and the sample size is n. Then, as Gelfand and Dey (1994) state, a possible choice for g to approximate the posterior would be a multivariate normal or Student t with mean of length q and covariance matrices of dimension $q \times q$ that are computed from the sampled $\theta_j^{(t)}, t = 1,.., T; j = 1,..q$. The formal basis for this assumption of multivariate normality of the posterior density, possibly after selective parameter transformation, rests with the Bayesian version of the central limit theorem (Kim and Ibrahim, 2000).

In practice, for complex models with large numbers of parameters, one might split the parameters into sets (Lenk and Desarbo, 2000), such as regression parameters, variances, dispersion matrices, mixture proportions, and so on. Suppose the first subset of parameters in a particular problem consists of regression parameters with sampled values $\beta_j^{(t)}$, $t = 1,.., T; j = 1,.., q_1$. For these the posterior density might be approximated by taking $g(\beta)$ to be multivariate normal or multivariate t, with the mean and dispersion matrices defined by the posterior means and the $q_1 \times q_1$ dispersion matrix taken from a long MCMC run of T iterations on the q_1 parameters. Geweke (1989) considers more refined methods such as split Normal or t densities for approximating skew posterior densities, as might occur in nonlinear regression.

The next set, indexed $j = q_1 + 1, \ldots, q_2$, might be the parameters of a precision matrix

$$T = \sum^{-1}$$

for interdependent errors. For a precision matrix T of order $r = q_2 - q_1$, with Wishart prior $W(Q_0, r_0)$, the importance density $g(T)$ may be provided by a Wishart with $n + r_0$ degrees of freedom and scale matrix $Q = \hat{S}(n + r_0)$, where $\hat{S}$ is the posterior mean of T^{-1}. The set indexed by $j = q_2 + 1,.., q_3$ might be variance parameters ϕ_j for independ-

[3] A spreadsheet is most suitable for very large or small numbers that often occur in this type of calculation.

ent errors. Since variances themselves are often skewed, the posterior of $\chi_j = \log(\phi_j)$ may better approximate normality. The parameters indexed $j = q_3 + 1, .., q_4$ might be components $\psi = (\psi_1, \psi_2, .., \psi_J)$ of a Dirichlet density[4] of dimension $J = q_4 - q_3$. Suppose $J = 2$, as in Example 2.2 below, then there is one free parameter ψ to consider with prior beta density. If the posterior mean and variance of ψ from a long MCMC run are k_ψ and V_ψ, then these may be equated to the theoretical mean and variance, as in $M_\psi = a_p/H$ and $V_\psi = a_p b_p / H^2[H+1]$, where $H = (a_p + b_p)$. Solving gives an approximation to the posterior density of ψ as a beta density with sample size

$$H = [k_\psi(1 - k_\psi) - V_\psi]/V_\psi$$

and success probability k_ψ.

So for the MCMC samples $\theta^{(t)} = \{\beta^{(t)}, T^{(t)}, \chi^{(t)}, \psi^{(t)}, ..\}$, the values taken by the approximate posterior densities, namely $g^{(t)}(\beta)$, $g^{(t)}(T)$, $g^{(t)}(\psi)$ and $g^{(t)}(\chi)$ and other stochastic quantities, are evaluated. Let the values taken by the product of these densities be denoted $g^{(t)}$. This provides the values of each parameter sample in the approximation to the posterior density $\pi(\theta|y)$ (Lenk and Desarbo, 2000, p. 117), and these are used to make the estimate $\hat{m}(y)$ in Equations (2.6) or (2.9). An example of how one might obtain the components of g using this approach, a beta-binomial mixture is considered in Example 2.2.

Chib (1995) proposes a method for approximating the posterior in analyses when integrating constants of all full conditional densities are known as they are in standard conjugate models. Suppose the parameters fall into B blocks (e.g. $B = 2$ in linear univariate regression, with one block being regression parameters and the other being the variance). Consider the posterior density as a series of conditional densities, with

$$\pi(\theta|y) = \pi(\theta_1|y)\ \pi(\theta_2|\theta_1, y)\ \pi(\theta_3|\theta_1, \theta_2, y) \ldots . \pi(\theta_B|\theta_{B-1}, \theta_{B-2}, \ldots \theta_1, y)$$

In particular,

$$\begin{aligned}\pi(\theta^*|y) = \pi(\theta_1{}^*|y)\ \pi(\theta_2{}^*|\theta_1{}^*, y)\ \pi(\theta_3{}^*|\theta_1{}^*, \theta_2{}^*, y) \ldots . . \\ \pi(\theta_B{}^*|\theta_{B-1}{}^*, \theta_{B-2}{}^*, \ \ldots \theta_1{}^*, y)\end{aligned} \tag{2.11}$$

where θ^* is a high density point, such as the posterior mean $\bar{\theta}$, where the posterior density in the marginal likelihood identity (2.4) may be estimated.

Suppose a first run is used to provide θ^*. Then the value of the first of these densities, namely $\pi(\theta_1{}^*|y)$ is analytically

$$\pi(\theta_1{}^*|y) = \int \pi(\theta_1{}^*|y, \theta_2, \theta_3, ..\theta_B)\ \pi(\theta_2, \theta_3, ..\theta_B|y)\ d\theta_2, d\theta_3, ..d\theta_B$$

and may be estimated in a subsequent MCMC run with all parameters free. If this run is of length T, then the average of the full conditional density of θ_1 evaluated at the samples of the other parameters provides

[4] In a model involving a discrete mixture with J classes, define membership indicators G_i falling into one of J possible categories, so that $G_i = j$ if individual subject i is assigned to class j. The assignment will be determined by a latent class probability vector $\psi = (\psi_1, \psi_2, ..\psi_J)$, usually taken to have a Dirichlet prior. The MCMC estimates $E(\psi_j)$ and $\text{var}(\psi_j)$ then provide moment estimates of the total sample size ν in the posterior Dirichlet and the posterior 'sample' sizes ν_j of each component. ν is estimated as $[1 - \sum_j E(\psi_j)^2 - \sum_j \text{var}(\psi_j)]/\sum_j \text{var}(\psi_j)$, and ν_j as $E(\psi_j)\nu$. More (less) precise estimates of ψ_j imply a better (worse) identified discrete mixture and hence a higher (lower) posterior total 'sample' size ν in the Dirichlet.

$$\hat{\pi}(\theta_1{}^*|y) = T^{-1} \sum_t \pi(\theta_1{}^*|\theta_2^{(t)}, \theta_3^{(t)}, \ldots \theta_B^{(t)})$$

However, the second density on the right side of (2.11) conditions on θ_1 fixed at $\theta_1{}^*$, and requires a secondary run in which only parameters in the $B - 1$ blocks apart from θ_1 are free to vary (θ_1 is fixed at $\theta_1{}^*$ and is not updated). The value of the full conditional $\pi(\theta_2{}^*|y, \theta_1, \theta_3, \ldots \theta_B)$ is taken at that fixed value of θ_1, but at the sampled values of other parameters, $\theta_k^{(t)}$, $k > 2$, i.e. $\pi(\theta_2{}^*|y, \theta_1{}^*, \theta_3^{(t)}, \ldots \theta_B^{(t)})$. So

$$\hat{\pi}(\theta_2{}^*|\theta_1{}^*, y) = T^{-1} \sum_t \pi(\theta_2{}^*|\theta_1{}^*, \theta_3^{(t)}, \theta_4^{(t)}, \ldots \theta_B^{(t)})$$

In the third density on the right-hand side of (2.11), both θ_1 and θ_2 are known and another secondary run is required where all parameter blocks except θ_1 and θ_2 vary freely, and so on. One may then substitute the logs of the likelihood, prior and estimated posterior at θ^* in Equation (2.6). Chib (1995) considers the case where latent data z are also part of the model, as with latent Normal outcomes in a probit regression; see Example 3.1 for a worked illustration.

2.2.4 Predictive criteria for model checking and selection

Another approach to model choice and checking is based on the principle of predictive cross-validation. In Bayesian applications, this may take several forms, and may lead to alternative pseudo Bayes factor measures of model choice.

Thus, predictions might be made by sampling 'new data' from model means for case i at each iteration t in an MCMC chain. The sampled replicates $Z_i^{(t)}$ for each observation are then compared with the observed data, y_i. For a normal model with mean $\mu_i^{(t)}$ for case i at iteration t, and variance $V^{(t)}$, such a sample would be obtained by taking the simulations

$$Z_i^{(t)} \sim N\left(\mu_i^{(t)}, V^{(t)}\right)$$

Such sampling is the basis of the expected predictive approaches of Carlin and Louis (2000), Chen *et al.* (2000) and Laud and Ibrahim (1995).

Predictions of a subset y_r of the data may also be made from a posterior updated only using the complement of y_r, denoted $y_{[r]}$; see also Section 2.2.5. A common choice involves jack-knife type cross-validation, where one case (say case i) is omitted at a time, with estimation of the model based only on $y_{[i]}$, namely the remaining $n - 1$ cases excluding y_i. Under this approach an important feature is that even if the prior π, and hence possibly $\pi(\theta|y)$ is improper, the predictive density

$$p(y_r|y_{[r]}) = m(y)/m(y_{[r]}) = \int f(y_r|\theta, y_{[r]})\pi(\theta|y_{[r]})d\theta$$

is proper because the posterior based on using only $y_{[r]}$ in estimating θ, namely $\pi(\theta|y_{[r]})$, is proper. Geisser and Eddy (1979) suggest the product

$$\hat{m}(y) = \prod_{i=1}^{n} p(y_i|y_{[i]}) \tag{2.12}$$

of the predictive densities derived by omitting one case at a time (known as Conditional Predictive Ordinates, CPOs) as an estimate for the overall marginal likelihood. The ratio of two such quantities under models M_1 and M_2 provides a pseudo Bayes Factor (sometimes abbreviated as PsBF):

$$\text{PsBF} = \prod_{i=1}^{n} \{p(y_i|y_{[i]}, M_1)/p(y_i|y_{[i]}, M_2)\}$$

Another estimator of the marginal likelihood extends the harmonic mean principle to the likelihoods of individual cases: thus the inverse likelihoods for each subject are monitored, and their posterior averages obtained from an MCMC run. Then the product over subjects of the inverses of these posterior averages, which (see Chapter 1) are estimates of the conditional predictive ordinates for case i, produces another estimator of $m(y)$. The latter may be called the CPO harmonic mean estimator

$$\hat{m}(y) = \prod_{i} \hat{p}(y_i|y_{[i]}) \tag{2.13a}$$

where

$$\hat{p}(y_i|y_{[i]}) = \left[T^{-1} \sum_{t=1}^{T} \frac{1}{L_i(\theta^{(t)})}\right]^{-1} \tag{2.13b}$$

A method supplying an Intrinsic Bayes factor is proposed by Berger and Perrichi (1996), and involves defining a small subset of the observed data, y_T as a training sample. For instance, with a logit regression with p predictors, these samples are of size $p+1$. The posterior for θ derived from such a training sample supplies a proper prior for analysing the remaining data $y_{[T]}$. The canonical form of this method stipulates completely flat priors for θ in the analysis on the training samples, but one might envisage just proper priors being updated by training samples to provide more useful priors for the data remainders $y_{[T]}$. In practice, we may need a large number of training samples, since for large sample sizes there are many such possible subsets.

2.2.5 Replicate sampling

Predictive checks based on replicate sampling – without omitting cases – are discussed in Laud and Ibrahim (1995). They argue that model selection criteria such as the Akaike Information Criterion and Bayes Information Criterion rely on asymptotic considerations, whereas the predictive density for a hypothetical replication Z of the trial or observation process leads to a criterion free of asymptotic definitions. As they say, 'the replicate experiment is an imaginary device that puts the predictive density to inferential use'.

For a given model k from K possible models, with associated parameter set θ_k, the predictive density is

$$p(Z|y) = \int p(Z|\theta_k)\pi(\theta_k|y)d\theta_k$$

Laud and Ibrahim consider the measure

$$C^2 = \sum_{i=1}^{n} [\{E(Z_i) - y_i\}^2 + \text{var}(Z_i)] \tag{2.14a}$$

involving the match of predictions (replications) to actual data, $E(Z_i) - y_i$, and the variability, var(Z) of the predictions. Better models will have smaller values of C^2 or its square root, C. In fact, Laud and Ibrahim define a 'calibration number' for model k as the standard deviation of C, and base model choice on them. If different models k and m provide predictive replicates Z_{ik} and Z_{im}, one might consider other forms of distance or separation measure between them, such as Kullback–Leibler divergence.

Gelfand and Ghosh (1998) generalise this procedure to a deviance form appropriate to discrete outcomes, and allow for various weights on the matching component $\Sigma_{i=1}^{n} \{E(Z_i) - y_i\}^2$. Thus, for continuous data and for any $w > 0$,

$$C^2 = \sum_{i=1}^{n} \text{var}(Z_i) + [w/(w+1)] \sum_{i=1}^{n} \{E(Z_i) - y_i\}^2 \tag{2.14b}$$

This criterion may also be used for discrete data, possibly with transformation of both y_i and z_i (Chen and Ibrahim, 2000).

Typical values of w at which to compare models might be $w = 1$, $w = 10$ and $w = 100\,000$. Larger values of w put more stress on the match between ν_i and y_i, and so downweight precision of predictions.

Gelman *et al.* (1995) provide an outline of another posterior predictive checking (rather than model choice) procedure. Suppose the actual data is denoted y_{obs} and that $D(y_{\text{obs}};\theta)$ is the observed criterion (e.g. a chi-square statistic); similarly, let the replicate data and the criterion based on them be denoted y_{new} and $D(y_{\text{new}};\theta)$. Then a reference distribution P_{R} for the chosen criterion can be obtained from the joint distribution of y_{new} and θ, namely

$$P_{\text{R}}(y_{\text{new}}, \theta) = P(y_{\text{new}}|\theta)\ \pi(\theta|y_{\text{obs}})$$

and the actual value set against this reference distribution. Thus a tail probability, analogous to a classical significance test, is obtained as

$$p_b(y_{\text{obs}}) = P_{\text{R}}[D(y_{\text{new}};\theta) > D(y_{\text{obs}};\theta)|y_{\text{obs}}] \tag{2.15}$$

In practice, $D(y^{(t)}_{\text{new}}, \theta^{(t)})$ and $D(y_{\text{obs}}, \theta^{(t)})$ are obtained at each iteration in an MCMC run, and the proportion of iterations where $D(y^{(t)}_{\text{new}}, \theta^{(t)})$ exceeds $D(y_{\text{obs}}, \theta^{(t)})$ calculated (see Example 2.2). Values near 0 or 1 indicate lack of fit, while mid-range values (between 0.2 and 0.8) indicate a satisfactory model. A predictive check procedure is also described by Gelfand (1996, p. 153), and involves obtaining 50%, 95% (etc.) intervals of the $y_{\text{new},\ i}$ and then counting how many of the actual data points are located in these intervals.

2.3 ENSEMBLE ESTIMATES: POOLING OVER SIMILAR UNITS

We now return to the modelling theme of this chapter, in terms of models for smoothing a set of parameters for similar units or groups in a situation which does not involve regression for groups or members within groups. Much of the initial impetus to development of Bayesian and Empirical Bayesian methods came from this problem, namely simultaneous inference about a set of parameters for similar units of observation (schools, clinical trials, etc.) (Rao, 1975). We expect the outcomes (e.g. average exam grades, mortality rates) over similar units (schools, hospitals) to be related to each other and drawn from a common density. In some cases, the notion of exchangeability may be modified: we might consider hospital mortality rates to be exchangeable within one group of teaching hospitals and within another group of non-teaching hospitals, but not across all hospitals in both groups combined. Another example draws on recent experience in UK investigations into cardiac surgery deaths: the performance of 12 centres is more comparable within two broad operative procedure types, 'closed' procedures involving no use of heart bypass during anaesthesia, and 'open' procedures where the heart is stopped and heart bypass needed (Spiegelhalter, 1999)

The data may take the form of aggregate observations y_j from the units, e.g. means for a metric variable or numbers of successes for a binomial variable, or be disaggregated to observations y_{ij} for subjects i within each group or unit of observation j. The data are seen as generated by a compound or hierarchical process, where the parameter λ_j relevant to the jth unit is sampled from a prior density at stage 2, and then at stage 1 the observations are sampled from a conditional distribution given the unit parameters.

A related theme but with a different emphasis has been in generalising the standard densities to allow for heterogeneity between sample units. Thus the standard densities (e.g. binomial, Poisson, normal) are modified to take account of heterogeneity in outcomes between units which is greater than postulated under that density. This heterogeneity is variously known as over-dispersion, extra-variation or (in the case of symmetric data on continuous scales) as heavy tailed data. Williams (1982) discusses the example of toxicological studies where proportions of induced abnormality between litters of experimental animals vary because of unknown genetic or environmental factors. Similarly in studies of illness, there is likely to be variation in frailty or proneness λ.

Under either perspective consider the first stage sampling density $f(y|\lambda)$, for a set of n observations, y_i, $i = 1, \ldots, n$, continuous or discrete, conditional on the parameter vector $\Lambda = \{\lambda_1, \ldots, \lambda_n)$. Often a single population wide value of λ (i.e. $\lambda_j = \lambda$ for all j) will be inappropriate, and we seek to model population heterogeneity. This typically involves either (a) distinct parameters $\lambda_1, \ldots, \lambda_n$ for each subject $i = 1, \ldots, n$ in the sample, or (b) parameters $\lambda_1, \ldots, \lambda_J$ constant within J sub-populations. The latter approach implies discrete mixtures (e.g. Richardson and Green, 1997; Stephens, 2000), while the first approach most commonly involves a parametric model, drawing the random effects λ_i from a hyperdensity, with form

$$\lambda \sim \pi(\lambda|\theta)$$

In this density the θ are sometimes called hyperparameters (i.e. parameters at the second or higher stages of the hierarchy, as distinct from the parameters of the first stage sampling density). They will be assigned their own prior $\pi(\theta)$, which may well (but not necessarily always) involve further unknowns. If there are no higher stages, the marginal density of y is then

$$m(y) = \int\int f(y|\lambda)\pi(\lambda|\theta)\pi(\theta)d\lambda d\theta \tag{2.16}$$

For example, consider a Poisson model $y \sim \text{Poi}(\lambda)$, where y is the number of nonfatal illnesses or accidents in a fixed period (e.g. a year), and λ is a measure of illness or accident proneness. Instead of assuming all individuals have the same proneness, we might well consider allowing λ to vary over individuals according to a density $\pi(\lambda|\theta)$, for instance a gamma or log-normal density to reflect the positive skewness in proneness. Since λ is necessarily positive, we then obtain the distribution of the number of illnesses or accidents (i.e. the marginal density as in Equation (2.16) above) as

$$\Pr(y = k) = \int\int [\lambda^k \exp(-\lambda)/k!]\pi(\lambda|\theta)\pi(\theta)d\lambda d\theta$$

where the range of the integration over λ is restricted to positive values, and that for θ depends upon the form of the parameters θ. In this case

$$E(y) = E(\lambda)$$

and

$$\text{Var}(y) = E(\lambda) + \text{Var}(\lambda) \tag{2.17}$$

so that $\text{Var}(\lambda) = 0$ corresponds to the simple Poisson. It is apparent from Equation (2.17) that the mixed Poisson will always show greater variability than the simple Poisson. This formulation generalises to the Poisson process, where counts occur in a given time t or over a given population exposure E. Thus, now $y \sim \text{Poi}(\lambda t)$ over time of observation period t, or $y \sim \text{Poi}(\lambda E)$, where y might be deaths in areas and E the populations living in them. The classic model for a mixed Poisson process (Newbold, 1926) assumes that λ for a given individual is fixed over time, and that there is no contagion (i.e. influence of past illnesses or accidents on future occurrences).

The model choice questions include assessing whether heterogeneity exists and if so, establishing the best approach to modelling it. Thus, under a discrete mixture approach, a major question is choosing the number of sub-populations, including whether one sub-population only (i.e. homogeneity) is the best option. Under a parametric approach we may test whether there is in fact heterogeneity, i.e. whether a model with var(λ) exceeding zero improves on a model with constant λ over all subjects, and if so, what density might be adopted to describe it.

2.3.1 Mixtures for Poisson and binomial data

Consider, for example, the question of possible Poisson heterogeneity or extravariation in counts O_i for units i with varying exposed to risk totals such that E_i events are expected. An example of this is in small area mortality and disease studies, where O_i deaths are observed as against E_i deaths expected on the basis of the global death rate average or more complex methods of demographic standardisation. Then a homogeneous model would assume

$$O_i \sim \text{Poi}(LE_i)$$

with L a constant relative risk across all areas, while a heterogeneous model would take

$$O_i \sim \text{Poi}(\lambda_i E_i)$$
$$\lambda_i \sim \pi(\lambda|\theta)$$

with $\pi(\lambda|\theta)$ a hyperdensity. For instance, if a gamma prior G(a,b) is adopted for the varying relative risks λ_i's, then $E(\lambda) = a/b$ and $\text{var}(\lambda) = a/b^2 = E(\lambda)/b$. The third stage might then be specified as

$$a \sim E(1)$$
$$b \sim G(1,\ 0.001)$$

that is in terms of relatively flat prior densities consistent with a and b being positive parameters.

Whatever mixing density is adopted for λ, an empirical moment estimator (Bohning, 2000) for $\tau^2 = \text{var}(\lambda)$, is provided by

$$\hat{\tau}^2 = 1/n[\sum_i \{(O_i - E_i\hat{L})^2/E_i^2\} - \hat{L}\sum_i \{1/E_i\}] \qquad (2.18)$$

and indeed, might be used in setting up the priors for a and b.

Heterogeneity may also be modelled in a transform of λ_i such as $\log(\lambda_i)$. This transformation extends over the real line, so we might add a normal or student t error u_i

$$\log(\lambda_i) = \kappa + u_i$$

This approach is especially chosen when λ is being modelled via a regression or in a multi-level situation, since one can include the fixed effects and several sources of extra-variability on the log scale (see Chapters 3 and 4).

For binomial data, suppose the observations consist of counts y_i where an event occurred in populations at risk n_i, with

$$y_i \sim \text{Bin}(p_i, n_i)$$

Rather than assume $p_i = p$, suppose the parameters for groups or subjects i are drawn from a beta density

$$p_i \sim \text{Beta}(\alpha, \beta)$$

The hyperparameters $\{\alpha, \beta\}$ may themselves be assigned a prior, $\pi(\alpha, \beta)$, at the second stage, though sometimes α and or β are assumed to be known. For instance, taking known hyperparameter values $\alpha = \beta = 1$ is the same as taking the p_i's to be uniform over (0, 1). If α and β are taken to be unknowns, then the joint posterior density of $\{\alpha, \beta, p_i\}$ is proportional to

$$\pi(\alpha, \beta)\Gamma(a+b)/\{\Gamma(a)\Gamma(b)\}\prod_{j=1}^{n} p_i^{\alpha-1}(1-p_i)^{\beta-1} \quad \prod_{i=1}^{n} p_i^{y_i}(1-p_i)^{n_i-y_i}$$

The full conditional density of the p_i parameters can be seen from above to consist of beta densities with parameters $\alpha + y_i$ and $\beta + n_i - y_i$.

An alternative approach to binomial heterogeneity is to include a random effect in the model for logit(p_i). This is sometimes known as the logistic-normal mixture (Aitchison and Shen, 1980); see Example 2.3.

Example 2.1 Hepatitis B in Berlin regions As an illustration of Poisson outcomes subject to possible overdispersion, consider the data presented by Bohning (2000) on observed and expected cases of Hepatitis B in 23 Berlin city regions, denoted $\{O_i, E_i\}$ i = 1, .., 23. Note that the standard is not internal[5], and so $\Sigma_i E_i = 361.2$ differs slightly from $\Sigma_i O_i = 368$. We first test for heterogeneity by considering a single parameter model

$$O_i \sim \text{Poi}(LE_i)$$

and evaluating the resulting chi-square statistic,

[5] An internal standardisation to correct for the impact of age structure differences between areas (on an outcome such as deaths by area) produces expected deaths or incidence by using age-specific rates defined for the entire region under consideration (e.g. Carlin and Louis, 2000, p. 307). Hence, the standard mortality ratio or standard incidence ratio for the entire region would be 100. An external standard means using a national or some other reference set of age-specific rates to produce expected rates for the region and areas within it.

$$\sum_i \{(O_i - \hat{L}E_i)^2 / \hat{L}E_i\}$$

The overall mean relative risk in this case is expected to be approximately 368/361.2, and a posterior mean $\hat{L} = 1.019$ is accordingly obtained. The chi square statistic averages 195, with median 193.8, and shows clear excess dispersion. The above moment estimator (2.18) for regional variability in hepatitis rates, $\hat{\tau}^2$, has mean 0.594.

A fixed effects model might be adopted to allow for such variations. Here the parameters λ_i are drawn independently of each other (typically from flat gamma priors) without reference to an overall density. In practice, this leads to posterior estimates very close to the corresponding maximum likelihood estimate of the relative incidence rate for the ith region. These are obtained simply as

$$R_i = O_i / E_i$$

Alternatively, a hierarchical model may be adopted involving a Gamma prior G(a,b) for heterogeneous relative risks λ_i, with the parameters a and b themselves assigned flat prior densities confined to positive values (e.g. Gamma, exponential). So with $\lambda_i \sim G(a, b)$ and

$$a \sim \mathrm{G}(J_1, J_2), \quad b \sim \mathrm{G}(K_1, K_2)$$

where J_1, J_2, K_1 and K_2 are known, then

$$O_i \sim \mathrm{Poi}(\lambda_i E_i)$$

Here take $J_i = K_i = 0.001$ for $i = 1, 2$. Running three chains for 20 000 iterations, convergence is apparent early (at under 1000 iterations) in terms of Gelman–Rubin statistics (Brooks and Gelman, 1998). While there is a some sampling autocorrelation in the parameters a and b (around 0.20 at lag 10 for both), the posterior summaries on these parameters are altered little by sub-sampling every tenth iterate, or by extending the sampling a further 10 000 iterations.

In terms of fit and estimates with this model, the posterior mean of the chi square statistic comparing O_i and $\mu_i = \lambda_i E_i$ is now 23, so extra-variation in relation to available degrees of freedom is accounted for. Given that the λ_i's are smoothed incidence ratios centred around 1, it would be anticipated that $E(\lambda) \approx 1$. Accordingly, posterior estimates of a and b are found that are approximately equal, with $a = 2.06$ and $b = 2.1$; hence the variance of the λ_i's is estimated at 0.574 (posterior mean of var(λ)) and 0.494 (posterior median). Comparison (Table 2.1) of the unsmoothed incidence ratios, R_i, and the λ_i, shows smoothing up towards the mean greatest for regions 16, 17 and 19, each having the smallest total (just two) of observed cases. Smoothing is slightly less for area 23, also with two cases, but higher expected cases (based on a larger population at risk than in areas 16, 17 and 19), and so more evidence for a low 'true' incidence rate.

Suppose we wish to assess whether the hierarchical model improves over the homogenous Poisson model. On fitting the latter an average deviance of 178.2 is obtained or a DIC of 179.2; following Spiegelhalter *et al.* (2002) the AIC is obtained as either (a) the deviance at the posterior mean $D(\bar{\theta})$ plus $2p$, or (b) the mean deviance plus p. Comparing $\bar{D}$ and $D(\bar{\theta})$ under the hierarchical model suggests an effective number of parameters of 18.6, since the average deviance is 119.7, but the deviance at the posterior mean (defined in this case by the posterior averages of the λ_i's) is 101.1. The DIC under the gamma mixture model is 138.3, a clear gain in fit over the homogenous Poisson model.

Table 2.1 Regional relative risks: simple maximum likelihood fixed effects and Poisson–Gamma mixture models

	Unsmoothed incidence ratios	Incidence ratios from hierarchical smoothing	2.5%	Median	97.5%
Region 1	2.66	2.42	1.62	2.39	3.34
Region 2	1.42	1.39	0.93	1.37	1.95
Region 3	2.92	2.77	2.07	2.75	3.59
Region 4	1.53	1.51	1.03	1.49	2.08
Region 5	0.71	0.75	0.44	0.74	1.15
Region 6	1.01	1.02	0.60	1.00	1.54
Region 7	0.61	0.69	0.30	0.66	1.23
Region 8	1.99	1.92	1.34	1.90	2.59
Region 9	0.89	0.91	0.55	0.90	1.35
Region 10	0.38	0.45	0.21	0.43	0.79
Region 11	1.31	1.31	0.95	1.30	1.72
Region 12	0.68	0.71	0.43	0.70	1.07
Region 13	1.75	1.62	0.95	1.59	2.52
Region 14	0.69	0.73	0.39	0.71	1.18
Region 15	0.91	0.94	0.49	0.91	1.53
Region 16	0.20	0.34	0.09	0.31	0.74
Region 17	0.18	0.32	0.08	0.29	0.70
Region 18	0.48	0.54	0.27	0.53	0.91
Region 19	0.38	0.55	0.14	0.51	1.17
Region 20	0.27	0.38	0.12	0.36	0.79
Region 21	0.54	0.59	0.31	0.58	0.95
Region 22	0.35	0.44	0.18	0.42	0.82
Region 23	0.15	0.28	0.07	0.25	0.62

Example 2.2 Hot hand in baseball This example considers data on shooting percentages in baseball, as obtained by Vinnie Jones over the 1985–89 seasons, and used by Kass and Raftery (1995) to illustrate different approximations for Bayes factors. The question of interest is whether the probability of successfully shooting goals p is constant over games, as in simple binomial sampling (model M_1), so that

$$y_i \sim \text{Bin}(p, n_i),$$

where n_i are attempts. Alternatively, under M_2 the hypothesis is that Vinnie Jones has a 'hot hand' – that is, he is significantly better in some games than would be apparent from his overall average. The latter pattern implies that p is not constant over games, and instead there might be extra-binomial variation, with successful shots y_i binomial with varying probabilities p_i in relation to all attempts, n_i (successful or otherwise):

$$\begin{aligned} y_i &\sim \text{Bin}(p_i, n_i) \\ p_i &\sim \text{Beta}(\alpha, \beta) \end{aligned} \tag{2.19}$$

Here the models (2=Beta-Binomial vs. 1=Binomial) are compared via marginal likelihood approximations based on importance sampling.

There are other substantive features of Jones' play that might be consistent with a hot hand, such as runs of several games with success rates y_i/n_i larger than expected under the simple binomial. Here, rather than global model hypothesis tests, a posterior predictive check approach might be used under the simple binomial sampling model. This entails using different test statistics applied to the observed and replicate data, y and y_{new}, and preferably statistics that are sensible in the context of application. In particular, Berkhof *et al.* (2000) test whether the maximum success rate $\max_i\{y_{i.\text{new}}/n_i\}$ in the replicate data samples exceeds the observed maximum.

In the binomial model, the prior $p \sim B(1, 1)$ is adopted, and one may estimate the beta posterior density $B(a_p, b_p)$ of p using moment estimates of the parameters. Thus, if k_p is the posterior mean of p and V_p its posterior variance, then $H = a_p + b_p$ is estimated as $[k_p(1 - k_p) - V_p]/V_p$. Thus with $k_p = 0.457$ and $V_p^{0.5} = 0.007136$, a posterior beta density with 'sample size' $H = 4872$ is obtained.

In the beta-binomial, Kass and Raftery (1995, p. 786) suggest reparameterising the beta mixture parameters in Equation (2.19). Thus $\alpha = \nu/\omega$, $\beta = (1 - \nu)/\omega$, where both ω and ν are assigned $B(1, 1)$ priors – equivalent to uniform priors on (0, 1). The posterior beta densities of the p_i's in Equation (2.19) are approximated using the moment estimation procedure, and similarly for the posterior beta densities of ω and ν.

It may be noted that the beta-binomial model is not especially well identified, and other possible priors such as vague gamma priors on the beta mixture parameters themselves, e.g.

$$\alpha \sim G(0.001, 0.001), \quad \beta \sim G(0.001, 0.001)$$

have identifiability problems. With the reparameterised version of the model convergence for ω is obtained from the second half of a three chain run with 10 000 iterations, with posterior mean of ω at 0.0024 and posterior standard deviation 0.002.

In a second run, iterations subsequent to convergence (i.e. after iteration 10 000) record the prior, likelihood and (approximate) posterior density values as in Equation (2.9), corresponding to the sampled parameters of the binomial and beta-binomial models, namely $\{p\}$ and $\{p_i, \omega, \nu\}$. Then with 1000 sampled parameter values and corresponding values of $\delta^{(t)} = \log(g^{(t)}) - [\log(L^{(t)}) + \log(\pi^{(t)})]$, the approximate marginal likelihoods under models 2 and 1 are -732.7 and -729.1, respectively. This leads to a Bayes factor in favour of the simple binomial of around 35.

Table 2.2, by contrast, shows that the beta-binomial has a slightly higher likelihood than the binomial. The worse marginal likelihood says in simple terms that the improved sampling likelihood obtained by the beta-binomial is not sufficient to offset the extra parameters it involves. Kass and Raftery (1995, p. 786) cite Bayes factors on M_1 between 19 and 62, depending on the approximation employed.

Features of the game pattern such as highest and lowest success rates, or runs of 'cold' or 'hot' games (runs of games with consistent below or above average scoring) may or may not be consistent with the global model test based on the marginal likelihood. Thus, consider a predictive check for the maximum shooting success rate under the simple binomial, remembering that the observed maximum among the y_i/n_i is 0.9. The criterion

$$\Pr(\max\{y_{\text{new}}/n\} > \max\{y/n\})$$

is found to be about 0.90 – this compares to 0.89 cited by Berkhof *et al.* (2000, p. 345). This 'significance rate' is approaching the thresholds which might throw doubt on the simple binomial, but Berkhof *et al.* conclude that is still such as to indicate that the observed maximum is not unusual or outlying.

Table 2.2 Posterior summary, baseball goals, binomial and beta-binomial parameters

Beta-binomial	Mean	St. devn.	2.50%	Median	97.50%
ω	0.00241	0.00205	0.00009	0.00193	0.00773
α	683.5	1599	58.49	236.6	5159
β	808.1	1878	70.61	282.7	6004
Log likelihood	−721.3	4.416	−727.5	−722.1	−711.4
SD(p) in beta-binomial	0.013				
Binomial					
p	0.457	0.007	0.441	0.457	0.474
Log likelihood	−725.6	0.7	−728.3	−725.3	−725.1

Example 2.3 Cycles to conception Weinberg and Gladen (1986) consider differences in the number of fertility cycles to conception according to whether the woman in each couple smoked or not. For $i = 1, \ldots, 100$ women smokers, 29 conceived in the first cycle, but from 486 non-smokers, 198 (or over 40%) conceived in this cycle. The full data, given in Table 2.3, consist of the number y of cycles required according to smoking status, with the last row relating to couples needing over 12 cycles.

Such an outcome is a form of waiting time till a single event, but in discrete time units only, and can be modelled as a geometric density. This is a variant of the negative binomial (see Chapter 1) which counts time intervals y until r events occur, when the success rate for an event is p. The negative binomial has the form

$$\Pr(y) = \binom{y-1}{r-1} p^r (1-p)^{y-r} \tag{2.20}$$

and it follows that the number of intervals until $r = 1$ (e.g. cycles to the single event, conception) is

Table 2.3 Cycles to conception

Cycle	Non-smokers	Cumulative proportion conceiving	Smokers	Cumulative proportion conceiving
1	198	0.41	29	0.29
2	107	0.63	16	0.45
3	55	0.74	17	0.62
4	38	0.82	4	0.66
5	18	0.86	3	0.69
6	22	0.90	9	0.78
7	7	0.92	4	0.82
8	9	0.93	5	0.87
9	5	0.94	1	0.88
10	3	0.95	1	0.89
11	6	0.96	1	0.9
12	6	0.98	3	0.93
Over 12	12	1.00	7	1
Total	486		100	

$$\Pr(y) = (1 - p)^{y-1}\, p$$

So p is equivalently the chance of conception at the first cycle (when $y = 0$).

Consider first a constant probability model (Model 1) for couples within the smoking group and within the non-smoking group, so that there are just two probabilities to estimate, p_1 for smokers and p_2 for non-smokers. Under this density, the probability of more than N cycles being required is

$$\Pr(y > N) = p \sum_{i=N+1}^{\infty} (1 - p)^{i-1} = (1 - p)^N \tag{2.21}$$

In the present example, $N = 12$. Note that this is an example of censoring (non-observation) of the actual cycles to conception; only the minimum possible cycle number for such couples is known.

The likelihood (2.20) and (2.21) may be modelled via the non-standard density option available in BUGS (the dnegbin option in BUGS might also be used but is complicated by the censoring). Thus, for a density not available to sample from in BUGS, an artificial data series Z_i of the same length N as the actual data is created, with $Z_i = 1$ for all cases. Then, if C_i are the number of conceptions at cycle i, $i = 1, .., N$, and if there are no groups to consider (such as the smoking and non-smoking groups here), the density for Z is

$$Z_i \sim \text{Bern}(L_i)$$

where L_i is the likelihood defined by

$$L_i = [(1 - p)^{i-1}\, p]^{C_i}$$

where Bern() denotes Bernoulli sampling. The corresponding coding in BUGS (with $N = 12$ and a $B(1, 1)$ prior for p) is

```
{for (i in 1:N) {Z[i] <- 1
                 Z[i] ~ dbern(L[i])
                 L[i] <- pow(r[i],C[i])
                 r[i] <- pow (1-p,i-1)*p}
                 p ~ dbeta(1, 1)}.
```

In the example here, this approach is extended so that at each cycle i there are J groups of cases (here $J = 2$ for smoking and non-smoking women), so that

$$Z_{ij} \sim \text{Bern}(L_{ij}) \quad i = 1, .., N;\ j = 1, .., J$$
$$L_{ij} = [(1 - p_j)^{i-1}\, p_j]^{C_{ij}}$$

Alternatively (Model 2), one might allow for extra-variability in conception chances with a beta mixture at the couple level. This is known as a beta-geometric mixture, and is analogous to the more commonly encountered beta-binomial. There are now distinct probabilities p_{ik} for couples $k = 1, \ldots C_i$ taking i cycles to conceive (with $i = 1, .. N$). Disregarding possible grouping, and continuing the artificial data device, the coding corresponds to the model

$$\begin{aligned} Z_{ik} &= 1 \\ Z_{ik} &\sim \text{Bern}(L_{ik}) \\ L_{ik} &= (1 - p_{ik})^{i-1} p_{ik} \\ p_{ik} &\sim \text{B}(\alpha, \beta) \quad i = 1, .., N;\ k = 1, .., C_i \end{aligned} \tag{2.22}$$

with the total likelihood for subjects taking i cycles being

$$L_i = \prod_{k=1}^{C_i} L_{ik}$$

Program 2.3 includes this approach with the data arranged as a single string of length ΣC_i.

Under the constant probability model (within the two groups), the posterior means for the conception probabilities are $p_1 = 0.23$ and $p_2 = 0.33$, with 95% credible intervals (0.19, 0.27) and (0.31, 0.36). The predicted cycle distribution under this model, denoted nhat[,] in Program 2.3, under-predicts both short and long cycles to conception frequencies (1 cycle and over 12 cycles) – see Table 2.4. This brings into doubt the constant probability model.

For the heterogeneous case, with a mixture of conception probabilities over couples, attention is confined to couples where the female partner smokes. The data are then defined as a vector of 100 couple level observations of cycles required, with 29 observations requiring one cycle, 16 requiring two and so on, with the couples needing over 12 cycles coded to 13. As well as the artificial data device, a reparameterisation as in Example 2.2 is employed for the beta mixture parameters.

A three chain run taken to 10 000 iterations shows apparent convergence from 1000 iterations (the beta mixture parameters are relatively slow to converge according to Gelman–Rubin criteria). The average conception probability from the beta parameters, namely

$$P = \alpha/(\alpha + \beta)$$

is 0.29 with 95% credible interval from 0.22 to 0.37. Short and long cycles to conception frequencies are much more closely reproduced under this model (Table 2.5). The credible interval for these frequencies includes the actual frequencies of 29 and 7, respectively. The benefits from allowing conception chances to vary over couples are evident.

Table 2.4 Cycles to conception, predictions under constant probability model

	Smokers			Non-Smokers		
Cycle	Mean	2.5%	97.5%	Mean	2.5%	97.5%
1	22.6	18.7	26.7	161.3	149.8	172.9
2	17.4	15.2	19.6	107.7	103.6	111.4
3	13.5	12.4	14.4	71.9	71.6	72.0
4	10.4	10.0	10.6	48.1	46.2	49.6
5	8.1	7.7	8.2	32.1	29.8	34.3
6	6.3	5.6	6.6	21.5	19.2	23.7
7	4.8	4.1	5.4	14.4	12.4	16.4
8	3.8	3.0	4.4	9.6	8.0	11.4
9	2.9	2.2	3.6	6.4	5.1	7.9
10	2.3	1.6	2.9	4.3	3.3	5.4
11	1.8	1.2	2.4	2.9	2.1	3.8
12	1.4	0.9	1.9	1.9	1.4	2.6
Over 12	4.9	2.4	8.4	3.9	2.5	5.8

Table 2.5 Couples with female partner as smokers: predicted cycle distribution under mixture model

	Mean	St. devn.	2.50%	97.50%
P	0.29	0.04	0.22	0.37
α	3.36	3.75	1.16	11.61
β	9.23	13.33	2.25	36.77
Cycle				
1	28.9	3.4	22.5	35.6
2	18.2	0.9	16.4	19.8
3	12.2	0.8	10.6	13.6
4	8.6	0.8	7.1	10.1
5	6.3	0.7	5.0	7.7
6	4.7	0.6	3.7	5.9
7	3.6	0.5	2.8	4.6
8	2.8	0.4	2.1	3.6
9	2.2	0.3	1.7	2.9
10	1.8	0.2	1.4	2.3
11	1.4	0.2	1.1	1.9
12	1.2	0.2	0.9	1.5
Over 12	8.2	2.2	4.2	12.9

In practice, couple characteristics (e.g. the woman's age) might be expected to affect the chance of conception, and one might then link the p_{ik}'s in Equation (2.22) to predictors via a logit regression. Remaining variability between couples might then be modelled via a Normal error in the logit link. In fact, this approach might be adopted for all 586 couples using the female smoking status S_{ik} (=1 for smokers, 0 otherwise) as a predictor, so that

$$\text{logit}(p_{ik}) = \gamma_0 + \gamma_1 S_{ik} + u_{ik}$$

with u_{ik} Normal.

2.3.2 Smoothing methods for continuous data

For metric data, assumed initially at least to be approximately Normal, a typical problem involves two way data with $i = 1, \ldots, n_j$ replicated observations y_{ij} within groups $j = 1, \ldots, K$. Sometimes the observed data may be provided only as group averages $\bar{y}_j$ aggregating over individual observations, though with details on the variability within groups (or on 95% intervals for the mean). Assuming the observed means are derived from similar observation settings or similar types of unit, they may be regarded as draws from an underlying common density for the unknown true means μ_j. This assumption leads to a hierarchical model, with the first stage specifying the density of the observations, and the second (and maybe higher) stages specifying hyperparameters which underlie the observations. Typically, the underlying cell means μ_j are taken to differ by group but the variance σ^2 (and so also the precision $\tau = \sigma^{-2}$) is assumed constant over groups. In analysis of variance situations, the goal may be to assess additionally whether the underlying group means are equal. For data in a one group y_i, $i = 1, \ldots, n$, the higher stage density involves a single mean and variance.

While a symmetric, unimodal density such as the Normal is often appropriate, some work has focused on skewed options to the Normal (Leonard, 1980; Fernandez and Steel, 1998), which may involve a different variance according to whether the observation is located above or below the mean. Other options model skewness via latent 'factor scores' introduced into the prediction of the regression mean (Branco and Dey, 2001). Other departures from symmetry or unimodality may be modelled by discrete mixtures (see Section 2.4).

Even if the assumption of symmetry is accepted, other features of the Normal density may be inappropriate. There are issues of robustness to outlier observations or parameters. The exceptional observation(s) may indicate a robust (heavy-tailed) alternative to the Normal so that estimates of summary parameters at the second stage are not distorted, while an exceptional parameter (e.g. one of the means μ_j) will indicate an alternative second stage prior to the Normal (Gelman *et al.*, 1995, Chapter 12).

However, suppose initially the observations are assumed to be Normally distributed such that

$$y_{ij} \mid \mu_j \sim \mathrm{N}(\mu_j, \sigma^2)$$

Equivalently, the likelihood can be specified in terms of the means of the observations within group j, namely $\bar{y}_j$, also with means μ_j but with variances σ^2/n_j (and precisions $n_j\tau$, where $\tau = 1/\sigma^2$), so that

$$\bar{y}_j|\mu_j \sim \mathrm{N}(\mu_j, \sigma^2/n_j) \tag{2.23}$$

Often – for example in meta analyses – the observed data will be available in the form of summary statistics y_j such as odds ratios, with sample sizes and sampling variances V_j for the summary statistics also provided. It may be noted that a Normal approximation to the distribution of the means or other summary measures may remain appropriate, even if the original data are non-Normal. Thus the y_j might be the logged odds ratios resulting from 2×2 tables of case versus exposure status, and the variances V_j would have been obtained by the usual formula in such applications (Woodward, 1999).

Suppose the second stage prior for the μ_j's is also assumed as Normal with mean M and variance ϕ^2. At the highest stage a gamma prior for the inverse variance $1/\phi^2$ might be adopted, and a flat (though proper) prior for the mean. Inferences about the μ_j are likely to be sensitive to the prior on the higher level variance or precision: to avoid over-smoothing, the prior for $1/\phi^2$ may be specified to avoid high precision (see Example 2.6).

Inferences for data assumed Normal, for instance regarding cell means in one or two-way analysis of variance situations, may be influenced by discordant observations. In these cases the heavier tailed alternatives to the univariate and multivariate Normal, such as the Student t in its univariate and multivariate forms (with additional degrees of freedom parameter, ν) may provide more robust inferences with regard to the location of the overall mean. Small values of ν (under 10) indicate that Normality of the data is doubtful, while values in excess of 50 are essentially equivalent to Normality. It may also be necessary (see Example 2.5) to account for outlier parameters at the second or higher stages of a hierarchical model, such as a discrepant μ_j. Heavy tailed departures from non-Normality may also be modelled by exponential power distributions, of which the double exponential is an example (Box and Tiao, 1973, Chapter 3).

As noted in Chapter 1, the Student t density is obtainable as a scale mixture of the Normal. Thus

$$y_i \sim \mathrm{N}(\mu^2/\lambda_i) \quad i = 1, .., n$$

with the variance scaling parameters or weights λ_i being drawn from a $\mathrm{G}(\nu/2, \nu/2)$ density. If additionally ν is treated as an unknown parameter, then a flat prior is not appropriate (Geweke, 1993), and an exponential prior is one option; also for the posterior mean of μ to exist, values of $\nu \in [0, 2]$ are not admissible. The scaling parameters are lowest for observations which are discrepant (i.e are potential outliers) from the main set; the scale mixture approach to outlier detection is exemplified by authors such as West (1984) and DeFinetti (1961).

In fact, specifying the λ_i's in this form is one option in a broader class of scale mixture models in which the prior for the weights does not necessarily refer to the degrees of freedom parameter (Geweke, 1993, p. S23). In general,

$$y_i \sim \mathrm{N}(\mu, \sigma^2/w_i) \tag{2.24}$$

where w_i are positive random variables with average 1 representing variance heterogeneity. Markedly lower values of w_i correspond to suspect values of y, only included in the density in Equation (2.24) by inflating the variance to accommodate them.

Example 2.4 Univariate Scale Mixture: the Darwin data An often cited example of distortion in the univariate Normal parameters caused by potentially outlying points is the data obtained by Darwin relating to the benefits of cross-fertilisation in plants. Specifically, he noted differences in heights (in eighths of an inch) within pairs of corn plants, one cross-fertilised and the other self-fertilised. Here variations on the first stage density are introduced, and their impact assessed on the summary measures (hyper-parameters) of location and variability.

The data y_i are differences from 15 plant pairs, as follows:
49, −67, 8, 16, 6, 23, 28, 41, 14, 29, 56, 24, 75, 60, −48
Suppose the first stage density for these data is taken as Normal. A run of 10 000 iterations with 1000 burn in provides a posterior mean (95% credible interval) for the Normal mean μ of 20.9 (1.6, 40.1) and a posterior median on σ^2 of 1300. The mean is thus imprecisely estimated, though its 95% credible interval is confined to positive values.

An alternative sampling density is a univariate[6] Student t, $y_i \sim t_\nu(\mu, \sigma^2)$, without an explicit scale mixture (Model 2 in Program 2.4), one aspect of which is to assess the density of the degrees of freedom parameter ν. Here an exponential prior for ν is assumed, $\nu \sim E(\kappa)$, with κ itself assigned a uniform prior $U(0.01, 0.5)$. This is approximately equivalent to assuming the degrees of freedom lies within the range 2–100. Assigning a preset value such as $\kappa = 0.1$ would be a more informative option. There is also a formal constraint on ν excluding the range $\nu \in [0, 2]$. A flat gamma prior for $\tau = 1/\sigma^2$ is adopted, namely $\tau \sim \mathrm{G}(0.0001, 0.0001)$, and an $\mathrm{N}(0, 10^7)$ prior on μ. One set of initial values for a three chain run is based on null start values, the others on the mean and 97.5% point of a trial run.

Early convergence of κ is apparent, but ν itself only converges after about 150 000 iterations. Table 2.6 is based on iterations 150 000–200 000. A median value of 10 for the

[6] In BUGS this involves the code

```
y[i] ~ dt(mu, tau, nu)
```

with tau as precision, and nu as degrees of freedom.

Table 2.6 Student t parameters for fertilisation data, degrees of freedom unknown

Parameter	Mean	St. devn.	2.5%	Median	97.5%
κ	0.11	0.12	0.01	0.06	0.47
Mean	23.90	9.84	−1.06	24.17	46.72
ν	29.57	52.66	2.07	10.17	289.30
Variance	1225.00	709.20	237.10	1088.00	3655.00

degrees of freedom is obtained (though the posterior mean for ν is 30). The posterior median of σ^2 is lower than in the Normal analysis, namely 1090, and the Normal mean is higher – though not more precisely identified – with an average of 24.

This analysis illustrates how inferences about the second stage parameters may be distorted by incorrectly assuming a Normal first stage density, but does not indicate which observations are the source of the heavy tails. In a third analysis (Model 3 in Program 2.4), the scale mixture version of the Normal is applied, with ν still an unknown. Taking the prior on the degrees of freedom as above we obtain – from iterations 50 000–100 000 of a three chain run – factors λ_i for each sample member that provide a measure of outlier status. It may be noted that sampling convergence is improved by this approach as compared to the standard Student t form above. The lowest λ_i (i.e. most suspect points) are for corn plant pairs 2 and 15, the observations for which are both highly negative (Table 2.7). The high positive value for pair 13 is also suspect. The Normal mean parameter μ is again estimated to have posterior mean 24, and the median variance is around 1050.

Table 2.7 Parameters for scale mixture density of fertilisation data

	Mean	St. devn.	2.5%	Median	97.5%
Mean	24.1	9.8	3.9	24.4	42.8
Degr. Freedom	23.7	40.8	2.2	8.7	140.8
Variance	1199	706	304	1058	2950
		Weights for Individual Cases			
	Mean	St. devn.	2.5%	Median	97.5%
λ_1	1.03	0.53	0.22	0.97	2.37
λ_2	0.59	0.36	0.05	0.57	1.29
λ_3	1.06	0.56	0.24	0.99	2.47
λ_4	1.11	0.60	0.26	1.01	2.68
λ_5	1.05	0.55	0.23	0.98	2.43
λ_6	1.14	0.63	0.27	1.02	2.78
λ_7	1.14	0.62	0.28	1.02	2.78
λ_8	1.09	0.58	0.24	1.00	2.57
λ_9	1.10	0.60	0.25	1.01	2.64
λ_{10}	1.14	0.63	0.27	1.02	2.78
λ_{11}	0.97	0.49	0.19	0.93	2.17
λ_{12}	1.14	0.62	0.27	1.03	2.78
λ_{13}	0.82	0.42	0.12	0.83	1.74
λ_{14}	0.94	0.47	0.17	0.91	2.07
λ_{15}	0.68	0.38	0.07	0.68	1.45

Example 2.5 Labour market rateable values As an example of the impact of adopting a multivariate t density at the first or higher stage, consider data on four types of Rateable Value (RV) change in four groups of Labour Market Area in England and Wales over 1966–71 (Kennett, 1983, Table 9.4). The $n = 4$ Labour Market groups in this example are either cores, rings, outer rings, or unclassified, with the $p = 4$ variables being the change in per cent in total RV, in domestic RV, in commercial RV and in industrial RV. These data contain extreme values on industrial RV change, which cast doubt on Normality and raise problems in estimating the mean for this variable.

It may be noted that in BUGS, the direct form of the multivariate t (i.e. not based on scale mixing) for a p-dimensional outcome y_i involves the command

```
y[i, 1: p] ~ dmt(mu[1: p], P[, ], nu),
```

where nu (the degrees of freedom) is usually[7] preset (Model A in Program 2.5). A profile of the log-likelihood or some predictive measure like the expected predictive deviance over different values of the degrees of freedom is then one approach to assessing both the extent of departures from multivariate normality and the most appropriate degrees of freedom. For illustration, a multivariate t likelihood at the first stage is compared between two options for the degrees of freedom, $\nu = 20$ and $\nu = 100$ (the latter effectively equivalent to multivariate normality). There is, in fact, very little to choose between these models in terms of likelihood. With runs of 50 000 iterations, median log-likelihoods are -53.7 for $\nu = 100$ and -53.9 for $\nu = 20$. The posterior for industrial RV change when $\nu = 20$ has 95% credible interval from -29 to 78, with mean 25.

One may obtain outlier indicators λ_i by specifying a univariate Normal measurement error model at the first stage, and a multivariate t (via scale mixing) in the second stage model for the means (Model 2 in Program 2.5):

$$
\begin{aligned}
&Y_{i,j} \sim \mathrm{N}(\mu_{ij}, \tau_j) \quad i = 1, \ldots, n; \; j = 1, \ldots, p \\
&\mu_{ij} = \gamma_j + \varepsilon_{ij} \\
&\varepsilon_{ij} = \eta_{ij}/\sqrt{\lambda_i} \\
&\lambda_i \sim \mathrm{G}(\nu/2, \nu/2) \\
&\eta_{i,\,1:p} \sim N_p(0, \Sigma)
\end{aligned}
$$

where γ_j are assigned a just proper Normal prior. The variances τ_j specify the expected level of measurement error, and one might assume relatively informative priors favouring lower error variances. The μ_{ij} are then the underlying multivariate form of the data. Conclusions about Σ and the outlier indicators may be sensitive to the assumed τ_j, especially for small samples as here. As above the degrees of freedom is assigned an exponential prior $E(\kappa)$, excluding values under 2, and with κ itself a free parameter with prior $\kappa \sim U(0.01, 0.5)$.

It may be noted that this approach might be used with joint discrete outcomes subject to extra-variation, where such extra-variation was suspected to be located in relatively few observations. For example, if Y_{ij} were pairs of Poisson outcomes such as deaths from cause A and from cause B, with expectations (in the demographic sense) E_{ij}, then one might assume

[7] Though a discrete prior on different values of d.f. is possible.

$$Y_{ij} \sim \text{Poi}(\mu_{ij} E_{ij})$$
$$\log(\mu_{ij}) = \gamma_j + \varepsilon_{ij}$$

with the ε_{ij}'s specified as above.

In the rateable values example, three chains are run for 10 000 iterations from over-dispersed initial values (two sets of which are based on a trial run, the other being a 'null' start point) with convergence after 2500 iterations. Informative priors on the inverse variances of the measurement error are based on such error accounting for approximately 10% of the total variance[8] of each variable. The outlier indicators show in fact that the core areas have lowest weight (around 0.55), though the posterior distributions of all the λ_i's have substantial spread. The degrees of freedom has median 4. Industrial RV change is more precisely identified under this mixture model approach, with 95% interval 2 to 52, and mean 28. Of course, other priors on the measurement error variances might well be applied to assess sensitivity.

Example 2.6 Coaching programs In this example we consider a cell means analysis of a metric outcome, defined by the effects y_j of coaching programs for the Scholastic Aptitude Test-Verbal in $K = 8$ US high schools. The data are available as the estimated coaching effects and their standard errors s_j after an analysis of covariance using scores from an earlier test administration. The goal is to estimate underlying means μ_j, the prior for which expresses a belief that such means are drawn from a common distribution. Assume the effects y_j are draws from a hierarchical model with the first stage sampling variances V_j provided by the squared standard errors s_j^2, so that

$$y_j \sim \text{N}(\mu_j, V_j)$$

with the second stage prior specifying a common Normal distribution

$$\mu_j \sim \text{N}(M, \phi^2).$$

As one route to establish a prior on ϕ^2 that avoids over-smoothing of the μ_j consider the harmonic mean of the observed V_j, given by

$$N/\left[\sum_{j=1}^{K} 1/V_j\right]$$

which is $s_0^2 = 139$, with square root $s_0 = 11.8$. This mean is the basis for a Pareto prior density for ϕ, as proposed by DuMouchel (1990), involving s_0 as a guide quantity in the prior. Specifically, the prior

$$\pi(\phi) = s_0/(s_0 + \phi)^2 \tag{2.25}$$

is still relatively vague but has median s_0.

Alternatively, a prior may be set on ϕ^2 or its inverse ϕ^{-2}, which incorporates (though downweights) the guide value s_0^2 or its inverse s_0^{-2}. Suppose a prior is set as follows:

$$\phi^{-2} \sim \text{G}(a, b)$$

where $a = 1/\kappa$ and $b = s_0^2/\kappa$, where κ is a downweighting constant in excess of 1. Large values of κ (e.g. $\kappa = 1000$) correspond to just proper priors.

[8] The moment variances V_j are 11,22,6 and 800, and G(10, V_j) priors on the inverse variances of the measurement errors are adopted.

With an arbitrary value of $\kappa = 10$, a three chain run (Model A in Program 2.6) shows early convergence and summaries are based on the last 24 500 iterations from 25 000. This leads to a posterior mean value for ϕ^2 of 62.5, and to smoothed program effects μ_j ranging from 5.3 to 11.7, as represented in terms of posterior means (see Table 2.8).

Then (Model B) uncertainty in κ is introduced via an exponential prior,

$$\kappa \sim E(0.01)$$

Table 2.8 Program effects: observed and smoothed

School Effectiveness Data

School	Y	s	s^2	$1/s^2$	
A	28.39	14.9	222	0.0045	
B	7.94	10.2	104	0.0096	
C	−2.75	16.3	265.7	0.0038	
D	6.82	11	121	0.0083	
E	−0.64	9.4	88.4	0.0113	
F	0.63	11.4	130	0.0077	
G	18.01	10.4	108.2	0.0092	
H	12.16	17.6	309.8	0.0032	
			Sum of $1/s^2$	0.0576	
			Harmonic Mean for Variance	138.81	= (8/0.0576)

Smoothed Effects Model A				Model B			
	Mean	2.5%	97.5%		Mean	2.5%	97.5%
μ_1	11.7	−5.3	34.8	μ_1	11.4	−4.8	34.8
μ_2	8.0	−7.7	23.8	μ_2	8.0	−7.5	23.5
μ_3	6.4	−14.6	24.4	μ_3	6.5	−14.2	24.0
μ_4	7.7	−8.4	23.8	μ_4	7.7	−8.5	23.5
μ_5	5.3	−11.4	19.5	μ_5	5.5	−11.3	19.5
μ_6	6.2	−11.5	21.8	μ_6	6.3	−11.8	21.8
μ_7	11.0	−3.9	29.3	μ_7	10.6	−3.9	28.9
μ_8	8.7	−10.4	29.1	μ_8	8.5	−10.3	29.0
ϕ^2	62.6	4.3	412.9	ϕ^2	58.5	0.9	423.0
M	8.1	−3.7	20.2	M	8.1	−3.8	20.3
				κ	47.3	0.2	304.8

Model C

	Mean	2.50%	97.50%
μ_1	10.19	−1.11	26.00
μ_2	8.15	−2.73	19.68
μ_3	7.20	−6.90	19.65
μ_4	8.05	−3.44	19.85
μ_5	6.61	−5.88	17.48
μ_6	7.20	−5.11	18.74
μ_7	9.72	−0.64	22.44
μ_8	8.60	−4.06	22.52
ϕ^2	4.23	0.06	14.15

A three chain run shows convergence for this parameter at around 2500 iterations, and from iterations 2500–25 000 a median κ of 24 and mean 47 are obtained, with the posterior mean for ϕ^2 now 58.5. The posterior density for κ has most mass on low values but has a positive skew. The smoothed effects are slightly more pulled to the grand mean than under Model A, and range from 5.5 to 11.4. Finally, Model C is defined via the Pareto prior in Equation (2.25). Note that initial values in three chains for the Pareto variable (in its BUGS parameterisation and denoted tau0 in Program 2.6) are randomly sampled from the prior. This model, after a run of 25 000 iterations, leads to greater smoothing with a posterior mean for ϕ^2 now 32 and the program effects ranging from 6.4 to 10.0.

Gelman *et al.* (1995, p. 147) cite a range in posterior median effects from 5–10 for these data. They conclude that one program effect, namely the first program with mean 28.4 and empirical standard error 14.9, is unlikely in terms of the smoothing density $\mu_j \sim N(M, \phi^2)$. Thus, in Model A one may compare y_j with μ_j using the standard normal cumulative distribution function Φ, and find that the median chance of a μ_j smaller than 28.4 is 0.998.

2.4 DISCRETE MIXTURES AND DIRICHLET PROCESSES

The priors considered above for the underlying population mixing density have a specific parametric form. However, to avoid being tied to particular parametric forms of prior, non-parametric options have been proposed, such as discrete mixtures of parametric densities (Laird, 1982) or Dirichlet process priors. The goal in choosing parametric or non-parametric mixing may be to improve fit and robustness of inferences by approximating more closely the true density of the sample (Robert, 1996; West, 1992). This is especially so for observations that are multimodal or asymmetric in form. Here a parametric model assuming a single component population is likely to be implausible.

For instance, in smoothing health outcomes over sets of small areas, especially when there may be different modes in subsets of areas, discrete mixtures may be used (Clayton and Kaldor, 1987). The mixture will generally involve a small but unknown number of sub-populations, as determined by an underlying latent set of group membership variables for each observation. There are then likely to be issues of discrimination between models involving different numbers of sub-populations, especially in smaller samples or if data likelihoods are relatively flat (see Böhning, 2000).

2.4.1 Discrete parametric mixtures

Discrete mixture models have wide flexibility in representing heterogeneous data, when a choice of parametric form for the heterogeneity is unclear or when inferences are sensitive to particular choices of parametric mixture. Assume a mixture of J sub-populations or groups, and let L_i denote a latent group membership indicator for sample member i; L_i can take any value between 1 and J. The latent data may also be expressed by multinomial indicators $Z_{ij} = 1$ if $L_i = j$ and $Z_{ij} = 0$ otherwise. If the missing data L_i were actually known, the density of the data $Y = (y_1, \ldots, y_n)$ could be written

$$y_i | L_i = j \sim f(y_i | \theta_j)$$

The densities conditional on L are generally taken to have the same form, e.g. all normal or all Poisson, with the parameters of groups j being denoted θ_j. For a Poisson the θ_j's would be means and for a Normal mixture the θ_j's could be group specific means and variances $\{\mu_j, \phi_j\}$, varying means and a common variance $\{\mu_j, \phi\}$ or a common mean but differing variances $\{\mu, \phi_j\}$. The marginal density of y is then

$$f(y_i) = \sum_{j=1}^{J} f(y_i|\theta_j)\pi_j \tag{2.26}$$

where π_j is the probability of belonging to sub-population j, and

$$\sum_{j=1}^{J} \pi_j = 1$$

In practice the L_i are unknown and the estimation problem relates to the mixing proportions π_j, the parameters of the separate sub-populations, and the posterior estimates of group membership L_i given the data and parameters. Typically, the vector π is assigned a Dirichlet prior with elements $\{\alpha_1, \ldots, \alpha_j\}$ with a typical choice being such as $\alpha_1 = .. = \alpha_J = 1$; the posterior for π is then provided by a Dirichlet with elements $K_j + \alpha_j$, where K_j is the number of sample members assigned to the jth group. The probabilities governing the assignment of subjects i to groups j are specified by

$$\mathrm{P_{ij}} = \Pr(Z_{ij} = 1) = \pi_j f(y_i|\theta_j)/f(y_i) \tag{2.27}$$

where $f(y_i)$ is as in Equation (2.26). If means differ by sub-population, then the smoothed mean ρ_i for the ith subject is the sum over j of the population means μ_j times the assignment probabilities

$$\rho_i = \sum_j \mathrm{P}_{ij}\mu_j \tag{2.28}$$

These means for subjects will often show shrinkage towards the global mean $\tilde{\mu} = \pi_1\mu_1 + \pi_2\mu_2 + ..\pi_J\mu_J$ (e.g. in disease mapping applications), even if the subjects have high posterior probabilities of belonging to groups with means much above or below the global mean.

The major issues in identifying mixture models using parametric densities $f(y|\theta_j)$ are the general question of identifiability in the face of possibly flat likelihoods (Bohning, 2000), and the specification of appropriate priors that are objective, but also effective in estimation. Thus, Wasserman (2000) cites the hindrance in mixture modelling arising from the fact that improper priors yield improper posteriors. More generally, vague priors even if proper, may lead to poorly identified posterior solutions, especially for small samples. Various approaches to prior specification in mixture modelling have been proposed and often mildly informative proper priors based on subject matter knowledge may be employed.

There are also issues of 'label switching' in MCMC estimation of mixture models. If sampling takes place from an unconstrained prior with J groups then the parameter space has $J!$ subspaces corresponding to different ways of labelling the states. In an MCMC run on an unconstrained prior there may be jumps between these subspaces. Constraints may be imposed to ensure that components do not 'flip over' during estimation. One may specify that one mixture probability is always greater than another, or that means are ordered, $\mu_1 > \mu_2 > \ldots > \mu_J$, or that variances are ordered. It remains problematic whether such constraints distort the final estimate. Depending on the problem, one sort of constraint may be more appropriate to a particular data set: constraining means may not be effective if subgroups with different means are not well identified, but groups with different variances are (Fruhwirth–Schattner, 2001). A sophisticated constraint scheme is proposed by Robert (1996) which for continuous Normal or Student data encompasses both mean and variance: for a two group model, this takes the form

$$Y \sim p\text{N}(\mu, \sigma^2) + (1 - p)\text{N}(\mu + \sigma\theta, \tau^2\sigma^2)$$

There is also the possibility of empty groups (e.g. only $J - 1$ groups of cases are chosen at a particular MCMC iteration when a J group model is being fitted); to avoid this, Wasserman (2000) proposes a data dependent prior that avoids a null group occurring. Chen *et al.* (2001) seek to overcome boundary and identifiability problems by introducing an extra penalty component to the likelihood that amounts to introducing JC extra 'observations', C with mean μ_1, C with mean μ_2, and so on. As an example, let there be $J = 2$ sub-groups with probabilities π and $(1 - \pi)$, differing only in their means μ_1 and μ_2, with $\mu_1 < \mu_2$. The number $2C$ of extra observations is related to the precision on the prior for the μ_j's. Thus if μ_1, μ_2 are uniform with priors $U(-R, R)$ then Chen *et al.* take $C = \log(R)$. For a normal prior on the μ_j, one might take $C = \log(3\sigma)$, where σ^2 is the prior variance of the μ_j. The likelihood[9] for $J = 2$ is penalised as follows:

$$\Sigma_i \log[(1 - \pi) f(y_i|\mu_2) + \pi f(y_i|\mu_1)] + C \log[4\pi(1 - \pi)]$$

This approach generalises to $J > 2$ sub-groups, with the same inequality constraint on the μ_j's, and with a penalty $C\Sigma_{j=1}^{J} \log(2\pi_j)$.

2.4.2 DPP priors

Dirichlet Process Priors (abbreviated as DPP priors) offer another approach avoiding parametric assumptions and, unlike a mixture of parametric densities, are less impeded by uncertainty about the appropriate number of sub-groups (see Dey *et al.*, 1999). The DPP method deals with possible clustering in the data without trying to specify the number of clusters, except perhaps a maximum conceivable number. Let y_i, $i = 1, \ldots, n$, be drawn from a distribution with unknown parameters ν_i, φ_i,

$$f(y_i|\nu_i, \varphi_i)$$

where a Dirichlet process prior is adopted for the ν_i's but a conventional parametric prior for the φ_i's. The Dirichlet process specifies a baseline prior G_0 from which candidate values for ν_i are drawn. For instance, in Example 2.1 most of the incidence ratios λ_i are likely to be between 0.25 and 4 in a hierarchical model, so a suitable baseline prior G_0 on the $\nu_i = \log(\lambda_i)$ might be a N(0, 0.5) density.

Suppose the ν_i's are unknown means or log means for each case, and that clustering in these values is expected. Then for similar groups of cases within a cluster, the same value θ of ν_i would be appropriate for them. Theoretically, the maximum number M of clusters could be n, but it will usually be much less. So a set of M potential values of θ, denoted θ_m, $m = 1, \ldots, M$, is drawn from G_0. The most appropriate value θ_m for case

[9] In BUGS, for $J = 2$ components in a normal mixture with equal variance across groups, and C rounded to an integer, the prior for the additional points might be specified with the coding:

```
for (i in 1:C) {y.aux[i] ~ dnorm(mu[1],tau);      y.aux[i+C] ~ dnorm(mu[2],tau)}
for (i in 1:n) {y[i] ~ dnorm(mu[L[i]],tau);       L[i] ~ dcat(pi[1:2])}
```

i is selected using multinomial sampling with M groups, and the Dirichlet prior on the group probabilities has a precision parameter α. So the cluster indicator for case i is chosen according to

$$L_i \sim \text{Categorical}(p)$$

where p is of length M, and has uniform elements determined by a precision parameter α. Then case i is assigned the parameter θ_{L_i}. The parameter α may be preset (typical values are $\alpha = 1$ or $\alpha = 5$), or itself assigned a prior. Following Sethuraman (1994) and Ishwaran and James (2001), one may generate the mixture parameters by considering the ν_i's as iid with density function q, where

$$q() = \sum_{j=1}^{\infty} p_j \; h(|\theta_j) \tag{2.29}$$

where h is the density of the θ_j under G_0. In practice, the mixture may be truncated at $M \leq n$ components with $\Sigma_{j=1}^{M} p_j = 1$.

The random mixture weights p_j at any iteration may be constructed by defining a sequence $r_1, r_2, \ldots, r_{M-1}$ of Beta(1, α) random variables, with $r_M = 1$. Thus, sample

$$r_j \sim B(1, \alpha) \quad j = 1, \ldots, M \tag{2.30}$$

and set

$$\begin{aligned} p_1 &= r_1 \\ p_2 &= r_2(1 - r_1) \\ p_3 &= r_3(1 - r_2)(1 - r_1) \end{aligned}$$

and so on. This is known as a stick-breaking prior, since at each stage what is left of a stick of unit length is broken and the length of the broken portion assigned to the current value p_j. At any iteration it is likely that some of the M potential clusters will be empty and one may monitor the actual number J of non-empty, though differing, clusters. Other sampling strategies for the r_j are discussed by Ishwaran and James (2001); one option takes the random weights as $r_j \sim B(a_j, b_j)$, where $a_j = 1 - \alpha$, $b_j = j\alpha$ and $0 < \alpha < 1$.

Example 2.7 SIDS deaths Consider the well known data on SIDS deaths S_i among births B_i in 100 counties of South Carolina during 1974–78. The statewide SIDS death rate per 1000 births is 2 per 1000 (667 deaths in 329 962 births), but some counties have (unsmoothed) rates approaching 10 per 1000. Symonds *et al.* (1983) consider a two group Poisson mixture for these events, such that

$$\begin{aligned} S_i &\sim \text{Poi}(R_i B_i) \\ L_i &\sim \text{Categorical}(\pi_{1:2}) \end{aligned}$$

where

$$R_i = \mu_{L_i}$$

is the Poisson mean in the group with index L_i. It is common practice to treat rare vital events as Poisson in this way, though strictly they are binomial; in more complex models adopting Poisson sampling, occasional inadmissible values may be generated for death rates R_i unless a constraint in the prior is included. An option is to reframe the problem in terms of expected deaths E_i, where $E_i = RB_i$, and R is here the statewide rate. Then $S_i \sim \text{Poi}(\nu_i E_i)$, where the ν_i's are relative risks with average close to 1 (and exactly equalling 1 if the sum of E_i equals the sum of the S_i).

The smoothed mean for the ith county is drawn from a mixture with two components, and will be determined by the average of formula (2.28) over a large number of iterations: the mean for county i will be smoothed towards the sub-population mean, depending on its highest assignment probabilities. In 'disease mapping' applications such as these, a primary object is the identification of counties with genuinely high risk. In this regard, simple maximum likelihood estimates (or 'crude' rates) $1000S_i/B_i$ may distort true high risks (e.g. giving spurious impressions of clustering), especially if the most extreme rates are based on small birth totals or small death totals.

For the two group parametric mixture model, $G(0.0, 0.01)$ priors on μ_j, $j = 1, 2$ are adopted and a Dirichlet D(1, 1) prior on the mixing probabilities π_j. An initial unconstrained run (with a single chain) suggested two well-separated means; a subsequent three chain run was therefore undertaken with a constraint on the means. Note that for county memberships L_i arbitrary initialising may be used, though preferably not all set to one subgroup. A run of 10 000 iterations identified a minority of areas with high SIDS death rates (mortality at around 3.8 per 1000 births, with a summary in Table 2.9 based on the second half of the run). In terms of the resulting estimates of SIDS rates at county level, broadly similar results to Symonds *et al.* are obtained though with different rankings within the high risk counties. The estimates of county rates are based on a further 1000 iterations subsequent to the first 10 000.

The shrinkage obtained under this model is exemplified by county 87 (Swain county) with a crude SIDS rate of 4.4 per 1000, but based on three deaths among only 675 births over five years. Despite its crude rate exceeding the higher subpopulation mean of 3.8 per 1000, it has a posterior probability of 0.24 of belonging to the low risk group, and a

Table 2.9 Mixture parameters (two groups) and highest posterior SIDS rates

	Mean	St. devn.	2.50%	Median	97.50%
SIDS Rates per 1000 Births					
Anson	3.70	0.52	2.71	3.64	5.02
Robeson	3.69	0.53	2.70	3.64	5.01
Halifax	3.69	0.53	2.67	3.64	5.02
Columbus	3.64	0.61	1.69	3.62	5.00
Northampton	3.57	0.69	1.62	3.59	5.00
Rutherford	3.42	0.77	1.62	3.51	4.86
Rockingham	3.42	0.80	1.58	3.52	4.89
Bladen	3.26	0.91	1.55	3.43	4.87
Hertford	3.21	0.93	1.52	3.40	4.89
Hoke	3.18	0.95	1.52	3.38	4.89
Washington	3.02	0.99	1.47	3.28	4.81
Bertie	2.98	1.02	1.46	3.25	4.87
Means of components					
μ_1	0.0017	0.0001	0.0014	0.0017	0.0019
μ_2	0.0038	0.0005	0.0028	0.0037	0.0049
Probabilities of components					
π_1	0.765	0.098	0.538	0.780	0.912
π_2	0.235	0.098	0.088	0.220	0.462

smoothed rate of 2.5 per 1000. It may be noted from Table 2.9 that this model provides essentially no discrimination between the counties with the three highest rates (counties 4, 42 and 78): they have a virtual unity probability of belonging to the higher risk group and identical median SIDS rates of 3.64.

One advantage of discrete parametric mixtures is that the number of model parameters is usually well defined, and so penalised fit measures such as the AIC or BIC present no problem in definition. Note that this is not the case for random effects mixtures, which is why the DIC approximation discussed above has been developed. In particular, one might base choice between a model with $J+1$ components (Model 2) against one with J components (Model 1) in terms of that model with minimum AIC or BIC. There is a case for using the BIC here since the 'sample size' might be defined not by the number of counties, but by the number of births in them (Upton, 1991), and also because the BIC approximation becomes more reliable for large samples.

For closely competing models, one might want to assess the probability that the AIC or BIC for Model 2 exceeds that for Model 1. If a Poisson mixture with three components is compared against one with two components, there are $p_1 = 3$ parameters in the simpler model and $p_2 = 5$ in the more complex one (three means, and two free mixture probabilities). If the models are run in parallel and at each iteration $D_2^{(t)} + 2p_2$ is compared with $D_1^{(t)} + 2p_1$ (where $D = -2\log L$), then accumulating over iterations gives the probability that model j has the minimum AIC (Model B in Program 2.7).

In this application, the priors on the means for both models are constrained, and an upper limit for the highest smoothed mean is set by the maximum crude rate of 9.6 per 1000. Without this limit there is a tendency for impossible means (mortality exceeding 100%) to be generated.

A three chain run of 25 000 iterations (and burn-in of 5000) with initial values for two chains provided by a trial run, shows some support for a three group solution. In approximate terms, a quarter of counties belong to a low risk sub-group, 50% to a medium risk group, and a quarter to a high risk group (Table 2.10). Comparison of the AICs at each iteration between the models shows the three group mixture to have a lower AIC with a probability of 60%.

Finally, a Dirichlet process prior approach is applied, with the goal of assessing the relative SIDS risks in Anson, Robeson and Halifax counties under a semi-parametric approach, and also whether the relative risk in Swain county is clearly above 1 or

Table 2.10 Comparison of two and three group models

		Mean	St. devn.	1%	Median	99%
2 sub-populations	AIC	10030	150	9669	10030	10360
Group Means	μ_1	0.0016	0.0001	0.0013	0.0017	0.0019
	μ_2	0.0037	0.0005	0.0026	0.0036	0.0050
Group Probabilities	π_1	0.745	0.105	0.429	0.759	0.923
	π_2	0.255	0.105	0.077	0.241	0.571
3 sub-populations	AIC	9968	198	9262	9989	10330
Group Means	μ_1	0.0008	0.0007	0.0000	0.0011	0.0018
	μ_2	0.0020	0.0005	0.0014	0.0018	0.0036
	μ_3	0.0040	0.0009	0.0027	0.0039	0.0080
Group Probabilities	π_1	0.243	0.244	0.000	0.184	0.818
	π_2	0.556	0.221	0.043	0.597	0.896
	π_3	0.202	0.104	0.013	0.189	0.507

whether alternatively it straddles 1 (and so is not definitively high risk). The reframed format describe above is adopted, with expected SIDS deaths defined as $E_i = RB_i$, and taking $S_i \sim$ Poi($\nu_i E_i$). The relative risks $\nu_i = \exp(\phi_i)$ are modelled via a DPP with a ceiling of $M = 10$ clusters, and with an N(0, 1) baseline prior on ϕ_i. The prior on r_j is as in (30) with $\alpha = 1$ preset.

A three chain run (with initial values randomly generated from the priors) shows convergence of the non-empty cluster total J at around iteration 1000 and the summary is based on iterations 1000–5000. This shows the relative risk clearly highest in Anson county, and with the Swain county risk straddling 1 (Table 2.11). The distribution of non-empty clusters shows six clusters as the most frequent (Figure 2.1). One may wish to assess sensitivity of inferences to different baseline densities (e.g. Student t with mean 0, variance 1 and low degrees of freedom) or to alternative values of α.

Example 2.8 Exponential mixtures for patient length of stay distributions The work of Harrison and Millard (1991) and McClean and Millard (1993) relates to lengths of stay of patients in hospital, with a particular focus on patient lengths of stay in geriatric departments. Lengths of stay of other classes of patients have also attracted discrete mixture analysis (especially psychiatric patients), as well as other patient characteristics such as age at admission (Welham *et al.*, 2000). Lengths of stay in geriatric departments exhibit pronounced skewness, but exponential mixture models with relatively few components have been found effective in modelling them. The analysis here shows the utility of Bayesian sampling estimation in deriving densities for structural or system parameters.

Table 2.11 DPP prior on SIDS deaths; selected parameters

Non-empty clusters	Mean	St. devn.	1%	Median	99%
J	5.99	1.55	3	6	9
Relative Risks					
Anson (county 4)	3.07	1.26	1.44	2.68	6.76
Halifax (county 42)	1.94	0.48	1.03	1.89	3.48
Robeson (county 78)	1.80	0.35	1.07	1.78	2.65
Swain (county 87)	1.27	0.54	0.55	1.09	3.01

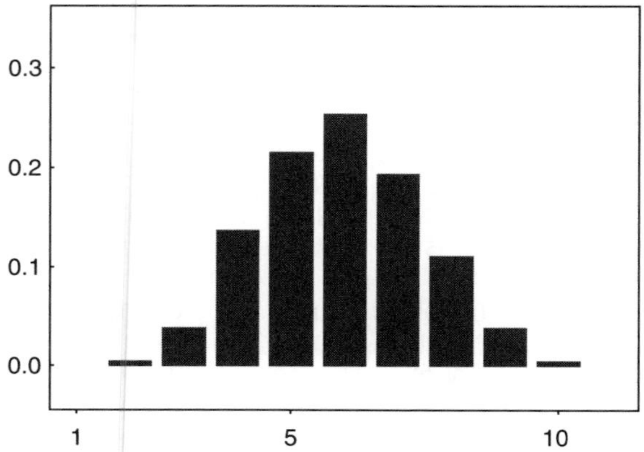

Figure 2.1 Number of DPP clusters in SIDS deaths analysis

Thus, let x denote length of stay for a patient in days, and suppose a two component exponential mixture, EM(2), is fitted

$$f(x) = \pi \exp(-Bx) + (1-\pi)\exp(-Dx) \tag{2.31}$$

where the identifiability constraint is that the average length of stay $1/B$ in group 1 is lower than that in group 2. Harrison and Millard (1991) draw on continuous time models for drug flow in pharmacokinetics to derive interpretations for the EM(2) parameters. Envisage two classes of patients, a standard stay group, of size $A(t)$ at calendar time t, and a long stay group of size $L(t)$ at t (these are distinct from the two groups in Equation (2.31)). The long stay group (were it possible to identify them on admission to hospital) form an unknown fraction k of patients admitted to the geriatric unit, but no one is formally admitted as a long stay patient. Instead, they are notionally converted to such a status at some time during their hospital stay. The standard group account for the remaining $1-k$ share of the patient population.

Harrison and Millard (1991) assume a steady state system operating at full capacity, with $A(t) = A$ and $L(t) = L$ constant and in equilibrium. The stock A in the standard group is kept constant by (a) patient discharges from hospital at a rate r, by (b) conversion from standard to long stay status at rate v, and by (c) a constant number of new patients per day A_0. Finally, long stay patients themselves have a low chance d of discharge (including mortality). Under the steady state assumption, it can be shown that

$$k = v/(v+r-d)$$

Under the exponential model for stay lengths, the rate of loss through discharges, deaths, and so on, is the inverse of the expected length of stay in days. So the expected length of stay for standard patients is

$$S_1 = 1/(\mathrm{v}+r)$$

while that for long stay patients is

$$S_2 = 1/d \tag{2.32}$$

The expected length of stay at admission under the steady state model is therefore

$$S = (1-k)S_1 + \mathrm{k}S_2$$

Let T be the total number of patients, or equivalently under equilibrium, the number of occupied beds in the hospital. Then, from Equation (2.31), the total of patients who have been in hospital for at least x days is given by

$$t(x) = T\pi \exp(-Bx) + T(1-\pi)\exp(-Dx)$$

while from the steady state model assumptions, it can be shown to be

$$t(x) = (1-k)A_0\ S_1(1-v-r)^x + k\ A_0\ S_2(1-d)^x \tag{2.33}$$

Under the steady state model, admissions equal exits so that A_0 is (Harrison and Millard, 1991, Eq. (17)) given by

$$(v+r)T\pi + dT(1-\pi)$$

Also, equating parameters in Equation (2.33), the proportion k of long stay patients may be derived from

$$kA_0S_2 = T(1-\pi)$$

Using Equation (2.32) gives

$$k = dT(1 - \pi)/A_0$$

So the parameters of the $EM(2)$ model may be used to estimate system parameters, resulting, say, from better discharge arrangements to non-hospital care for long stay patients. For example, if the daily discharge rate d for such patients is increased by $1/(365L)$, then $(r + v)/(v + d)$ more patients per year can be admitted.

To illustrate the practical working of this approach, and the derivation of system parameters from basic model parameters such as those in Equation (2.31), consider length of stay observations for $T = 469$ patients in a psychiatric hospital in North East London in January 1991. These are incomplete lengths of stay, but following the work discussed above, there is no need to model the censoring mechanism to obtain the structural parameters. Following the steady state model of Harrison and Millard as discussed above, we consider a two group mixture. One may, however, generalise these concepts for, say, $J = 3$ group mixtures and higher by invoking conversions from a short stay group to a standard group at rate v_1, from a standard group to long stay at rate v_2, etc.

The exponential mixture model (2.31) in Program 2.8A is parameterised so that B is larger than D, i.e. so that lengths of stay in group 1 (with prior probability π) are shorter than those in group 2. From a three chain run of 25 000 iterations, with convergence after a 500 iteration burn-in, a median estimate for π of 0.565 is obtained (Table 2.12). This larger group has an exponential parameter of 0.0016 and average length of stay of 620 days, compared to over 7500 days in the smaller group. To informally assess model fit, one may monitor the 'complete data' likelihood, taking the latent group membership indices ($L[i]$ in the coding in Program 2.8A) at each iteration as known. (Note that Example 2.7 used the 'incomplete data' likelihood.) This likelihood has a posterior median of −3999. So there is a clear gain in fit over a single group exponential model, which has median likelihood −4324 and estimates an average length of stay of 3700 days.

The proportion k of long stay patients under the steady state model is just under 6%, and the median number of such patients is then $L = 27$. Increasing the discharge rate

Table 2.12 Structural length of stay parameters: steady state patient flow model

	Parameter	Mean	St. devn.	2.5%	97.5%
A_0	Admission Rate (New patients per day)	0.457	0.041	0.383	0.547
	Complete Data Log-likelihood	−3999	11.5	−4023	−3977
	Expected LOS, Mixture Group 1	620.9	74.1	478.2	769.7
	Expected LOS, Mixture Group 2	7810	720	6543	9336
L	Number of long stay patients	27.23	4.41	19.58	36.94
S	Expected LOS on admission, all patients	1034	92.5	857.4	1225
	Benefit from higher discharge rate of long stay patients	7.596	0.813	6.105	9.298
π	Mixture Probability	0.565	0.040	0.487	0.642
K	Proportion long stay	0.058	0.009	0.042	0.079
B	Exponential Mean, Group 1	0.0016	0.0002	0.0013	0.0021
D	Exponential Mean, Group 2	0.00013	0.000012	0.00011	0.00015
R	Standard group discharge rate	0.0015	0.00019	0.0012	0.0020
v	Conversion rate, standard to long stay	0.000088	0.000019	0.000056	0.00013

among long stay patients by $1/(365L)$, or by one patient a year, means that on averaged 7.6 more patients per year can be admitted; this is the posterior mean of $(r+v)/(v+d)$. The uncertainty attaching to some of the structural parameters, such as the length of stay in group 2 (95% interval from 6500 to 9300 days) feeds through to the structural parameters. Thus, the number of new patients per day, A_0, is estimated with 95% interval $\{0.38, 0.55\}$, and the number of long stay patients has a 95% interval between 20 and 37.

We consider next mixtures with more groups (though without estimating the structural parameters). Thus for $J = 3$, and running three chains for 25 000 iterations yields convergence after 2500 iterations. The two extra parameters produce a further gain in fit (Model C in Program 2.8A). The median complete data likelihood is −3916, and a short stay group with average LOS around 54 days, exponential mean 0.0203, and relative probability of 12%, is identified.

A DPP version of this mixture analysis for the patient lengths of stay (Program 2.8B) is also applied. A maximum possible $M = 5$ alternative exponential means is set, i.e. up to five potential clusters in which the 469 lengths of stay can be arranged. Trial analysis suggested that most of the mass was concentrated on 3–5 clusters, even for larger α (the average number of non-empty clusters J is expected to be increase with the Dirichlet precision parameter). As well as monitoring the number of non-empty clusters and the complete data likelihood, one may track the largest cluster mean (th.R[M] or th.R[5] in the code), corresponding to the shortest stay patients. Monitoring convergence is complicated by the fact that parameters may have different meanings according to whether the number of non-empty clusters at a particular iteration is $J = 3$, $J = 4$, $J = 5$, etc.

A three chain run with $\alpha = 1$ is run for 5000 iterations, with the second half giving an average for non-empty clusters J of 3.1, and complete likelihood averaging −3918. The two cluster solution was not selected in any iteration. The exponential parameter corresponding to the short stay group has an average value 0.37 when five clusters are chosen (in 27 of 7500 iterations), 0.085 when four clusters are selected (which occurs in about 12.5% of iterations) and an average value 0.0213 at iterations when three clusters are chosen. To assess this patterning, it is necessary to use the CODA facility in BUGS and sort iterations according to the number of clusters selected. Increasing the Dirichlet parameter (e.g. to $\alpha = 5$) does raise the proportion of iterations when four or five clusters are selected, but there is no overwhelming evidence to suggest the existence of four or more clusters in the data.

2.5 GENERAL ADDITIVE AND HISTOGRAM SMOOTHING PRIORS

Many types of smoothing problems involve a series of observations $y_1, \ldots, y_n$ at equally spaced observation or design points (e.g. consecutive ages, income groups, etc.). While often considered as time series methods, it is worth emphasising the scope for more general applications. Among the approaches which have figured in the Bayesian literature, we consider here two relatively simple methodologies which are suitable for a wide range of problems. One is based on smoothness priors derived from differencing an underlying series of true values (Kitagawa and Gersch, 1996). The other, histogram smoothing, is applicable to frequency plots, including those deriving from originally metric data aggregated to equal length intervals – an example involving weight gains in pigs is considered by Leonard and Hsu (1999).

2.5.1 Smoothness priors

For the smoothness prior approach, it is assumed initially that the y_t are metric with normal errors. Extensions to discrete outcomes and unequally spaced observation intervals are also considered below, as are extensions to grouped observations – multiple observations at a single design point t.

The model for metric observations is then $y_t = f_t + e_t$, where $e_t \sim \mathrm{N}(0, \sigma^2)$, and where the f_t's may be taken as an underlying series, free of the measurement error in the observed series. The measurement error assumption leading to an underlying series which is the object of smoothing is especially frequent for continuous data. For discrete outcomes, the focus of smoothing is typically on the mean count, risk or probability. Thus, if the y_t were Poisson, then one might have

$$\begin{aligned} y_t &\sim \mathrm{Poi}(\mu_t) \\ \log(\mu_t) &= f_t = f_{t-1} + u_t \end{aligned}$$

while if they were binomial with populations n_t, then the model might be

$$\begin{aligned} y_t &\sim \mathrm{Bin}(\pi_t, n_t) \\ \mathrm{logit}(\pi_t) &= f_t = f_{t-1} + u_t \end{aligned}$$

A widely used smoothing model assumes Normal or Student distributed random walks in the first, second or higher differences of the f_t's. For example, a Normal Random Walk (RW) in the first difference Δf_t is equivalent to the smoothness prior

$$f_t \sim \mathrm{N}(f_{t-1}, \tau^2) \tag{2.34}$$

and a random walk in the second difference $\Delta^2 f_t$ leads to the prior

$$f_t \sim \mathrm{N}(2f_{t-1} - f_{t-2}, \tau^2) \tag{2.35}$$

These two priors can be expressed in the alternative forms

$$f_t = f_{t-1} + u_t$$

and

$$f_t = 2f_{t-1} - f_{t-2} + u_t$$

respectively. These models have to be set in motion by assumptions about the initial parameter(s): this involves separate priors on the initial parameters, such as f_1 in the case of a first difference RW, f_1 and f_2 in the case of a second order RW, and so on. A common practice is to ascribe vague priors with large variances to these parameters.

A rationale for this is provided by the measurement error model for continuous Normal data. Here the ratio $\lambda = \sigma^2/\tau^2$ is a smoothing parameter, which for a second order random walk, RW(2), appears in the penalised least squares criterion

$$\Sigma_t\, e_t^2 + \lambda \sum_{t=3}^{n} \Delta^2 f_t$$

With flat initial priors on the initial f_t's in the series the posterior modes for f_t are equivalently those that minimise this criterion. Thus, one might set initialising priors

$$f_j \sim N_r(0, kI), \quad j = 1, \ldots, r$$

where r is the order of differencing in the smoothness prior and k is large.

It may be noted that one is not confined to asymmetric priors like (2.34) and (2.35), which make most sense if t does denote time; if the index t were something like income group, then it might make sense for f_t to depend upon adjacent points at either side. Thus, a generalisation of the first order prior might involve dependence on both f_{t-1} and f_{t+1}. For example,

$$f_t = 0.5f_{t-1} + 0.5f_{t+1} + u_t \quad 2 \leq t \leq n-1$$
$$f_1 = f_2 + u_1$$
$$f_n = f_{n-1} + u_n$$

Further issues concern specialised priors in the event of unequal spacing or tied observations. For unequal spacing, it is necessary to weight each preceding point differently, and to change the precision such that wider spaced points are less tied to their predecessor than closer spaced points. Thus, suppose the observations were at points $t_1, t_2, \ldots, t_n$ with $\delta_1 = t_2 - t_1, \delta_2 = t_3 - t_2, \ldots, \delta_{n-1} = t_n - t_{n-1}$. The smoothness prior (2.34) would become

$$f_t \sim \mathrm{N}(f_{t-1}, \delta_t \tau^2)$$

so that wider gaps δ_t translates into larger variances. Similarly, the prior (2.35) would become

$$f_t \sim \mathrm{N}(\nu_t, \delta_t \tau^2)$$

where $\nu_t = f_{t-1}(1 + \delta_t/\delta_{t-1}) - f_{t-2}(\delta_t/\delta_{t-1})$. ν_t reduces to the standard form if $\delta_t = \delta_{t-1}$. For multiple observations at one point in the series, we have one model for the n observations and a prior for the G distinct values of the series index. For instance, suppose the observations y_t were 1, 3, 4, 7, 11, 15 ($n = 6$) at observation points 1, 1, 2, 2, 3, 4 (i.e. $G = 4$). Then if g_t denotes the grouping index to which observation t belongs, and a first order random walk in the underlying series is assumed, we would have

$$y_t \sim \mathrm{N}(f[g_t], \sigma^2), \quad t = 1, \ldots, n$$
$$f_k \sim \mathrm{N}(f_{k-1}, \tau^2), \quad k = 2, \ldots, G$$

2.5.2 Histogram smoothing

Suppose observations of an originally continuous variable y are grouped into frequency counts, with f_j being the number of observations between cut points K_{j-1} and K_j. For example, Leonard and Hsu (1999) discuss the pig weight gain data of Snedecor and Cochran (1989), where weight gains of 522 pigs are aggregated into 21 intervals of equal width. One may also consider counts f_j of events (e.g. migrations, deaths) arranged in terms of a monotonic classifier, at levels $j = 1, \ldots, J$, such as age band or income level. Often, sampling variability will mean the observed frequencies f_j in the jth interval will be ragged in form when substantive considerations would imply smoothness.

Let prior beliefs about such smoothness be represented by an underlying density $h(y)$, and let π_j be the probability that an observation is in the jth interval

$$\pi_j = \int_{K_{j-1}}^{K_j} h(u)du$$

The observed frequencies are assumed to be multinomial with parameters $\{\pi_1, \pi_2, .., \pi_J\}$, and to imperfectly reflect the underlying smooth density. A prior structure on the π_j's that includes smoothness considerations is appropriate. Specifically, express π_j via a multiple logit model

$$\pi_j = \exp(\phi_j)/\sum_k \exp(\phi_k)$$

where the ϕ_k are taken to be multivariate normal with means $a_1, a_2, .., a_J$ and $J \times J$ covariance matrix V. A noninformative prior on the π_j's would take them to be equal in size, i.e.

$$\pi_j = 1/J$$

which is equivalent to

$$a_j = -\log(J)$$

In the case of frequencies arranged by age, income, etc., there are likely to be total population or exposure totals E_j, and the appropriate prior on the π_j's then would involve taking

$$\pi_j = E_j/\sum_k E_k \qquad (2.36a)$$

or

$$a_j = \log(E_j/\sum_k E_k) \qquad (2.36b)$$

The covariance matrix **V** is structured to reflect dependence between frequencies neighbouring on the histogram classifier. Thus, one option is a first order dependence of order 1 with correlation ρ, as discussed in Chapter 5. Thus

$$V_{ij} = \rho^{|i-j|}\sigma^2 \qquad (2.37)$$

Then $\rho = 0$ leads to exchangeability in the histogram probabilities, that is to a joint distribution that is unaffected by permutation of the suffixes $1, 2, .., J$. This is contrary to expectation in many situations where greater similarity is anticipated between π_j and π_{j+1} than between π_j and π_{j+2}. Note that defining $\kappa = 1/(\sigma^2 - \sigma^2\rho^2)$, the elements of $T = V^{-1}$ in the case of the prior (2.37) are given by

$$\begin{aligned}
&T_{11} = T_{JJ} = \kappa, \\
&T_{jj} = \kappa(1+\rho^2) \quad j = 2, .., J-1 \\
&T_{j,j+1} = T_{j+1,j} = -\kappa\rho \quad j = 1, .., J-1 \\
&T_{jk} = 0 \text{ elsewhere}
\end{aligned}$$

Example 2.10 Hemorrhagic Conjunctivitis As an example of the smoothness prior approach, though from a maximum likelihood perspective, Kashigawi and Yanagimoto (1992) consider smoothing of serial count data through a state-space model, specifically weekly data y_i on new cases of acute hemorrhagic conjunctivitis in the Chiba-prefecture of Japan in 1987.

Thus, if $y_t \sim \text{Poi}(\lambda_t)$, $t = 1, .. n$, with the data sequenced in time, they assume that the λ_t's change gradually through time. They therefore impose a difference prior of order r on the log link $v_t = \log \lambda_t$, with

$$\Delta^r v_i \sim N(0, \sigma^2) \tag{2.38}$$

The case $r = 1$ with $\sigma^2 = 0$ leads to a trendless model with constant incidence of new cases, $\lambda_1 = \lambda_2 = .. = \lambda_n$. This leads on to a test for homogeneity of the means through time (with $r = 1$ in Equation (2.38)), namely $\sigma^2 = 0$ vs. $\sigma^2 > 0$. While these data are time series, the approach is applicable to any series arranged against a monotonic index (e.g. age, income, weight). Kashigawi and Yanagimoto detected a significant departure from a stable mean weekly incidence in these series with an increased level apparent between about the 28th and 38th weeks.

Here a flat N(0, 1000) prior on f_1 is assumed with an RW(1) smoothness prior, and initially a G(1, 0.001) prior on $1/\sigma^2$. The findings of Kashigawi and Yanagimoto are replicated, and Figure 2.2 shows the underlying series of mean incidence. The mean number of new cases peaks at around 1.21 in the weeks 35–38, then falls back to around 0.7 at the end of the year. This smooth gives an appearance of over-smoothing, and Figure 1 of Kashigawi and Yanagimoto in fact shows a smooth where there is a more pronounced variation in mean incidence than in Figure 2.2. Their smoothed series can be replicated more closely with a G(1, 1) prior on σ^{-2}, since this places lower weight on high precisions (lower variances σ^2 which produce less smoothing). The means vary now from 0.25–2.4. To resolve the problem of the most appropriate smoothing variance, several alternative priors might be adopted and their appropriateness compared with simple fit measures (e.g. log-likelihoods) or formal Bayes choice methods. With the G(1, 1) prior on the precision, the log-likelihood of a trend model (with non-zero variance σ^2) averages -56, as against -69 for a constant mean model.

To estimate the effective number of parameters in the model with a G(1, 1) prior on the precision, the DIC procedure suggests six effective parameters based on comparing the mean deviance of 33.6 with the deviance at the posterior mean of 27.8. Note that deriving this (in a non-iterative program) involves the posterior estimates of λ_i as these

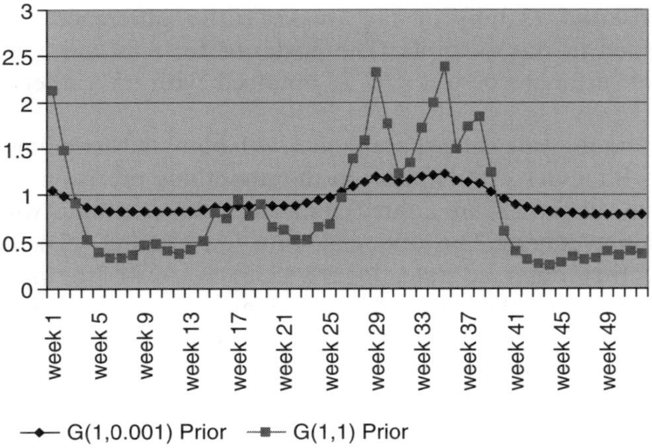

Figure 2.2 Smoothed plots of mean conjunctivitis incidence

alone define the likelihood. On this basis, the reduction in average deviance of around 26 is offset by five extra parameters.

Another option is to set a higher stage prior on the gamma parameters a_1, b_1 in $1/\sigma^2 \sim G(a_1, b_1)$. For example, the choice

$$a_1 \sim \exp(1) \text{ and}$$
$$b_1 \sim G(0.001, 0.001)\, I(0.001,)$$

is satisfactory in terms of identifiability, and yields a smooth series intermediate between the two in Figure 2.2. The DIC is worse than under the preset G(1, 1) prior on $1/\sigma^2$. Fidelity to the original data is not the only criterion governing choice of a smoothing model, and genuine cross-validation (involving omission of cases and prediction by the remaining data) might also be investigated.

Example 2.11 Mental illness hospitalisations As an example comparing histogram and state space smoothing, consider hospitalisations f_j for mental illness at age j over two London boroughs in the year 1st April 1998 to 31st March 1999. These are classified by single year of age from ages 15–84, and may be seen as binomial outcomes in relation to age-specific population estimates P_j at mid-1999.

Fitting a histogram smoothing model entails setting (see Equation (2.36))

$$a_j = \log(P_j / \sum_k P_k)$$

where P_j is the population at risk at age j. A U(0, 1) prior on the correlation parameter ρ is assumed, and a gamma G(1, 0.001) prior on $1/\sigma^2$; this defines the covariance matrix $\mathbf{V}$ as in Section 2.5.2. A three chain run shows convergence of Gelman–Rubin criteria for these two parameters at around iteration 1500, and the remaining iterations of a run of 5000 provide the summary.

Table 2.13 shows the smoothing of ragged frequencies to reflect the operation of the prior in adjacent categories of the histogram. Thus, the observed schedule shows 19 hospital cases at age 21, six at age 22 and 25 at age 23. The smoothed version shows much less discontinuity with 16.5, 13.6, and 19 as the estimated frequencies of hospitalisation at these ages. Taking account of exposures to risk as in the prior parameterisation (2.36b) in fact makes little difference to the degree of smoothing obtained in this example. The posterior for ρ is concentrated above 0.9. A posterior mean estimate of $\rho = 0.97$ is obtained with 95% interval from 0.92 to 0.995.

A first order random walk state space model with binomial sampling at the first stage is then applied, with a G(1, 0.001) prior on the smoothing precision and an N(0, 1000) prior on the initial value f_1. From a three chain run (5000 iterations with 500 burn-in) a posterior precision averaging 22 is obtained, with f_1 averaging -6.4. This compares to the logit of the first observed rate, logit(5/4236) $= -6.74$, so the first underlying probability has been smoothed towards the average. Values for remaining smoothed frequencies are close to those obtained under the histogram smoothing approach. The random walk and histogram smooth model are expected to be broadly similar for values of ρ in Equation (2.37) close to 1.

Table 2.13 Smooth histogram for mental illness hospital cases

Age	Popn.	Actual Hospitalisations	Obs. Rate × 1000	Histogram smoothing Mean	Histogram smoothing 2.5%	Histogram smoothing 97.5%	Random Walk Mean	Random Walk 2.5%	Random Walk 97.5%
15	4236	5	1.2	7.4	4.3	11.5	7.1	4.2	10.9
16	4293	7	1.6	8.2	5.3	11.9	7.9	5.1	11.6
17	4378	8	1.8	9.6	6.4	13.5	9.3	6.2	13.1
18	4760	13	2.7	12.9	8.9	17.6	12.5	8.7	17.0
19	4940	17	3.4	16.1	11.6	21.9	15.8	11.2	21.4
20	4907	21	4.3	18.2	13.1	24.5	17.9	13.0	24.2
21	4603	19	4.1	16.6	12.0	22.4	16.4	11.8	21.8
22	4596	6	1.3	13.9	9.7	18.8	14.0	9.9	18.8
23	4861	25	5.1	19.0	13.6	25.0	18.9	13.7	24.9
24	5116	11	2.2	18.3	13.0	24.1	18.4	13.3	24.1
25	5375	32	6.0	26.2	19.5	34.2	25.9	19.4	33.5
26	5709	29	5.1	27.5	20.9	35.7	27.5	21.0	35.5
27	5932	24	4.0	26.3	19.6	34.1	26.7	20.1	34.2
28	5942	27	4.5	28.3	21.3	36.3	28.5	21.6	36.4
29	5705	30	5.3	31.0	23.7	39.5	31.0	23.8	39.4
30	5678	42	7.4	37.1	29.3	46.5	36.8	28.5	46.8
31	5583	34	6.1	33.6	26.1	42.9	33.7	26.0	42.7
32	5617	31	5.5	31.1	24.0	39.4	31.3	23.9	39.8
33	5569	32	5.7	29.0	22.0	37.4	28.8	21.8	37.1
34	5610	19	3.4	23.4	17.5	30.6	23.2	17.1	29.9
35	5932	25	4.2	24.0	17.9	31.2	24.0	17.7	31.0
36	5758	19	3.3	21.6	15.7	28.3	21.9	16.1	28.5
37	5600	24	4.3	22.7	16.8	29.6	23.0	17.0	30.1
38	5414	23	4.2	22.6	17.0	29.6	22.7	16.8	29.9
39	5176	22	4.3	21.6	16.1	28.1	22.0	16.3	28.5
40	5099	21	4.1	21.2	15.8	28.4	21.6	16.1	28.0
41	4950	23	4.6	21.2	15.7	27.8	21.5	15.9	28.0
42	4782	14	2.9	19.0	13.9	25.4	19.4	14.2	25.6
43	4646	25	5.4	22.7	16.9	30.2	22.8	16.9	29.6
44	4491	26	5.8	24.0	17.7	31.6	23.9	17.8	31.2
45	4626	24	5.2	24.2	17.8	31.3	24.1	18.2	31.3
46	4591	27	5.9	23.7	17.4	31.0	23.6	17.6	30.9
47	4502	16	3.6	19.1	14.0	25.0	19.3	14.0	25.3
48	4565	20	4.4	19.1	13.9	25.3	19.1	13.9	25.3
49	4689	19	4.1	18.4	13.0	24.5	18.5	13.4	24.5
50	4856	18	3.7	17.5	12.5	23.8	17.5	12.6	23.3
51	5188	15	2.9	16.5	11.7	22.4	16.6	11.8	22.2
52	5662	15	2.6	17.2	12.4	23.1	17.4	12.6	23.3
53	4426	20	4.5	14.6	10.6	19.9	14.7	10.7	19.9
54	4339	4	0.9	11.6	7.9	15.7	11.8	8.1	16.1
55	4329	16	3.7	14.3	10.2	19.5	14.3	10.2	19.5
56	4098	19	4.6	15.4	11.1	21.0	15.2	10.9	20.6
57	3711	13	3.5	13.0	9.0	17.9	13.0	9.1	17.7

(*continues*)

Table 2.13 (*continued*)

Age	Popn.	Actual Hospitalisations	Obs. Rate × 1000	Histogram smoothing			Random Walk		
				Mean	2.5%	97.5%	Mean	2.5%	97.5%
58	3411	9	2.6	11.3	7.7	15.5	11.3	7.9	15.7
59	3605	15	4.2	12.7	8.6	17.5	12.5	8.8	17.2
60	3731	7	1.9	12.0	8.2	16.4	12.1	8.3	16.6
61	3680	19	5.2	14.4	10.1	19.7	14.2	10.1	19.4
62	3580	12	3.4	13.0	9.1	18.3	13.0	9.1	17.8
63	3480	9	2.6	12.3	8.6	16.6	12.4	8.5	16.9
64	3355	13	3.9	13.8	9.8	18.5	13.7	9.7	18.6
65	3330	23	6.9	16.3	11.6	22.1	16.1	11.5	21.9
66	3259	10	3.1	13.4	9.4	18.1	13.3	9.5	18.0
67	3291	11	3.3	13.5	9.5	18.2	13.5	9.6	18.3
68	3277	16	4.9	15.3	11.0	20.9	15.3	10.9	20.7
69	3209	12	3.7	16.1	11.3	21.7	16.4	11.7	21.7
70	3144	24	7.6	21.7	16.0	28.4	21.5	16.0	28.3
71	3033	25	8.2	25.0	18.7	32.2	24.7	18.6	32.2
72	2985	30	10.1	29.0	22.3	36.9	28.7	21.9	36.9
73	2929	24	8.2	30.4	23.1	38.4	30.8	23.4	39.1
74	2831	47	16.6	45.1	35.6	56.7	45.2	35.8	56.1
75	2753	71	25.8	61.9	50.0	75.4	61.2	49.1	74.8
76	2688	49	18.2	53.2	42.8	65.7	53.4	42.6	65.2
77	2718	59	21.7	59.4	48.1	71.7	59.5	48.2	72.2
78	2738	69	25.2	67.5	56.1	80.9	67.5	55.2	81.5
79	2710	63	23.2	69.8	56.8	83.5	69.9	56.7	84.6
80	1626	64	39.4	61.1	49.7	74.9	60.9	49.5	73.9
81	1446	69	47.7	68.7	56.4	82.4	68.3	56.0	82.3
82	1502	90	59.9	86.7	72.2	101.9	86.8	72.2	102.5
83	1489	86	57.8	89.5	74.5	106.4	90.2	75.7	106.2
84	1444	115	79.6	109.2	91.5	128.5	110.0	92.7	129.5

2.6 REVIEW

The above discussion and examples have emphasised parametric smoothing methods based on exchangeable sample members. The fully Bayes approach to combining information over exchangeable units using exponential family densities is exemplified by George *et al.* (1994), and stresses the benefits (e.g. in fully expressing uncertainty) as compared to parametric empirical Bayes smoothing (see, for example, Morris 1983). The fully Bayes method implemented through repeated sampling allows the derivation of complex inferences concerning the relationships among the units, such as the density of the maximum or the density of the rank attached to each sample unit (Marshall and Spiegelhalter, 1998, p. 237).

If the simplest model based on exchangeable units assumes smoothing to a common global mean and a unique variance readily made modifications may be more realistic, for example allowing asymmetric skewed densities (Branco and Dey, 2001) or allowing the data to be distributed as a mixture of two or three distributions with different means and/or variances. Adopting discrete mixtures of parametric densities leads into semi and non-parametric Bayesian methods. Further flexibility is provided by the range of approaches based on the Dirichlet process priors (illustrated in Examples 2.7 and 2.8), and discussed by authors such as Dey *et al.* (1999) and Walker *et al.* (1999). These are more natural approaches if clustering of sub-groups within the sample is expected, or as providing a sensitivity analysis against baseline unimodal smoothing model. Sometimes the latter may suffice: for example, Marshall and Spiegelhalters analysis of 33 transplant centres failed to confirm a two cluster division of the centres. More specialised departures from exchangeability occur in the analysis of spatially correlated data where the clustering is based on spatial contiguity (Chapter 7).

Whether a unimodal symmetric density is appropriate or not as a basis for combining information, a further major element to the process of joint inferences about sample units is the presence of further relevant information, possibly over different levels of data hierarchies (pupils, schools, etc.). Hence inferences about means or ranks for sample units may need to take account of covariates: for instance, severity or casemix indices may be relevant to rankings of medical institutions (see Example 4.6). The next two chapters accordingly consider the modelling of covariate effects in single and multi-level data.

REFERENCES

Akaike, H. (1973) Information theory and an extension of the maximum likelihood principle. *2nd Internat. Sympos. Inform. Theory*, Tsahkadsor 1971, 267–281.

Aitchison, J. and Shen, S. M. (1980) Logistic-normal distributions: Some properties and uses. *Biometrika* **67**, 261–272.

Aitkin, M. (1991) Posterior Bayes factors. *J. Roy. Stat. Soc., Ser. B* **53**(1), 111–114.

Berger, J. and Pericchi, L. (1996) The intrinsic Bayes factor for model selection and prediction. *J. Am. Stat. Assoc.* **91**(433), 109–122.

Berkhof, J., van Mechelen, I. and Hoijtink, H. (2000) Posterior predictive checks: principles and discussion, *Computational Stat.*, **15**(3), 337–354.

Besag, J. (1989) A candidates formula: a curious result in Bayesian prediction. *Biometrika* **76**, 183.

Bohning, D. (2000) *Computer Assisted Analysis of Mixtures and Applications.* Monographs on Statistics and Applied Probability, 81. London: Chapman & Hall.

Box, G. and Tiao, G. (1973) *Bayesian Inference in Statistical Analysis.* Addison-Wesley.

Bozdogan, H. (2000) Akaike's Information Criterion and recent developments in information complexity. *J. Math. Psychol.*, **44**(1), 62–91.

Branco, M. and Dey, D. (2001) A general class of multivariate skew elliptical distributions. *J. Multivariate Analysis* **79**, 99–113.

Brooks, S. and Gelman, A. (1998) General methods for monitoring convergence of iterative simulations. *J. Comp. Graph. Statist.* **7**, 434–455.

Carlin, B. and Chib, S. (1995) Bayesian model choice via Markov chain Monte Carlo methods. *J. Roy. Stat. Soc., Ser. B* **57**(3), 473–484.

Carlin, B. and Louis, T. (2000) *Bayes and Empirical Bayes Methods for Data Analysis*, 2nd edn. Texts in Statistical Sciences. Boca Raton: Chapman and Hall/ RCR.

Chen, M., Shao, Q. and Ibrahim, J. (2000) *Monte Carlo Methods in Bayesian Computation.* Springer Series in Statistics. New York, NY: Springer.

Chen, H., Chen, J. and Kalbfleisch, J. (2001) A modified likelihood ratio test for homogeneity in finite mixture models. *J. Roy. Stat. Soc, Ser B*, **63**(1), 19–30.

Chib, S. (1995) Marginal likelihood from the Gibbs output. *J. Am. Stat. Assoc.* **90**, 1313–1321.

Clayton, D. and Kaldor, J. (1987) Empirical Bayes estimates of age-standardised relative risks for use in disease mapping. *Biometrics* **43**, 671–681.

Davison, A. C. and Hinkley, D. (1997) *Bootstrap Methods and their Application.* Cambridge Series on Statistical and Probabilistic Mathematics. Cambridge: Cambridge University Press.

De Finetti, B. (1961) The Bayesian approach to the rejection of outliers. *Proc. 4th Berkeley Symp. Math. Stat. Probab.* **1**, 199–210.

DeSantis, F. and Spezzaferri, F. (1997) Alternative Bayes factors for model selection. *Can. J. Stat.* **25**, 503–515.

Dey, D., Muller, P. and Sinha, D. (1999) *Practical Nonparametric and Semiparametric Bayesian Statistics*. Lecture Notes in Statistics 133. New York, NY: Springer.

DiCiccio, T., Kass, R., Raftery, A. and Wasserman, L. (1997) Computing Bayes factors by combining simulation and asymptotic approximations. *J. Am. Stat. Assoc.* **92**(439), 903–915.

DuMouchel, W. (1990) Bayesian meta-analysis. In: Berry, D. (ed.), *Statistical Methodology in the Pharmaceutical Sciences*. Marcel Dekker.

Fernandez, C. and Steel, M. (1998) On Bayesian modelling of fat tails and skewness. *J. Am. Stat. Assoc*, **93**, 359–367.

Fruhwirth-Schattner, S. (2001) Markov Chain Monte Carlo estimation of classical and dynamic switching and mixture models. *J. Am. Stat. Assoc.* **96**, 194–209.

Geisser, S. and Eddy, W. (1979) A predictive approach to model selection. *J. Am. Stat. Assoc.* **74**, 153–160.

Gelfand, A. (1996) Model determination using sampling based methods. In: W. Gilks, S. Richardson and D. Spieglehalter (eds.), *Markov Chain Monte Carlo in Practice*, Boca Raton: Chapman & Hall/CRC.

Gelfand, A. and Dey, D. (1994) Bayesian model choice: asymptotics and exact calculations. *J. Roy. Stat. Soc., Ser. B* **56**(3), 501–514.

Gelman, A., Carlin, J., Stern, H. and Rubin, D. (1995) *Bayesian Data Analysis*. CRC Press.

George E., Makov, U. and Smith, A. (1994) Fully Bayesian hierarchical analysis for exponential families via Monte Carlo Computation. In: Freeman, P. and Smith, A. (eds.), *Aspects of Uncertainty – A Tribute to D. V. Lindley*. New York: Wiley.

Geweke, J. (1993) Bayesian treatment of the independent Student-t linear model. *J. Appl. Econometrics* **8**, S19–S40.

Green, P. (1995) Reversible jump Markov Chain Monte Carlo computation and Bayesian model determination. *Biometrika* **82**(4), 711–732.

Harrison, G. and Millard, P. (1991) Balancing acute and long-term care: the mathematics of throughput in Departments of Geriatric Medicine. *Meth. Infor. Medicine* **30**, 221–228.

Hsiao, C., Tzeng, J. and Wang, C. (2000) Comparing the performance of two indices for spatial model selection: application to two mortality data sets. *Stat. in Medicine* **19**, 1915–1930.

Ishwaran, H. and James, L. (2001) Gibbs sampling methods for stick-breaking priors. *J. Am. Stat. Assoc.* **96**, 161–173.

Kashiwagi, N. and Yanagimoto, T. (1992) Smoothing serial count data through a state-space model. *Biometrics* **48**, 1187–1194.

Kass, R. and Raftery, A. (1995) Bayes factors. *J. Am. Stat. Assoc.* **90**, 773–795.

Kass, R. and Steffey, D. (1989) Approximate Bayesian inference in conditionally independent hierarchical models. *J. Am. Stat. Assoc.* **84**, 717–726.

Kennett, S. (1983) Migration within and between labour markets. In: Goddard, J. and Champion, A. (eds.), *The Urban and Regional Transformation of Britain*. London: Methuen.

Kim, S. and Ibrahim, J. (2000) Default Bayes factors for generalized linear models. J. Stat. Plan. Inference **87**(2), 301–315.

Kitagawa, G. and Gersch, W. (1996) *Smoothness Priors Analysis of Time Series*. Lecture Notes in Statistics 116. New York, NY: Springer.

Laird, N. (1982) Empirical Bayes estimates using the nonparametric maximum likelihood estimate for the prior. *J. Stat. Comput. Simulation* **15**, 211–220.

Laud, P. and Ibrahim, J. (1995) Predictive model selection. *J. Roy. Stat. Soc. Ser. B* **57**(1), 247–262.

Lee, P. (1997) *Bayesian Statistics: An Introduction. 2nd ed.* London: Arnold.

Leonard, T. (1980) The roles of inductive modelling and coherence in Bayesian statistics. In: Bernardo, J., DeGroot, M., Lindley, D. and Smith, A. (eds.), *Bayesian Statistics I*. Valencia: University Press, pp. 537–555.

Lenk, P. and Desarbo, W. (2000) Bayesian inference for finite mixtures of generalized linear models with random effects. *Psychometrika* **65**, 93–119.

Leonard, T. (1973) A Bayesian method for histograms. *Biometrika* **60**, 297–308.

Leonard, T. and Hsu, J. (1999) *Bayesian Methods: An Analysis for Statisticians and Interdisciplinary Researchers*. Cambridge: Cambridge University Press.

Lindley, D. (1957) A statistical paradox. *Biometrika* **44**, 187–192.

McClean, S. and Millard, P. (1993) Patterns of length of stay after admission in geriatric-medicine – an event history approach. *The Statistician* **42**(3), 263–274.

McCullagh, P. and Nelder, J. (1989) *Generalized Linear Models*. Boca Raton: Chapman & Hall/CRC.

Marshall, E. and Spiegelhalter, D. (1998) Comparing institutional performance using Markov chain Monte Carlo methods. In: Everitt, B. and Dunn, G. (eds.), *Recent Advances in the Statistical Analysis of Medical Data*. London: Arnold, pp. 229–250.

Morgan, B. (2000) *Applied Stochastic Modelling*. London: Arnold.

Morris, C. A. (1983) Parametric empirical Boyesian inference: theory and applications. *J. Am. Stat. Assoc.* **78**, 47–65.

Newbold, E. (1926) A contribution to the study of the human factor in the causation of accidents. Industrial Health Research Board, Report 34, London.

Raftery, A. (1995) Bayesian model selection in social research. *Sociological Methodology* **25**, 111–163.

Rao, C. (1975) Simultaneous estimation of parameters in different linear models and applications to biometric problems. *Biometrics* **31**, 545–554.

Richardson, S. and Green, P. (1997) On Bayesian analysis of mixtures with an unknown number of components. *J. Roy. Stat. Soc. Ser. B* **59**(4), 731–758.

Robert, C. (1996) Mixtures of distributions: inferences and estimation. In: *Markov Chain Monte Carlo in Practice*, Gilks, W., Richardson, S. and Spieglehalter, D. (eds.), Boca Raton: Chapman & Hall/CRC.

Sethuraman, J. (1994) A constructive definition of Dirichlet priors. *Stat. Sin.* **4**(2), 639–650.

Schwarz, G. (1978) Estimating the dimension of a model. *Ann. Stat*, **6**, 461–464.

Snedecor, G. and Cochran, W. (1989) *Statistical Methods*, 8th ed. Ames, IA: Iowa State University Press.

Spiegelhalter, D. (1999) An initial synthesis of statistical sources concerning the nature and outcomes of paediatric cardiac surgical services at Bristol relative to other specialist centres from 1984 to 1995. Bristol Royal Infirmary Inquiry (http://www.bristol-inquiry.org.uk/brisdsanalysisfinal.htm# Background Papers).

Spiegelhalter, D., Best, N., Carlin, B., and van der Linde, A. (2002) Bayesian measures of model complexity and fit, *J. Royal Statistical Society*, **64B**, 1–34.

Stephens, M. (2000) Bayesian analysis of mixture models with an unknown number of components – An alternative to reversible jump methods. *Ann. Stat.* **28**(1), 40–74.

Symons, M., Grimson, R. C. and Yuan, Y. (1983) Clustering of rare events. *Biometrics* **39**, 193–205.

Tsutakawa, R. (1985) Estimation of cancer mortality rates: A Bayesian analysis of small frequencies. *Biometrics* **41**, 69–79.

Upton, G. (1991) The exploratory analysis of survey data using log-linear models. *The Statistician* **40**, 169–182.

Walker, S., Damine, P., Laud, P. and Smith, A. (1999) Bayesian nonparametric inference for random distributions and related functions (with discussion). *J. Roy. Stat. Soc., Ser. B*, **61**, 485–527.

Wasserman, L. (2000)Asymptotic inference for mixture models using data-dependent priors. *J. Roy. Stat. Soc., Ser. B, Stat. Methodol.* **62**(1), 159–180.

Weinberg, C. and Gladen, B. (1986) The Beta-geometric distribution applied to comparative fecundability studies. *Biometrics* **42**(3), 547–560.

Welham, J., McLachlan, G. and Davies, G. (2000) Heterogeneity in schizophrenia; mixture modelling of age-at-first-admission, gender and diagnosis. *Acta Psychiat. Scand.* **101** (4), 312–317.

West, M. (1984) Outlier models and prior distributions in Bayesian linear regression. *J. Roy. Stat. Soc., Ser. B* **46**, 431–439.

West, M. (1992) Modelling with mixtures. pp 503–524 In: Bernardo, J., Berger, J., David, A. and Smith, A. (eds.), *Bayesian Statistics 4*. New York, NY: OUP.

Williams, D. (1982) Extra-binomial variation in logistic linear models. *J. Roy. Stat. Soc., Ser. C* **31**, 144–148.

Woodward, M. (1999) *Epidemiology*. London: Chapman & Hall.

EXERCISES

1. Using the data in Example 2.1, consider comparing model fit (e.g. via the DIC approach) between the fixed effects and gamma-Poisson mixture models. The fixed effects model allows the underlying relative risks λ_i to be different, but does not relate them to an overall hyperdensity. Identify the largest rate under each approach and the probability that it exceeds the average rate (by using the sample to assess this probability).

2. Also in Example 2.1, again try to analyse via random effects, but using Normal and Student t mixtures applied in the log scale for λ_i. How far does the robust Student *t* alternative (with degrees of freedom an unknown parameter) make a difference to the smoothed relative risks?

3. In Example 2.2, apply the DIC procedure to discriminate between binomial and beta-binomial models.

4. Repeat the beta-geometric mixture analysis of Example 2.3 for couples with non-smoking female partners. Calculate the chi square statistic for comparing actual and predicted cycles to conception counts (as in Table 2.4). Also, consider how to use this statistic in a predictive check fashion (see Equation (2.15)).

5. In Example 2.8, try the DPP analysis with a Dirichlet precision parameter α of 5. How does this compare with the results when taking $\alpha = 1$, and what are the implications for the number of sub-groups apparent in the data. Also, try to identify a four group mixture by 'conventional' discrete mixture methods and consider how identifiability is compromised.

6. In Example 2.11, program the sampling of replicate frequencies Z_i for ages 15–84, and so compare the predictive criterion G^2 in Equation (2.14) between the two models.

CHAPTER 3

Regression Models

3.1 INTRODUCTION: BAYESIAN REGRESSION

Methods for Bayesian estimation of the Normal linear regression model, whether with univariate or multivariate outcome, are well established. Assuming the predictors are exogenous and measured without error (i.e. not random), they may be conditioned on as fixed constants. For a univariate outcome the parameters are then the regression coefficients, linking the mean outcome for case i to predictors $x_{i1}, x_{i2}, .. x_{ip}$ for that case, and the conditional or residual variance. With an inverse gamma prior on the variance, and conjugate Normal prior on the regression coefficients, conditional on the sampled variance, analytic formulae for the posterior densities of these coefficients and other relevant quantities (e.g. predictions for new explanatory variable values) are available. These permit direct estimation with no need for repeated sampling.

However, generalisations which include discrete outcomes, non-linear or varying coefficient relationships, non-conjugate priors (Carlin and Polson, 1991), or priors constraining coefficients to substantively expected ranges, may all be facilitated by a sampling-based approach to estimation. Similar advantages apply in assessing the density of structural quantities defined by functions of parameters and data. The Bayesian approach may also be used to benefit with regression model selection, in terms of priors adapted to screening out marginally important predictors, or in comparisons between non-nested models (see Example 3.3). Recent methods tackle some of the identification problems (such as label-switching) in discrete mixture regression (Frühwirth-Schnatter, 2001). Bayesian methods have also been proposed (see the book by Dey *et al.*, 1999) for discrete outcome data which are over-dispersed in relation to standard densities such as the Poisson.

The development below is selective among the wide range of modelling issues which have been explored from a Bayes perspective, but intended to illustrate some potential benefits of the Bayes approach. The first case studies of Bayesian regression techniques include questions of predictor and model choice, for instance comparing marginal likelihood approximations with approaches with methods based on external validation (Section 3.2). Outcomes are binary, counts or univariate continuous. The focus is then on models for multiple category outcomes, which includes ordinal data, and discrete mixture regressions (Sections 3.3 and 3.4). These models illustrate the central role of prior parameter specification to reflect the form of the data. As one possible

Applied Bayesian Modelling P. Congdon
 ISBN: 0-471-48695-7

methodology for nonlinear regression, Section 3.5 discusses random walk and state space priors. A final theme (Section 3.6) is possible approaches to robust estimation in the event of departures from standard densities, and the related problem of outlier or influential observations. First, though, we consider questions around the specification of priors on regression model parameters. Among the aspects to consider are that priors express accumulated knowledge or subject matter constraints, may be defined by the form of outcome (e.g. in ordinal regression), and may be central to identification (e.g. in discrete mixture regressions).

3.1.1 Specifying priors: constraints on parameters

Thus, prior constraints on parameters may be defined by subject matter as in certain econometric models. As a subject matter example, Griffith *et al.* (1993) discuss the aggregate consumption function in economics (with C=consumption, and Y=disposable income),

$$C = \beta_1 + \beta_2 Y$$

where β_1 is autonomous consumption (consumption when there is no disposable income), and β_2 is the marginal propensity to consume. Economic principles suggest that of every dollar earned as disposable, some will be consumed (spent) and some will be saved. Hence, β_2 should lie between 0 and 1. Similarly, baseline consumption is expected to be positive, so that $\beta_1 > 0$.

Constraints on parameters can usually be dealt with in the prior for those parameters. Formally, let Ω be the unconstrained parameter space of β and let R be a constraint, expressed as

$$R{:}\, \beta \in \Omega_R$$

where Ω_R is a subspace of Ω. The posterior density of β given the constraint, outcomes y and regressors X is then, assuming $P(R|y) \neq 0$,

$$P(\beta|R, y, X) = P(\beta|y, X)P(R|\beta, y)/P(R|y)$$

This is the posterior density that would have been obtained in the absence of a constraint, namely $P(\beta|y, X)$, multiplied by a factor proportional to the conditional probability of the constraint given β. Such constraints on parameters imply that MCMC parameter samples $\beta_j^{(t)}$ at iteration t may be generated from non-truncated densities, but if they violate the constraint they are rejected with probability 1.

This principle also finds expression in applications where the constraint reflects the form of the outcome. Thus for ordinal outcomes, an underlying continuous scale may be postulated on which cutpoints – which are constrained to be increasing – correspond to increasingly ranked categories. Specifically, the cumulative odds model (McCullagh, 1980) is framed in terms of the cumulative probability γ_{ij} of subject i being located in the first j categories of a ranked outcome. Then logit(γ_{ij}) – or maybe some other link function of γ_{ij} – is related to a series of cutpoint parameters θ_j on the underlying scale. The utility of Bayesian approaches to ordinal outcomes is illustrated below in terms of applications to diagnostic tests (Example 3.9).

3.1.2 Prior specification: adopting robust or informative priors

Specification of the prior may be less clearly defined in certain applications by the subject matter or outcome form than in Section 3.1.1, but is still often central to model identifiability, fidelity to the observed data, and robustness in inference. Often, the structure of residuals suggests alternatives to standard density assumptions, in terms of heavier tails than under the Normal, as well as outlier and influential observations. To model heavy tails Dickey (1976), Chib *et al.* (1990) and Geweke (1993) are among those adopting univariate or multivariate Student t densities as sampling densities for the data, or as priors for linear regression parameters. A contaminated Normal or Student t model, in which outliers have shifted location and or variances, provides one Bayesian approach to outliers (e.g. Verdinelli and Wasserman, 1991). For instance, assuming a small probability such as $\varepsilon = 0.05$ that an outlier occurs, the density of a metric y_i may take the form

$$f(y_i|\mu, \sigma^2, \varepsilon) = (1-\varepsilon)\phi(y_i|\mu, \sigma^2) + \varepsilon\phi(y_i|\mu + D_i, \sigma^2)$$

where D_i is the shift in location for case i if it is selected as an outlier. One may adopt similar methods for shifts in regression coefficients.

Other options to protect against the influence of outliers or influential cases may be particular to the form of outcome. With a binary outcome, for instance, a prior may be set on a 'transposition', namely when $y_i = 1$ is the actual observation but the regression model provides much higher support for $\Pr(y_i = 0)$ than for $\Pr(y_i = 1)$. In terms of the contaminated data approach of Copas (1988, p. 243) this would be a measure of 'the evidence that each individual observation has been misrecorded' (see Example 3.14). Discrete mixture regressions also intend to improve robustness (to assumptions of a single homogenous density across all data points), as well as providing for substantively based differences in the regression coefficients between sub-populations (Wedel *et al.*, 1993).

Model identifiability, and credibility in terms of existing knowledge, may be improved by adopting an informative prior for one or more regression coefficients. Such a prior might refer to results from similar studies. For instance, if the estimate of a regression parameter β from previous study had mean β_0 and covariance Σ_0, we might inflate Σ_0 by a factor $a > 1$ (e.g. $a = 5$) to provide a prior covariance on β in the current study (Birkes and Dodge, 1993, p. 156; Dellaportas and Smith, 1993, p. 444). More formal procedures for reflecting historical data may actually include such data in the likelihood, though down-weighted relative to the current data (Ibrahim and Chen, 2000).

One might also consider some summary statistic from the observed data to inform the specification of the prior, though with suitable down-weighting, so that priors become 'gently data determined' (Browne and Draper, 2000). Note, however, that strictly this is departing from fully Bayes principles. Sometimes it is preferable to avoid completely flat priors, especially if flat priors lead to effectively improper posterior densities or poor identifiability. Thus, Fernandez and Steel (1999) show how choosing a proper Wishart prior for the conditional precision matrix Σ^{-1} improves identifiability of multivariate Student t errors regression. In mixture regressions (Section 3.5), some degree of prior information may be essential to identifiability.

3.1.3 Regression models for overdispersed discrete outcomes

Below we consider regression applications both for metric and discrete outcomes. General Linear Models (GLMs) have been proposed as a unified structure for both types of outcomes. As elaborated by McCullagh and Nelder (1989, p. 28) several discrete densities (together with the Normal and Student t densities) can be encompassed within the exponential family. The exponential family density (Gelfand and Ghosh, 2000) has the form

$$f(y_i|\theta_i) = \exp\{\varphi_i^{-1}[y_i\theta_i - b(\theta_i)] + c(y_i, \varphi_i)\} \tag{3.1}$$

where φ_i are scale parameters, and the means are obtained as

$$E(y_i) = \mu_i = b'(\theta_i)$$

and variances as

$$V(y_i) = \varphi_i b''(\theta_i) = \varphi_i V(\mu_i)$$

Poisson and binomial densities have a fixed scale factors $\varphi_i = \varphi = 1$. So for the Poisson, with $b(\theta_i) = \exp(\theta_i)$, the mean and variance are both μ_i.

Often, though, as also noted in Chapter 2, residual variability under these densities exceeds that expected under the postulated variance-mean relationship. The variance function remains as above, but now $\varphi > 1$ for all subjects, or possibly dispersion factors vary by subject with $\varphi_i > 1$, and are themselves modelled as functions of covariates.

One approach to handling such heterogeneity in general linear models involves random effects modelling in the model for the mean response. Thus for a Poisson outcome, $y_i \sim \text{Poi}(\mu_i)$ we might stipulate a model for the mean μ_i containing both fixed and random effects:

$$\log(\mu_i) = \beta X_i + \varepsilon_i$$

where the ε_i are parametric (e.g. Normal) or possibly semi-parametric (e.g. where a DPP prior for the ε_i is used with a Normal baseline density). This effectively means adding a set of parameters which increase in number with the sample size, technically making the likelihood nonregular and raising the question about how many parameters are actually in the model (see Chapter 2 on assessing the number of effective parameters).

Alternative approaches to overdispersion include reparameterisations of the variance function, and generalised density and likelihood forms such as the double exponential that model both mean and variance functions (Carroll and Ruppert, 1988; Efron, 1986). Nelder and Pregibon (1987) propose a 'quasi-likelihood', which in logged form for one observation is

$$-0.5D(y_i, \mu_i)/\varphi_i - 0.5\log[2\pi\varphi_i\ V(\mu_i)]$$

where $D(,)$ is the deviance function[1]. These are quasi-likelihoods because they do not correspond to any complete likelihood[2] within the exponential family (Albert and Pepple, 1989). West (1985) has, however, proposed a Bayesian approach involving a scaled, and hence complete, likelihood

$$f(y_i|\mu_i, \phi_i) = \gamma_i^{0.5} \exp[-\gamma_i D(y_i, \mu_i)] \tag{3.2}$$

where $\gamma_i = \lambda/(\lambda + \phi_i)$, and λ reflects the level of over-dispersion. As λ tends to infinity, and γ_i approaches 1, the scaled likelihood approaches the baseline exponential density. Various parameterisations of the variance function are possible: for instance, Engel (1992) has suggested

$$V(y_i) = \varphi_i V(\mu_i) = \varphi_i \mu_i^\kappa \tag{3.3}$$

where

$$\log(\varphi_i) = \eta W_i$$

and W_i contains covariates relevant to explaining extra-variability.

Breslow (1984) and others have proposed a related approach where a pseudo-likelihood for each observation has the logged form

$$-0.5(y_i - \mu_i)^2/[\varphi_i V(\mu_i)] - 0.5\log[2\pi\varphi_i \; V(\mu_i)] \tag{3.4}$$

The pseudo-likelihood approach is thus equivalent to assuming a Normal outcome but with modified variance function. Ganio and Schafer (1992) and Dey *et al.* (1997) are among those adopting these forms or extensions of them, and relating the variance inflators ϕ_i to predictors; Nelder and Lee (1991) term such models joint GLMs.

Another methodology in instances of over-dispersion involves generalising the standard variance function for particular densities. Thus, for a count regression with mean $\mu_i = \exp(\beta x_i)$, Winkelmann and Zimmerman (1991) propose a variance function

$$V(y_i) = (\omega - 1)\mu_i^{\kappa+1} + \mu_i \tag{3.5}$$

where $\omega > 0$ and $\kappa \geq -1$. Then $\omega = 1$ corresponds to the Poisson, and $\omega > 1$ to over-dispersion. It can be seen that the particular value $\kappa = 0$ means the Poisson variance μ_i is inflated by the factor ω. This type of 'linear' adjustment has been proposed by McCullagh and Nelder (1989) as a correction procedure for over-dispersion.

At its simplest, a correction for linear over-dispersion involves dividing the fitted deviance or chi-square for a Poisson regression by the degrees of freedom and deriving a

[1] For example, in the Poisson the deviance for case i is

$$2\{y_i \log(y_i/\mu_i) - (y_i - \mu_i)\}$$

while in the binomial, with $y_i \sim \text{Bin}(p_i, \; T_i)$ and $\mu_i = p_i T_i$, it is

$$2\{y_i \log(y_i/\mu_i) + (T_i - y_i)\log([T_i - y_i]/[T_i - \mu_i])$$

[2] It may be noted that in BUGS introducing synthetic data Z to represent densities not included does not necessarily require a complete likelihood. For instance, a quasi likelihood, with variance as in (3.3) and covariate $W[i]$, could be coded as

```
for (i in 1:N) {Z[i] <- 1; Z[i] ~ dbern(q[i])
   q[i] <- exp(Q[i])
Q[i] <- -0.5*log(6.28*phi[i]*pow(mu[i],kappa)) - 0.5*D[i]/phi[i]
log(phi[i]) <- eta[1] + eta[2]*W[i]}
```

with $D[i]$ being the relevant deviance.

scaling factor (i.e. estimating ω when $\kappa = 0$) to adjust the standard errors of the β_j for over-dispersion. The scaling factor ω can also be obtained by a pseudo-likelihood model assuming Normal errors, with means μ_i, and variances $\omega\mu_i$. Another variance function with $\kappa = 1$ occurs in the event of a specific type of gamma-Poisson mixture (as considered in Chapter 2), when y_i is marginally a negative binomial (see McCullagh and Nelder, 1989, Chapter 11). In fact, both the linear and quadratic variance forms (with $\kappa = 0$ and $\kappa = 1$, respectively) can be derived from mixing a gamma with the Poisson, but differ in the way the gamma is parameterised.

3.2 CHOICE BETWEEN REGRESSION MODELS AND SETS OF PREDICTORS IN REGRESSION

Many methods have been discussed for comparing the structural form and specification of regression models, including Bayesian methods, likelihood based methods, penalised likelihoods (Akaike, 1978), supermodel methods (Atkinson, 1969), bootstrap methods and cross-validation techniques (Browne, 2000). Very often, model selection schemes combine elements of different basic approaches.

Model uncertainty in regression analysis may involve different aspects of model specification:

(a) the error structure of the residuals (e.g. Normal vs. Student *t*);
(b) whether or not transformations are applied to predictors and outcome, which may be appropriate, for example, if metric data are skewed before transformation or to induce approximate Normality in a count outcome;
(c) which link to adopt for discrete outcomes in general linear modelling, since the standard links are not necessarily always appropriate;
(d) whether a general additive or non-linear regression term be used or just a simple linear expression;
(e) which is the best subset of predictors in the regression term, regardless of its form;
(f) for discrete mixture regressions, there is an issue of optimal choice of the number of components in the mixture.

Many of these questions can be addressed by the model choice techniques considered in Chapter 2, for example using approximations to the marginal likelihood, using joint model-parameter space methods, using penalised measures of fit such as the BIC or DIC or using predictive (cross-validatory) criteria.

In addition to the marginal likelihood approximations discussed in Chapter 2 may be mentioned the Laplace approximation. This is useful for standard regressions, such as Poisson, logistic or Normal linear models, since it requires differentiation of the log-likelihood with regard to all model parameters, and so may be difficult for complex nonlinear or random effect models. The Laplace approximation is based on the Taylor series expansion for a q-dimensional function $g(x_1, x_2, .. x_q)$, such that

$$\int \exp[g(x)]dx \approx (2\pi)^{q/2}|S|^{0.5}\exp[g(\hat{x})] \quad (3.6)$$

where $\hat{x}$ is the value of x at which g is maximised, and S is minus the inverse Hessian of $g(x)$ at the point $\hat{x}$. The Hessian is the matrix of double differentials $\partial^2 g/\partial x_i \partial x_j$. For a model with parameters ϕ, the marginal likelihood has the form

$$m(y) = \int f(y|\phi)\pi(\phi)d\phi$$
$$= \int \exp\,[\log f(y|\phi) + \log \pi(\phi)]d\phi$$

and in this case, the Laplace approximation may be obtained at the posterior mode $\bar{\phi}$ (though it may also be defined at the maximum likelihood point $\hat{\phi}$). For a univariate regression, with $\phi = (\sigma^2, \beta)$, and $q = p + 2$, where p is the number of regressors (excluding the constant), the approximation is therefore

$$m(y) \approx (2\pi)^{q/2}|S|^{0.5}f(y|\bar{\phi})\pi(\bar{\phi})$$

In this approximation, $\pi(\bar{\phi})$ is the probability of the values in $\bar{\phi}$ evaluated in terms of the assumed prior densities $\pi(\phi)$, and S is minus the inverse Hessian of

$$g(\phi) = \log f(y|\phi) + \log \pi(\phi)$$

evaluated at the posterior mean $\bar{\phi}$ of the conditional variance and regression coefficients. Then the log of the marginal likelihood is approximated as

$$\log\,[m(y)] \approx 0.5q\log\,(2\pi) + 0.5\log|S| + \log f(y|\bar{\phi}) + \log \pi(\bar{\phi})$$

In practice (see Raftery, 1996), S may be estimated by minus the inverse Hessian of the log-likelihood at the mode, $L(\bar{\phi}|y) = \log f(y|\bar{\phi})$, rather than that of $g(\phi)$. An approximation might also be based on the variance-covariance matrix of ϕ based on a long MCMC run, though this may be less valid for if regression parameters have non-Normal posterior densities (Diciccio *et al.*, 1997); other options include multivariate interpolation (Chib, 2000).

3.2.1 Predictor selection

While approaches based on the marginal likelihood have been successfully applied to regression model selection, the structure of regression models is such that distinctive methods have been proposed for some of the particular model choice questions occurring in regression. One of the major motivations for seeking optimal or parsimonious models occurs if there is near collinearity between predictors, $X_1, X_2, \ldots, X_p$. Taken as a single predictor, the coefficient β_j for predictor X_j may have a clearly defined posterior density in line with subject matter knowledge (i.e. a 95% credible interval confined to positive values, assuming X_j was expected to have a positive effect on y). However, with several predictors operating together coefficients on particular X_j may be reduced to 'insignificance' (in terms of a 95% interval neither clearly positive nor negative), or even taking signs opposite to expectation.

Choice among p predictors to reduce such parameter instability can be seen as a question of including or excluding each predictor, giving rise to choice schemes specifying priors on binary indicators γ_j relating to the probabilities of inclusion $\Pr(\gamma_j = 1)$ or exclusion $\Pr(\gamma_j = 0) = 1 - \Pr(\gamma_j = 1)$ of the jth predictor. Note that exclusion may be defined as 'exclusion for all practical purposes'.

Suppose a Poisson density with mean μ_i is assumed for count outcome y_i, so that if all predictors are included in a log link

$$\log\,(\mu_i) = \beta_0 + \beta_1 X_{i1} + \beta_2 X_{i2} + \ldots + \beta_p X_{ip}$$

Then in a predictor selection model, we introduce between 1 and p binary variables γ_j relating to predictors about which inclusion is uncertain (the intercept is presumed always necessary). If uncertainty is applied to all predictors, the model becomes

$$\log(\mu_i) = \beta_0 + \gamma_1\beta_1 X_{i1} + \gamma_2\beta_2 X_{i2} + \ldots + \gamma_p\beta_p X_{ip} \tag{3.7}$$

George and McCullough (1993) propose a stochastic Search Variable Selection Scheme (SVSS), whereby β_j has a vague prior centred at zero (or some other value) when $\gamma_j = 1$, but when $\gamma_j = 0$ is selected the prior is centred at zero with high precision (i.e. β_j is zero for all practical purposes). When $\gamma_j = 0$ one might make[3] the variance τ_j^2 small, but multiply this by a large constant c_j^2 when $\gamma_j = 1$. Typically, Bernoulli priors with probability 0.5 are assigned to the probability of each selection index γ_j being 1. An MCMC analysis would then be carried out and the frequency of different combinations of retained predictors enumerated; there are 2^p possible combinations in Equation (3.7).

Another scheme which takes the priors for binary indicators γ_j and coefficients β_j as independent is presented by Kuo and Mallick (1998). Thus in Equation (3.7) one option is that the β_j are assigned conventional priors, such as

$$\beta_j \sim N(0, V_j)$$

with V_j large, rather than mixture priors as in the SVSS approach. This independent priors approach allows the possibility of more particular model elements than just whether predictors should be included or not. Thus for a log-linear regression of counts y_{ij} on factors A_i, B_j and their interaction C_{ij}, one would not generally include C_{ij} unless both A_i and B_j were included. So coding the γ_j to reflect this, with $y_{ij} \sim \text{Poi}(\mu_{ij})$, leads to

$$\log(\mu_{ij}) = \alpha + \max(\gamma_1, \gamma_3)\ \beta_1[A_i] + \max(\gamma_2, \gamma_3)\ \beta_2[B_j] + \gamma_3\beta_3[C_{ij}]$$

so that $\gamma_3 = 1$ corresponds to including A_i, B_j and C_{ij}. Kuo and Mallick (1998, p. 72) also suggest a dependent priors scheme, whereby $\beta_j \sim N(0, V_j)$ when $\gamma_j = 1$, but $\beta_j \sim N(0, k_j V_j)$ when $\gamma_j = 0$, such that $k_j V_j$ is close to zero.

3.2.2 Cross-validation regression model assessment

In general, cross-validation methods involve predictions of a subset y_r of cases when only the complement of y_r, denoted $y_{[r]}$ (i.e. the remaining observations) is used to update the prior on parameters ϕ. Here ϕ for a linear regression might consist of regression parameters and a variance term. Suppose an $n \times p$ matrix $\mathbf{X}$ of predictors is also partitioned into $\mathbf{X}_r$ and $\mathbf{X}_{[r]}$. A discussion of cross-validation principles in Bayes regression model selection is provided by Geisser and Eddy (1979). These authors suggested a marginal likelihood approximation based on leaving one observation out at a time. The links (in terms of asymptotic equivalence) between this single-case omission strategy and penalised fit measures such as the AIC – where such measures are based on fitting to the complete sample – are discussed in earlier work by Stone (1974, 1977).

Consider a procedure where cases are omitted singly, and let $y_{(r)}$ denote the data excluding a single case r. Then the cross-validatory predictions have the form

[3] George and McCulloch suggest c_j between 10 and 100 and $\tau_j \approx \delta_j/(2\log c_j)^{0.5}$, where δ_j is the largest value for $|\beta_j|$ that would be considered unimportant. One might take $\delta_j = \Delta y/\Delta X_j$, where Δy is say under 0.5 of s.d.(y) but ΔX_j is the range of X_j.

$$p(y_r|y_{(r)}) = \int f(y_r|\phi, y_{(r)}, \mathbf{X}_{(r)})\pi(\phi|y_{(r)})d\phi$$

with $p(y_r|y_{(r)})$ often known as the Conditional Predictive Ordinate (CPO). A proxy for the marginal likelihood $m(y)$ is defined by the product of these terms

$$m(y) = \prod_{r=1}^{n} p(y_r|y_{(r)})$$

The ratio of $m_1(y)$ and $m_2(y)$ for two models M_1 and M_2 is then a surrogate for the Bayes factor B_{12}. Cross-validation methods also play a major role in regression model checking in identifying influential cases and other model discrepancies (see Example 3.15). Gelfand *et al.* (1992) propose several checking functions involving comparison of the actual observations with predictions from $p(y_r|y_{(r)})$.

Rust and Schmittlein (1985) and Fornell and Rust (1989) propose a cross-validatory scheme for regression analysis including Bayesian model selection criteria, but drawing on split-half and other sample splitting methods also used in frequentist inference. Rust and Schmittlein emphasize that the choice of validation function may be based on subject matter choices (e.g. predicted vs. actual marketing share in the sales applications they consider). They consider cross-validation to compare models $j = 1, .. J$ via random splitting of datasets. Let D_1 and D_2 be such a split, with D_1 being the pre-sample on which parameter estimation and D_2 is the post or validation sample, used to assess the model. Let ϕ_j denote the parameter estimate under model j and $\hat{\phi}_{j1}$ the estimate of ϕ_j using just the data in D_1. Then Rust and Schmittlein propose that the posterior probabilities of each model be approximated as

$$\begin{aligned} P(M_j|D_2) &= P(D_2|M_j)\ \pi(M_j)/P(D_2) \\ &= \pi(M_j)\int \pi(\phi_j)\ f(D_2|\phi_j)d\phi_j/ \\ &\left\{\sum_{k=1}^{J}\pi(M_k)\int \pi(\phi_k)\ f(D_2|\phi_k)d\phi_k\right\} \\ &\approx f(D_2|\hat{\phi}_{j1})\ \pi(M_j)/\left\{\sum_{k=1}^{J}\pi(M_k)\ f(D_2|\hat{\phi}_{k1})\right\} \end{aligned} \tag{3.8}$$

Rust and Schmittlein take $\hat{\phi}_{j1}$ to be the maximum likelihood estimator, but one might also take the posterior average of the parameters under model j, $\bar{\phi}_{j1}$. Another option, given equal prior model probabilities, would involve comparing the average log-likelihoods of the validation data (evaluated over a long run of T iterations) under model j

$$\bar{f}_j = T^{-1}\sum_{t=1}^{T} f(D_2|\phi_{j1}^{(t)})$$

Fornell and Rust consider the measure (3.8) for a single random split of the original data into sets $\{D_1, D_2\}$, but one might consider a large number $r = 1, .. R$ of such splits and carry out such comparisons by averaging over the sets $\{D_{1r}, D_{2r}\}$. Thus, Alqallaf and Gustafson (2001) consider cross-validatory checks based on repeated two fold data splits into training and validation samples. Another option is k fold validation, where the data is split into a small number of groups (e.g. $k = 5$) of roughly equal size, and cross-validation is applied to each of the k partitions obtained by leaving each group out at a time. Kuo and Peng (1999) use such an approach to obtain the predictive likelihood for the gth omitted group (the validation group), and suggest a product of these likelihoods over the k partitions as a marginal likelihood approximation.

Example 3.1 Nodal involvement To illustrate model choice under marginal likelihood and cross-validation approaches, consider binary outcomes on prostatic cancer nodal involvement and four possible predictors (Collett, 1992). These data illustrate a frequent problem in model fitting: simple unpenalised fit measures such as the deviance may show a slight gain in fit as extra predictors (or other parameter sets) are added but penalised fit measures, or marginal likelihood type measures, show that any improvement is offset by extra complexity. Cross-validation approaches may also favour the less complex model in such situations.

In the nodal involvement example, the predictors are $x_1 = \log$(serum acid phosphate), $x_2 =$ result of X-ray ($1 = +$ve, $0 = -$ve), $x_3 =$ size of tumour ($1 =$ large, $0 =$ small) and $x_4 =$ pathological grade of tumour ($1 =$ more serious, $0 =$ less serious). A probit regression model including all predictors is then

$$y_i \sim \text{Bern}(\pi_i)$$
$$\text{probit}(\pi_i) = \beta_0 + \beta_1 x_{1i} + \ldots + \beta_4 x_{4i}$$

The model may be fitted directly or by introducing latent continuous and Normally distributed responses[4] as in Albert and Chib (1993). Using the latter method, Chib (1995) shows that a model with $x_1 - x_3$ only included has a worse log likelihood than a model including all four predictors (-24.43 as against -23.77) but a better marginal likelihood (-34.55 vs. -36.23). Letting M_2 denote the full model (four predictors) and M_1 the reduced model, then the approximate Bayes factor B_{12} obtained by Chib is $\exp(1.68) = 5.37$. By conventional criteria on interpreting Bayes factors (Kass and Raftery, 1996), this counts as 'positive evidence' for the reduced model, though far from conclusive.

Here alternative approaches to marginal likelihood approximation and model choice, as discussed above and in Chapter 2, are considered. As noted in Chapter 2 the marginal likelihood identity

$$m(y|M_j) = p(y|\beta, M_j)\pi(\beta|M_j)/\pi(\beta|y, M_j)$$

applies at any value of β, and in particular at points such as the posterior mean $\bar{\beta}$. So (omitting M_j), one may estimate $\log[m(y)]$ as

$$\log\{m[y]\} = \log\{p(y|\bar{\beta}) + \log\{\pi(\bar{\beta})\} - \log\{\pi(\bar{\beta}|y)\} \tag{3.9}$$

[4] Thus, suppose X_i denotes a set of predictors, with regression parameter β. If y_i is 1 then the latent response z_i is constrained to be positive and sampled from a Normal with mean βX_i and variance 1. This is equivalent to assuming a probit link for the probability that $y_i = 1$. For the observed outcome $y_i = 0$, z_i is sampled in the same way, but constrained to be negative (with 0 as a ceiling value). The latent variable approach is especially advantageous for the probit link, and has benefits (e.g. for residual analysis). In BUGS, the following code (for observations y[] and one covariate x[]) might be used

```
  for (i in 1:n) {z[i] ~ dnorm(mu[i],1) I(low[y[i] + 1],high[y[i] + 1]);
    mu[i] <- b[1] + b[2]*x[i]}
# sampling bounds
    low[1] <- - 20; low[2] <- 0; high[1] <- 0; high[2] <- 20;
```

A logit link is approximated by sampling z from a Student t with 8 degrees of freedom; this is equivalent to Normal sampling, as in the above code, but with the precision of 1 replaced by case-specific precisions sampled from a Gamma density with shape and index both equal to 4.

The main issue is then how $\pi(\bar{\beta}|y)$ is approximated. If augmented data z is part of the model definition, and there is just one other parameter block (e.g. as here where there are latent Normal responses underlying an observed binary response, and the only other parameters are regression coefficients), then one may estimate $\pi(\bar{\beta}|y)$ using draws $z^{(t)}$ from a sampling run where all parameters are updated, and the probability of the fixed value $\bar{\beta}$ (obtained from an earlier run) is evaluated against the full conditional density for β given $z^{(t)}$. The latter density is derived essentially by considering z as dependent variable in a Normal errors model (Chib, 1995, 2000). If the prior for $\beta = (\beta_0, \beta_1, \ldots, \beta_4)$ is denoted $\beta \sim N_p(c_0, C_0^{-1})$, then the relevant full conditional is

$$N_p(\hat{\beta}, C^{-1})$$

where $C = C_0 + X'X$, with X of dimension $n \times p$, and

$$\hat{\beta} = C^{-1}[C_0 c_0 + X'z] \tag{3.10}$$

with z being the vector of sampled z_i. If C_0 is diagonal, then the prior reduces to a set of univariate Normals.

Multivariate Normal or Student approximations to the posteriors $\pi(\bar{\beta}|y)$ may also be adopted to estimate the posterior ordinate in Equation (3.9), with covariance matrix based on correlations between the sampled $\beta_k^{(t)}$, where $k = 0, 1, \ldots, 4$ for model 2 and $k = 0, \ldots, 3$ for model 1. Then $\pi(\bar{\beta}|y)$ is the value of the multivariate Normal or Student density evaluated at its mean $\bar{\beta}$.

Harmonic mean and importance sample estimators of the marginal likelihood (see Chapter 2) may also be considered. To examine stability in these estimators, the directly estimated probit model (without augmented data z) is fitted, and seven batches of 2000 iterations are taken from a single run of 15 000 after excluding the first 1000 (see Program 3.1(A), Model A). Priors are as in Chib (1995). The importance sample estimate is as in (2.9)

$$\begin{aligned}\hat{m}(y) &= 1/[T^{-1} \sum_t g^{(t)}/\{L^{(t)}\pi^{(t)}\}] \\ &= T/[\sum_t g^{(t)}/\{L^{(t)}\pi^{(t)}\}]\end{aligned} \tag{3.11}$$

where $L^{(t)}$ and $\pi^{(t)}$ are the likelihood and prior densities evaluated at iterations $t = 1, \ldots, T$, and $g^{(t)}$ is the value of an importance function intended to approximate the posterior density $\pi(\beta|y)$. This function is provided by a multivariate Normal approximation.

Table 3.1 accordingly shows the sort of fluctuations that occur in the harmonic mean estimates (Equation (2.8) in Chapter 2) of the marginal likelihood. This involves monitoring the total log likelihood (L in Program A, Model 3.1A), using the coda facility to extract runs of 2000 values, exponentiating $L^{(t)}$ to give $H^{(t)}$, taking the average $\bar{h}$ of $h^{(t)} = 1/H^{(t)}$, and then taking minus the log of $\bar{h}$. Comparison of the average marginal likelihood estimates (averaging over batches) gives $B_{12} = 1.81$. The importance sample estimates of the marginal likelihoods, as calculated in Equation (3.11), are -35.20 (Model 1), and -36.89 (Model 2), and so $B_{12} = 5.42$. These estimates of the log marginal likelihood are virtually identical to those obtained from the application of the marginal likelihood identity (3.9) in Model C of Program 3.1(A) (at posterior mean) using the MVN approximation to the posterior of β. At the posterior mean, the estimates are -35.19 and -36.89 for Models 1 and 2.

Table 3.1 Probit models for nodal involvement, regression coefficients and marginal likelihood approximations

Model 1 with 3 predictors	Mean	St. devn.	2.50%	Median	97.50%
β_0	−0.74	0.40	−1.50	−0.74	0.07
β_1	1.42	0.67	0.17	1.40	2.79
β_2	1.30	0.48	0.40	1.29	2.22
β_3	1.08	0.43	0.25	1.08	1.92
Model 2 with 4 predictors					
β_0	−0.79	0.42	−1.64	−0.78	0.01
β_1	1.63	0.70	0.30	1.63	3.07
β_2	1.26	0.49	0.32	1.26	2.21
β_3	0.96	0.45	0.10	0.96	1.84
β_4	0.55	0.45	−0.33	0.54	1.45

Marginal likelihood estimates by iteration batch

	Harmonic Mean		CPO Method	
Iterations	Model 1	Model 2	Model 1	Model 2
1001–3000	−29.86	−28.85	−29.20	−29.85
3001–5000	−28.41	−30.83	−29.18	−30.03
5001–7000	−28.89	−28.47	−28.90	−29.88
7001–9000	−28.55	−30.23	−29.07	−29.87
9001:11000	−28.47	−28.60	−29.15	−29.50
11000:13000	−28.33	−29.35	−28.91	−29.78
13000–15000	−29.25	−29.58	−29.28	−29.86

Program 3.1(B) follows Albert and Chib (1993), and takes the latent data z from the distribution $z|y$. Then $\bar{\beta}$ is evaluated against $N_p(\hat{\beta}, C^{-1})$ for the samples $z^{(t)}$ by substituting in Equation (3.10). This gives a marginal likelihood estimate of −34.04 for Model 1 and −35.28 for Model 2, a Bayes factor of 3.46; see Model B in Program 3.1(B). Model A in Program 3.1(B) illustrates the basic truncated Normal sampling needed to implement the Albert–Chib algorithm for probit regression.

The CPO estimate (Equation (2.13) in Chapter 2) is obtained by taking minus logs of the posterior means of the inverse likelihoods (the quantities $G[]$ in Program 3.1A, Model A), and then totalling over all cases. This estimator leads to $B_{12} = 2.06$, and is more stable over batches than the harmonic mean estimate.

A Pseudo Bayes Factor (PsBF) is also provided by the Geisser–Eddy cross-validation method based on training samples of $n - 1$ cases and prediction of the remaining case. Program 3.1(C) (Nodal Involvement, Cross-Validation) evaluates 53 predictive likelihoods of cases 1, 2, 3,.., 53 based on models evaluated on cases {2,.., 53}, {1, 3, 4,.., 53}, {1, 2, 4,.., 53),....{1, 2,..., 52}, respectively. Each omitted case y_i provides a predictive likelihood and Bayes factor under model j (where here $j = 1, 2$), based on the components

$$f_j(y_i|\beta),\ \pi_j(\beta|y_{(-i)})$$

where $y_{(i)}$ denotes the sets of cases omitting the ith, namely

$$y_1, y_2, .. y_{i-1}, y_{i+1}, .. y_n$$

The ratios of the predictive likelihoods for case i (when it is the omitted case) under Models 1 and 2 may be denoted

$$b_{12}^{[i]}$$

and their product b_{12} is a form of Bayes factor. Since b_{12} is skew, one may sample $\log(b_{12})$ in an MCMC run with validation models running in parallel, and then take the exponential of the posterior mean of $\log(b_{12})$ as an estimate of the PsBF. Additionally, if it is possible to run the n validation models in parallel and monitor $\log(b_{12})$ to find it exceeds 0 with a high probability, then one may say Model 1 is preferred to Model 2. For numeric stability, a logit rather than probit link is used in Program 3.1C.

From iterations 500–2000 of a run of 200 iterations (pooling over three chains), $\Pr(\log(b_{12}) > 0)$ converges to around 67%, while the posterior mean of $\log(b_{12})$ stabilises at around 1.77 (i.e. $\text{PsBF}_{12} = 5.87$). One may also monitor the total predictive likelihoods involved (TL[] in Program 3.1C), and compare their posterior averages or medians to obtain an alternative estimate of B_{12}. The mean values of these likelihoods are -32.33 and -34.07, giving $\text{PsBF}_{12} = 6.4$. Overall, the cross-validation methods provide slightly greater support for the reduced model than the approximate marginal likelihood methods do[5], but both approaches support a more parsimonious model, whereas a simple deviance comparison might lead to adopting the complex model.

Example 3.2 Hald data A data set from Hald (1952) refers to the heat evolved in calories per gram of cement, a metric outcome for $n = 13$ cases. The outcome is related to four predictors describing the composition of the cement ingredients. These data may be used to illustrate predictor selection in regression modelling. Options include

- the SSVS scheme of George and McCulloch (1993), and related approaches;
- true cross-validation methods based on omitting portions of the data as validation samples; and
- predictive model assessment based on sampling 'new data' Z from a model fitted to all cases, and seeing how consistent the new data are with the observations.

Here we adopt the third approach, and undertake a simultaneous analysis of the eight models by declaring repeat versions of the outcome, and evaluate them using a predictive criterion derived from the work of Laud and Ibrahim (1995). As discussed in Chapter 2, for model k and associated parameter set θ_k, the predictive density is

$$p(Z|y, M = k) = \int f(Z|\theta_k)\pi(\theta_k|y)d\theta_k$$

The predictive criterion of Laud and Ibrahim is then

$$C^2 = \sum_{i=1}^{n} [\{E(Z_i) - y_i\}^2 + \text{var}(Z_i)]$$

[5] A test of the cross-validation approach with the probit link provided similar differentials between Models 1 and 2 (e.g. in terms of predictive likelihoods), but is subject to occasional numerical errors when the linear regression term becomes too large.

Better models will have smaller values of C^2 or its square root, C. Table 3.2 compares the eight (2^3) possible models in terms of C.

Laud and Ibrahim employ a specialised prior, namely a guess at the fitted regression outcome for each case in terms of the available predictors; this is then translated into priors on the regression coefficients. Here non-informative N(0, 10^5) priors are assumed on the regression coefficients themselves, and a G (1, 0.001) prior for the conditional variance. Despite this, similar C measures to those reported by Laud and Ibrahim (1995, Table 1, p. 255) are obtained. Taking into account variability in the criterion C, there is no overwhelming evidence against any model.

However, on the Occam's Razor principle that a Model 1 which is embedded in Model 2 (i.e. is nested within it and less heavily parameterised) and also has a better average C than Model 2, we can eliminate the models $\{x_1, x_2, x_3, x_4\}$ and $\{x_2, x_3, x_4\}$.

Comparisons of the predictive criteria C_i and C_j between models i and j over repeated samples is carried out in Program 3.2 via the step() command; thus Comp[i, j] is 1 if model j has a worse criterion than model i in a particular iteration. On this basis, it can be confirmed that there is essentially no clear advantage in fit for any model in Table 3.2. For example, Comp[1, 2] averages 0.51 from the second half of a run of 20 000 iterations over three chains, so that Model 2 has a worse fit than Model 1 in 51% of 30 000 iterations. Of the $28 = 8 \times 7/2$ model comparisons, the highest (comparing models 1 and 7) is 0.70 and the lowest is 0.46.

Example 3.3 Ship damage Both Winkelmann and Zimmerman (1991) and Dey et al. (1997) consider the ship damage data of McCullagh and Nelder (1989, p. 205). These data illustrate approaches to overdispersion in discrete outcomes, and the questions of model selection involved in introducing extra parameters to account for overdispersion. While a gamma-Poisson mixture is a standard option, this is not the only choice, and the goal is to use as parsimonious model as possible to ensure overdispersion is effectively allowed for. Hence model choice is better based on penalised measures of fit than unmodified likelihood and deviances.

Excluding null observations, the data consists of counts y_i of wave damage to 34 cargo ships according to ship type t_i (A–E), year of construction c_i (1960–64, 1965–69, 1970–74, 1975–79), and period of observation o_i (1960–74, 1975–79). Most analyses

Table 3.2 Predictive fit of models for Hald data (iterations 10 000–20 000, three chains)

Model (included predictors)	Mean C	s.d. C	Median C
1,2,4	11.40	2.73	11.04
1,2,3	11.45	2.71	11.09
1,3,4	11.79	2.82	11.41
1,2,3,4	11.85	2.89	11.45
1,2	11.98	2.72	11.65
1,4	13.60	3.07	13.22
2,3,4	14.11	3.33	13.68
3,4	13.57	3.08	13.20

treat these as categorical factors, and take the months of service total s_i as an offset with coefficient 1. Thus, the model for damage counts is

$$y_i \sim \text{Poi}(\mu_i)$$
$$\log(\mu_i) = \log(s_i) + \alpha + \beta(t_i) + \gamma(c_i) + \delta(o_i)$$

First, consider a standard Poisson regression, with N(0, 10^3) priors on the regression coefficients $\phi = \{\alpha, \beta, \gamma, \delta\}$, and taking appropriate corner constraints on $\{\beta, \gamma, \delta\}$. Unpenalised fit measures are not ideal for model choice, but do indicate whether over-dispersion is effectively corrected for. They include the mean Poisson deviance $\bar{D}_1$, and a chi-square measure based on standardised residuals,

$$D_2 = \sum R(y_i, \mu_i) = \sum [(y_i - \mu_i)^2/V(y_i)]$$

We can also calculate the deviance at $\bar{\phi}$ and obtain the effective parameters p_e, and so obtain the DIC

$$D_3 = D_1(\bar{\phi}) + 2p_e = \bar{D}_1 + p_e$$

Another penalised fit measure is provided by the deviance criterion (for Poisson data) under minimum predictive loss (Gelfand and Ghosh, 1998), which, like the criterion used in Example 3.2, involves[6] sampling new data Z_i from the model means μ_i.

The baseline Poisson regression model, with nine parameters (and 25 degrees of freedom) is estimated using three chains[7] taken to 100 000 iterations and with the summary excluding 5000 burn-in iterations. This model suggests over-dispersion, albeit not pronounced, with the mean value of D_1 standing at 42 and that of D_2 at 55. The deviances are highest for ships 19, 20, and 31 with 6, 2 and 7 damage incidents, respectively; two of these ships (19, 31) have relatively short exposure periods. The evaluation of the deviance at $\bar{\phi}$ gives $D_1(\bar{\phi}) = 33.6$, so $p_e = 8.4$ and $D_3 = 33.6 + 2(8.4) = 50.4$. The deviance criterion under minimum predictive loss is 55.4.

The Poisson regression (see Table 3.3) shows lower damage rates for ship types B and C (type A is the first level in the categorical factor and has coefficient zero). Damage rates are higher for later construction and observation periods.

McCullagh and Nelder (1989) suggest a variance inflator (based implicitly on a linear variance model with $\kappa = 0$ in (3.5)) of 1.69, namely $\omega = 1.69$. To formalise this, one may adopt the Normal errors model of the pseudo-likelihood (3.4), as set out in Model B in Program 3.3. Thus the damage counts are taken as Normal with mean μ_i and variance

$$\omega\mu_i$$

where ω is assigned an $E(1)$ prior. This formulation in fact allows for under-dispersion,

$$V(y_i) < \mu_i$$

as well as for overdispersion. ω then has a posterior mean of 1.73, a median of 1.59 and 95% interval 0.96 to 3.1, consistent with the analysis of McCullagh and Nelder. This

[6] Let M_i denote the posterior average of Z_i and T_i the posterior average of the Poisson deviance component $t(Z_i) = Z_i \log Z_i - Z_i$. Define $Q_i = (M_i + ky_i)/(1 + k)$, where k is positive; then the Gelfand–Ghosh measure is $2\Sigma_i[T - t(M_i)] + 2(k+1)\Sigma_i[\{t(M_i) + ky_i\}/\{1 + \text{k}\} - \text{t}(\text{Q}_\text{i})]$.

[7] Starting values are provided by null values, values based on the posterior mean of a trial run, and values based on the upper 97.5% point of the trial run.

model enhances the coefficients for year of construction, but diminishes the size and 'significance' of the coefficients for ship types B and C. The mean value of D_1 is slightly increased at 49, while the mean of D_2 is now 36.

Then in Model C of Program 3.3 a general variance[8] function (3.5) is assumed, in conjunction with a Poisson-gamma mixture. Following Winkelmann and Zimmerman (1991), this is achieved as

$$y_i \sim \text{Poi}(\mu_i)$$
$$\mu_i \sim \text{G}(\eta_{1i}, \eta_{2i})$$
$$\eta_{1i} = \exp(\beta x_i)^{1-\kappa}/(\omega - 1)$$
$$\eta_{2i} = \exp(\beta x_i)^{-\kappa}/(\omega - 1)$$

Three parallel chains are taken to 50 000 iterations, with different starting values for κ and ω. An $E(1)$ prior on $\kappa + 1$ and a gamma prior, G (1, 0.01), on

$$\tau = \omega - 1$$

are adopted. Note that the latter prior amounts to 'forcing' overdispersion. Convergence is relatively slow in τ, and hence ω, and posterior summaries are based on iterations 15 000–50 000.

This approach leads to a posterior means for the statistics D_1 and D_2 of around 26 and 19, respectively. Effective parameters are estimated as approximately 14, the deviance at the posterior mean being 11.9, so $D_3 \approx 11.9 + 28 = 39.9$. This improves over the simple Poisson regression. The deviance criterion under minimum predictive loss is reduced to 49.8.

There is clear evidence of overdispersion in the sense that ω averages 2.94, though in this model the departure from the Poisson assumption is explicitly modelled. The posterior median of the power parameter κ is -0.59, though with a long tail of values that straddle the value of zero; this corresponds approximately to a square root transform in Equation (3.5). Figure 3.1 shows how the posterior for this parameter

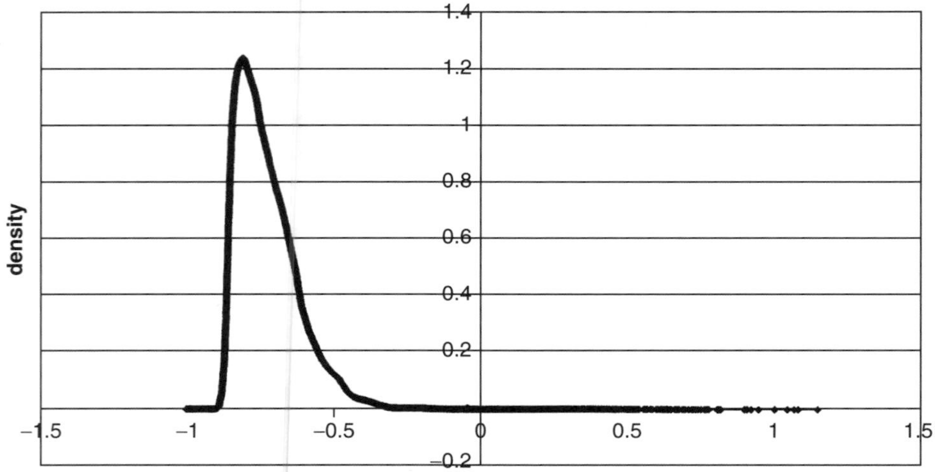

Figure 3.1 Smooth for power parameter, κ

[8] Fitted in version 1.2 of WINBUGS.

(based on iterations 20 000–50 000 over three chains, with sub-sampling every tenth-iteration) focuses on negative values near the minimum threshold.

The credible intervals for the regression coefficients are little affected by this reformulation as compared to the Poisson regression, despite the allowance for overdispersion (Table 3.3). Note that Winkelmann and Zimmerman obtain a value for κ of -0.74 and for ω of 1.85.

Another possible modelling strategy, considered by Dey *et al.* (1997) and others, introduces a dependence of the variance for case i on predictors W_i. These models may be substantively of interest, but may pose identifiability and interpretation problems if, say, there are common predictors of both the mean and variance. Following Dey *et al.*, the variance is a function of the log of the months of observation (Model D in Program 3.3). This variable is also in the model for the mean, and usually taken to have coefficient of unity (this means it becomes an 'offset', in line with a Poisson process).

Here a gamma-Poisson mixture is again assumed with the inverse variances φ_i of the gamma mixture linked to months of service. Thus, Model D involves the steps

$$y_i \sim \text{Poi}(\mu_i)$$
$$\mu_i = \nu_i \delta_i$$
$$\delta_i \sim \text{G}(\varphi_i, \varphi_i)$$
$$\log \phi_i = \alpha_0 + \alpha_1 \log(\text{months service})$$
$$\log \nu_i = \log(\text{months service}) + \text{effects of ship type, etc.}$$

This is consistent with a quadratic variance function, with

$$V(y_i) = \mu_i + \mu_i^2/\varphi_i$$

In fitting this model, N(0, 1) priors are specified on α_0 and α_1 for numerical stability and to improve convergence.

Table 3.3 Ship damage data models

	Poisson (model A)			General variance Function (model C)			Variance regression (model D)		
	Mean	2.5%	97.5%	Mean	2.5%	97.5%	Mean	2.5%	97.5%
Regression parameters									
Intercept	−6.424	−6.863	−6.013	−6.367	−6.732	−5.920	−6.43	−7.16	−5.72
Type B	−0.536	−0.876	−0.167	−0.574	−0.953	−0.091	−0.53	−1.14	0.10
Type C	−0.714	−1.398	−0.080	−0.751	−1.601	0.042	−0.65	−1.60	0.31
Type D	−0.098	−0.683	0.470	−0.170	−1.005	0.551	−0.16	−1.11	0.79
Type E	0.320	−0.143	0.807	0.252	−0.335	0.844	0.39	−0.43	1.28
Built 1965–69	0.699	0.416	0.991	0.676	0.298	1.047	0.69	0.24	1.16
Built 1970–74	0.820	0.499	1.147	0.787	0.380	1.222	0.86	0.38	1.43
Built 1975–79	0.445	−0.024	0.900	0.430	−0.142	0.966	0.49	−0.19	1.20
Period of obs'n 1975–79	0.388	0.150	0.624	0.387	0.119	0.683	0.39	0.01	0.76
Other parameters									
κ				−0.52	−0.99	0.57			
ω				2.94	1.09	9.16			
α_0							1.25	0.27	2.35
α_1							0.88	0.07	1.71

The last 15 000 of a 20 000 iteration run with three chains show that there is a positive (negative) effect α_1 of months service on the precision (variance). The most marked over-dispersion (as expressed also in the deviances at observation level in the simple Poisson regression above) occurs when damage incidents are relatively high in relation to exposure periods. Posterior means on coefficients in the model for the means are similar to in the baseline Poisson model, though with wider credible intervals. The average deviance D_1 and chi-squared sum D_2 are reduced to 25.5 and 18, and all damage counts are well predicted.

So in this example different model approaches with similar fit (in simple deviance terms) have effectively accounted for, or modelled, the over-dispersion but have different implications for the form of the variance function and the regression effects. A suggested exercise is to assess, (a) the effective parameters for the DIC, and (b) the deviance based criteria under minimum predictive loss, in Models C and D, and so attempt to discriminate between them. It should also be noted that the Poisson regression analysis may be improved by allowing an interaction between ship type and year of construction (Lawless, 1987).

Example 3.4 Sales territory While many goodness of fit criteria are 'internal' approaches comparing predictions with actual data, cross-validation methods involve out-of-sample predictions as a way of model assessment. Rust and Schmittlein (1985) consider the application of the split sample Bayesian Cross Validation (BCV) method to 40 observations relating to marketing territory data for sales of click ball point pens (y, in \$000s), advertising spots (X_1), number of sales representatives (X_2), and a wholesale efficiency index (X_3). Two fold cross-validation is one option; other options are n-fold cross-validation, or k-fold cross-validation (k small but exceeding two).

Two alternative sales prediction models are to be assessed via the split sample or two fold cross validation, namely linear and linear-in-logs forms:

$$Y = \alpha_0 + \alpha_1 X_1 + \alpha_2 X_2 + \alpha_3 X_3 + e_1$$
$$Y = \beta_0 + \beta_1 \log(X_1) + \beta_2 \log(X_2) + \beta_3 \log(X_3) + e_2$$

As Rust and Schmittlein note, these are non-nested models, the relative performance of which may be difficult to assess with classical methods. Note that we take sales in thousands of dollars divided by 100 provide a scaled outcome Y. This means likelihoods will differ from those cited by Rust and Schmittlein.

Assuming Normal errors e, we compare penalised likelihood criteria for model choice involving a model fitted to all observations with a BCV check based on a single random split of the 40 observations into an estimation sample of 20 and a validation sample of 20. This split is provided by Rust and Schmittlein, though Program 3.4 includes code to derive alternative random split halves of 40 observations.

The BCV approach yields results as in Table 3.4, based on a fit to the 20 training cases and a log-likelihood comparing actual and predicted values of the outcome for the remaining 20 validation cases. The latter suggests a small advantage to the linear model: a posterior model probability of $2.46/(1+2.46) = 0.71$, where $2.46 = \exp(-21.7 + 22.6)$. A suggested exercise is to:

(a) estimate and validate in the 'reverse' direction, noting that there is no reason in particular to regard one sub-sample as the training sample and one the validation sample; and

(b) split the data randomly into four sub-samples and estimate the two models four times over cases in three sub-samples, validating each time over cases in the remaining sub-sample, and accumulating the predictive log-likelihoods.

To derive standard penalised fit measures, the log-likelihoods at the posterior means of the parameters (i.e. the regression parameters and error variance) are calculated under the two models and penalised according to $p = 5$ parameters under the AIC or $5p/2 \times \log(40) = 9.22$ under the BIC. The respective log-likelihoods, following runs of 10 000 iterations under the linear and log-linear models, are −22.2 and −25.7, so the linear model is also preferred on the basis of a full data set analysis. Since both models have the same number of parameters, the use of AIC or BIC makes no difference to this conclusion. Hence, there is unanimity between model choice methods in this example. It is apparent from Table 3.4 that the inclusion of X_3 is doubtful under either model, and the approach illustrated in Example 3.2 might be used to compare reduced models.

Example 3.5 Laplace approximation to marginal likelihood: counts of schizophrenia Example 3.3 considered simple and penalised deviance measures of fit for a Poisson count outcome, analogous to classical approaches. Here we consider the Laplace approximation as applied to model choice for Poisson regression, approximating formal Bayes model choice based on marginal likelihood. In this approximation, the parameter dispersion matrix S in Equation (3.6) is estimated using the second order differentials of the log-likelihood of the data.

Consider counts y_i of inpatient hospitalisations of schizophrenic patients for 44 small areas (electoral wards) in two London boroughs with expected cases denoted E_i (see Congdon *et al.*, 1998). These are to be 'explained' in terms of measures of social deprivation and community structure. Thus, Model 1 relates psychiatric morbidity to a single predictor, the Townsend deprivation index (X_1), and Model 2 combines the Townsend and anomie indices (X_1 and X_2, respectively). Univariate Normal priors are adopted for the regression coefficients, namely $b_j \sim N(0, 100)$.

Then $y_i \sim \text{Poi}(\mu_i)$, where the Poisson means μ_i are related to predictors via a log-link:

$$\log(\mu_i) = \log(E_i) + b_0 + b_1X_{1i} + b_2X_{2i} + \ldots$$

Table 3.4 Click ball points: BCV analysis

Linear	Mean	St. devn.	2.5%	Median	97.5%
log-likelihood	−21.7	3.0	−28.7	−21.4	−16.8
A_0	0.18	0.98	−1.81	0.19	2.10
A_1	0.16	0.09	−0.02	0.16	0.35
A_2	0.35	0.26	−0.16	0.35	0.87
A_3	0.10	0.23	−0.36	0.10	0.56
Linear-in-logs					
log-likelihood	−22.6	3.1	−29.8	−22.2	−17.6
B_0	−2.93	1.36	−5.60	−2.93	−0.26
B_1	1.87	0.97	−0.01	1.86	3.75
B_2	1.63	1.27	−0.86	1.66	4.13
B_3	0.06	0.55	−1.04	0.06	1.16

Omitting constants, the log likelihood for area i is

$$L(b|y) = -\mu_i + y_i \log(\mu_i)$$

and the second-order differentials of L with respect to b_j and b_k have the form

$$-X_{ij}X_{ik}\mu_i$$

We fit Model 1 and obtain posterior means and standard deviations $b_0 = -3.1$ (0.35), $b_1 = 1.27$ (0.14). The total Poisson log-likelihood (F[1] in program 3.5, Model B) is -137.3, while the marginal likelihood is approximated via the Laplace method[9] is -147.2.

Adding the predictor anomie produces a gain in likelihood which would be judged significant by usual methods, with L now at -133.8. The posterior means and standard deviations are

$$b_0 = -3.24(0.33),\ b_1 = 0.81(0.21) \text{ and } b_2 = 0.55(0.21)$$

The size of the Townsend score coefficient is much reduced and its precision lessened, and the marginal likelihood suggests in fact no gain in fit, standing at -147.5. So there is in effect nothing to choose between the models in Bayesian terms, with the (approximate) Bayes factor close to unity.

Support for a simple model including X_1 only is provided by applying the Kuo–Mallick dependent priors predictor selection method (Kuo and Mallick, 1998, p.72). Here the model is extended to include a third index, the mental illness needs index or mini[] in Program 3.5. Priors on the three regression coefficients are set so that a value $\gamma_j = 0$ on the binary index for the jth coefficient results in an effectively zero coefficient. With three chains, convergence on the coefficients is apparent from around iteration 9000. The second half of a run of 40 000 iterations shows γ_1 to have value 1 (X_1 included in every iteration), whereas γ_2 is 0.09 and γ_3 is only 0.006.

3.3 POLYTOMOUS AND ORDINAL REGRESSION

Many outcomes relating to political or religious affiliation, labour force or social status, or choice (e.g. of travel mode to work) involve ordered or unordered polytomous variables (Amemiya, 1981). The underlying model for such outcomes typically involves a latent continuous variable, which may be conceptualised as an attitudinal, prestige

[9] The BUGS coding for the Laplace approximation involves the inverse( ) and logdet( ) commands. For a Poisson outcome y[], covariates X[,1:p], posterior mean parameter estimate b[1:p] from an earlier fit involving N(0,100) priors, and expected counts *E*[], the coding for a non–iterative program could be:

```
for (i in 1:N) { # Poisson mean
    log(mu[i]) <- log(E[i]) + sum(C[i,1:p])
# log-likelihood
    L[i] <- −mu[i] + y[i]*log(mu[i]) – logfact( y[i])
  for (j in 1:p) { C[i,j] <- b[j]*X[i,j]
  for (k in 1:p) { D[i,j,k] <- −X[i,j]*X[i,k]*mu[i]}}}
# inverse Hessian
for (j in 1:p) { for (k in 1:p) { H[j,k] <- – sum(D[,j,k])
    S[j,k] <- inverse(H[,],j,k)}}
# log probs of posterior mean values under Normal prior densities
for (j in 1:p) { tau.b[j] <- 0.01; mu.b[j] <- 0;
Pr[j] <- 0.5*log(tau.b[j]/6.28) – 0.5*tau.b[j]*pow(b[j] – mu.b[j],2)}
# Laplace approxn
F <- 0.5*p*log(6.28) + 0.5*logdet(S[,]) + sum(L[]) + sum(Pr[])
```

or utility scale, that determines the observed rating or category of belief, status or choice.

The multinomial logit and multinomial probit models generalise their binomial equivalents and are suitable – perhaps with modifications – to modelling multicategory outcomes which do not involve ordered categories. So just as the logit model for a binary outcome involves the log odds of a positive to a negative response, so a multinomial logit involves stipulating a baseline category (say the first of K possible outcomes) and comparing the probabilities of outcomes 2, 3,.., K against that of the first.

3.3.1 Multinomial logistic choice models

Covariates may be defined for individuals i, according to different choices j, or in terms of particular features of choice j which are unique to individual i. Thus, in a travel mode choice example, the first type of variable might be individual income, the second might be the generic cost of alternative modes, and the third might be individual costs attached to different modes. Consider a vector of covariates X_i specific to individuals i alone, and let $y_{ij} = 1$ if option j is chosen. Then for K possible categories in the outcome, we may specify a multiple logit model

$$\begin{aligned} \Pr(y_{ij} = 1) = p_{ij} &= \exp(\beta_j + \gamma_j X_i) \Big/ \left\{ 1 + \sum_{k=2}^{K} \exp(\beta_k + \gamma_k X_i) \right\} \quad j > 1 \\ \Pr(y_{i1} = 1) = p_{i1} &= 1 \Big/ \left\{ 1 + \sum_{k=2}^{K} \exp(\beta_k + \gamma_k X_i) \right\} \end{aligned} \tag{3.12}$$

or equivalently,

$$\log\{p_{ij}/p_{i1}\} = \beta_j + \gamma_j X_i$$

Also, for j and k both exceeding 1,

$$\log\{p_{ij}/p_{ik}\} = (\beta_j - \beta_k) + (\gamma_j - \gamma_k)X_i$$

so that choice probabilities are governed by differences in coefficient values between alternatives. Strictly, the above formulation (involving covariates constant across alternatives j) is known as a 'multinomial logit', and focuses on the individual as a unit of analysis.

If instead we consider attributes W_{ij} of the jth alternative specific for individual i, then a conditional logit model is obtained with

$$p_{ij} = \exp(\delta W_{ij}) \Big/ \sum_{k=1}^{K} \exp(\delta W_{ik}) \tag{3.13}$$

Dividing through by $\exp(\delta W_{ij})$ gives

$$p_{ij} = 1 \Big/ \sum_{k=1}^{K} \exp(\delta[W_{ik} - W_{ij}])$$

In the conditional logit model, the coefficients δ are usually constant across alternatives, and so choice probabilities are determined by differences in the values of characteristics between alternatives. A mixed model, combining features of both Equations (3.12) and

(3.13) would include both individual level attributes X_i and alternative specific characteristics W_{ij}. Thus

$$\log(p_{ij}/p_{ik}) = (\beta_j - \beta_k) + (\gamma_j - \gamma_k)X_i + \delta(W_{ij} - W_{ik}) \tag{3.14}$$

Multiple logit models can be expressed in terms of a model for individual choice behaviour. Thus, let z_{ij} be the unobserved value or utility of choice j to individual i, with

$$z_{ij} = U(X_i, S_j, W_{ij}, \varepsilon_{ij})$$

where S_j are known predictors for choice j (e.g. climate in state j for potential migrants to that state), and X_i and W_{ij} are as above. The ε_{ij} are random utility terms. Assuming additivity and separability of stochastic and deterministic components leads to

$$z_{ij} = \nu_{ij} + \varepsilon_{ij} \tag{3.15}$$

with a regression function such as

$$\nu_{ij} = \beta_j + \gamma_j X_i + \delta W_{ij} + \phi S_j$$

Then the choice of option j means

$$z_{ij} > z_{ik} \quad k \neq j$$

and so

$$p_{ij} = \Pr(z_{ij} > z_{ik})$$

Equivalently

$$y_{ij} = 1 \text{ if } z_{ij} = \max(z_{i1}, z_{i2}, \ldots . z_{iK})$$

Assume the ε_{ij} follow a type I extreme value (double exponential) distribution with cdf:

$$F(\varepsilon_{i.}) = \prod_j \exp(-\exp(-\varepsilon_{ij}))$$

and if the assumption in (3.15) holds also,

$$\Pr(y_{ij} = 1 | X_i, W_{ij}, S_j) = \exp(\nu_{ij}) / \sum_k \exp(\nu_{ik})$$

with $\beta_1 = \gamma_1 = 0$ as in Equation (3.9) for identifiability.

3.3.2 Nested logit specification

A feature of the conditional model (3.10) is that the relative probability of choosing the jth alternative as compared to the kth is independent of the presence or absence of other alternatives (the so-called independence of irrelevant alternatives or IIA axiom). In practice, the presence of other alternatives may be far from irrelevant, and there may be similar alternatives (e.g. with similar utilities to the consumer) between which substitution may be made – see Congdon (2000) for an application involving patient flows to hospitals. We may consider a simple form of nested logit model adapted to account for real world departures from IIA (Poirier, 1996; Fischer and Aufhauser, 1988). Thus, suppose there is one nest of $K - 1$ alternatives within which the simple conditional form (3.10) holds, and a single alternative m from the original set of K which is isolated from the nest. Let $K[-m]$ denote the set of alternatives excluding the mth. Then for choice by individuals and predictors W_{ij} specific to i and j

$$y_{i,\,1:K} \sim \mathrm{Mult}(p_{i,\,1:K},\ 1)$$

For the isolated alternative

$$p_{im} = \exp(\delta W_{im})/\{\exp(\delta W_{im}) + \exp(\alpha \mathrm{I}_i)$$

where I_i is known as the inclusive value

$$I_i = \log\left\{\sum_{j\varepsilon K[-m]} \exp(W_{ij}\delta/\alpha)\right\}$$

Within the nest defined by $K[-m]$, the second level choice probabilities are given by

$$p_{ij} = (1 - p_{im})q_{ij}$$

where

$$q_{ij} = \exp(W_{ij}\delta/\alpha)/\sum_{k\varepsilon K[-m]} \exp(W_{ik}\delta/\alpha)$$

is the standard conditional logit confined to alternatives in the set $K[-m]$. The coefficient α has null value 1, in which case the model reduces to the standard conditional logit over all K alternatives and with unrestricted IIA applying. If α is between 0 and 1 then substitution is greater within a nest than between nests. If substitution among nests exceeds substitution within nests, then α exceeds 1. In the example below we assume α is positive, though negative values of α are not invalid numerically. Poirier (1996, p. 172) suggests the reparameterisation $\lambda = \delta/\alpha$, and so $\delta = \alpha\lambda$, with priors taken on λ and α.

3.3.3 Ordinal outcomes

The multinomial and conditional logit models make no assumptions about the ordering of a categorical outcome. However, ordinal response data are frequently encountered in the social and health sciences. In opinion surveys, respondents are often asked to grade their views on a statement on scales from 'strongly agree' to 'strongly disagree'. Health status or diagnostic gradings of disease are often measured on multicategory scales as 'normal' to 'definitely abnormal' (diagnosis) or as 'good' to 'poor' (health).

Among statistical issues raised by such data are the modelling of ordinal regression relationships (e.g. with predictors as well as outcome ordinal); whether an underlying latent scale assumption need be invoked (Armstrong and Sloan, 1989); and whether or not the ordering of responses is relevant to stratifying regression relationships, for example with different slopes according to each ordinal response category (Anderson, 1984).

The cumulative odds model is often expressed in terms of an underlying continuous response, though this is not strictly necessary. Thus McCullagh (1980) outlined the regression model for an observed ordinal response variable T_i (with possible values $1, 2, \ldots K$) taken to reflect an underlying continuous random variable. For $i = 1, \ldots N$ respondents this model has a systematic component in the form of a cumulative probability

$$\gamma_{ij} = \Pr(T_i \leq j) = F(\theta_j - \mu_i) \qquad j = 1, \ldots K - 1$$

or

$$F^{-1}(\gamma_{ij}) = \theta_j - \mu_i$$

Here $\mu_i = \beta X_i$ is a regression term defined by covariates X_i, θ_j represents the cut point corresponding to the *j*th rank, and F is a distribution function. If the X are all categorical then i typically indexes subject groups defined by combinations of covariates. Special cases of F include the logistic (leading to a proportional odds model) and the extreme value distribution leading to a proportional hazards form.

So if p_{ij} is the probability of lying in the *j*th ranked group, then $\gamma_{ij} = p_{i1} + \ldots . + p_{ij}$. Conversely, the probabilities of an observation lying in the *j*th rank are given by differencing the cumulative probabilities:

$$p_{i1} = \gamma_{i1}$$
$$p_{ij} = \gamma_{ij} - \gamma_{i,\, j-1}$$
$$p_{iK} = 1 - \gamma_{i,\, K-1}$$

If F is a logistic, with

$$C_{ij} = \text{logit}(\gamma_{ij}) = \theta_j - \beta X_i \qquad (3.16)$$

and β is uniform across response categories j, then the θ_j are the logits of belonging to categories up to and including $1, .. j$ (as against categories $j + 1, .. K$) for subjects with $X = 0$. The difference in cumulative logits for different values of X, for example X_1 and X_2, is independent of j. This is known as the 'proportional odds' property[10], with

$$C_{1j} - C_{2j} = \beta(X_2 - X_1)$$

3.3.4 Link functions

Several authors have considered the issue of more general link functions for binary or ordinal outcomes than the logistic or probit links. Lang (1999) proposes a mixture of the symmetric logistic form for F and two asymmetric forms. The latter are the Left Skewed Extreme Value (LSEV) distribution

$$F_1(t) = 1 - \exp(-\exp(t)) \qquad (3.17)$$

and the Right Skewed Extreme Value (RSEV) distribution

$$F_3(t) = \exp(-\exp(-t))$$

[10] In BUGS the essential coding for the proportional odds model applied to individual choice (and for one covariate *X*[*i*]) has the form, together with specimen priors:

```
model{ for (i in 1:N) { for (j in 1:K-1) { # logit of cumulative probability of rank j or lower
    logit(gamma[i,j]) <- theta[j] – mu[i] }
# probability of jth rank
    p[i,1] <- gamma [i,1];
    for (j in 2:K-1) { p[i,j] <- gamma [i,j] - gamma [i,j-1] }
    p[i,K] <- 1 – gamma [i,K – 1];
    mu[i] <- b[1] + x[i]*b[2]
    y[i] ~ dcat(p[i,1:K])}
b[1] ~ dnorm(0,0.001); b[2] ~ dnorm(0,0.001)
theta[1] ~ dnorm(0,1) I(0,theta[2])
for (j in 2:K-1) {theta[j] ~ dnorm(0,1) I(theta[j-1],)}}
```

Note that sampling of y[] could also be represented, with input response data appropriately changed, as y[i,1:K] ~ dmulti(p[i,1:K],1).

with $F_2(t)$ being the logistic distribution

$$F_2(t) = \exp(t)/(1 + \exp(t))$$

The distribution in (3.17) corresponds to the complementary log- log form of link for probability p:

$$F^{-1}(p) = \log(-\log(1 - p))$$

The mixture has the form

$$F_\lambda(t) = w_1(\lambda)F_1(t) + w_2(\lambda)F_2(t) + w_3(\lambda)F_3(t) \tag{3.18}$$

The mixture proportions on the LSEV and RSEV links in the model of Lang are given by

$$w_1(\lambda) = \exp(-\exp(3.5\lambda + 2))$$

and

$$w_3(\lambda) = \exp(-\exp(-3.5\lambda + 2))$$

respectively. They depend upon an additional parameter λ which has a Normal density with mean zero and known variance σ^2_λ. The mixture proportion on the straightforward (i.e. conventional) logit link, intermediate between the asymmetric forms, is obtained as $w_2(\lambda) = 1 - w_1(\lambda) - w_3(\lambda)$.

Lang outlines how the mixture proportions on the three possible links relate to the value of λ, and its variance. Negative values of λ are obtained when the LSEV form is preferred, and positive values when the RSEV is preferred; $\lambda = 0$ corresponds to the logit link. Most variation in the mixture proportions $\{w_1(\lambda), w_2(\lambda), w_3(\lambda)\}$ occurs when λ is in the interval $[-3, 3]$, and this leads to a prior on λ with variance 5 as being essentially non-informative with regard to the appropriate link out of the three possible.

For an ordinal response as in Equation (3.16), we then have

$$p_{ij} = F_\lambda(\theta_j - \beta X_i) - F_\lambda(\theta_{j-1} - \beta X_i).$$

Other Bayesian approaches for general linear models with link functions taken as unknown include Mallick and Gelfand (1994).

Example 3.6 Infection after caesarian birth We first consider a multinomial logistic model applied to data from Fahrmeier and Tutz (1994). In this example the outcome is multinomial but unordered and the predictors are categorical. Specifically, the outcome is a three fold categorisation of infection in 251 births involving Caesarian section; namely no infection, type I infection, and type II infection. The risk factors are defined as NOPLAN = 1 (if the Caesarian is unplanned, 0 otherwise), FACTOR = 1 (if risk factors were present, 0 otherwise), and ANTIB=1 if antibiotics were given as prophylaxis. Of the eight possible predictor combinations, seven were observed. The numbers of maternities under the possible combinations range from 2 (for NOPLAN=0, ANTIB=1, FACTOR=0) to 98 (NOPLAN=1, ANTIB=1, FACTOR=1). N(0, 10) priors are assumed for the impacts of different levels of these categorical variables.

The estimates for this MNL model, following the structure of Equation (3.9), show antibiotic prophylaxis as decreasing the relative risk of infection type I more than type II. Hence γ_{22} is more highly negative than γ_{23}. By contrast, the presence of risk factors in a birth increases the risk of type II infection more: the odds ratio of type II vs type I infection when a risk factor is present, measured by p_{i3}/p_{i2} is $\exp(-2.54+2.14)/\exp(-2.61+1.78)$ or 1.54 (see Table 3.5).

In fact we can use sampling to provide a density for such structural parameters, rather than obtaining 'point estimates' based on the posterior means of constituent parameters. This involves defining new quantities $s_1 = \exp(\beta_3 + \gamma_{33})$ and $s_2 = \exp(\beta_2 + \gamma_{32})$ and monitoring them in the same way as basic parameters. One might assess the probability that $s_1 > s_2$ by repeated sampling also – for example, by using the step() function in BUGS.

If there were not strong evidence that $s_1 > s_2$, the coefficients on NOPLAN, ANTIB and FACTOR might be equalised since Table 3.5 suggests homogeneity across the two infection types. This would involve three fewer parameters, and may provide a better fit after allowing for reduced parameterisation.

Example 3.7 Travel choice As an example of multicategory choice and possible departures from the IIA axiom discussed in Section 3.3.2, consider the travel mode choice data from Powers and Xie (2000). These relate to three choices (1=train, 2=bus, 3=car) among $N = 152$ respondents, with

$$y_{i,\,1:3} \sim \text{Multi}(p_{i,\,1:3},\ 1)$$

where, for example, $y_{i1} = 1$ if a subject chooses the train and zero otherwise. Choice is modelled as a function of respondent income (an X variable, as in Section 3.3.1) and of destination attributes which differ by respondent (W variables). The latter are terminal

Table 3.5 Risk factors for infection after Caesarian delivery

	Mean	St. devn.	2.5%	Median	97.5%
β_2	−2.610	0.605	−3.963	−2.565	−1.595
β_3	−2.537	0.519	−3.627	−2.507	−1.638
NOPLAN					
γ_{12}	1.075	0.492	0.112	1.085	2.066
γ_{13}	0.927	0.448	0.013	0.929	1.828
ANTIB					
γ_{22}	−3.410	0.650	−4.722	−3.398	−2.208
γ_{23}	−3.012	0.535	−4.119	−2.999	−1.991
FACTOR					
γ_{32}	1.775	0.658	0.596	1.746	3.227
γ_{33}	2.140	0.556	1.108	2.127	3.317

waiting time for bus and train (denoted TIME[] in Program 3.7, and zero by definition for car users), in-vehicle time (IVT[]), and in-vehicle cost (IVC[]). Log transforms of the four predictors are taken, though the analysis of PX was in their original scale. The log transformed predictor model gives a better log-likelihood than the untransformed predictor model.

A model with common regression coefficients across choices as in the conditional logit specification (3.13) is fitted first. This provides evidence that the probability of choosing a particular mode decreases as in-vehicle costs and in-vehicle time increase. Note that the effect of terminal waiting time is wrongly signed in substantive terms. Thus the log-odds of bus (Model 2) against the baseline category train (mode 1) may be obtained (see Table 3.6) as

$$\log(p_{i2}/p_{i1}) = 0.31(\log[(\mathrm{TIME}_{i2}+1)/(\mathrm{TIME}_{i1}+1)] \\ - 13.3\log(\mathrm{IVC}_{i2}/\mathrm{IVC}_{i1}) - 0.58\log(\mathrm{IVT}_{i2}/\mathrm{IVT}_{i1})$$

where TIME2 is bus terminal time and TIME1 is train waiting time. The same coefficients apply for the log odds outcome $\log(p_{i3}/p_{i1})$.

Because of the unexpected direction of one coefficient, one might consider whether prior constraints or specialised priors to tackle the collinearity in the predictors. Constraining the coefficients to be negative while retaining all respondents results in a worsening of fit, with the average log-likelihood falling from −87 to −92. The absolute size of the coefficient on invehicle costs is reduced. While assessment of outlier or influential observations, or perhaps variable selection, might seem natural steps as well, it is necessary to consider whether specification might be improved (e.g. to allow for interdependent destinations or the impact of omitted individual level influences).

As an alternative specification option, nesting of choices might first be considered. For illustration, the nested logit model with bus and train as alternatives within a nest of $K-1=2$ choices, and with the car option taken as an isolated single alternative. Unconstrained priors on TIME, IVC and IVT are assumed. This leads (with 5000 iterations over three chains and 1000 burn in) to a posterior mean on α around 0.38. The 95% credible interval is entirely below 1, suggesting substitution within the nest.

Table 3.6 MNL model for travel choice (Non-nested (IIA) and Nested (Non-IIA))

		Mean	2.5%	97.5%
IIA				
	Log likelihood	−87.3	−90.4	−85.9
TIME	δ_1	0.31	0.10	0.53
IVC	δ_2	−13.3	−16.8	−10.2
IVT	δ_3	−0.58	−1.21	0.06
Non-IIA				
	Log Likelihood	−82.5	−86.1	−80.7
	α	0.38	0.21	0.61
TIME	δ_1	0.26	0.10	0.44
IVC	δ_2	−10.2	−13.7	−7.2
IVT	δ_3	−0.41	−0.86	0.00

The coefficients γ are reduced (absolutely) as compared to the IIA alternative where $\alpha = 1$ by default. While there is a clear gain in fit as measured by log-likelihood, the unexpected sign on TIME remains.

We then allow, in a mixed model approach of (3.14) and assuming IIA, for individual specific characteristics X_i. This involves creating dummy variables for $K - 1$ alternatives with train as the base category and multiplying each individual covariate by the $K - 1$ dummies. Here household income provides a single individual level covariate and this involves four extra predictors, with dummies defined for the bus and car options. This analysis is based on 5000 iterations over three chains, with 1500 burn-in, and is shown in Table 3.7. It can be seen that while the attractiveness of bus and car as against train, both increase with income, only the car vs. train choice (measured by γ_2) shows a clear positive impact of income.

One may also at this stage assess whether certain observations are influential or outlying, and distorting the model estimates for all observations combined. For assessing outliers, a procedure suggested by Gelfand and Dey (1994), Weiss (1994) and others, avoids the computing burden of single case omission by considering an estimate for the CPO statistic using output from an MCMC run of length T. This is obtained as

$$\mathrm{CPO}_i^{-1} = T^{-1} \sum_{t=1}^{T} [f(y_i|\theta^{(t)}]^{-1}$$

where $f(y_i|\theta)$ is the total likelihood for each respondent, totalling over destinations. Thus, we find in the mixed model that the CPO statistics to be especially low for respondent numbers 5, 59, 61 and 80, and one might consider an analysis excluding them.

To assess the influence of an individual case i, one might compare regression parameter estimates or samples of new data Z based on (a) the posterior $p(\theta|y^{(N)})$ using the full data $y^{(N)}$, as against (b) the posterior $p(\theta|y_{[i]})$ using all the data set except case i, $y_{[i]}$. One might, for example, compare a certain regression parameter β^* by the discrepancy or checking function (Weiss, 1994)

$$D_{\beta_i^*} = E(\beta^*|y_{[i]}) - E(\beta^*|y^{(N)})$$

In the transport choice analysis, we might consider the two parameters γ_1 and γ_2 describing interaction between mode and income. Other types of discrepancy involve predictions of replicate data Z, with distributions under options (a) and (b)

$$p(Z|y^{(N)}) = \int p(Z|\theta)p(\theta|y^{(N)})d\theta$$

Table 3.7 Mixed MNL model, parameter summary

		Mean	2.5%	97.5%
Log Likelihood		−58.9	−63.5	−56.2
δ_1	TIME	−3.2	−4.8	−1.9
δ_2	IVC	−11.0	−14.7	−7.7
δ_3	IVT	−0.88	−1.86	0.11
β_1	BUS	−3.32	−6.97	−0.09
γ_1	BUS × INCOME	0.75	−0.29	1.90
β_2	CAR	−19.2	−27.0	−12.6
γ_2	CAR × INCOME	1.80	0.69	3.06

$$p(Z|y_{[i]}) = \int p(Z|\theta)p(\theta|y_{[i]})d\theta$$

One might compare the predictions Z_i and observations y_i for the suspect case i. For instance, in the multinomial example here, one might compare the relative distribution of the $Z_{i,\,1:3}$ (i.e. proportions in the three travel modes) between the two sampling options (a) and (b). One might also compare the likelihoods of Z_i under the two options, with a discordancy index, such as (Geisser, 1990)

$$D_{Z_i} = f(Z_i|y_{[i]}) \log \{f(Z_i|y_{[i]})/f(Z_i|y^{(N)})\}$$

Such comparisons may be computationally intensive, and here we consider the impacts of omitting the cases with the lowest and highest CPO as defined above. These are, respectively, case 80 and 63, with case 80 opting for car despite lower than average income. Using single long runs of 20 000 iterations (see Model D in Program 3.7) we find, as in Table 3.8, that case 80 has a discordancy index clearly higher than case 63. There is also more difference in the relative proportions over the three modes (comparing in Program 3.7 the samples y.new[80,] and y.new.d[80,] of new data when estimation is based on all cases $y^{(N)}$ and on $y_{[i]}$, respectively). There seems less influence in terms of the income by mode interaction parameters, though excluding case 80 tends to slightly raise the income by car effect.

Example 3.8 O-ring failures The logit and probit links are the default options for categorical choice models. To illustrate the potential impact of alternative links, the model of Section 3.3.4 is applied to 23 observations relating to thermal problems on the US Challenger missions in 1986. Let $y_i = 1$ if a primary O-ring showed thermal distress, and $y_i = 0$ otherwise. The single covariate is the temperature S in Fahrenheit at launch. Following Lang (1999), we take this covariate as uncentred, and with $\pi_i = P_r(y_i = 1)$, obtain a logistic model with posterior mean (and s.d.) parameter estimates

$$\text{logit}(\pi_i) = \underset{(9.3)}{19.2} - \underset{(0.14)}{0.29}\,\text{S}$$

Following Lang, a Normal prior with mean 0 and variance 5 on λ is taken in the general link model (3.18). Because of the relatively sparse data, a clear choice of link is unlikely,

Table 3.8 Checking functions under case omission, cases 63 and 80

	Omitting Case 80		Omitting Case 63	
	Mean	St. devn.	Mean	St. devn.
New Data Discrepancy D_{z_i}	0.0433	0.1520	0.0000	0.0001
Difference in Bus by Income $D_{\gamma 1_i}$	0.0237	0.7473	−0.0330	0.7845
Difference in Car by Income $D_{\gamma 2_i}$	0.1529	0.9156	0.0211	0.8809
Relative distribution between modes				
Full data, Z = Train	0.6561	0.4750	0.9999	0.0079
Full data, Z = Bus	0.3330	0.4713	0.0000	0.0000
Full data, Z = Car	0.0109	0.1037	0.0001	0.0079
Data omitting case, Z = Train	0.6442	0.4788	0.9999	0.0079
Data omitting case, Z = Bus	0.3525	0.4778	0.0000	0.0000
Data omitting case, Z = Car	0.0033	0.0573	0.0001	0.0079

but the analysis leans towards the complementary log-log link, corresponding to the left skewed extreme value distribution.

Three chains are taken to 100 000 iterations (with convergence of λ apparent at around 7000 iterations). The weights on the three components are 0.53, 0.35 and 0.11, and the average of λ is close to -1, with $\beta_0 = 16.5$ and $\beta_1 = -0.26$ (Table 3.9). So there is support for a non-logit form of link. Figure 3.2 shows the posterior density of λ, based on 9000 iterations (every tenth iterate over 30 000 iterations on three chains), with most mass between -2 and 0.

On this basis, we can summarise the relation as

$$F_\lambda(\beta_0 + \beta_1 S) = F_{-1}(16.5 - 0.26S)$$

By contrast, Lang obtains

$$F_\lambda(\beta_0 + \beta_1 S) = F_{-1.43}(14.7 - 0.23S)$$

As a check to model form, a Bayes predictive check (Gelman *et al.*, 1995) is used, based on comparing a chi-square measure for the observed data $\chi^2(Z_i, \pi_i)$ and for replicated data, namely $\chi^2(Z_{i.\text{rep}}, \pi_i)$. This is estimated at 0.27, and shows no evidence of lack of fit.

Example 3.9 Ordinal ratings in diagnostic tests As an illustration of Bayesian ordinal regression, we consider an issue in medical diagnosis. The performance of diagnostic tests is measured by various probabilities: the probability of correctly assessing a diseased individual is known as the sensitivity (i.e. true positive rate) of the test, and that

Table 3.9 Choice of link: O-ring data, parameter summary

	Mean	St. devn.	2.5%	97.5%
β_0	16.52	8.19	3.65	36.12
β_1	−0.26	0.12	−0.55	−0.06
λ	−0.99	1.78	−4.69	2.72

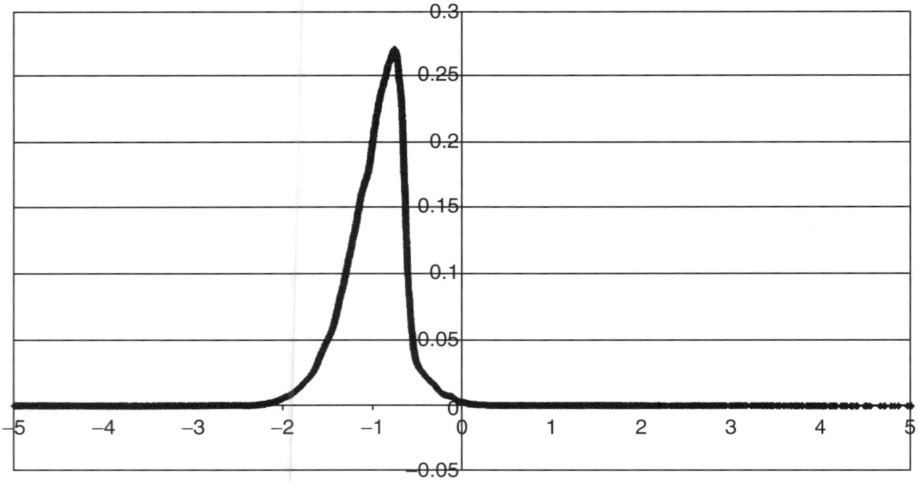

Figure 3.2 Smooth of λ

of correctly classifying a healthy individual is the specificity. The false positive rate (chance of classifying someone healthy as ill) is 1 minus the specificity.

Assume outcomes T_i on subjects are available as a rating scale with K levels, with category 1 corresponding to the lowest indication of disease and category K to the highest (e.g. for $K = 3$ we might have 1=normal, 2=equivocal, 3=abnormal). As in Section 3.3.3, $K - 1$ cut points θ_j are assumed so that an individual will fall in category 1 if the underlying scale value for that subject z_i is less than θ_1, in category 2 if $\theta_1 < z_i \leq \theta_2$, and so on, and in category K if $z_i > \theta_{K-1}$.

The performance of a test depends upon the threshold chosen on the latent scale z. Lowering the threshold will improve sensitivity, but at the expense of more false positives. The relation between the true and false positive rate as the threshold varies is known as the Receiver Operating Characteristic (ROC) curve. A parametric approach to estimating this can be derived from signal detection theory, and leads to a smooth ROC curve. However, this approach does not adapt to include covariates (e.g. age or disease stage) which might affect the test accuracy.

Assume, therefore, that p covariates are available on each subject, and that the first of these is a dummy index for true disease state ($x_{i1} = 1$ for diseased cases i), so that there is no intercept. Then following Tosteson and Begg (1988), we may specify the cumulative probability $\gamma_j(x_i)$ of response up to and including category j for a subject with covariates x_i.

$$F^{-1}\{\gamma_j(x_i)\} = (\theta_j - \beta x_i)/\exp(\delta x_i) \tag{3.19}$$

Taking F as the Normal distribution function amounts to assuming the latent variable z is Normal with mean βx_i and standard deviation $\exp(\delta x_i)$. Other links are possible, such as the logit, but interpretation of the β parameters changes. When the first covariate x_1 is true disease status and x_1 is set at 1, the true positive rate or sensitivity for a subject with rating j is

$$1 - \gamma_j(x_{1i})$$

since $\gamma_j(x_{1i})$ with $x_{1i} = 1$ is the probability that a diseased subject is ranked in one of the diagnostic groups $1, . . j$.

The basis for making the variance of the latent scale depend upon covariates is likely to be governed by the application area. Tosteson and Begg (1988) consider the detection of hepatic metastases by ultrasonography, and cite evidence that the spread of responses in radiological rating data is wider in diseased than non-diseased subjects. This difference would be modelled by the parameter δ_1. Their analysis and a reanalysis by Peng and Hall (1996) concern the accuracy of ultrasonography ratings on a five point scale (see Appendix Table 1). These ratings are used in assessing the presence or otherwise of metastases in patients. Ratings are related to true metatstasis x_1, and cancer site (either breast cancer, with $x_2 = 1$, or colon cancer, with $x_2 = 0$). A third predictor is the interaction $x_3 = x_1 x_2$ between true metastasis status and cancer type.

A 'full model' includes all three covariates in both β and δ in Equation (3.19). A logit link is adopted for numerical stability – a probit link would best be implemented with the Albert–Chib (1993) sampling method. The priors on the cut points are constrained to be increasing, with

$$\theta_j \sim N(j, 10)\ I(\theta_{j-1}, \theta_{j+1})$$

The logit link full model is not well identified (cf. Tosteson and Begg), but suggests that variability is greater for metastasis cases (for whom $x_1 = 1$) and breast cancer cases for whom $x_2 = 1$ (see Table 3.10). These findings are in line with Tostesan and Begg (1988, Table 2), though they report different signs on β_2 and β_3 than obtained here (though neither effect is significant in their results). The sensitivity of the scale runs at 0.75 or more across the possible ranks.

A reduced model, involving just x_1, the true status, in defining β and δ, shows a clear association between true status and the diagnostic classification. With a run of 10 000 iterations from three chains (and 2500 burn-in), the parameter β_1 averages 1.9 and has a 95% interval consisting only of positive values. However, the effect of metastasis on variability in ultrasonography ratings (as measured by δ_1), is no longer present. As in the full model, the latent scale cut points θ_j suggest that upper ranks on the scale could be merged.

3.4 REGRESSIONS WITH LATENT MIXTURES

Chapter 2 considered finite mixtures to describe heterogeneity in which latent classes $j = 1, \ldots, J$ differ only in their means μ_j and other population parameters (e.g. variances

Table 3.10 Ultrasonography ratings parameters

Full model	Mean	St. devn.	2.5%	Median	97.5%
Influences on diagnostic rank					
β_1	3.86	1.61	1.60	3.53	7.79
β_2	−0.62	1.55	−4.34	−0.23	1.48
β_3	1.00	2.45	−3.09	0.73	6.32
Influences on variability					
δ_1	0.76	0.60	−0.51	0.79	1.84
δ_2	1.30	1.25	−0.52	1.15	4.52
δ_3	−1.65	1.67	−5.35	−1.46	1.19
Cutpoints					
θ_1	0.35	0.21	0.02	0.32	0.81
θ_2	0.99	0.17	0.65	0.99	1.33
θ_3	1.07	0.16	0.75	1.07	1.39
θ_4	1.11	0.16	0.79	1.10	1.43
Reduced Model					
β_1	1.88	0.44	1.18	1.82	2.88
δ_1	−0.26	0.41	−1.12	−0.25	0.51
θ_1	0.31	0.19	0.02	0.29	0.71
θ_2	0.87	0.15	0.59	0.87	1.16
θ_3	0.94	0.14	0.68	0.94	1.23
θ_4	0.98	0.14	0.71	0.97	1.26

in a Normal example). Finite mixture regressions introduce covariates into either the determination of the latent class indicators or to describe the relation between the mean of subject i on each latent class μ_{ij} and that subject's attribute profile. Mixture regressions have been applied to modelling the behaviour or attitudes of human subjects so that each individual has their overall mean determined by their membership probabilities (Wedel *et al.*, 1993).

Thus, for univariate Normal observations y_i, a p-dimensional vector of predictors X_i, define latent indicators z_i of class membership among possible classes $j = 1, .. \mathrm{J}$. Were the indicators known,

$$y_i|z_i = j \sim \mathrm{N}(\beta_j X_i, \tau_j) \tag{3.20}$$

where β_j is a class specific regression vector of length p, and τ_j is the conditional variance. If $\lambda_{ij} = \Pr(z_i = j)$ then the overall mean for subject i is

$$\lambda_{i1}\mu_{i1} + \lambda_{i2}\mu_{i2} \ldots . + \lambda_{iJ}\mu_{iJ}$$

where $\mu_{ij} = \beta_j X_i$. The indicators z_i may be sampled from a multinomial without additional covariates, so that the multinomial has parameters $\lambda_{ij} = \lambda_j$. Alternatively, an additional regression seeks to model the z_i as functions of covariates W_i such that the multinomial is defined by parameters

$$\lambda_{ij} = \exp(\phi_j W_i)/[1 + \sum_{k=2} \exp(\phi_k W_i)] \quad j > 1$$

as in Section 3.3.1.

Several applications of regression mixtures have been reported in consumer choice settings: Jones and McLachlan (1992) consider metric outcomes y, namely consumer preference scales for different goods, which are related to product attributes (appearance, texture, etc.). They find sub-populations of consumers differing in the weight they attach to each attribute. Here the multinomial logit regressions for z_i might involve covariates W such as consumers' age, class, or type of area of residence, while the modelling of the y_i might involve covariates X describing the quality or price of goods.

Binomial, ordinal or multinomial mixture regressions have utility both in representing departures from the baseline model assumptions (e.g. overdispersion), as well as differential regression slopes between sub-populations (Cameron and Trivedi, 1986). For example, Wedel *et al.* (1993) argue for using a latent class mixture in an application involving Poisson counts y, both because of its advantage in modelling differential purchasing profiles among customers of a direct marketing company, and its potential for modelling over-dispersion in relation to the Poisson assumption.

As mentioned in Chapter 2, there are the usual problems in Bayesian analysis (as in frequentist analysis) concerning the appropriate number of components. Additionally, Bayesian sampling estimation may face the problems of empty classes at one or more iterations (e.g. no subjects are classified in the second of $J = 3$ groups) and the switching of labels unless the priors are constrained. On the other hand, the introduction of predictors provides additional information that may improve identifiability. To counter label switching we might apply a constraint to one or more of the intercepts, regression coefficients, variance parameters, or mixture proportions that ensures a consistent labelling. In some situations one may be able to specify informative priors consistent with widely separated, but internally homogenous groups (Nobile and Green, 2000). These ensure (a) that different groups are widely separated, for instance a prior on intercepts β_{0j} when $J = 2$ might be $\beta_{01} \sim \mathrm{N}(-5, 2)$, $\beta_{02} \sim \mathrm{N}(0, 2)$, effectively ensuring separation, and (b) that if cases do fall in the same group, they are expected to be similar.

An alternative to constrained priors involves re-analysis of the posterior MCMC sample, for example by random or constrained permutation sampling (Fruhwirtth-Schattner, 2001). Suppose unconstrained priors in model (3.20) are adopted, and parameter values $\theta_j^{(t)} = \{\beta_j^{(t)}, \tau_j^{(t)}\}$ are sampled for the nominal group j at iteration t. We may investigate first whether – after accounting for the label switching problem – there are patterns apparent on some of the parameter estimates which support the presence of sub-populations in the data. Thus, if there is only $p = 1$ predictor and the model is

$$\begin{aligned} y_i &\sim \Sigma_j \mathrm{N}(\mu_{ij}, \tau_j) \\ \mu_{ij} &= \beta_{0j} + \beta_{ij} x_i \end{aligned} \tag{3.21}$$

then a prior constraint which produces an identifiable mixture might be $\beta_{01} > \beta_{02}$, or $\beta_{11} > \beta_{12}$ or $\tau_1 > \tau_2$. (Sometimes more than one constraint may be relevant, such as $\beta_{01} > \beta_{02}$, *and* $\beta_{11} > \beta_{12}$). Fruhwirtth-Schattner proposes random permutations of the nominal groups in the posterior sample from an *un*constrained prior to assess whether there are any parameter restrictions apparent empirically in the output that may be associated with sub-populations in the observations.

From the output of an unconstrained prior run with $J = 2$ groups, random permutation of the original sample labels means that the parameters nominally labelled as 1 at iteration t are relabelled as 2 with probability 0.5, and if this particular relabelling occurs then the parameters at iteration t originally labelled as 2 are relabelled as 1. Otherwise, the original labelling holds. If $J = 3$ then the nominal group samples ordered {1, 2, 3} keep the same label with probability 1/6, change to {1, 3, 2} with probability 1/6, etc.

Let $\tilde{\theta}_{jk}$ then denote the relabelled group j samples for parameters $k = 1,.., K$. (A suffix for iteration t is understood.) The parameters relabelled as 1 (or any other single label among the $j = 1,..J$) provide a complete exploration of the unconstrained parameter space, and one may consider scatter plots involving $\tilde{\theta}_{1k}$ against $\tilde{\theta}_{1m}$ for all pairs k and m. If some or all the plots involving $\tilde{\theta}_{1k}$ show separated clusters, then an identifying constraint may be based on that parameter. To assess whether this is an effective constraint, the permutation method is applied based not on random reassignment, but on the basis of reassignment to ensure the constraint is satisfied at all iterations.

Example 3.10 Viral infections in potato plants Turner (2000) considers experiments in which viral infections in potato plants were assessed in relation to total aphid exposure counts. The experiment was repeated 51 times. The data are in principle binomial, recording numbers of infected plants in a 9×9 grid with a single plant at each point. However, for reasons of transparency, a Normal approximation involving linear regression was taken. The outcome is then just the totals of plants infected y, which vary from 0 to 24. A plot of the infected plant count against the number of aphids released (x) shows a clear bifurcation, with one set of (y, x) pairs illustrating a positive impact of aphid count on infections, while another set of pairs shows no relation.

Here we consider two issues: the question of possible relabelling in the MCMC sample, so that samples for the nominal group 1 (say) in fact are a mix of parameters from more than one of the underlying groups. Secondly, we consider possible assessments of the number of groups.

A two group mixture, with means and conditional variances differing by group, was accordingly well identified by Turner and his bootstrap analysis showed marked improvement in fit over a single group model. As discussed in Chapter 2, the bootstrap is often applied when the usual likelihood ratio tests do not follow standard asymptotic densities. Suppose one wished to assess the gain in fit in a $J+1$ vs. J component model, with Q° denoting the observed likelihood ratio comparing (say) a three group to two group model, with parameters $\hat{\theta}_{(3)}$ and $\hat{\theta}_{(2)}$, respectively. Then one might sample S repeated data sets from the parameters $\hat{\theta}_{(2)}$ and derive the likelihood ratio statistic Q_s for each, and find the proportion of sets where Q° exceeds Q_s. This is the bootstrap test criterion, as opposed to the nominal test based on asymptotic results.

First we run the model (3.21) for $J=2$ and $K=3$ (intercepts β_{0j}, slopes β_{1j}, and variances τ_j). Using the output from 20 000 iterations with a single chain, 500 iteration burn-in, and starting from null starting values, we apply (e.g. in a spreadsheet) the random permutation sampler of Fruhwirtth-Schattner (2001). The relabelled group 1 parameters (every tenth sample) are then plotted against each other (Figures 3.3a–c). Both plots involving the slope, namely Figures 3.3a and 3.3b, show well separated clusters of points, suggesting an identifiability constraint on the β_{1j}, such as $\beta_{11} > \beta_{12}$. Applying the constrained permutation sampler, again to the output from the unconstrained prior, shows that this is an effective way of identifying sub-populations, and one can either use the output from the constrained permutation to derive parameter estimates or formally apply the constraint in a new MCMC run. From the constrained permutation sampler, the following characteristics of the two groups are obtained:

	β_{01}	β_{02}	β_{11}	β_{12}	$1/\tau_1$	$1/\tau_2$
Mean	0.868	3.324	0.002	0.055	0.900	0.096
2.5%	0.115	1.047	−0.003	0.040	0.361	0.043
97.5%	1.733	5.593	0.007	0.069	2.117	0.170

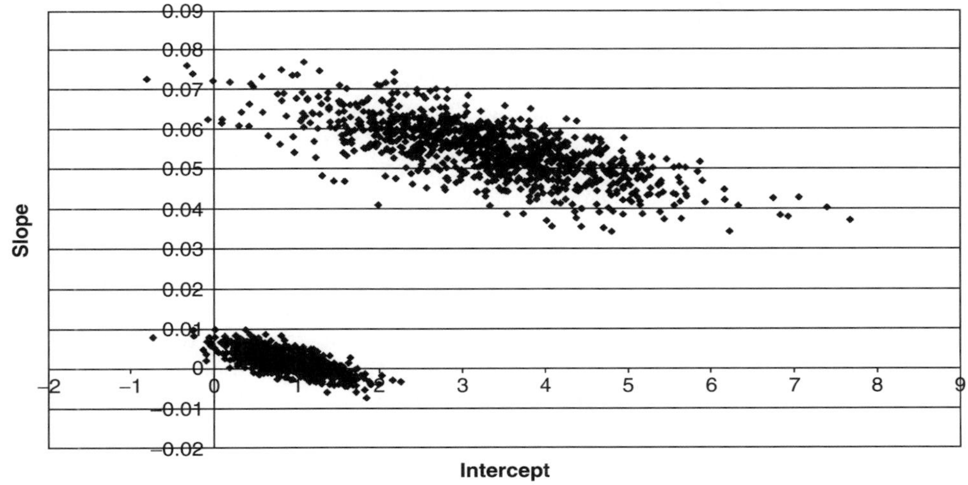

Figure 3.3a Plot of intercept vs. slope for relabelled first group iterations

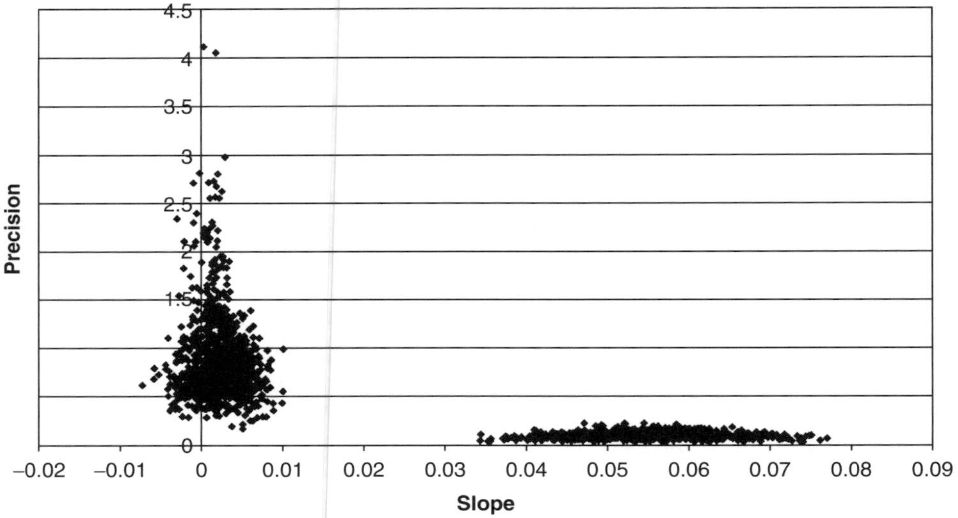

Figure 3.3b Slope against precision for relabelled first group iterations

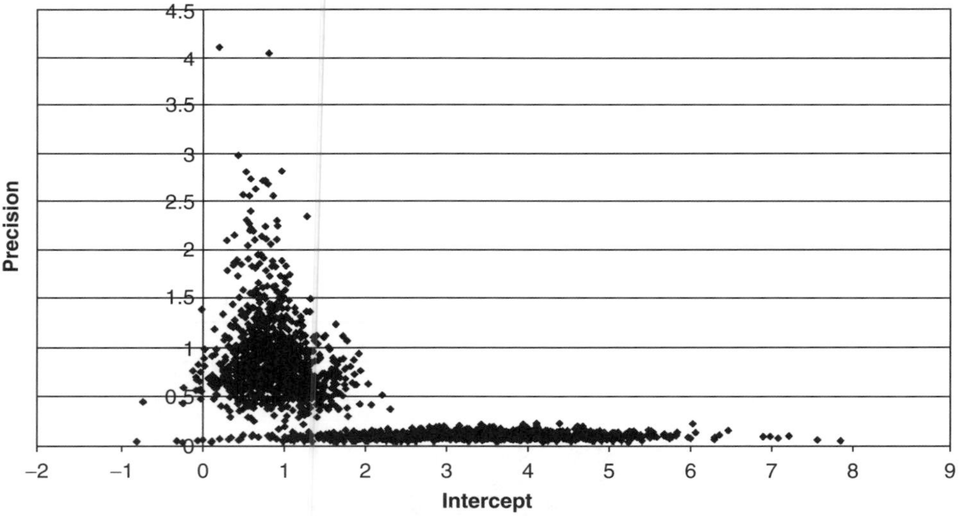

Figure 3.3c Intercept against precision for relabelled group 1 iterations

In the present application, it was not possible to identify a stable solution with three groups that provided an improvement in log-likelihood over the two group solution. This was apparent both from MCMC sampling with a constraint on the slopes β_{1j} and using the permutation sampler on the output from a three group model without a prior constraint. Plots of the intercepts on the slopes from the relabelled parameter iterates $\tilde{\theta}_{1k}(k = 1, 3)$ showed no distinct groups for $J = 3$, but instead just 'scatter without form'.

It remains to demonstrate the gain in a two over a one group model. This might be possible to demonstrate using penalised fit measures such as the AIC statistics. Thus, parallel sampling of a one and two group model (the latter with constrained prior) could be carried out, and the proportion of iterations assessed where the AIC for the two group model improved over that for the one group model. Note that the number of parameters are usually taken as $p_1 = 3$ and $p_2 = 7$ in the one and two group models, respectively. Other options (see Raftery, 1995) might involve marginal likelihood approximations leading to approximate Bayes factors. Another possibility is Bayes sampling analogues of bootstrap and Monte Carlo test procedures (see Dey *et al.*, 1998).

Here a marginal likelihood approximation is obtained by monitoring the inverse likelihood $g_i^{(t)} = 1/f(y_i|\theta^{(t)})$ for each case. The inverse of the posterior mean of $g_i^{(t)}$ is an estimate of the CPO for case i, and the sum of the logarithms of these estimated CPO_i is an estimate of the log marginal likelihood of the model. Here this procedure gives log marginal likelihood estimates of -162.9 and -140.5 under the one and two group models, and a clear preference for the two group model.

3.5 GENERAL ADDITIVE MODELS FOR NONLINEAR REGRESSION EFFECTS

A generalisation of the smoothing prior approach of Chapter 2 is to generalised additive models in regression. Such models provide an approach to modelling possible nonlinearity, but avoiding the need to specify complex algebraic forms. Thus for a metric outcome $y_1, .. y_n$, assume there are corresponding values of a regressor variate $x_1, ... x_n$ ordered such that

$$x_1 < x_2 < .... < x_n$$

The model for the observations may then be

$$y_t = \beta_0 + f(x_t) + \varepsilon_t$$

where $\varepsilon_t \sim N(0, \sigma^2)$. Let $g_t = f(x_t)$ be the smooth function representing the changing, possibly nonlinear, impact of x on y as it varies over its range. As in Chapter 2, it is common to assume Normal or Student random walks in the first, second or higher differences of the g_t. A variant on this is when the smooth in the variable x modifies the effect of a predictor z, with

$$y_t = \beta_0 + z_t f(x_t) + \varepsilon_t$$

If $x_t = t$ denotes time and y_t is a time sequenced response, then a dynamic coefficient or state space model is obtained, with

$$y_t = \beta_0 + z_t \beta_{1t} + \varepsilon_t$$

It will commonly be the case that the x_t are unequally spaced, and it is then necessary in specifying the prior for g_t (or β_{1t}) to weight each preceding point differently. This means adjusting the precision such that wider spaced points are less tied to their predecessor than closer spaced points. Thus, suppose the x_t were irregularly spaced and that the spaces between points are $\delta_1 = x_2 - x_1$, $\delta_2 = x_3 - x_2, \ldots$ $\delta_{n-1} = x_n - x_{n-1}$. A first order random walk smoothness prior, with Normal errors, would then be specified as

$$g_t \sim N(g_{t-1}, \delta_t \tau^2)$$

and a second order one would be

$$g_t \sim \mathrm{N}(\nu_t,\ \delta_t\tau^2)$$

where $\nu_t = g_{t-1}(1 + \delta_t/\delta_{t-1}) - g_{t-2}(\delta_t/\delta_{t-1})$ (see Fahrmeir and Lang, 2001). If there is equal spacing then the first and second order random walk priors are just

$$g_t \sim \mathrm{N}(g_{t-1},\ \tau^2)$$
$$g_t \sim \mathrm{N}(2g_{t-1} - g_{t-2},\ \tau^2)$$

With smooths on several regressors x_{1t}, x_{2t}, x_{3t} .. it will be necessary to supply an ordering index on each one at observation level O_{1t}, O_{2t}, O_{3t}, etc. A frequent situation is the semi-parametric additive form, with smooths on a subset of p regression variables, with the remainder modelled conventionally. Quite often there would be just a single regressor with a general additive form and the remainder included in a conventional linear combination. It may be noted that results may be sensitive to the priors assumed for the initial smoothing values (e.g. g_1 in a first order random walk) and for the evolution variance τ^2. This is especially so for sparse data, such as for binary outcomes y_t. In BUGS convergence may benefit from sampling τ^2 directly from its full conditional density.

Other smoothness priors have been proposed. For example, Carter and Kohn (1994) consider the signal plus noise model

$$y_t = g_t + \varepsilon_t$$

where the g_t are generated by a differential equation

$$\frac{d^2 g_t}{dt^2} = \tau \frac{dW_t}{dt}$$

with W_t a Weiner process, and τ^2 the smoothing parameter. This leads to a bivariate state vector $s_t = (g_t,\ \frac{dg_t}{dt})$ in which

$$s_t = F_t s_{t-1} + u_t \qquad (3.22)$$

where, with δ_t defined as above,

$$F_t = \begin{pmatrix} 1 & \delta_t \\ 0 & 1 \end{pmatrix}$$

Also, the u_t are bivariate Normal with mean zero and covariance $\tau^2 U_t$, where

$$U_t = \begin{pmatrix} \delta_t^3/3 & \delta_t^2/2 \\ \delta_t^2/2 & \delta_t \end{pmatrix}$$

Wood and Kohn (1998) discuss the application of this prior with binary outcomes, which involves applying the latent variable model of Albert and Chib (1993).

Example 3.11 Kyphosis and age Consider the kyphosis data from Hastie and Tibshirani (1990) on 81 patients receiving spinal surgery (Appendix Table 2). The binary outcome y_t relates to the post-surgical presence or otherwise of forward flexion of the spine from the vertical. As mentioned above, this is a sparse data form and so results of additive smoothing may be sensitive to priors assumed.

Risk factors taken to influence the outcome are x_{1t} = the number of vertebrae level involved, x_{2t} = the starting vertebrae level of the surgery and x_{3t} = age in months. Initially, a probit model with linear effects in $X_t = (x_{1t}, x_{2t}, x_{3t})$ is applied using the latent dependent variable method. Thus

$$z_t \sim N(\mu_t, 1)\, I(0,) \qquad \text{if } y_t = 1$$
$$z_t \sim N(\mu_t, 1)\, I(, 0) \qquad \text{if } y_t = 0$$

with mean

$$\mu_t = \beta_0 + \beta_1 x_{1t} + \beta_2 x_{2t} + \beta_3 x_{3t}$$

N(0, 10) priors are set on β_0, β_1, and β_2, but an N(0, 1) prior on β_3 as it is applied to ages which exceed 200, and too vague a prior may lead to numeric problems as a large coefficient is applied to a large age value. This analysis shows clear positive effects for x_1 and a negative effect for x_2, but a less clear positive (linear) effect of age.

To clarify possible nonlinear effects, the impact of age is instead modelled via a general additive form. Thus, sampling of the z_t is defined as above, but with mean now

$$\mu_t = \beta_0 + \beta_1 x_{1t} + \beta_2 x_{2t} + \beta_3 x_{3t} + f(x_{3t})$$

In defining the smoothness prior for $f()$, the age variable is grouped according to the $n_g = 64$ distinct values, as in Appendix Table 2, which includes both the original age and the age group index. The differential equation prior (3.22) is used, sampling $\sigma^2 = 1/\tau^2$ from the full conditional specified by Carter and Kohn (1994, p. 546). Prior values of 0.001 for are used for index and shape parameters in the gamma full conditional density for σ^2. Using the initial parameter values as in Program 3.11 convergence in τ^2 is apparent after 5000 iterations, and the posterior summary is based on iterations 5000–50 000.

The evidence from the smooth (Figure 3.4) is of a clear nonlinear effect in age, reaching a maximum at ages between 90 and 100 months. The linear age effect is eliminated, but the values of the regression coefficients on the other covariates are somewhat enhanced in the general additive model.

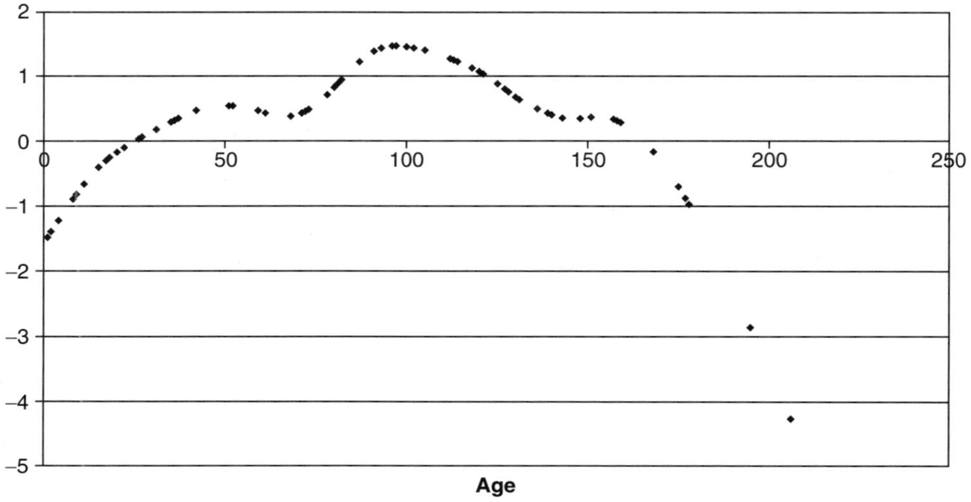

Figure 3.4 Smooth for age

Table 3.14 Kyphosis outcome, standard Probit and general additive models

Probit regression		Mean	St. devn.	2.5%	Median	97.5%
	Deviance	64.9	2.7	61.6	64.3	71.8
	Intercept	−1.130	0.744	−2.615	−1.102	0.277
	Number	0.229	0.110	0.020	0.227	0.448
	Start	−0.125	0.038	−0.202	−0.125	−0.053
	Age	0.006	0.004	−0.001	0.006	0.013
Additive form on age						
	Deviance	59.4	4.1	52.1	58.8	70.8
	Intercept	−1.337	1.206	−4.301	−1.295	1.270
	Number	0.249	0.127	−0.033	0.247	0.553
	Start	−0.147	0.045	−0.255	−0.146	−0.046
	Age (Linear)	0.004	0.013	−0.024	0.004	0.033
	τ^2	0.0013	0.0011	0.0002	0.0009	0.0057

The deviance of the GAM model improves only slightly on the linear model, despite being more heavily parameterised; the penalised fit (as may be verified) therefore deteriorates. The plot of the smooth suggests a low order polynomial might be sufficient to model the nonlinearity, and this would be less heavily parameterised than a GAM.

3.6 ROBUST REGRESSION METHODS

Statistical modelling based on the Normal distribution is often the default option in regression with metric outcomes, or in models including hierarchical random effects, when outcomes are discrete. One approach seeks to identify outliers in terms of posterior probabilities that an absolute value of an estimated residual is surprisingly large (see Chaloner, 1994 and Example 3.12). However, instead of adopting Normality and then seeking possible outlier observations inconsistent with Normality, an alternative is to embed the Normal in a broader model set. This larger set may be defined by an extra parameter (or parameters) that afford resistance or robustness to non-Normality. Alternatively, the Normal can be set within a range of alternative densities, allowing heavier or (less commonly) lighter tails than the Normal. Thus Bayesian approaches to robustness in metric outcomes include the *t* density, and contaminated Normal densities that explicitly allow some observations to be outliers (Verdinelli and Wasserman, 1991).

Under the Student *t* density, resistance to outliers is accommodated by varying the degrees of freedom parameter. As considered in Chapter 2, introducing this extra parameter is in fact still equivalent to the Normal but with a variable scale parameter for each observation. The associated weight that may be used to indicate outlier status in relation to the overall assumed density or regression model.

Suppose the data consists of univariate metric outcomes y_1, $i = 1, .., n$ and an $n \times p$ matrix of predictors. Then consider a Student *t* regression model for the means $\mu_i = \beta x_i$ with variance σ^2 and known degrees of freedom ν. Assuming the reference prior (Gelman *et al.*, 1995)

$$\pi(\beta, \sigma^2) \propto \sigma^{-1}$$

the posterior density is proportional to

$$\sigma^{-(n+1)} \prod_{i=1}^{n} [1 + (y_i - \beta x_i)^2/\nu\sigma^2]^{-(\nu+1)/2}$$

Similarly, if the outcome y is multivariate Student t of dimension q with dispersion Σ, and $\pi(\beta, \Sigma) \propto |\Sigma|^{-1}$, the posterior is proportional to

$$|\Sigma|^{-(n+1)} \prod_{i=1}^{n} \left[1 + \frac{1}{\nu}(y_i - \beta x_i)\Sigma^{-1}(y_i - \beta x_i)\right]^{-(\nu+q)/2}$$

where Σ is a $q \times q$ dispersion matrix.

The equivalent scale mixture specification in either case involves unknown weight parameters ω_i that scale the overall variance or dispersion parameter(s) of the Normal. Thus, for a univariate outcome, the Student t model may be expressed as

$$\begin{aligned} y_i &= \beta x_i + e_i \\ e_i &\sim \mathrm{N}(0, \sigma^2/\omega_i) \\ \omega_i &\sim G(\nu/2, \nu/2) \end{aligned} \tag{3.23}$$

The multivariate version of this takes again ω_i as $G(\nu/2, \nu/2)$, and takes the ith vector observation y_i to be sampled from a multivariate Normal with dispersion matrix

$$\Sigma_i = \Sigma/\omega_i$$

Suspect observations (i.e. potential outliers) with small weights ω_i and large distances $y_i - \beta x_i$ from the regression model have effects

$$(y_i - \beta x_i)\Sigma_i^{-1}(y_i - \beta x_i)$$

on the posterior density down-weighted, with the degree of down-weighting usually being enhanced for smaller values of ν.

Other densities with heavier (or possibly lighter) tails than the Normal are obtained with alternative densities for the ω_i or other types of mixing of densities. For instance, Smith (1981) considers a model for calculating marginal likelihoods, which involves choosing between the Normal, Double Exponential or Uniform densities, with prior probabilities of 1/3 on each density. The M-estimators of Huber (1981), which form the base for much work on robust estimation, are based on densities

$$P(y|\mu, \sigma, \kappa) \quad \propto \quad \exp[-U(d)]$$

where $d = (y - \mu)/\sigma$, and

$$\begin{aligned} U(d) &= 0.5d^2 \quad \text{if} \quad |d| < \kappa \\ U(d) &= \kappa|d| - 0.5\kappa^2 \quad \text{if} \quad |d| \geq \kappa \end{aligned}$$

As κ tends to infinity, this form tends to the Normal while for κ near zero it approximates the double exponential.

3.6.1 Binary selection models for robustness

A variant of the scale mixture approach for metric dependent variables takes ω as binary with probability $\lambda = \Pr(\omega = 1)$ of selecting a Normal density with mean μ and standard deviation σ. On the other hand, if $\omega = 0$, then an overdispersed alternative is selected with the same mean but standard deviation $\kappa\sigma$, where κ considerably exceeds 1. For example, taking λ to be small, e.g. $\lambda \sim U(0, 0.1)$ and $\kappa \sim U(2, 3)$ allows protection against a low level of contamination (of up to 10% of the observations) and variance inflation in that contaminated component of between four and nine times the overall level. Setting λ to a very low level, e.g. $\lambda = 0.01$, and κ positive with unrestricted ceiling, allows for a small number of extreme outliers.

Outlier resistant models may be seen as allowing for measurement error or (for categorical outcomes) misclassification. Thus in regression for binary outcomes, an outlier in the response y can be seen as possibly due to a transposition $0 \rightarrow 1$ or $1 \rightarrow 0$. These two cases occur when (a) the observed outcome $y_i = 1$ despite a model probability for $y = 1$ being close to zero, i.e.

$$\Pr(y_i = 1|x_i) = \pi_i \approx 0$$

and (b) when $y_i = 0$ despite π_i being close to 1. If residuals are defined as $\eta_i = y_i - \pi_i$, then an outlier is indicated if the absolute residual is close to 1. The sensitivity of a binary regression to outliers in part depends on the assumed link (e.g. the logit and complementary log-log links have heavier tails than the probit).

Regardless of link, a model for resistant binary regression may also be specified, including a mechanism for transposition between 0 and 1. Let y_i be the recorded response, and also define $\tilde{\pi}_i = \Pr(\tilde{y}_i = 1|x_i)$, where $\tilde{y}_i$ is the true response. Following Copas (1988), assume a transposition[11] occurs with a small probability γ such that the probability of the actually recorded response being 1 is given by

$$\begin{aligned}\pi_i &= \Pr(y_i = 1|x_i)\\ &= \Pr(y_i = 1|\tilde{y}_i = 0)\Pr(\tilde{y}_i = 0|x_i) + \Pr(y_i = 1|\tilde{y}_i = 1)\Pr(\tilde{y}_i = 1|x_i)\\ &= (1 - \gamma)\tilde{\pi}_i + \gamma(1 - \tilde{\pi}_i) \qquad (3.24)\end{aligned}$$

where (for example)

$$\text{logit}(\tilde{\pi}_i) = \beta x_i$$

The likelihood ratio that y is an outlier (or more particularly, misrecorded as a result of transcription) as against it belonging to the data (or being a genuine observation) is then given by

$$R_i = y_i(1 - \tilde{\pi}_i)/\tilde{\pi}_i + (1 - y_i)\tilde{\pi}_i/(1 - \tilde{\pi}_i) \qquad (3.25)$$

3.6.2 Diagnostics for discordant observations

The search for robust or resistant fits in general linear models extends to consider outlying points in the design space (of the X variables) as well as outlying responses (y). Logistic models, binary or multiple, may be especially sensitive to such outliers. Adjusting for such outliers may help to avoid unjustified rejection of models in their entirety, or of particular explanatory variates within models, because of distortions due to a few unusual data points.

[11] In BUGs this involves the coding, for a single predictor $x[i]$:

```
y[i] ~ dbern(pstar[i])
pstar[i] <- (1-gamma)*p[i] + gamma*(1-p[i])
logit(p[i]) <- beta[1] + x[i]*beta[2]
```

Methods for outlier detection or for down-weighting influential observations may be based on appropriately defined residual terms (Copas, 1988). For example, for a logit regression for binary outcomes with probability π_i, the components of deviance are

$$d_i = -2\log(1-\hat{\pi}_i) \quad \text{when } y_i = 0$$
$$-2\log(\hat{\pi}_i) \quad \text{when } y_i = 1$$

and taking $\hat{\varepsilon}_i = y_i - E(y_i|x_i) = y_i - \hat{\pi}_i$ in line with a general definition of residuals gives

$$d_i = -2\log(1 - |\hat{\varepsilon}_i|) \tag{3.26}$$

One may modify the usual likelihoods or deviances to downweight influential observations, via influence functions $g(u)$ which penalise large values of u (Pregibon, 1982). Whereas maximum likelihood estimation is equivalently minimisation of $\Sigma_i d_i$, where d_i is the deviance component of the ith case, robust estimation instead minimises $\Sigma_i g(d_i)$, where $g(d_i) < d_i$ for large d_i, so lessening the influence of cases with large deviances. The $g(d_i)$ may be based on the estimated residuals, as in Equation (3.26).

Another approach to obtaining residuals from a binary regression involves the latent variable method of Albert and Chib (1993), which is equivalent either to probit or logit regression: for example, if cases have different and known weights under the probit option, then a homogenous variance of 1 may be replaced by variances (inverse of the weights) averaging 1.

One may also consider the effect of each observation on the fitted model; for example, regression coefficients may be sensitive to particular points with unusual configurations of design variables, $x_{i1}, x_{i2}, \ldots x_{ip}$. Thus estimates of a coefficient β when all cases are included may be compared with the same coefficient estimate $\beta[i]$ when case i is excluded (Weiss, 1994; Geisser, 1990). The differences

$$\Delta\beta_i = \beta[i] - \beta$$

may then be plotted in order of the observations. A cross-validatory approach to model assessment omitting a single case at a time therefore has the advantage not just of providing a pseudo marginal likelihood and pseudo Bayes factor, but of providing a measure of the sensitivity of the regression coefficients to exclusion of certain observations. One may obtain posterior summaries of the $\Delta\beta_i$, ascertain which are most clearly negative or positive, and so produce the most distortion of the all cases estimate β.

Example 3.12 Group means from contaminated sampling Chaloner (1994) and Chaloner and Brant (1988) discusses the identification of outliers at two levels within a one way analysis of variance problem. Specifically, Chaloner (1994) considers an outlier as an observation with residual ε_i exceeding a threshold κ appropriately defined. For Normal data with n observations, and setting

$$\kappa = \Phi^{-1}\{0.5 + 0.5(0.95^{1/n})\}$$

ensures that the prior probability of no outliers is 0.95. Chaloner then analyses data from Sharples (1990) using both inflated variance and mean shift contaminated mixture models for outliers (see Table 3.15 for the data concerned). For a one-way data set with $I = 5$ groups (and $n_i = 6$ observations within the ith group), within group errors were sampled with probability 0.1 from a gamma distribution with mean 5.5. In addition the group means used to generate the data were 25 for groups 1–4, but 50 for group 5.

Table 3.15 Data generated with contamination process (Sharples, 1990) (Outliers starred)

	Observations in Group						
Group	1	2	3	4	5	6	Average
1	24.80	26.90	26.65	30.93	33.77	63.31*	28.61
2	23.96	28.92	28.19	26.16	21.34	29.46	26.34
3	18.30	23.67	14.47	24.45	24.89	28.95	22.46
4	51.42*	27.97	24.76	26.67	17.58	24.29	24.25
5	34.12	46.87	58.59*	38.11	47.59	44.67	42.27

Three of the 30 observations are identified as *a priori* outliers (known to be sampled from the contamination gamma density). The question is then to identify the probability of outliers among the data $y_{ij}(i = 1,.., n_j; j = 1,.., J)$ and among the group means. Specifically, suppose

$$y_{ij} \sim \mathrm{N}(\theta_j, \tau_w^2)$$
$$\theta_j \sim \mathrm{N}(\mu, \tau_b^2)$$

Then the first and second stage residuals are defined as

$$\varepsilon_{ij} = (y_{ij} - \theta_j)/\tau_w$$

and

$$\varepsilon_j = (\theta_j - \mu)/\tau_b$$

and compared to κ above. If $\Pr(\varepsilon_{ij} > \kappa)$ or $\Pr(\varepsilon_j > \kappa)$ exceeds the prior probability of 0.05, then an outlier is indicated. Prior specification of τ_w^2 and τ_b^2 is important, as certain priors may allow excessive shrinkage (τ_b^2 too small). Note that the observed between group variance is around 60.

Consider first a prior (prior A) for the overall variance $\tau^2 = \tau_w^2 + I\tau_b^2$ and then the proportion assigned to within group variation decided by a ratio π with a flat beta prior. Prior B is used by Chaloner, namely

$$\Pr(\tau_w^2, \tau_b^2) \propto \tau_w^{-2}(\tau_w^2 + I\tau_b^2)^{-1}$$

which in BUGS involves a double grid prior scaled to ensure total mass of one. The grid takes account of the actual value of the between group variance to the extent that small overall variances (below 5) are excluded. Finally, prior C is a double grid prior with equal probability over the pairs of values of τ_w^2 and τ_b^2 in the grid.

Whatever the prior, it is clear that the likelihood for τ_b^2 is relatively flat, but prior C results in a higher mean estimate for between group variance than the other two. Despite this, probabilities of individual observation outliers are relatively similar regardless of the prior adopted as are the estimates of the true group means $\theta_1, \theta_2, .., \theta_5$. The probabilities that $y_{1,6}$ and $y_{4,1}$ are outliers are, respectively, {0.38, 0.028), {0.42, 0.037} and {0.35, 0.031} under the three priors, whereas Chaloner, who uses a Normal approximation in conjunction with a Laplace approximation, finds values of 0.44 and 0.05. Priors A and B give a posterior probability of group 5 being an outlier of around 0.05 to 0.06; this compares to 0.019 cited by Chaloner.

Example 3.13 Stack loss To illustrate the effect of robust alternatives to the Normal for metric outcomes based on the Student t density, consider $n = 21$ points in the classic data set for stack loss, y and predictors x_1 = air flow, x_2 = temperature and x_3 = acid. A simple Normal errors model gives an estimated equation, with posterior means and standard deviations

$$y = -43.6 \; + 0.72x_1 + 1.28x_2 - 0.11x_3$$
$$(10.5) \quad (0.12) \quad (0.32) \quad (0.13)$$

Lange *et al.* (1989) show a small improvement in likelihood in moving from a Normal (degrees of freedom $\nu = \infty$) to a Cauchy density ($\nu = 0.5$), with the maximum likelihood estimate of the degrees of freedom provided by $\nu = 1.1$. Here an exponential prior for ν is assumed with mean η, which is itself assigned a uniform prior[12] between 0.01 and 1. It may be noted that a proper prior is needed on ν to avoid relapsing to the Normal (see Geweke, 1993, p. S27).

Adopting this approach (Model B in Program 3.13) gives a median value of around 1.7 for ν, with the credible interval ranging from 0.5–60. The estimated equation is now

$$y = -39.4 + 0.81x_1 + 0.71x_2 - 0.09x_3$$
$$(7.6) \quad (0.12) \quad (0.37) \quad (0.11)$$

so that the coefficient on x_2 is considerably reduced.

We then take the scale mixture approach to the Student t but with ν known (as 1.7). This shows the lowest weights (ω_i as in (3.23)), namely 0.78 and 0.67, for observations 4 and 21. Adopting the MLE value of 1.1 instead gives the lowest weights (all under 0.25) on observations 3, 4 and 21 (cf. Lange *et al.*, 1989, p. 883).

Finally Model D in[13] Program 3.13 also uses a scale mixture approach, but with ν also unknown. The outcome of this model, from a run of 10 000 iterations, is a median for ν of 1.3, and weights of 0.14 and 0.10 on observations 4 and 21. The coefficient on x_2 is further reduced to average 0.61 (compare Lange *et al.*, 1989, Table 1).

Example 3.14 Leukaemia survival To illustrate the transposition model (3.24) for a binary outcome, we consider the leukaemia data of Fiegl and Zelen (1965), but with response for $n = 33$ subjects being defined according to whether they survived for a year or more ($y = 1$ for deaths at more than a year). The two covariates are x_1 = white blood cell count and x_2 = positive or negative AG (presence or absence of certain morphologic characteristics in the white cells). A standard logistic regression shows both covariates negatively related to extended survival (Table 3.20), with the WBC coefficient being −0.04.

In adopting the alternative model allowing for possible contamination, it is to be noted that Copas (1988, p. 245) found by maximum likelihood methods that a value of $\gamma = 0.003$ in (3.21) leads to a large increase in the absolute size of the WBC coefficient. This means the predicted longer term survival chances of patients with low WBC counts are

[12] This is broadly equivalent to assuming the degrees of freedom is between 1 and 100.
[13] Run in version 1.2.

Table 3.20 Leukaemia data, logit model and logit under transposition, parameter summary

	Mean	St. devn.	2.5%	Median	97.5%
Standard Model					
Intercept	1.073	0.708	−0.231	1.049	2.538
WBC	−0.041	0.022	−0.092	−0.038	−0.005
Negative AG	−2.49	1.036	−4.688	−2.432	−0.601
Transposition Model					
Intercept	3.68	2.04	0.32	3.39	11.07
WBC	−0.32	0.18	−0.89	−0.30	−0.02
Negative AG	−3.38	1.89	−10.47	−3.12	−0.30
γ	0.0066	0.0049	0.0003	0.0055	0.0228

Likelihood Ratio (Misrecording vs. Genuine) (see equation (3.22))

	Mean	Median		Mean	Median
D_1	0.1268	0.0689	D_{18}	53.7300	2.6940
D_2	0.0969	0.0433	D_{19}	48.4700	1.7380
D_3	0.1906	0.1263	D_{20}	0.6457	0.4211
D_4	0.1342	0.0753	D_{21}	1.8150	0.9181
D_5	18.82	4.73	D_{22}	0.1486	0.0917
D_6	1.2400	0.7816	D_{23}	0.4155	0.2796
D_7	1.0180	0.6772	D_{24}	0.1177	0.0680
D_8	0.5933	0.1937	D_{25}	0.0260	0.0047
D_9	−39.61	5.65	D_{26}	0.0120	0.0005
D_{10}	0.3770	0.2835	D_{27}	0.0111	0.0004
D_{11}	0.8145	0.5736	D_{28}	0.0091	0.0002
D_{12}	0.0970	0.0022	D_{29}	0.0130	0.0007
D_{13}	0.0809	0.0010	D_{30}	0.0206	0.0026
D_{14}	0.0114	0.0001	D_{31}	0.0017	0.0001
D_{15}	0.0114	0.0001	D_{32}	0.0011	0.0001
D_{16}	0.0383	0.0001	D_{33}	0.0011	0.0001
D_{17}	8966	9999			

increased. However, further increases in γ had a relatively small impact. Therefore, a prior on γ may be set drawing on this analysis, and specifically

$$\gamma \sim E(300)$$

A three chain run of 25 000 iterations shows convergence from around 5000 iterations, and the summary is based on iterations 5000–25 000. The posterior mean for the WBC coefficient now stands at −0.34, with a 95% interval confined to negative values. The posterior mean for γ itself is just under 0.007. The likelihood ratios (3.25) are highest (hence chance of outlier status greatest) for observation 17, which has $y = 1$ (the patient survived 65 weeks) but a WBC measure of 100. The coefficient on the dummy index for AG status is also increased in absolute size, though like the WBC coefficient is estimated less precisely (i.e. the posterior standard deviation is increased over the standard logit).

Example 3.15 Travel for shopping To illustrate alternative approaches to robustness with binary outcomes to the contamination/misclassification model, consider data from Guy *et al.* (1983) for a panel survey of shopping behaviour. This involved 84 family households in suburban Cardiff, approximately equidistant from the city centre, with the response y_i being whether or not the household used a city centre store during a particular week. The predictors are income (Inc), household size (Hsz, for number of children) and whether the wife was working (WW). The first two covariates are respectively ordinal (with levels 1 for income under £1000 up to 8 for incomes over £15 000) and discrete – but both are taken as continuous.

Wrigley and Dunn (1986) consider issues of resistant and robust logit regression against a substantive background, whereby positive effects of income and working wife on central city shopping are expected, but a negative effect of household size. They argue that it is preferable to work with all the explanatory variables considered relevant and that exclusion of predictors because of one or two outliers or influential observations should be avoided. Let

$$\pi_i = \Pr(y_i = 1)$$

Wrigley and Dunn cite estimates from a maximum likelihood fit as follows (with standard errors in brackets)

$$\begin{aligned} \text{logit}(\pi_i) = &-0.72 + 0.14\,\text{Inc} - 0.56\,\text{Hsz} + 0.83\,\text{WW} \\ &\;(0.91)\quad (0.23)\qquad (0.19)\qquad (0.54) \end{aligned} \tag{3.27}$$

So the significance of the working wife variable is only marginal (i.e. income is significant at 5% only if a one-tail test is used).

Here we adopt mildly informative priors (cf. Chib, 1995) in a logit link model: N(0.75, 25) priors on β_{Inc} and β_{WW} are taken in line with an expected positive effect on central city shopping of income and female labour activity, while a N(−0.75, 25) prior on β_{Hsz} reflects the expected negative impact of household size. From a 10 000 iteration three chain run a stronger effect of income is obtained than in Equation (3.27). The analogous equation to that above, with posterior standard deviations in brackets, is

$$\begin{aligned} \text{logit}(\pi_i) = &-0.56 + 0.39\,\text{Inc} - 0.61\,\text{Hsz} + 1.07\,\text{WW} \\ &\;(0.97)\quad (0.25)\qquad (0.20)\qquad (0.59) \end{aligned}$$

The 90% credible intervals on both the income and working wife variables are entirely confined to positive values, though this is not true for the 95% intervals. The highest deviance components are obtained for observations 5, 55, 58, 71 and 83. These points account for about 20% of the total deviance (minus twice the log likelihood, which averages about −50). The highest deviance is for case 55. The conditional predictive ordinates are lowest for cases 55 and 71.

Similarly, case 55 is the highest average residual under the latent utility logit model of Albert and Chib (1993) – see Model (B) in Program 3.15. The logit is matched by a latent utility following Student t errors with 8 degrees of freedom (logistic errors are 'heavy tailed' like the Student t density). Trying Student t sampling with smaller degrees of freedom (e.g. $t(2)$ errors) makes little difference to any conclusions about credible intervals for income and working wife coefficients – the 95% intervals still straddle zero.

Note that either the Normal or Student *t* latent utility approach of Albert and Chib allow different known weights for each point. This approach might therefore be used to downweight certain observations using an influence function based on deviance or leverage contributions (see the Exercises).

As an illustration of potential distortion from particular data points Model C in Program 3.15 applies the full cross-validation methodology based on single case omission. The differences between β_{Inc} and $\beta_{Inc[i]}$ for income (Del.beta.Inc[] in Model C) show that major changes in this coefficient are caused by exclusion of particular points. The average value of β_{Inc} from the standard logit link model is 0.39 with posterior standard deviation of 0.25. Exclusion of case 71 raises this coefficient by over half this standard deviation, to around 0.58, while exclusion of case 58 raises it to around 0.47. By contrast excluding case 29 lowers the coefficient to around 0.26. There might therefore be grounds for excluding case 71 at least, as it figures as an outlier and is influential on the regression. As discussed above, other options to assess robustness of inferences may be used, which retain the suspect case(s), but model them via contaminated priors or discrete mixture regressions.

3.7 REVIEW

The Bayesian approach to linear regression with continuous outcomes, general linear modelling (including discrete outcomes), and nonlinear regression offers several advantages. These include the facility with which repeated sampling estimation may be used to select the most appropriate subsets of regressors (and hence best regression model) in situations where there is multicollinearity. Recent developments include modifications of the univariate selection procedures (such as in Kuo and Mallick, 1998) to allow multivariate responses, or to vary hyperparameters so that more or less parsimonious models are chosen (Brown *et al.*, 1998). The facility with which varying levels of prior information regarding regression slopes is another advantage. Prior information, such as from historical data or previous studies of similar data, may allow one to set a prior expected value of β based on past information (Birkes and Dodge, 1993), or to assign some weight to prior data D_0 relative to that assigned to likelihood of the data D of the current study (Chen *et al.*, 2000). A range of non-parametric regression methods – though some are possible to fit without a full Bayesian methodology – allows flexible regression modelling where effects are non-constant and this may be combined with predictor selection (Smith and Kohn, 1996). Both non-parametric regression and other approaches to regression (e.g. determining outliers via scale mixture Student *t* regression) allow for a range of sensitivity checks to standard modelling assumptions of linear effects and constant variances. On the other hand regression in certain settings, e.g. discrete mixture regression, is quite complex under a Bayesian approach with repeated sampling estimation leading to problems of identifiability. Issues such as prior specification and related questions of identification, model choice and robust inference, recur in the following chapters which consider more specialised data structures (multi-level data, time series data and panel data).

APPENDIX: DATA TABLES

Table 1 Ultrasonography ratings

		Diagnostic ratings					Predictors		
Tumor Site	Status (Metastasis)	R1	R2	R3	R4	R5	X1	X2	X3
Colon	N	27	17	2	0	0	0	0	0
Colon	Y	4	1	2	2	13	1	0	0
Breast	N	6	5	2	1	0	0	1	0
Breast	Y	0	2	0	2	5	1	1	1

Table 2 Kyphosis data

Sequence No	Kyphosis	Age	Number	Start	Age Group
1	absent	71	3	5	24
2	absent	158	3	14	57
3	present	128	4	5	47
4	absent	2	5	1	2
5	absent	1	4	15	1
6	absent	1	2	16	1
7	absent	61	2	17	22
8	absent	37	3	16	17
9	absent	113	2	16	40
10	present	59	6	12	21
11	present	82	5	14	30
12	absent	148	3	16	54
13	absent	18	5	2	9
14	absent	1	4	12	1
16	absent	168	3	18	59
17	absent	1	3	16	1
18	absent	78	6	15	27
19	absent	175	5	13	60
20	absent	80	5	16	28
21	absent	27	4	9	13
22	absent	22	2	16	11
23	present	105	6	5	38
24	present	96	3	12	34
25	absent	131	2	3	49
26	present	15	7	2	7
27	absent	9	5	13	5
29	absent	8	3	6	4
30	absent	100	3	14	36
31	absent	4	3	16	3
32	absent	151	2	16	55
33	absent	31	3	16	14
34	absent	125	2	11	45

(continues)

Table 2 *(continued)*

Sequence No	Kyphosis	Age	Number	Start	Age Group
35	absent	130	5	13	48
36	absent	112	3	16	39
37	absent	140	5	11	52
38	absent	93	3	16	33
39	absent	1	3	9	1
40	present	52	5	6	20
41	absent	20	6	9	10
42	present	91	5	12	32
43	present	73	5	1	26
44	absent	35	3	13	15
45	absent	143	9	3	53
46	absent	61	4	1	22
47	absent	97	3	16	35
48	present	139	3	10	51
49	absent	136	4	15	50
50	absent	131	5	13	49
51	present	121	3	3	44
52	absent	177	2	14	61
53	absent	68	5	10	23
54	absent	9	2	17	5
55	present	139	10	6	51
56	absent	2	2	17	2
57	absent	140	4	15	52
58	absent	72	5	15	25
59	absent	2	3	13	2
60	present	120	5	8	43
61	absent	51	7	9	19
62	absent	102	3	13	37
63	present	130	4	1	48
64	present	114	7	8	41
65	absent	81	4	1	29
66	absent	118	3	16	42
67	absent	118	4	16	42
68	absent	17	4	10	8
69	absent	195	2	17	63
70	absent	159	4	13	58
71	absent	18	4	11	9
72	absent	15	5	16	7
73	absent	158	5	14	57
74	absent	127	4	12	46
75	absent	87	4	16	31
76	absent	206	4	10	64
77	absent	11	3	15	6
78	absent	178	4	15	62
79	present	157	3	13	56
80	absent	26	7	13	12
81	absent	120	2	13	43
82	present	42	7	6	18
83	absent	36	4	13	16

REFERENCES

Akaike, H. (1978) A Bayesian analysis of the minimum AIC procedure. *Ann. Inst. Stat. Math.* **30**, 9–14.

Albert, J. and Chib, S. (1993) Bayesian regression analysis of binary and polychotomous response data. *J. Am. Stat. Assoc.*, **88**, 657–667.

Albert, J. and Pepple, P. (1989) A Bayesian approach to some overdispersion models. *Can. J. Stat.* **17**(3), 333–444.

Alqallaf, F. and Gustafson, P. (2001) On cross-validation of Bayesian models. *Canad. J. Stat.* **29**, 333–340.

Amemiya, T. (1981) Qualitative response models – a survey. *J. Economic Literature* **19**(4), 1483–1536.

Anderson, J. (1984) Regression and ordered categorical variables (with discussion), *J. Roy. Stat. Soc., Ser. B* **46**, 1–30.

Armstrong, B. and Sloan, M. (1989) Ordinal regression models for epidemiologic data. *Am. J. Epidemiology* **129**, 191–204.

Atkinson, A. (1969) A test for discriminating between models. *Biometrika* **56**, 337–347.

Bianco, A. and Yohai, V. (1996) Robust estimation in the logistic regression model. In: Rieder, H. (ed.), *Robust Statistics, Data Analysis, and Computer Intensive Methods.* Springer-Verlag, pp. 17–34.

Birkes, D. and Dodge, Y. (1993) *Alternative Methods of Regression.* Chichester: Wiley.

Breslow, N. (1984) Extra-Poisson variation in log-linear models. *Appl. Stat.*, **33**, 38–44.

Browne, M. (2000) Cross-validation methods. *J. Math. Psych.* **44**, 108–132.

Browne, W. and Draper, D. (2000) Implementation and performance issues in the Bayesian and likelihood fitting of multilevel models. *Computational Stat.* **15**, 391–420.

Brown, P., Vannucci, M. and Fearn, T. (1998) Multivariate Bayesian variable selection and prediction. *J. Roy Stat Soc., B*, **60**, 3.

Cameron, A. C. and Trivedi, P. (1986) Econometric models based on count data: comparisons and applications of some estimators and tests. *J. Appl. Econometrics* **1**, 29–54.

Carlin, B. and Polson, N. (1991) Inference for nonconjugate Bayesian models using the Gibbs sampler. *Canad. J. Stat.*, **19**(4), 399–405.

Carroll, R. and Ruppert, D. (1988) *Transformation and Weighting in Regression.* London: Chapman & Hall.

Carter, C. and Kohn, R. (1994) On Gibbs sampling for state space models. *Biometrika* **81**(3), 541–553.

Chaloner, K. (1994) Residual analysis and outliers in Bayesian hierarchical models. In: *Aspects of Uncertainty. A Tribute to D. V. Lindley*. New York: Wiley, pp. 149–157.

Chaloner, K. and Brant, R. (1988) A Bayesian approach to outlier detection and residual analysis. *Biometrika* **75**(4), 651–659.

Chatuverdi, A. (1996) Robust Bayesian analysis of the linear regression model. *J. Stat. Planning and Inference* **50**, 175–186.

Chen, M., Ibrahim, J. and Shao, Q. (2000) Power prior distributions for generalized linear models. *J. Stat. Planning and Inference* **84**, 121–137.

Chib, S., Jammalamadaka, S. and Tiwari, R. (1990) On Bayes Smoothing in a Time Varying Regression Model, Bayesian and likelihood methods. In: *Statistics and Econometrics: Essays in Honor of George A. Barnard*. New York: Elsevier/North-Holland, pp. 105–119.

Chib, S. (1995) Marginal likelihood from the Gibbs output. *J. Am. Stat. Assoc.* **90**, 1313–1321.

Chib, S. (2000) Bayesian analysis of correlated binary data. In: Dey, D., Ghosh, S. and Mallick, B. (eds.), *Generalized Linear Models: A Bayesian Perspective*. New York: Marcel-Dekker, pp. 113–131.

Collett, D. (1991) *Modelling Binary Data*. London: Chapman & Hall.

Congdon, P. (2000) A Bayesian approach to prediction using the gravity model, with an application to patient flow modelling. *Geographical Anal.*, **32**, 205–224.

Congdon, P., Smith, A. and Dean, C. (1998) Assessing psychiatric morbidity from a community register: methods for Bayesian adjustment. *Urban Stud.*, **35**(12), 2323–2352.

Copas, J. (1988) Binary regression models for contaminated data. *J. Roy. Stat. Soc., Ser. B* **50**, 225–265.

Dellaportas, P. and Smith, A. (1993) Bayesian inference for generalized linear and proportional hazards models via Gibbs sampling. *J. Roy. Stat. Soc., Ser. C* **42**(3), 443–459.
Dey, D., Gelfand, A. and Peng, F. (1997) Overdispersed generalized linear models. *J. Stat. Planning and Inference* **64**, 93–107.
Dey, D., Gelfand, A., Swartz, T and Vlachos, P. (1998) A simulation-intensive approach for checking hierarchical models. *Test* **7**(2), 325–346.
Dey, D., Ghosh, S. K. and Mallick, B. K. (eds.) (1999) *Generalized Linear Models: A Bayesian Perspective*. New York: Marcel Dekker.
DiCiccio, T., Kass, R., Raftery, A. and Wasserman, L. (1997) Computing Bayes factors by combining simulation and asymptotic approximations. *J Am. Stat. Assoc.* **92**, 903–915.
Dickey, J. (1976) A new representation of Student's *t* as a function of independent t's, with a generalization to the Matrix *t*. *J. Multivariate Anal.* **6**, 343–346.
Efron, B. (1986) Double exponential families and their use in generalized linear regression. *J. Am. Stat. Assoc.*, **81**, 709–721.
Engel, J. (1992) Modelling variation in industrial experiments. *J. Roy. Stat. Soc., Ser. C* **41**(3), 579–593.
Fahrmeir, L. and Tutz, G. (1994) *Multivariate Statistical Modelling Based on Generalized Linear Models*. Berlin: Springer-Verlag.
Fahrmeir, L. and Lang, S. (2001) Bayesian inference for generalized additive mixed models based on Markov random field priors. *Appl. Stat.* **50**, 201–220.
Feigl, P. and Zelen, M. (1965) Estimation of exponential survival probabilities with concomitant information. *Biometrics* **21**, 826–838.
Fernandez, C. and Steel, M. (1999) Multivariate Student-t regression models: pitfalls and inference. *Biometrika* **86**(1), 153–167.
Fischer, M. and Aufhauser, E. (1988) Housing choice in a regulated market – a nested multinomial logit analysis. *Geographical Anal.* **20**, 47–69.
Fornell, C. and Rust, R. (1989) Incorporating prior theory in covariance structure analysis: a Bayesian approach. *Psychometrika* **54**, 249–259.
Frühwirth-Schnatter, S. (2001) MCMC Estimation of classical and dynamic switching and mixture models. *J. Am. Stat. Assoc.* **96**, 194–209.
Gamerman, D. (1997) *Markov Chain Monte Carlo*. London: Chapman & Hall.
Ganio, L. and Schafer, D. (1992) Diagnostics for overdispersion. *J. Am. Stat. Assoc.* **87**, 795–804.
Geisser, S. and Eddy, W. (1979) A predictive approach to model selection. *J. Am. Stat. Assoc.* **74**, 153–160.
Geisser, S. (1990) Predictive approaches to discordancy testing. In: Geisser, S. *et al.* (eds.), *Bayesian and Likelihood Methods in Statistics and Econometrics*. Amsterdam: North Holland, pp. 321–336.
Gelfand, A. and Dey, D. (1994) Bayesian model choice: Asymptotics and exact calculations. *J. Roy. Stat. Soc., Ser. B* **56**(3), 501–514.
Gelfand, A. and Ghosh, S. (1998) Model choice: A minimum posterior predictive loss approach. *Biometrika* **85**(1), 1–11.
Gelfand, A. and Ghosh, S. (2000) Generalized linear models: a Bayesian view. In: Dey, D., Ghosh, S. and Mallick, B. (eds.), *Generalized Linear Models: A Bayesian Perspective*. New York: Marcel-Dekker, pp. 1–22.
Gelman, A., Carlin, J., Stern, H. and Rubin, D. (1995) *Bayesian Data Analysis. Texts in Statistical Science Series*. London: Chapman & Hall.
George, E. and McCullough, R. (1993) Variable selection via Gibbs sampling. *J. Am. Stat. Assoc.* **88**, 881–889.
Geweke, J. (1993) Bayesian treatment of the independent Student-t linear model. *J. Appl. Econometrics* **8S**, 19–40.
Griffiths, W., Hill, R. and Judge, G. (1993) *Learning and Practicing Econometrics*. New York: Wiley.
Guy, C., Wrigley, N., O'Brien, L. and Hiscocks, G. (1983) *The Cardiff Consumer Panel*. UWIST papers in Planning Research 68.
Hastie, T. and Tibshirani, R. (1990) *Generalized Additive Models*. London: Chapman & Hall.
Huber, P. (1981) *Robust Statistics*. New York: Wiley.
Ibrahim, J. G. and Chen, M. (2000) Power prior distributions for regression models. *Stat. Sci.*, **15**, 46–60.

Jones, P. N. and McLachlan, G. (1992) Fitting finite mixture models in a regression context. *The Australian J. Stat.* **34**, 233–240.

Kass, R. and Raftery, A. (1995) Bayes factors. *J. Am. Stat. Assoc.* **90**, 773–795.

Kuo, L. and Mallick, B. (1998) Variable selection for regression models. *Sankhya* **60B**, 65–81.

Kuo, L. and Peng, F. (1999) A mixture model approach to the analysis of survival data. In: Dey, K. D., Ghosh, S. K. and Mallick, B. K. (eds.), *Generalized Linear Models: A Bayesian Perspective*. New York: Marcel Dekker.

Lang, J. (1999) Bayesian ordinal and binary regression models with a parametric family of mixture links. *Computational Stat. and Data Anal.* **31**(1), 59–87.

Lange, K., Little, R. and Taylor, J. (1989) Robust Statistical Modeling Using the t-distribution. *J. Am. Stat. Assoc.* **84**, 881–896.

Laud, P. and Ibrahim, J. (1995) Predictive model selection. *J. Roy. Stat. Soc., Ser. B* **57**(1), 247–262.

Lawless, J. (1987) Negative binomial and mixed Poisson regression. *Can. J. Stat.* **15**(3), 209–225.

Lindley, D. V. and Smith, A. F. M. (1972) Bayes estimates for the linear model (with discussion). *J. Roy. Stat. Soc., Ser. B* **34**, 1–41.

McCullagh, P. (1980) Regression models for ordinal data. *J. Roy. Stat. Soc., Ser. B* **42**, 109–142.

McCullagh, P. and Nelder, J. (1989) *Generalized Linear Models* London: Chapman & Hall/CRC.

Mallick, B. and Gelfand, A. (1994) Generalized linear models with unknown link functions. *Biometrika* **81**, 237–245.

Nelder, J. and Pregibon, D. (1987) An extended quasi-likelihood function. *Biometrika* **74**, 221–232.

Nelder, J. and Lee, Y. (1991) Generalized linear models in the analysis of Taguchi-type experiments. *Appl. Stochastic Models in Data Anal.* **7**, 103–120.

Peng, F. and Hall, W. (1996) Bayesian analysis of ROC curves using Markov-chain Monte Carlo Methods. *Med. Decis. Making* **16**(4), 404–411.

Poirier, D. (1996) A Bayesian analysis of nested logit models. *J. Econometrics* **75**(1), 163–181.

Powers, D. and Xie, Y. (1999) *Statistical Methods for Categorical Data Analysis*. London: Academic Press.

Pregibon, D. (1982) Resistant fits for some commonly used logistic models with medical applications. *Biometrics* **38**, 485–498.

Raftery, A. (1995) Bayesian model selection in social research. In: Marsden, P. (ed.), *Sociological Methodology 1995*. Oxford: Blackwell.

Raftery, A. (1996) Approximating Bayes factors and accounting for model uncertainty. *Biometrika* **83**, 251–266.

Rust, R. T. and Schmittlein, D. (1985) A Bayesian cross-validated likelihood method for comparing alternative specifications of quantitative models. *Marketing Sci.*, **4**, 20–40.

Sharples, L. (1990) Identification and accommodation of outliers in general hierarchical models. *Biometrika* **77**, 445–453.

Smith, M. and Kohn, R. (1996) Nonparametric regression using Bayesian variable selection. *J. Econometrics* **75**, 317–334.

Stone, M. (1974) Cross-validatory choice and assessment of statistical predictions (with discussion). *J. Roy. Stat. Soc., Ser. B* **36**, 111–147.

Stone, M. (1977) An asymptotic equivalence of choice of model by cross-validation and Akaike's criterion. *J. Roy. Stat. Soc., Ser. B* **39**, 44–47.

Tosteson, A. and Begg, C. (1988) A general regression methodology for ROC curve estimation. *Med. Decision Making* **8**, 204–215.

Turner, T. (2000) Estimating the propagation rate of a viral infection of potato plants via mixtures of regressions. *J. Roy. Stat. Soc., Ser. C, Appl. Stat.* **49**(3)

Verdinelli, I. and Wasserman, L. (1991) Bayesian analysis of outlier problems using the Gibbs sampler. *Stat. and Comput.* **1**, 105–117.

Wedel, M., DeSarbo, W., Bult, J. and Ramaswamy, V. (1993) A latent class Poisson regression model for heterogeneous count data. *J. Appl. Econometrics* **8**, 397–411.

Weiss, R. (1994) Pediatric pain, predictive inference, and sensitivity analysis. *Evaluation Rev.* **18**, 651–677.

West, M. (1985) Generalized linear models: Scale parameters, outlier accomodation and prior distributions. *Bayesian Statistics 2, Proc. 2nd Int. Meet.* 531–558.

Winkelmann, R. and Zimmermann, K. F. (1995) Recent developments in count data modeling: theory and applications. *J. Econ. Surv.* **9**, 1–24.

Wood, S. and Kohn, R. (1998) A Bayesian approach to robust binary nonparametric regression. *J. Am. Stat. Assoc.* **93**(441), 203–213.

Wrigley, N. and Dunn, R. (1986) Diagnostics and resistant fits in logit choice models. In: Norman, G. (ed.), *Spatial Pricing and Differentiated Markets*. London Papers in Regional Science 16. London: Pion.

EXERCISES

1. For the ship damage data of Example 3.3, try allowing for extravariability by adding a random effect in the log link for the regression mean. How do regressor effects then compare with the Poisson model? How might robustness to outliers be gained under this approach? Try assessing gain in fit by the effective parameter method using the Poisson deviance (Chapter 2).

2. Also in Example 3.3, assess the expanded variance function (Model C) against the regression variance function (Model D) via the effective parameters method and predictive loss methods. Under the predictive loss approach try alternate values of k such as $k = 5$, $k = 10$, and assess stability of inferences on model choice.

3. Following Example 3.5 compare models for schizophrenia cases in small areas involving (a) the deprivation index as sole predictor and (b) the deprivation and anomie indices. Evaluate model choice using classical fit measures (i.e. deviances), the DIC method, and pseudo marginal likelihood based on single case omission.

4. In Example 3.6 add code for the quantities s_1 and s_2 comparing risks of infection when additional risks are present, and assess the probability that s_1 exceeds s_2.

5. Consider the influence discrepancy functions in Example 3.7 if the four cases 5, 59, 61 and 80 are excluded simultaneously.

6. Suggest how the binomial link mixture model (Section 3.3.4) could be adapted to modelling an unknown extra-variation parameter κ in the Poisson link as in Equation (3.5), with $-1 \leq \kappa \leq 1$. Try this approach with the ship damage data of Example 3.3.

7. In Example 3.9, try the probit model for the cumulative probability of diagnosis of metastasis, using the Albert–Chib sampling method.

8. In Example 3.11 consider a logit link and direct Bernoulli sampling (i.e. without referring to an underlying continuous variable) to reproduce the linear and GAM models. Also, try a model with a random walk prior on the age effect, such that the Bernoulli probability is

$$\mu_t = \beta_0 + \beta_1 x_{1t} + \beta_2 x_{2t} + \beta_{3t} x_{3t}$$

The prior on β_{3t} will need to take account of the varying gaps between (grouped) ages.

9. In Example 3.15, the code for Model A includes the influence function of Bianco and Yohai (1996), where

$$\begin{aligned} g(d) &= d - d^2/c \quad (d \leq c) \\ &= c/2 \qquad (d > c) \end{aligned}$$

d_i is the deviance contribution of case i, and $c = -\log(\kappa)$, and κ is small (between 0.01 and 0.05). In Program 3.15, c is set to $3.6 \approx -\log(0.03)$. Try using the posterior means on the weights $g(d_i)/d_i$, scaled to average 1, as alternative precisions in the Albert–Chib method with Student t sampling via scale mixing. How does varying κ to 0.01 affect inferences?

CHAPTER 4

Analysis of Multi-Level Data

4.1 INTRODUCTION

The behaviour and experience of, and outcomes for, individual actors in society is defined by the institutions or social groups to which they belong. Failing to take account of such contextual effects on individual level outcomes may lead to what are known as atomic or atomistic errors of inference (Courgeau and Baccaini, 1997). However, analysis of aggregates of individual outcomes (e.g. studies linking area health status to area income) may be subject to a so-called ecological fallacy by neglecting to include the mode of operation of risk factors at the individual level. Hence, an analytic approach simultaneously taking account of different levels of aggregation but considering the outcome at an individual level may be preferred and less subject to either fallacy. The statistical task is then 'to partition the variation in a response variable as a function of levels in the hierarchy and relate this variability to descriptions of the data structure' (Browne *et al.*, 2000).

Examples where this type of approach is relevant include pupil attainment within schools, psychopathological behaviour within families (Martinius, 1993), or illness or mortality rates among residents classified by county or local authority (Brodsky *et al.*, 1999). In these cases, pupils, offspring and individual community residents define the lower level, known as level 1, of the data hierarchy, and the groups they are members of define the higher level or level 2. Further levels of aggregation are possible, with lower level clusters or groups (e.g. classes of pupils) arranged within schools at level 3, which are further arranged according to local education authority at level 4. Another multi-level structure is defined by repeat observations on an individual subject (e.g. in growth curve models), in which the observations on the same individual constitute the first or most basic level of analysis and the subjects themselves are at level 2.

The model will need to reflect typical features of multi-level data. Thus, individuals within groups tend to be more alike than individuals across groups. Similarities between pupils taught in the same school, or between residents of the same neighbourhood, then generate a correlation at group level. The schools or neighbourhoods constitute 'clusters' of the level 1 units and the correlation of responses within clusters is denoted intra-cluster correlation (Goldstein, 1995; Humphrey and Carr-Hill, 1991). A multi-level analysis may then seek to identify the effects of both cluster and individual level variables on the individual level outcome.

Applied Bayesian Modelling P. Congdon
 ISBN: 0-471-48695-7

Such an analysis will take account of the positive correlation within clusters which otherwise will result in understating the standard deviation of the effects of cluster variables. As well as introducing known influences at cluster level, we may introduce shared unobserved variables within clusters which to some degree model the intra-cluster correlation among the responses. We are obviously also interested in the impact of individual attributes on the outcome, and whether they vary across clusters. A multi-level model will provide estimates of cluster-specific regression estimates, namely varying intercepts and slopes, that use the information from subjects within each cluster but also borrow strength from other clusters.

Consider the case of spatial epidemiology, where both individual and contextual factors are relevant in explaining variations in individual health outcomes. The context is provided by various levels of geographic aggregation, and effects of area variables have been termed ecologic effects by Blakely and Woodward (2000), Morgenstern (1998) and others. Health inequalities are associated with many individual risk factors (e.g. marital status, ethnicity, social class), but are also known to show wide spatial variation beyond that which can be attributed to differences in the social composition of the population. An example is the North-South contrast in both mortality and illness rates in Britain, which is not wholly explained by differences in social and demographic structure (see Example 4.2).

At the cluster level therefore some contrasts in mean rates may be *compositional* – merely reflecting the aggregate effect of the socio-economic composition of each area's or region's residents.[1] However, there may additionally be genuine *contextual* effects: for instance, if the health experience or behaviour of an individual of a given type (e.g. as defined by age, class, etc.) varies across regions (Duncan *et al.*, 1999). Morgenstern (1998) gives some interesting examples of contextual effects: thus, residence in a mainly Protestant area may raise the suicide risk among non-Protestants, so that there is an interaction between individual religion and the religious composition of areas of residence. If, for instance, x denoted individual religion (Protestant or not), X the proportion Protestant in areas of residence, then prediction of y (suicide risk) might involve a model $y = b_0 + b_1 x + b_2 X + b_3 xX$, where b_2 reflects a direct contextual effect and b_3 an indirect one, through the interaction just mentioned.

One possible approach to such contrasts across contextual settings may be denoted the tabulation method: this involves separate regression analyses for each cluster and comparison of the resulting effect estimates across clusters. Separate analyses for each cluster are, however, not the best way to study the interplay of contextual and compositional effects, since they neglect within cluster dependence.

Multi-level analysis often raises further complex statistical issues. For instance, cluster effects (such as the effect of school variables on pupil attainment) may involve small sample sizes and possible clusters which are outliers. Hence, robust methods may be needed to assess inferences and here Bayes methods (e.g. for heavy tailed alternatives to say normally distributed cluster effects) come into play. There may be issues of heteroscedasticity where for example the level one variance is not constant but depends on explanatory variables – an example might be a residual variance at level 1 differing between boy and girl pupils. Multilevel outcomes may well also be multivariate (e.g. two types of exam score for pupils), and the repeated observations of scores on the same

[1] Note that this term does not refer to a decomposition of variance in the sense of spatial vs unstructured errors in spatial outcomes as will be considered in Chapter 7. However, spatial correlation of cluster effects (e.g. among varying intercepts and slopes) may be relevant when the clusters are geographic areas.

subject constitute the first level in this instance. Further, while many multi-level applications consider only inference within the sample or survey, one may also use multi-level approaches in making inferences beyond the survey, specifically for small domain estimation. The goal may be to make inferences about the average outcome in a domain (formed, say, by cross-classifying demographic attributes with relatively low scale geographic level) where the survey data themselves contain only a few units, perhaps none, from that domain. This may involve referring to other sources of data on the domains (e.g. Census data as well as the survey results themselves).

A Bayesian estimation perspective, especially one based on MCMC methods, may have advantages over maximum likelihood estimation based on (say) iterative generalised least squares. Thus, GLS estimates may understate random effect variances because not all sources of sampling uncertainty in the relevant parameters are allowed for in their derivation. This is particularly so for small numbers of clusters (e.g. the number of classes or schools), small cluster sizes, and in unbalanced hierarchical data sets, where cluster sizes differ between clusters. GLS methods may also suffer from convergence problems in these situations. A Bayesian model based on iterative sampling may be useful in providing simplified 'significance tests' on parameters or derived quantities, based on counting samples where the hypothesis holds. By contrast, a maximum likelihood approach typically rests on asymptotic normality and relevant standard errors for derived quantities (e.g. for a difference in level 1 variances if these are made functions of regressors) may be difficult to derive. However, although the Bayesian approach may be advantageous in some respects, there may be the sensitivity to prior specifications, especially for small samples, or small numbers of clusters at higher levels. This may be the case, for instance, regarding covariation of the random effects at different levels (Daniels, 1999; Browne and Draper, 2000; Daniels and Kass, 1999).

This chapter adopts a Bayes perspective in terms of the application of multi-level concepts to both continuous and discrete data (Section 4.2) Examples of different outcome types are provided by two level models applied to continuous pupil attainment and binary health outcome data. Then Section 4.3 considers multi-level models including heteroscedasticity at one or more levels, with variances dependent on continuous or categoric regressors. This is illustrated with the pupil attainment data set and with a binary attitude outcome from a survey. Questions of robust inference are raised in multi-level models, where multivariate normality of cluster and higher level effects is a typical default, but may be problematic if cluster sample sizes are small (Seltzer, 1993). This question is considered in Section 4.4. The chapter concludes in Sections 4.5 and 4.6 by considering multivariate outcomes within a multi-level context, and the application of multi-level concepts to derive population wide predictions within domains defined by survey variables.

4.2 MULTI-LEVEL MODELS: UNIVARIATE CONTINUOUS AND DISCRETE OUTCOMES

The nesting of observations allows considerable scope for differentiating or indexing regression and or error variance effects, guided both by subject matter indications and by statistical criteria such as model parsimony and identifiability. For example, suppose we have pupils arranged by class and school (levels 2 and 3, respectively); then the effect of the level 2 variable, classroom size, on a continuous attainment score (the level 1 outcome) may be affected by school resources (at level 3). The effect on attainment of a level 1 variable, such as pupil ability, may be differentiated both by class and school.

In terms of formal model specification, consider simply the nesting of pupils within classes. Suppose we have a two level data set with continuous observations Y_{ij} in classes $j = 1, .. J$ at level 2, and pupils $i = 1, .. n_j$ within classes at the lowest level. Predictors may be defined at each level, say $X_{hij}(h = 1, .. p)$ at level 1 and $Z_{hj}(h = 1, .. q)$ at level 2; an example of the first type of predictor might be pupil ability or gender, and of the second, class size or teacher style. In a multi-level regression model, covariate effects and errors may be specified at each level and are potential cumulative sources of explanation or error in the outcome.

Thus, in predicting the outcome at level 1, one may define error terms v_j at level 2, and u_{ij} at level 1 (both with zero mean). Suppose $p = 2$ with X_{1ij} = ability and X_{2ij} = gender (1 = girls, 0 = boys). Then a two-level model for continuous outcomes with fixed impacts of the level 1 predictors might take the form

$$Y_{ij} = \alpha + \beta_1 X_{1ij} + \beta_2 X_{2ij} + v_j + u_{ij}$$

Combining the intercept and level 2 error gives

$$Y_{ij} = \alpha_j + \beta_1 X_{1ij} + \beta_2 X_{2ij} + u_{ij} \tag{4.1}$$

so that the α_j are varying regression intercepts over the level 2 units. The mean response at level 1 would be conditional on the random intercept:

$$E(Y_{ij}|\mathbf{X}_{ij}, \alpha_j) = \beta_1 X_{1ij} + \beta_2 X_{2ij} + \alpha_j$$

So the centred Normal version of this two level continuous data model, where $u_{ij} \sim \text{N}(0, \sigma^2)$ would be

$$Y_{ij} \sim \text{N}(\beta_1 X_{1ij} + \beta_2 X_{2ij} + \alpha_j, \sigma^2)$$

Note that this is the form for Gaussian or Student t data used in BUGS (remembering of course that BUGS uses a different parameterisation of the Normal).

In this model the impact of level 1 predictors is constant across higher level contexts and the effect of the cluster is reflected only in different intercepts. Typically, the cluster effects (random intercepts) α_j would be exchangeable errors with no correlation structure, that is unstructured 'white noise'; however, if the clusters were geographic areas one might envisage them being spatially correlated. One may well seek to explain varying intercepts in terms of the characteristics of clusters (or characteristics of higher level groupings of the clusters themselves).

The interpretation of the regression coefficient β_h for the hth predictor X_{hij} in (4.1) is then as a change in the expected response for a unit change in that predictor, with the error term at level 2 held constant. By contrast, multi-level models commonly differentiate the effects of the level 1 predictors according to clusters (here school classes) $j = 1, .. J$. So the effect of pupil ability or gender may be differentiated as follows

$$Y_{ij} = (\beta_1 + \delta_{1j})X_{1ij} + (\beta_2 + \delta_{2j})X_{2ij} + \alpha_j + u_{ij}$$

where δ_{1j} and δ_{2j} express the differential effect (with mean zero) that a pupil's class has on the impacts on attainment of pupil ability or gender. One might also, conflating the fixed and zero-centred random effects δ_{hj}, write this model as

$$Y_{ij} = \beta_{1j} X_{1ij} + \beta_{2j} X_{2ij} + \alpha_j + u_{ij} \tag{4.2}$$

As noted by Clayton (1996), the Bayesian viewpoint means that there is no longer a need to partition a parameter effect into fixed and random components. So we have a model with varying intercepts and slopes. To enable pooling of strength typically involves assuming these effects are random, namely drawn from a parametric population wide distribution over school classes – though fixed effects approaches may arguably have more validity in certain circumstances (Rice *et al.*, 2000).

A common assumption is for univariate or multivariate normal errors for the higher level effects, namely α_j in (4.1), and $\{\alpha_j, \beta_{hj}\}$ in (4.2). Thus, the priors for varying intercepts and slope effects for schools might be taken as independent Normals

$$\alpha_j \sim N(A, \tau_\alpha)$$

$$\beta_{hj} \sim N(B_h, \tau_{\beta \cdot h}) \quad h = 1, .. p$$

where A and the B_h are known constants. Alternatively, a $p + 1$ multivariate normal density could be used allowing for covariation between intercepts and slopes. As mentioned above, more informative from a subject matter viewpoint may be to relate variations in the parameters $\{\alpha_j, \beta_{hj}\}$ to cluster predictors Z_j. As considered further below, there may also be gains from modelling the variances as functions of predictors. In applications where the clusters are geographic areas, the multivariate density for $\{\alpha_j, \beta_{hj}\}$ might allow for spatial correlation (see, for example, Leyland *et al.*, (2000), and the discussion in Chapter 7).

While the above notation implies a nested arrangement of the data, it is often convenient, especially with unequal n_j in each cluster, to arrange the data in terms of a single subject index, so that a univariate outcome y would be arranged in terms of a vector of length $T_n = \Sigma n_j$. A vector of cluster membership indices, with values between 1 and J would also be of length N. This type of arrangement is also useful for crossed rather than nested data structures, for example, pupil data by school and area of residence.

4.2.1 Discrete outcomes

Multi-level analysis of discrete outcomes may be carried out in the appropriate linked regression (e.g. via a log link for a Poisson dependent variable or logit link for a binomial variable) (Goldstein, 1991). This may involve introducing into the regression structure such forms of random variation that are needed both to describe the hierarchical data structure and account for heterogeneity beyond that expected under the model form. For example, consider a two level Poisson count Y_{ij} for $i = 1, .. n_j$ units within $j = 1, .. J$ groups, and with $T_n = \Sigma n_j$. Denoting the Poisson means as μ_{ij} and assuming a single level 1 predictor X_{ij}, a log-linear regression may be specified

$$\log(\mu_{ij}) = \alpha_j + \beta_j X_{ij}$$

or with a level 1 error term

$$\log(\mu_{ij}) = \alpha_j + \beta_j X_{ij} + u_{ij} \tag{4.3}$$

The first form assumes heterogeneity will be largely accounted for by the group specific intercepts and slopes, while the second allows an unstructured error with variance σ^2 to account for residual heterogeneity beyond that associated with the Poisson regression (Congdon, 1997, Ghosh *et al.*, 1998).

This sort of over-dispersion is apparent in deviances at the posterior mean exceeding the available degrees of freedom (McCullagh and Nelder, 1989). For a binary or binomial outcome, there is a similar choice. Thus, suppose Y_{ij} is binomial, with $Y_{ij} \sim B(\pi_{ij}, R_{ij})$, where R_{ij} is the total number of events, and Y_{ij} is the number of 'successes'. Then a logit link model with cluster only intercepts

$$\text{logit}(\pi_{ij}) = \beta X_{ij} + \alpha_j$$

may need (if there is still over-dispersion) to be expanded to include level 1 variability, so that

$$\text{logit}(\pi_{ij}) = \beta X_{ij} + \alpha_j + u_{ij} \tag{4.4}$$

It may be noted though that a suitable prior for the level 1 variance in a Poisson or binomial model, as in Equations (4.3) or (4.4) may require care to avoid large values of σ^2 since otherwise the random effects tend to produce too close fit to the data (Johnson and Albert, 1999, p. 113). Thus Johnson and Albert suggest for their application an informative inverse gamma prior for the variance, with parameters 5 and 1.5, which result in most of the mass for the variance being concentrated between values 0.1 and 1.5.

One advantage of simulation based Bayesian estimation and inference is the ease of direct modelling of Poisson or binomial outcomes without resorting to weighted least squares approximations. For example, over-dispersion may be modelled via Normal or possibly Student t random errors in the log-link, as in Equation (4.3). One might also adopt conjugate forms to model the level 1 variation, with a Poisson outcome for instance being modelled as

$$Y_{ij} \sim \text{Poi}(\mu_{ij}\gamma_{ij})$$

where the γ_{ij} are taken as gamma variables with mean 1. Alternatively, for suitably large counts Normal approximations to the binomial or Poisson may be used. This implies Normal sampling (or possibly Student t sampling) but with a variance function appropriate to the form of the data. Thus for a binomial outcome, with Y_{ij} events occurring in R_{ij} at risk

$$Y_{ij} \sim \text{N}(\pi_{ij}R_{ij}, V_{ij})$$

with

$$V_{ij} = \phi^2 R_{ij}\pi_{ij}(1 - \pi_{ij}) \tag{4.5}$$

and with the regression for the π_{ij} involving a logit or probit link. In Equation (4.5) $\phi^2 = 1$ would be expected (approximately) if the level 1 variation were binomial, whereas heterogeneity beyond that expected under the binomial yields ϕ^2 well above 1.

Example 4.1 Language scores in dutch schools As an example of the two level situation for continuous data, consider data on language scores in 131 Dutch elementary schools for $T_n = 2287$ pupils in grades 7 and grade 8, and aged 10 and 11 (Snijders and Bosker, 1999). In each school a single class is observed, and so the nesting structure is of pupils within $J = 131$ classes. We are interested in the impact on language scores of pupil level factors such as IQ, and pupils social status (SES). Also relevant to explaining possible differences in intercepts and slopes (on IQ and SES) are class level variables: these include the class size, the average IQ of all pupils in the class, and whether the class is mixed over grades: thus COMB = 1 if the class includes both grade 7 and grade 8 pupils and COMB = 0 if the class contains only grade 8 pupils. These variables are denoted ClassSize[], IQ.class[]) and comb[] in Program 4.1. Following Snijders and Bosker, the pupil variables and the class variables are centred, and this aids in achieving earlier MCMC sampling convergence. Note that in Program 4.1 the data are arranged by subject in a 'single string' vector, with class memberships 1, 2... 131 in a vector of length 2287.

A model with random intercepts but fixed impacts of the two pupil level variables (Model A) is estimated first. The intercepts are explained by three class level variables: average IQ, mixed grade class or not and class size. So Model A is defined as

$$Y_{ij} \sim N(\mu_{ij}, \phi^{-1}) \quad i = 1, ..n_j; j = 1, ..J$$

$$\mu_{ij} = \alpha_j + \beta_1(X_{1ij} - \bar{X}_1) + \beta_2(X_{2ij} - \bar{X}_2)$$

$$\alpha_j \sim N(\nu_j, \phi_\alpha^{-1})$$

$$\nu_j = \gamma_1 + \gamma_2(Z_{1j} - \bar{Z}_1) + \gamma_3(Z_{2j} - \bar{Z}_2) + \gamma_4(Z_{3j} - \bar{Z}_3)$$

where X_1 and X_2 are pupil IQ and SES and Z_1, Z_2 and Z_3 are, respectively, class IQ, class grade type, and class size. The assumed priors are $\phi \sim G(0.001, 0.001)$, $\phi_\alpha \sim G(0.001, 0.001)$, $\beta_k \sim N(0, 10^7)$, $k = 1, 2$ and $\gamma_k \sim N(0, 10^7)$, $k = 1, 4$.

Posterior estimates are based on the second half of a three chain run of 5000 iterations. Initial values for β and γ in the three chains are provided by null start values (chain 1), start values in chain 2 equal to the posterior means from the trial run, and values in the third chain equal to the 97.5% point from the trual run. Starting values for the inverse variances are also based on the test run. Convergence by the Brooks–Gelman criteria (Brooks and Gelman, 1998) is evident at well under 1000 iterations, with mixing satisfactory in terms of rapid decline in the autocorrelations between parameter iterates at successive lags.

The parameter estimates for model A (Table 4.1) show that pupil IQ and SES both have clearly positive impacts on attainment. The intercepts vary between classes in such a way that classes with higher average IQs and containing only grade 8 pupils have higher attainments. There is also a weak negative impact γ_4 of class size, with 95% interval (from −0.14 to 0.06) biased towards negative values. One option here would be to try a transform of the class size variable, since a negative value might be expected on subject matter grounds.

A second model adopts random slopes on pupil IQ and SES; the same class level predictors as in Model A are used to explain variations in the slopes on these two pupil predictors. Thus in Model B

$$Y_{ij} \sim N(\mu_{ij}, \phi^{-1}) \quad i = 1, ..n_j; j = 1, ..J$$

$$\mu_{ij} = \beta_{1j} + \beta_{2j}(X_{1ij} - \bar{X}_1) + \beta_{3j}(X_{2ij} - \bar{X}_2)$$

$$\beta_j \sim N_3(\nu_j, V \cdot \beta) \tag{4.6a}$$

$$\nu_{kj} = \gamma_{1k} + \gamma_{2k}(Z_{1j} - \bar{Z}_1) + \gamma_{3k}(Z_{2j} - \bar{Z}_2) + \gamma_{4k}(Z_{3j} - \bar{Z}_3) \tag{4.6b}$$

The varying class level parameters $\beta_j = (\beta_{1j}, \beta_{2j}, \beta_{3j})$ are taken to be trivariate Normal with means $\nu_j = (\nu_{1j}, \nu_{2j}, \nu_{3j})$. The initial values for the three chains are provided by the

Table 4.1 Dutch language tests

Model A					
	Mean	St. devn.	2.5%	50%	97.5%
Deviance Precisions	14930	16.98	14900	14930	14960
ϕ	0.025	0.001	0.024	0.025	0.027
$\phi.\alpha$	0.131	0.022	0.093	0.129	0.178
Level 1 Effects					
Pupil IQ	2.21	0.07	2.06	2.21	2.35
Pupil SES	0.16	0.01	0.13	0.16	0.19
Level 2 Model for Level 1 Intercepts					
Intercept	41.58	0.44	40.72	41.58	42.46
Class IQ	1.10	0.32	0.48	1.10	1.72
Combined Class	−2.10	0.82	−3.75	−2.10	−0.50
Class Size	−0.038	0.053	−0.142	−0.037	0.065

Model B					
	Mean	St. devn.	2.5%	50%	97.5%
Deviance	14860	24.26	14810	14860	14910
Level 2 Covariance Elements					
$V.\beta_{11}$	7.658	1.343	5.373	7.533	10.57
$V.\beta_{12}$	−0.537	0.295	−1.145	−0.528	0.023
$V.\beta_{13}$	−0.001	0.067	−0.132	−0.002	0.134
$V.\beta_{22}$	0.251	0.087	0.121	0.238	0.449
$V.\beta_{23}$	−0.013	0.014	−0.044	−0.012	0.012
$V.\beta_{33}$	0.030	0.005	0.021	0.029	0.042
Level 1 Precision					
ϕ	0.026	0.001	0.024	0.026	0.027
Level 2 Parameters (Equation (4.6))					
γ_{11}	41.58	0.49	40.63	41.57	42.54
γ_{12}	2.13	0.15	1.84	2.14	2.43
γ_{13}	0.166	0.036	0.096	0.166	0.238
γ_{21}	0.896	0.361	0.171	0.902	1.586
γ_{22}	−0.013	0.091	−0.189	−0.013	0.166
γ_{23}	−0.012	0.026	−0.064	−0.012	0.040
γ_{31}	−1.947	0.886	−3.696	−1.935	−0.228
γ_{32}	0.380	0.282	−0.139	0.374	0.915
γ_{33}	0.016	0.064	−0.110	0.016	0.142
γ_{41}	−0.044	0.055	−0.154	−0.044	0.062
γ_{42}	−0.006	0.017	−0.041	−0.006	0.025
γ_{43}	−0.002	0.004	−0.010	−0.002	0.006

same procedure as in model A except that the starting values for the precision matrix on the parameters $\{\beta_{j1}, \beta_{j2}, \beta_{j3}\}$ are obtained by random draws from the Wishart prior for this matrix. This prior is taken to have three degrees of freedom and an identity scale matrix.

A 5000 iteration run with three chains is used for estimation, with the summary based the second half of the run. Note that Bayes estimation suggests that there does appear to be variation in the impact of SES across the classes, and some suggestion of covariation between SES and IQ: by contrast, Snijders and Bosker, using a weighted least squares method, failed to gain convergence when this variation and covariation was allowed for.

However, the extended model calls into question the relevance of the full set of cluster variables Z_j in predicting variability in slopes, and suggests over-parameterisation; the exception is the parameter γ_{32} which represents the additional impact of child IQ on attainment in mixed classes where COMB $= 1$. Note that these findings point to reformulation of the regression model (4.6b) to explain variable slopes, and not necessarily to a drawback in the random intercepts and slopes assumption (4.6a). As Snijders and Bosker (1999, p. 77) point out, it is not necessary to use all the cluster variables Z_{hj} in explaining variability in the coefficients $\beta_{kj}, k = 1, .., p + 1$.

Example 4.2 Long term illness As an example of binary outcome at individual level, this example follow Shouls, Congdon and Curtis (1996) in considering variation in individual level chances of the binary outcome, namely being long term ill. The data used draw on a 2% sample of anonymised individual records (abbreviated as the SAR) from the 1991 UK Census, and nested within 278 local authority areas. The full analysis focused on the age group 15–64 and on a 10% sample of the SAR data itself, i.e. a 0.2% sample of the full Census, amounting to around 90 thousand males and females in this age group. We focus here on females aged 45–59, excluding cases with missing covariates; the covariates are age, non-white ethnicity, being married and being in lower skill manual occupations (social classes IV and V). Additionally, an indicator of multiple deprivation is included for each individual, and is a tally of yes/no responses according to whether the individual is unemployed or the household of the individual does not own their home, does not own their car, has no access to a separate bathroom, or lives at over 1 person per room. This tally S appears as the variable sumdep[] in Program 4.2, which is then transformed to the regressor $X = \log(1 + S)$. A quadratic term in age is used (see Program 4.2); other options might be linear or log-linear terms, though a linear effect does not reflect the fact that the chance of long term illness increases steeply towards the end of the 45–59 age band.

The study of Shouls *et al.* (1996) found that there were contextual impacts of individual deprivation: its effect was greater in (more affluent) areas with lower average illness levels. This conclusion was based on a bivariate normal model at local authority area level for the intercepts and slopes on individual deprivation. Note that the analysis of Shouls *et al.* was based on iterative weighted least squares. Here three models are considered:

- one with intercept only variation (model A);
- one which replicates the Shouls *et al.* analysis, but with a Bayes specification of the contextual effect just described (Model B); and
- Model C, in which area variability in the effect of social class is added.

Further to the work of Shouls *et al.*, a cluster level variable Z_1 is introduced into Models B and C: thus $Z_{1j} = 1$ if local authority j is in the North of England. This variable may assist in explaining intercept and slope variation.

It may be noted that varying effects across clusters implies two types of ecologic effect on health status: thus define $Y_{ij} = 1$ if individual i in local authority j is long term ill (and $Y_{ij} = 0$ otherwise). Then Model B states the following:

$$y_{ij} \sim \text{Bern}(\pi_{ij})$$

$$\begin{aligned}\text{logit}(\pi_{ij}) = \beta_{0j} + \beta_{1j}\text{Dep}_{ij} + \beta_2\text{AgeSq}_{ij} + \beta_3\text{LowSkill}_{ij} \\ + \beta_4\text{NonWhite}_{ij} + \beta_5\text{Married}_{ij}\end{aligned} \tag{4.7a}$$

where Dep is the transformed (and then centred) deprivation score, and the age squared variable is also centred. Model A has varying intercepts as in Equation (4.7), but a fixed effect of individual deprivation, $\beta_{1j} = \beta_1$.

The level 2 models for the variable intercepts and slopes in Model B are

$$\beta_{0j} = \gamma_{00} + \gamma_{01}Z_{1j} + \varepsilon_{0j} \tag{4.7b}$$

$$\beta_{1j} = \gamma_{10} + \gamma_{11}Z_{1j} + \varepsilon_{1j} \tag{4.7c}$$

where the $\varepsilon_j = (\varepsilon_{0j}, \varepsilon_{1j})$ are bivariate normal. (In model A, β_{0j} is univariate Normal.) Substituting (4.7b) and (4.7c) into (4.7a) then gives

$$\begin{aligned}\text{logit}(\pi_{ij}) = \gamma_{00} + \gamma_{01}Z_{1j} + \gamma_{10}\text{Dep}_{ij} + \gamma_{11}Z_{1j}\text{Dep}_{ij} + \beta_2\text{AgeSq}_{ij} \\ + \beta_3\text{LowSkill}_{ij} + \beta_4\text{NonWhite}_{ij} + \beta_5\text{Married}_{ij} + \varepsilon_{0j} + \varepsilon_{1j}\text{Dep}_{ij}\end{aligned}$$

Thus there is both a direct 'cross-level effect' of Z_1 on the individual outcome (with coefficient γ_{01}) and an indirect cross-level effect with coefficient γ_{11} produced by the interaction of individual deprivation and the cluster variable (here the North-South divide in England).

In both Models A and B, three chains[2] are run for 2500 iterations, with convergence in the fixed effects and elements of the cluster precision matrix (cluster variance in Model A) apparent at under 500 iterations and with satisfactory mixing (fast decay in lagged dependence in sampled parameter values); the posterior summaries are based on iterations 500–2500.

Model A shows the anticipated positive impacts of age on the chance of illness and a lower rate for married people. It is notable that social class effects on illness exist even after allowing for individual deprivation. Model A shows significant variation between local authorities in long term illness rates after accounting for these observed characteristics. One advantage of sampling based estimation is the maximum and minimum intercept may be monitored; these average -1.33 and -3.02.

This suggests we consider either (a) simply allowing for variable effects on illness over area of the individual characteristics (age, etc.), and modelling covariation of the variable slopes with the illness rate, or (b) additionally modelling the variability in local authority intercepts and slopes in terms of area attributes, such as area deprivation,

[2] Initial values are provided by a null (zero values) on the regression coefficients, by the posterior mean of a test run, and by the upper 97.5% point of that test run.

position in terms of the North-South divide in the UK, and so on. Only option (b) allows for the cross-level effects just described.

Here Model B adopts a step along such a broader perspective in considering a bivariate model for varying intercepts and varying effects of individual deprivation, and in introducing a simple cluster variable. The results for Model B show a higher level of illness in the North than the south of England (a clearly positive γ_{01}), but no effect of this variable on the slope variation. There is only a weak negative correlation, averaging -0.13, between the illness level and the slope on individual deprivation. Model C shows more evidence suggesting a contextual effect, with the varying slopes on social class IV and V tending to be higher in low illness areas (Table 4.2).

4.3 MODELLING HETEROSCEDASTICITY

Regression models for continuous outcomes, whether single or multi-level, most frequently assume that the error variance is constant. In a multi-level analysis, for instance, this means that the level 1 variance is independent of explanatory variables at this level. It is quite possible, however, that the variance (and so the precision also) are related systematically to explanatory variables or other characteristics of the subjects. In discrete data models (e.g. Poisson or binomial) random effects at level 1 may be introduced if there is over-dispersion (see Equations (4.3) and (4.4)), and such errors may have a variance which depends on the explanatory variates. Heteroscedasticity may also be modelled at higher levels, as the abortion attitudes example below illustrates.

Recent work comparing marginal and conditional regression specifications in multi-level modelling emphasises the need to model heteroscedasticity where it is present (Heagerty and Zeger, 2000). Goldstein *et al.* (1991), and more recently Browne *et al.* (2000), have argued that proper specification of the random part of a multi-level model (i.e. allowing for possible non-homogenous variances at one or more levels) may be important in inferences on the mean regression coefficients. Therefore, one way towards more robust inference in multi-level, and potentially better fit also, is to model the dependence of variation on relevant factors; these might well be, but are not necessarily, among the main set of regressors.

If the differences in variance are specified according to a categorical variable C_{ij} observed at level 1, then one might simply take variances specific to the levels $1, .. K$ of C_{ij}. For instance, if ϕ_k denotes the inverse variance for the kth level of C_{ij}, then one might adopt a series of gamma priors

$$\phi_1 \sim G(a_1, b_1), \phi_2 \sim G(a_2, b_2), \ldots \quad \phi_c \sim G(a_k, b_k) \tag{4.8}$$

Equivalently, $\log(\phi_{ij})$ can be regressed on a factor defined by the levels $1, .., K$ of C_{ij}.

One might also model the heterosecdasticity by relating log variances to a general function of relevant factors or the entire regression term. An explicit random error derivation can be illustrated by a two level model, with

$$Y_{ij} = \beta_0 + \beta_1 X_{1ij} + \beta_2 X_{2ij} + .. + u_{ij}$$

Then the level 1 random effect is written as

$$u_{ij} = R_{0ij} + R_{1ij} W_{ij}$$

Table 4.2 Risk factors for long term illness

	Mean	St. devn.	2.5%	97.5%
Model A				
Log-Likelihood	−2710	12	−2737	−2689
Covariates modelled as random				
Average Intercept	−2.18	0.09	−2.38	−2.03
Intercept variance	0.14	0.04	0.08	0.22
Fixed Effects of Individual (Level 1) Covariates				
Age Squared	0.00059	0.00008	0.00043	0.00075
Social Classes IV,V	0.34	0.08	0.17	0.49
Non-White	0.19	0.18	−0.15	0.56
Married	−0.17	0.09	−0.34	0.02
Deprivation	0.68	0.09	0.50	0.86
Model B				
Log-Likelihood	−2687	11	−2707	−2665
Covariates modelled as random				
Average Intercept	−2.34	0.09	−2.52	−2.16
Average Effect of Individual Deprivation	0.64	0.13	0.40	0.89
Effect of N-S divide on Intercept	0.42	0.09	0.23	0.60
Interaction between N-S divide and Individual Deprivation	−0.01	0.20	−0.42	0.37
Intercept variance	0.14	0.04	0.08	0.22
Slope Variance	0.22	0.09	0.09	0.45
Correlation between Intercepts and Deprivation Slopes	−0.13	0.22	−0.55	0.31
Fixed Effects of Individual (Level 1) Covariates				
Age Squared	0.00060	0.00008	0.00044	0.00075
Social Classes IV,V	0.34	0.08	0.17	0.50
Non-White	0.30	0.19	−0.07	0.65
Married	−0.18	0.09	−0.36	−0.01
Model C				
Log-Likelihood	−2673	13	−2697	−2648
Covariates modelled as random				
Average Intercept	−2.32	0.10	−2.52	−2.14
Effect of N-S divide on Intercept	0.37	0.11	0.16	0.58
Average Effect of Individual Deprivation	0.66	0.13	0.42	0.90
Interaction between N-S divide and Individual Deprivation	−0.06	0.19	−0.43	0.33
Average Effect of Individual Social Class	0.23	0.12	−0.01	0.46
Interaction between N-S divide and Individual Social Class	0.19	0.19	−0.19	0.57
Intercept variance	0.13	0.04	0.07	0.23
Deprivation Slope Variance	0.25	0.12	0.09	0.53
Social Class Slope Variance	0.25	0.11	0.09	0.52
Correlation between Intercepts and Deprivation Slopes	0.00	0.24	−0.47	0.48
Correlation between Intercepts and Social Class Slopes	−0.32	0.22	−0.68	0.15
Correlation between Class Slopes and Deprivation Slopes	−0.20	0.27	−0.67	0.33
Fixed Effects of Individual (Level 1) Covariates				
Age Squared	0.00060	0.00008	0.00044	0.00076
Non-White	0.31	0.19	−0.07	0.66
Married	−0.19	0.09	−0.37	−0.01

where $W_{ij} = X_{ij}\beta$ is the total linear regression term[3], and where

$$\text{var}(R_{0ij}) = \sigma_0^2, \quad \text{var}(R_{1ij}) = \sigma_1^2$$

and $\text{cov}(R_{0ij}, R_{1ij}) = \sigma_{01}$. Hence,

$$V_{ij} = \text{var}(u_{ij}) = \sigma_0^2 + 2\sigma_{01} W_{ij} + \sigma_1^2 W_{ij}^2$$

As Snijders and Bosker (1999, p. 114) note, such a formula can be used without the interpretation that σ_0^2 and σ_1^2 are variances and σ_{01} a covariance; the formula may simply be used to imply that the level 1 variance is a quadratic function of W_{ij}. Note that in BUGS, one must transform back from precisions to variances to find how variances differ between groups or subjects defined by different predictors.

Example 4.3 Language score variability by gender This example continues Example 4.1 in terms of language scores of Dutch school children. However, instead of assuming a constant level 1 variance, we consider possible heteroscedasticity according to pupil characteristics. As a simple illustration of heteroscedasticity, consider the pupils in terms of grouped IQs, and the resulting averages and variances of the scores (Table 4.3).

It is apparent that at IQs above 12, there is a lesser variability in test scores (as well as higher average attainment).

Snijders and Bosker (1999, p. 111) consider the same phenomenon, but by pupil gender. We follow their analysis and include a 'Bayesian significance test' to assess whether in fact the gender specific variances do differ in terms of conventional significance levels. Let G_{ij} denote the gender of pupil i in class j (= 1 for girls, 0 for boys). Variable regression slopes for child IQ are assumed, but a homogenous regression effect of SES and gender. A single cluster attribute (Z_1 = class IQ) is used to explain variation in intercepts and the child IQ slopes. The model, coded in Program 4.3, may then be set out as follows:

$$Y_{ij} \sim \text{N}\left(\mu_{ij}, \phi_{ij}^{-1}\right) \quad i = 1, .. n_j; j = 1, .. J$$

$$\mu_{ij} = \beta_{1j} + \beta_{2j}(\text{IQ}_{ij} - \bar{\text{IQ}}) + \beta_3(\text{SES}_{ij} - \bar{\text{SES}}) + \beta_4 \text{G}_{ij}$$

$$\beta_j \sim \text{N}_2(\nu_j, V_\beta)$$

$$\nu_{kj} = \gamma_{1k} + \gamma_{2k}(Z_{1j} - \bar{Z}_1) \quad k = 1, 2$$

$$\log \phi_{ij} = c_1 + c_2 G_{ij}$$

[3] Alternatively, a single regressor might be used, with

$$u_{ij} = R_{0ij} + R_{1ij} X_{ij}$$

Hence,

$$V_{ij} = \text{var}(u_{ij}) = \sigma_0^2 + 2\sigma_{01} X_{ij} + \sigma_1^2 X_{ij}^2.$$

Table 4.3 Means and variances of scores by IQ group

IQ group	Average language score	St. devn. of language score
4–5.99	28.3	8.1
6–7.99	28.8	8.5
8–9.99	32.3	7.7
10–11.99	37.7	8.1
12–13.99	43.9	6.8
14–15.99	48.5	5.5
16+	50.2	4.7

Priors with large variances are assumed for the β and γ coefficients, but for numerical stability c_1 and c_2 are assigned relatively informative N(0, 1) priors.

Analysis is based on three parallel chains and a 5000 iteration run with 500 burn-in. Mixing and convergence are satisfactory. In Table 4.4, $R.\beta[1, 2]$ denotes the correlation between intercepts and IQ slopes, and shows a clear contextual effect: classes with lower than average attainment have higher impacts of individual IQ. The coefficient c_2 shows that girls have higher precision (and hence lower variance) in their language scores, and the posterior mean variances for boys and girls are, respectively, 38.6 and 36.2.

It may be noted that the coefficient c_2 straddles zero throwing doubt on a clear difference in variances. To assess whether the variance for boys exceeds that of girls, a significance test based on the proportion of iterations where the condition holds, shows a significance rate around 85%. Furthermore, a reduced model with equal gender

Table 4.4 Heteroscedasticity in Level 1 language score variances

	Mean	St. devn.	2.5%	97.5%
Deviance	14780	20.64	14740	14820
R.β[1, 2]	−0.55	0.15	−0.79	−0.22
Gender specific level 1 variances				
σ^2 (boys)	38.60	1.68	35.45	41.97
σ^2 (girls)	36.23	1.66	33.10	39.60
c_1	−3.65	0.04	−3.74	−3.57
c_2	0.064	0.064	−0.061	0.191
Fixed slopes				
β_3	0.15	0.01	0.12	0.18
β_4	2.47	0.25	1.97	2.96
Cluster model for variable intercept and IQ slope				
γ_{11}	39.57	0.32	38.94	40.19
γ_{12}	2.28	0.08	2.12	2.44
γ_{21}	1.07	0.32	0.42	1.70
γ_{22}	−0.11	0.08	−0.27	0.05

variances provides no increase in average deviance, and so one may conclude that there is in fact no pronounced evidence of different gender variances.

Example 4.4 Attitudes to abortion Heagerty and Zeger (2000) consider data from four waves of the British Social Attitudes Survey (1983–86) on attitudes to abortion. Level 1 is then the survey wave, level 2 is the survey respondent and level 3 is district of residence. Of particular interest in their analysis are the modelling of heterogeneity at levels 2 and 3 (subjects and clusters), and of heterogeneity at levels 2 and 3, and resulting impacts on fixed regression coefficients. The $T_n = 1056$ responses are classified by district j ($J = 54$), subject i (with 264 subjects), and by year t within subject ($t = 1, 4$).

Heagerty and Zeger consider a dichotomisation of abortion attitudes, namely the views encompassed under 'no legal restriction' as against 'possible legal restriction'. Thus, $Y_{ijt} = 1$ if a person believes abortion should be permitted in a range of hypothetical situations, while $Y_{ijt} = 0$ if a person believes abortion should not be allowed in one or more of these circumstances. At subject level the covariates are year, social class (1 = middle, 2 = upper working, 3 = lower working), sex (1= male, 2 = female), and religion (Roman Catholic = 1; Protestant or Church of England = 2; other religion = 3; no religion = 4). Because these covariates are fixed over time, they are denoted X_{ij}. They are modelled as categorical factors, with the first level in each having a null effect. Note that, for a factor with F levels, if $\beta_1 = 0$ and only $\beta_2, .. \beta_F$ are free coefficients, the profile of centred effects (averaging zero rather than with a corner constraint) can be obtained by monitoring the transformed coefficients

$$\kappa_k = \beta_k - \bar{\beta}$$

At cluster level (i.e. district of residence), there are an intercept and an overall percentage Protestant (Z_j): this variable measures the religious context or social environment governing attitudes as opposed to the individual's creed.

Thus a three level model (times within respondents within districts) with a logit link for the binary outcome, and with no random effects, may be specified as

$$Y_{ijt} \sim \text{Bern}(\pi_{ijt})$$

$$\text{logit}(\pi_{ijt}) = \varphi + \eta_t + \beta X_{ij} + \gamma Z_j$$

This independence model (Model A) therefore includes effects for year, social class, religion, gender and the district variable. φ is the overall intercept and the η_t are year effects with year 1 effect being zero. Early convergence in a three chain run is apparent in terms of Gelman–Rubin summaries for $\{\varphi, \eta, \beta, \gamma\}$, with fast decay in the autocorrelations at successive lags in the MCMC iterations. The posterior summary in Table 4.5 is based on 5000 iterations with 500 burn in.

The parameter estimates for Model A show that working class respondents are less likely to give the unrestricted view; and that women as compared to men, and Catholics and 'other religions' as compared to Protestants and non-believers are also less likely to give the unrestricted view. Living in a 'Protestant' area also boosts the chance of giving a 'no legal restriction' response.

A second model (Model B) introduces random variation at subject and district levels (levels 2 and 3). So

$$Y_{ijt} \sim \text{Bern}(\pi_{ijt})$$

Table 4.5 British social attitudes

	Mean	St. devn.	2.50%	97.50%
Model A (independence model)				
Log Likelihood	−628	2	−633	−624
Intercept	−0.79	0.28	−1.34	−0.24
Year 2	−0.42	0.20	−0.80	−0.04
Year 3	0.04	0.19	−0.33	0.41
Year 4	0.18	0.19	−0.19	0.56
Subject Effects				
Upper Working Class	−0.31	0.19	−0.68	0.06
Lower Working Class	−0.42	0.16	−0.74	−0.09
Gender	−0.27	0.14	−0.55	0.00
Protestant	−0.43	0.32	−1.08	0.18
Other Religion	−0.60	0.24	−1.09	−0.13
No religion	0.70	0.18	0.36	1.05
Cluster (District) Effects				
Protestant	0.80	0.29	0.25	1.37
Model B (Level 2 & 3 homogenous errors)				
Log Likelihood	−388	12	−412	−365
Intercept	−1.28	0.55	−2.39	−0.20
Year 2	−0.73	0.26	−1.24	−0.23
Year 3	0.07	0.24	−0.41	0.54
Year 4	0.30	0.24	−0.16	0.78
Subject Effects				
Upper Working Class	−0.51	0.35	−1.19	0.20
Lower Working Class	−0.38	0.34	−1.05	0.29
Gender	−0.53	0.35	−1.26	0.15
Protestant	−0.56	0.63	−1.82	0.66
Other Religion	−0.94	0.50	−1.93	0.02
No religion	1.07	0.40	0.30	1.86
Cluster (District) Effects				
Protestant	0.95	0.61	−0.27	2.11
Random Effects St devns				
Level 2	2.28	0.25	1.82	2.81
Level 3	0.53	0.40	0.01	1.33
Model C (Level 2 heteroscedasticity)				
Log Likelihood	−387	12	−411	−365
Intercept	−1.25	0.59	−2.38	−0.06
Year 2	−0.74	0.25	−1.23	−0.24
Year 3	0.07	0.24	−0.39	0.54
Year 4	0.31	0.24	−0.15	0.78
Subject Effects				
Upper Working Class	−0.48	0.39	−1.25	0.30
Lower Working Class	−0.22	0.37	−0.95	0.51
Gender	−0.65	0.35	−1.33	0.02
Protestant	−0.55	0.59	−1.72	0.61
Other Religion	−1.14	0.51	−2.16	−0.13

Model C (Level 2 heteroscedasticity) (*continued*)

No religion	0.77	0.41	−0.02	1.58
Cluster (District) Effects				
Protestant	1.07	0.61	−0.15	2.22
Level 2 St devns				
Middle	2.95	0.61	1.91	4.29
Skilled Working	2.94	0.57	1.98	4.22
Unskilled	1.77	0.29	1.27	2.40
Level 3 (District) Standard devn	0.40	0.35	0.01	1.22

$$\text{logit}(\pi_{ijt}) = \varphi + \eta_t + \beta X_{ij} + \gamma Z_j + u_{ij} + \alpha_j$$

where u_{ij} and α_j are normal random effects with homogenous variances τ_u and τ_α. Following Heagerty and Zeger, none of the slopes β are taken to vary randomly. Initially, G(0.0001, 0.0001) priors are assumed on $1/\tau_u$ and $1/\tau_\alpha$. N(0, 10) priors are taken on the intercept and the effects β, γ and η relating to year, percents Protestant, class, etc. Three parallel chains are run for 5000 iterations; convergence is apparent at under 500 iterations, and the posterior summary in Table 4.5 is based on iterations 500–5000.

Under the prior assumptions just mentioned, there is a pronounced gain in likelihood over Model A, though a full assessment would require penalising for the additional random effects and variance parameters (see the Exercises). The absolute effects β, in terms of posterior means, of most of the categorical variables (class, sex, religion) are enhanced, though generally with lower precision (i.e. posterior standard deviations are higher for the β coefficients in Model B than Model A). Little difference was made running a more informative G(5, 1.5) prior for $1/\tau_u$ following the example of Johnson and Albert (1999). The posterior mean estimate for τ_u was reduced slightly to 2.15.

A third model (Model C) allows for heteroscedasticity at respondent level such that the u_{ij} have variances at level 2 which are specific for the social class C_{ij} of the subjects. Since C_{ij} has three levels there are three possible values for σ^2_{ij}. So $\log(\phi_{ij})$, namely the logs of the level 1 precisions $\phi_{ij} = \sigma^{-2}_{ij}$, could depend upon social class via a regression containing an intercept and effects for the second and third social class groups. Alternatively, we here adopt the strategy in Equation (4.8), with

$$\phi_1 \sim G(a_1, b_1),\ \phi_2 \sim G(a_2, b_2),\ \phi_3 \sim G(a_3, b_3)$$

and $a_i = b_i = 0.0001$ for all i.

The analysis is based on three parallel chains over 5000 iterations, and shows that the unskilled working class respondents as most homogenous (least variable) in their views over districts. The direct regression effect of social class on attitudes loses further definition as against Models A and B, with the posterior standard deviation for the lower working class now 50% larger than the posterior mean of −0.22.

4.4 ROBUSTNESS IN MULTI-LEVEL MODELLING

Issues of robustness in multi-level modelling occur especially in comparing results of fully Bayesian estimation and ML estimation involving generalised least squares techniques (Langford and Lewis, 1998). The latter may condition on estimates of variance components which are treated as known, so obscuring the potential impact of outliers at different levels. Consider a standard two level hierarchical model with $j = 1,..J$ clusters at level 2, and $i = 1,..n_j$ cases within each cluster. Then for a continuous outcome Y, consider a level 1 model with cluster specific regression effects

$$Y_{ij} = \boldsymbol{\beta}_j X_{ij} + u_{ij} \tag{4.9}$$

where the u_{ij} have mean 0 and variance σ^2, and $\boldsymbol{\beta}_j$ and X_{ij} are of length $p+1$, with $X_{1ij} = 1$. At level 2 each of the stochastic coefficients from level 1 are related to q cluster level predictors Z_j. For the hth such coefficient ($h = 1,..p+1$), we might have

$$\beta_{jh} = Z_j \gamma_h + e_{jh} \quad j = 1,..J \tag{4.10}$$

where γ_h is a vector of regression coefficients.

Under typical assumptions that u_{ij} and e_{jh} are Normal, both the level 1 coefficients and the regression coefficients at level 2 may be sensitive to outliers. Estimates and inferences regarding the effects γ_h are especially subject to this if J is small. Thus, consider the posterior for γ, under these standard assumptions, conditional on the variance σ^2 at level 1, and on the $(p+1) \times (p+1)$ dispersion matrix V for the errors e at level 2, and assuming a uniform prior for γ. Then the posterior is multivariate normal of dimension $q \times (p+1)$

$$\gamma | y, V, \sigma^2 \sim \text{N}(G, D)$$

where

$$D^{-1} = \sum_{j=1}^{J} W_j (C_j + V)^{-1} W_j$$

$$G = D \sum_{j=1}^{J} W_j (C_j + V)^{-1} \hat{\beta}_j$$

$$C_j = \sigma^2 (X_j' X_j)^{-1}$$

and where $\hat{\beta}_j$ is the least squares estimate of β_j, namely

$$\hat{\beta}_j = (X_j' X_j)^{-1} X_j y_j$$

Many multi-level procedures involve maximum likelihood or EM procedures for estimating V and σ^2 which are then treated as known. The resulting estimates of G and D, and hence γ, will then not account for the uncertainty in estimating V and σ^2, so that the intervals on the components of γ will tend to be too narrow (Seltzer, 1993).

There may also be sensitivity regarding the assumed form of the cluster effect covariation and the observational errors u, even if uncertainty in their estimation is allowed for. Under the standard Normal assumptions regarding error terms at different levels, outlying data points, especially at level 2 and above, may unduly influence model parameter estimates and distort credible intervals tending to make them too wide. Options for robust estimation, especially of cluster disturbances, include discrete mixtures (Rabe-Hesketh and Pickles, 1999) and Student t errors with unknown degrees of freedom (Seltzer, 1993).

Browne and Draper (2000) consider alternatives to the standard non-informative choices for the prior on the precision matrix of variable cluster effects (intercepts and

p slopes varying at level 2 when the level 1 relates to subjects). Thus, for $p+1$ effects varying randomly at cluster level, a standard choice under multivariate Normal or Student t sampling for such effects is that V^{-1} is Wishart with $p+1$ degrees of freedom and an identity scale matrix. Browne and Draper consider an option whereby the degrees of freedom are $(p+1)+2$ and an estimated scale matrix (e.g. from an iterative GLS procedure) replaces the default identity matrix. Daniels and Kass (1999) consider further options, for instance where the degrees of freedom is an unknown parameter, or the element on the diagonal of the prior scale matrix is allowed to be a free parameter.

It may be noted that individual membership probabilities in the discrete mixture model may be related to explanatory variates, in a way that parallels regressions of (continuously varying) cluster slopes and intercepts on cluster variables. Thus Carlin *et al.* (2000) consider robustness from a Bayes perspective in a multi-level model for panel data on smoking in adolescent subjects: such subjects are then 'clusters' at level 2. With a binary outcome, Y_{it} at time t for subject i, they contrast the 'logistic-normal' model, namely normal subject effects, with a two group mixture of subjects. Group membership is related to a set of fixed subject level covariates W_i. In general, one might specify

$$Y_{it} \sim \text{Bern}(\pi_{it})$$

$$\text{logit}(\pi_{it}) = \beta_{L_i} x_{it}$$

$$\text{L}_i \sim \text{Categorical}(P_i)$$

$$\text{logit}(P_{i1}) = \phi W_i$$

Here L_i is the latent group membership aiming to distinguish between a high risk or susceptible group (in terms of smoking level) and a low risk group so P_i is of dimension 2. In fact, Carlin *et al.* take the probability of smoking in one 'non-susceptible' group as zero, and only apply a logit regression for π_{it} for subjects falling in a susceptible group.

Example 4.5 JSP Project: Maths over time To illustrate sensitivity issues in the specification of cluster effect covariation, consider the Joint Schools Project data, also analysed by Mortimore *et al.* (1988), and more recently in an extensive sensitivity analysis by Browne and Draper (2000). Here the pupils are level 1 and the clusters are schools; the data include differential weights on pupils, which affect the specification of the variance term. The model considers Maths attainment at year 5 in relation to such attainment at year 3, with

$$\text{MATH5}_{ij} = \alpha_j + \beta_j \text{MATH3}_{ij} + u_{ij} \tag{4.11}$$

The attainment continuity effects β_j and the intercepts are initially taken as bivariate normal over the 48 schools (Model A), with dispersion matrix **V**, while the level 1 errors u_{ij} are taken as normal with variance σ^2. To illustrate sensitivity with regard to prior assumptions on random cluster effects, we consider two variations towards robust inference. The first is a discrete mixture of intercept and attainment effects (Model B), with the α_j and β_j in Equation (4.11) being determined by a latent categorisation of the 48 schools into M groups. Following Rabe-Hesketh and Pickles (1999), a discrete mixture with $M=3$ groups of schools is assumed. The second is nonparametric hierarchical model using a truncted Dirichlet process prior (Ishwaran and Zarepour, 2000).

Estimation in Model A is based on three parallel chains with the summaries based on the last 24 000 from 25 000 iterations, since early convergence was apparent. In Model B there is later convergence and the last 15 000 of 25 000 iterations (over three chains) are used for the summary. In Model C the summary is based on the second half of a 10 000 iteration run with three chains.

The estimates for Model A show a negative correlation in the bivariate Normal for the school effects, averaging −0.46, between average school attainment and persistence in attainment. The persistence effect itself (the impact of Maths 3 scores on Maths 5 scores) is demonstrated by an average coefficient $\bar{\beta}$ of 0.62, with 95% credible interval from 0.50 to 0.74. The school persistence effects, measured by posterior means, vary between 0.24 (school 37) and 1.20 (school 43).

As to the discrete mixture model, this may be stated as

$$\text{MATH5}_{ij} = \alpha L_i + \beta L_i \text{MATH3}_{ij} + u_{ij} \tag{4.12}$$

where L_i is the group membership of the school. Rabe-Hesketh and Pickles estimate group specific persistence coefficients $\beta_1 = 0.68$, $\beta_2 = 0.77$ and $\beta_3 = 0.34$. In their analysis, these groups have respective probabilities 0.54, 0.19 and 0.28. Similar findings are obtained here in the analysis of Model B, adopting a mildly informative Dirichlet prior on the mixture proportions for $M = 3$, with prior weights of 2.5 on each component. Also, the prior on the attainment effects β_j in the three latent school groups is constrained to be increasing to improve identifiability (so the groups are in ascending order of persistence effect). With a relatively small number of schools at level 2, less informative priors may lead to identifiability problems unless a data dependent prior (see Chapter 2) is employed.

The largest of the $M = 3$ attainment coefficients stands at $\beta_3 = 0.80$, and the smallest at 0.35 (Table 4.6). In estimation there was some delay in the convergence of the intercepts α in Equation (4.12). There were also high sampling autocorrelations for the group intercept parameters, especially those for the second and third groups (those with the larger persistence effects), suggesting the group intercepts are less well separated than the persistence effects. Subsampling (every fiftieth iteration) makes little

Table 4.6 Parameter summaries, Maths attainment

	Mean	St. devn.	2.5%	Median	97.5%
Model A					
Log Likelihood	−2603	6.6	−2618	−2603	−2592
Corr(α_j, β_j)	−0.46	0.16	−0.73	−0.47	−0.10
σ^2	28.1	1.4	25.4	28.0	31.0
Level 2 covariance					
V_{11}	4.45	1.40	2.28	4.25	7.72
V_{12}	−0.31	0.15	−0.66	−0.30	−0.06
V_{22}	0.10	0.03	0.06	0.10	0.17
Average effects					
Intercept	30.59	0.37	29.86	30.60	31.31
Math3	0.62	0.06	0.50	0.62	0.74

Table 4.6 (*continued*)

	Mean	St. devn.	2.5%	Median	97.5%
Model B					
Actual Log Likelihood	−2660	7	−2673	−2660	−2648
Complete Log Likelihood	−2613	5	−2625	−2612	−2604
Math3 effects					
Group 1	0.35	0.09	0.22	0.35	0.47
Group 2	0.66	0.06	0.51	0.67	0.76
Group 3	0.80	0.11	0.66	0.78	1.05
Intercept					
Group 1	33.4	6.2	32.4	33.5	34.6
Group 2	29.4	1.6	26.5	30.0	32.9
Group 3	28.6	1.6	26.0	28.3	30.9
Mixture Proportions					
π_1	0.30	0.08	0.16	0.30	0.47
π_2	0.41	0.15	0.11	0.44	0.66
π_3	0.29	0.15	0.08	0.25	0.61
Model C					
Log Likelihood	−2609	7	−2625	−2609	−2597
Corr(α_j, β_j)	−0.71	0.15	−0.99	−0.71	−0.38
σ^2	28.43	1.44	25.70	28.39	31.37
Mixture Parameters κ (Dirichlet presicion)	3.16	3.22	0.29	2.00	12.03
Average Number of Clusters	6.3	1.8	3	6	9
Average effects					
Intercept	30.67	0.20	30.29	30.67	31.06
Math3	0.62	0.03	0.55	0.62	0.69

difference to the posterior means on the group intercepts, though the standard errors are altered to some degree.

The likelihood under Model B deteriorates as compared to the bivariate Normal random effects model, though it is likely to be less heavily parameterised. The 'complete data' likelihood (assuming group memberships known) is more comparable to the bivariate Normal random effects model.

The DPP nonparametric model assumes a maximum 10 clusters for the intercepts and slopes and adopts a baseline bivariate Normal prior for the covarying cluster intercepts and slopes. The baseline dispersion matric has prior as in model A. The Dirichlet precision parameter κ is updated as in Ishwaran and Zarepour (2000). Convergence is obtained in a two chain run after 1250 iterations with the mean number of clusters in a run of 5000 iterations being 6.3, and the Dirichlet parameter estimated at 3.16. The smoothed school effects on continuity (Table 4.7) are intermediate between the bivariate Normal parametric model and three-group discrete mixture.

4.5 MULTI-LEVEL DATA ON MULTIVARIATE INDICES

Frequently, profiling or performance rankings of public sector agencies (schools, hospitals, etc.) will involve multiple indicators. Inferences about relevant summary parameters such as comparative ranks, or the probability that a particular institution exceeds the average, are readily obtained under the Bayes sampling perspective (Deely and Smith, 1998). Such inferences will often be improved by allowing for the interplay between the indicators themselves, and also for features of the institutions (e.g. the case-mix of patients in health settings or intake ability of pupils in school comparisons) which influence performance on some or all of the indicators used. Similar gains in precision may occur in small area health profiling, where multiple mortality or morbidity outcomes provide a firmer basis for defining health problem areas than a single outcome.

Suppose individual level data Y_{ijh} are available for variables $h = 1, .. H$, clusters $j = 1, \ldots J$ and subjects $i = 1, .. n_j$ within each cluster. Then the measurements on the different variables can be envisaged as the lowest level (level 1) of the data hierarchy, in the same way as repeated measures on the same variable are treated. The subjects are at level 2, and the clusters (agencies, areas, etc.) at level 3. Thus for a set of H metric outcomes, taken to be multivariate Normal, and with M predictors X_{ijm}, $m = 1, .. M (X_{ij1} = 1)$ at subject level, one might propose

$$Y_{ij} \sim N_H\left(\mu_{ij}, \Sigma\right)$$

$$\mu_{ijh} = \beta_{h1} + \beta_{h2}X_{ij2} + \beta_{h3}X_{ij3} + .. \beta_{hM}X_{ijM} \tag{4.13}$$

with $Y_{ij} = (Y_{ij1}, Y_{ij2}, \ldots, Y_{ijH})$ and Σ an $H \times H$ dispersion matrix[4].

In Equation (4.13) intercept and predictor effects are specific to dependent variable h, and do not pool strength over clusters. Another option, therefore, allows random variability over clusters, with

$$\mu_{ijh} = \beta_{jh1} + \beta_{jh2}X_{ij2} + .. \beta_{jhM}X_{ijM}$$

The random effects β_{jhm} might then be related to cluster attributes within a multivariate density of dimension $H \times M$, or separate densities (for each outcome h) of dimension M.

In the same vein, multivariate data aggregated over individuals within clusters (i.e. averaged over the n_j subjects) constitute a form of multi-level data. The lowest level of the analysis models the observed outcomes are vectors of length H for agencies, or more generally clusters, $j = 1, .. J$

$$y_j = (y_{j1}, y_{j2}, \cdots y_{jH})$$

[4] In BUGS this might be expressed in the code

```
for (j in 1:J) {for (i in 1:n[j]) {Y[i,j,1:H] ~ dmnorm(mu[i,j,1:H],T[1:H,1:H])}}
```

where T[1:H,1:H] is the inverse dispersion.

Table 4.7 School effects on attainment continuity ($\beta_1, \beta_2, \ldots \beta_{48}$)

School no.	Mean effect Model A	2.5%	97.5%	Mean effect Model B	2.5%	97.5%	Mean effect Model C	2.5%	97.5%
S1	0.60	0.32	0.88	0.74	0.59	0.90	0.68	0.48	0.86
S2	0.45	0.06	0.80	0.59	0.28	0.78	0.56	0.25	0.80
S3	0.53	0.04	1.01	0.49	0.25	0.77	0.51	0.25	0.87
S4	0.89	0.54	1.26	0.77	0.62	0.98	0.77	0.56	1.24
S5	0.59	0.20	0.99	0.66	0.35	0.80	0.65	0.33	0.88
S6	0.50	0.12	0.89	0.56	0.27	0.76	0.54	0.26	0.82
S7	0.65	0.22	1.08	0.52	0.26	0.77	0.55	0.27	0.97
S8	0.70	0.40	1.00	0.67	0.52	0.79	0.68	0.44	0.92
S9	0.83	0.42	1.28	0.78	0.60	0.99	0.72	0.51	1.05
S10	0.71	0.32	1.10	0.65	0.35	0.79	0.66	0.35	0.99
S11	0.71	0.39	1.04	0.69	0.57	0.82	0.70	0.50	0.93
S12	0.36	0.03	0.69	0.65	0.34	0.77	0.56	0.18	0.78
S13	0.68	0.29	1.09	0.69	0.57	0.84	0.69	0.46	0.97
S14	0.48	−0.03	0.98	0.50	0.25	0.80	0.50	0.23	0.84
S15	0.82	0.42	1.22	0.70	0.58	0.85	0.74	0.53	1.20
S16	0.77	0.29	1.25	0.65	0.32	0.87	0.70	0.32	1.27
S17	0.65	0.30	1.00	0.68	0.55	0.82	0.68	0.43	0.89
S18	0.44	0.06	0.83	0.45	0.24	0.74	0.44	0.23	0.75
S19	0.67	0.18	1.18	0.68	0.34	0.88	0.67	0.32	1.04
S20	0.67	0.33	1.01	0.79	0.61	0.99	0.70	0.48	0.95
S21	0.77	0.36	1.19	0.74	0.59	0.94	0.73	0.52	1.08
S22	0.98	0.67	1.31	0.72	0.60	0.91	0.86	0.61	1.38
S23	0.37	0.03	0.71	0.35	0.23	0.48	0.37	0.22	0.52
S24	0.30	0.00	0.59	0.36	0.23	0.50	0.36	0.21	0.53
S25	0.44	0.09	0.79	0.51	0.26	0.75	0.48	0.24	0.76
S26	0.75	0.34	1.19	0.67	0.40	0.80	0.70	0.38	1.10
S27	0.74	0.31	1.14	0.74	0.60	0.93	0.72	0.52	1.05
S28	0.73	0.30	1.17	0.78	0.60	0.99	0.71	0.46	1.00
S29	0.51	0.15	0.88	0.56	0.29	0.76	0.53	0.26	0.81
S30	0.47	0.19	0.75	0.35	0.23	0.48	0.37	0.23	0.54
S31	0.63	0.33	0.92	0.67	0.47	0.79	0.65	0.40	0.84
S32	0.60	0.36	0.83	0.69	0.57	0.81	0.66	0.48	0.82
S33	0.46	0.10	0.83	0.35	0.23	0.48	0.37	0.22	0.54
S34	1.00	0.55	1.44	0.73	0.59	0.95	0.91	0.57	1.64
S35	0.41	0.05	0.78	0.35	0.23	0.48	0.37	0.22	0.54
S36	0.49	0.11	0.88	0.58	0.28	0.77	0.55	0.26	0.82
S37	0.24	−0.14	0.60	0.35	0.23	0.48	0.36	0.18	0.50
S38	0.60	0.12	1.11	0.71	0.52	0.90	0.68	0.37	0.97
S39	0.78	0.49	1.09	0.77	0.61	0.95	0.73	0.55	1.00
S40	0.37	0.00	0.74	0.37	0.23	0.65	0.38	0.22	0.64
S41	0.60	0.31	0.89	0.68	0.56	0.79	0.67	0.47	0.84
S42	0.70	0.23	1.19	0.67	0.32	0.89	0.67	0.32	1.07
S43	1.20	0.81	1.61	0.80	0.64	1.01	1.29	0.68	1.91
S44	0.56	0.09	1.03	0.58	0.26	0.81	0.57	0.26	0.91
S45	0.35	0.08	0.61	0.36	0.23	0.60	0.37	0.22	0.57
S46	0.77	0.56	0.97	0.69	0.58	0.81	0.72	0.57	0.93
S47	0.90	0.60	1.20	0.69	0.58	0.85	0.79	0.58	1.21
S48	0.26	−0.16	0.64	0.38	0.23	0.68	0.38	0.19	0.68

conditional on parameters of a hyperprior. For example, for continuous data one might be able to assume multivariate normality

$$y_j \sim N_H(\theta_j,\ V_j) \tag{4.14}$$

where θ_j is a vector of true averages on the H outcomes. Often the measurement dispersion matrices V_j will be known, and take account of varying sample sizes n_j.

A random effects model might then be used to predict the 'true' mean outcome θ_{jh} on index h in agency j on the basis of a multivariate population hyperprior. This might involve cluster level covariates

$$\{z_{jk},\ k = 1,..K, j = 1,..J\}$$

with $z_{j1} = 1$ for all j, to predict the means in this density. Suppose, following Everson and Morris (2000), the θ_{jh} are assumed to be functions of covariates z in a two stage multivariate model, with the V_j in Equation (4.14) known. Then at stage 2

$$\theta_j \sim N_H\left(\nu_j,\ \Sigma\right) \tag{4.15}$$

and the regression means ν_{jh} of the θ_{jh} are specified as

$$\nu_{j1} = \gamma_{11} + \gamma_{12}z_{j2} + \gamma_{13}z_{j3} + ..\gamma_{1K}z_{jK}$$
$$...$$
$$\nu_{jH} = \gamma_{H1} + \gamma_{H2}z_{j2} + \gamma_{H3}z_{j3} + ...\gamma_{HK}z_{jK}$$

Example 4.6 Hospital profiling In the hospital profiling application of Everson and Morris (2000), there are $H = 2$ percentage rates of patient-reported problems for $J = 27$ hospitals: one rate relates to surgical issues, the other to non-surgical issues. There is a severity index z_{j2} for each hospital, with higher levels of the index denoting more complex patient case-mix. Following Everson and Morris, a Normal approximation to the binomial is assumed and the measurement dispersion matrices V_j are known. Further it is assumed that measurement errors in Equation (4.14) are uncorrelated (so that V_j is diagonal), so restricting the modelling of correlations to the latent effects θ_j.

The variances, across hospitals and patients, for y_{j1} and y_{j2} are taken as supplied by Everson and Morris (2000, p. 405), namely 148.9 and 490.6, so that the first stage model is

$$y_{j1} \sim \mathrm{N}(\theta_{j1},\ V_{j1})$$
$$y_{j2} \sim \mathrm{N}(\theta_{j2},\ V_{j2})$$

where $V_{j1} = 149/n_j$, $V_{j2} = 491/n_j$. The latent means are modelled according to Equation (4.15).

Running three parallel chains for 50 000 iterations, early convergence is apparent (at under 500 iterations) on the hyperdensity parameters and underlying means; however, sampling autocorrelations are quite high on the γ parameters in Equation (4.15). However, comparing summaries based on 25 000 and 50 000 iterations shows little difference on posterior summaries for these parameters. Table 4.8, based on 50 000 iterations, shows that the impact of the severity index on the surgical problem rate for hospitals γ_{22} is clearly positive, but that on the non-surgical rate γ_{12} is less clear. There

Table 4.8 Bivariate hospital outcomes analysis ($h = 1$, non-surgical; $h = 2$, surgical)

Parameter covariance matrix	Mean	St. devn.		2.5%	97.5%
Σ_{11}	3.78	1.79		1.21	8.29
Σ_{12}	2.43	1.61		0.07	6.17
Σ_{22}	2.84	2.31		0.34	9.05
Intercepts and Severity effects					
γ_{11}	12.29	1.23		9.96	14.97
γ_{21}	12.68	1.47		9.52	15.46
γ_{12}	1.76	2.34		−3.29	6.40
γ_{22}	5.68	2.80		0.24	11.46
Smoothed rates (posterior means and standard deviations)					
Non-surgical	Mean	St. devn.	Surgical	Mean	St. devn.
$\theta_{1,1}$	12.3	1.6	$\theta_{1,2}$	16.1	1.6
$\theta_{2,1}$	12.8	1.4	$\theta_{2,2}$	16.0	1.4
$\theta_{3,1}$	14.3	1.5	$\theta_{3,2}$	16.7	1.5
$\theta_{4,1}$	12.5	1.3	$\theta_{4,2}$	14.1	1.4
$\theta_{5,1}$	13.3	1.4	$\theta_{5,2}$	17.5	1.7
$\theta_{6,1}$	12.7	1.3	$\theta_{6,2}$	15.2	1.3
$\theta_{7,1}$	14.1	1.3	$\theta_{7,2}$	15.9	1.3
$\theta_{8,1}$	13.1	1.2	$\theta_{8,2}$	15.6	1.3
$\theta_{9,1}$	12.4	1.2	$\theta_{9,2}$	14.1	1.3
$\theta_{10,1}$	10.2	1.3	$\theta_{10,2}$	13.1	1.6
$\theta_{11,1}$	16.7	1.4	$\theta_{11,2}$	18.6	2.0
$\theta_{12,1}$	12.5	1.2	$\theta_{12,2}$	14.5	1.3
$\theta_{13,1}$	12.6	1.2	$\theta_{13,2}$	14.1	1.3
$\theta_{14,1}$	14.0	1.2	$\theta_{14,2}$	16.5	1.3
$\theta_{15,1}$	14.1	1.2	$\theta_{15,2}$	14.6	1.3
$\theta_{16,1}$	14.5	1.2	$\theta_{16,2}$	17.1	1.3
$\theta_{17,1}$	11.7	1.2	$\theta_{17,2}$	12.5	1.5
$\theta_{18,1}$	15.2	1.2	$\theta_{18,2}$	16.2	1.3
$\theta_{19,1}$	15.4	1.2	$\theta_{19,2}$	17.1	1.3
$\theta_{20,1}$	11.7	1.1	$\theta_{20,2}$	13.8	1.2
$\theta_{21,1}$	14.1	1.1	$\theta_{21,2}$	16.4	1.2
$\theta_{22,1}$	11.5	1.2	$\theta_{22,2}$	15.6	1.5
$\theta_{23,1}$	14.7	1.1	$\theta_{23,2}$	16.2	1.2
$\theta_{24,1}$	12.8	0.9	$\theta_{24,2}$	16.0	1.2
$\theta_{25,1}$	11.5	1.0	$\theta_{25,2}$	14.8	1.1
$\theta_{26,1}$	10.6	0.9	$\theta_{26,2}$	11.5	1.3
$\theta_{27,1}$	13.8	0.8	$\theta_{27,2}$	17.4	1.1

is a clear positive correlation between the two latent rates θ_{jh} and taking account of this will improve precision in the estimated rates. Compared to the original surgical rate data, which range from 9–27%, the smoothed surgical problem rates vary from 11.5–18.5%.

Taking account of the uncertainty in estimating Σ may mean that Bayes estimates are less precise than REML estimates that assume a known dispersion matrix. Everson and Morris present evidence that the REML estimates of θ_{jh} show undercoverage in the sense of being too precise, and that the Bayes estimates show better coverage of the true distribution of these latent rates.

On the other hand, Bayes estimates are more precise than the classical intervals based on no pooling, namely via fixed effects maximum likelihood: the classical intervals for θ_{j2} (surgical problem rates) are 51% wider than the Bayesian estimates obtained by Everson and Morris. Here variances on θ_{j2} are obtained which are intermediate between the REML results and the estimates obtained by Everson and Morris (2000, Figure 2).

Example 4.7 Lung cancer death trends: bivariate Poisson outcomes This example uses data for 508 State Economic Areas in the US relating to male and female lung cancer deaths in the periods 1950–69 and 1970–94. The SEAs are nested within 51 states (including the District of Columbia) and eight regions. Of interest in explaining variability in the latter period are the persistence of area mortality differences from the earlier period, the impact on cancer mortality of economic indicators, and mortality differences by state and regional location of the SEAs. As economic indicators average incomes per head in 1982 (relative to the US average) are used, and the possible lagged impact of the same variable defined for 1960 may also be considered.

We define separate Poisson means for the two outcomes and relate them to state varying intercepts. Extensions to this model allow for the influence of past mortality rates and include regression slopes on the log(SMR) in 1950–69. The first model (Model A) for the correlation between the two SEA mortality outcomes in the later period is then (with $h = 1, 2$ for males and females)

$$Y_{ih} \sim \text{Poi}(\mu_{ih})$$

$$\log(\mu_{ih}) = \log(E_{ih}) + \beta_{S_i,\, h}$$

where E_{ih} = expected deaths (from demographic standardisation) and S_i = state to which SEA i belongs. This is a two-level model for each mortality outcome with intercepts varying according to the state that the SEA is located in. This arrangement of the data involves a vector $S[\,]$ of length $T_n = 508$ containing the state indicators for the SEAs. A nested arrangement might also be used, but would entail a square array of dimension [51, Q], where Q is the largest number of SEAs in a state. In a BUGS program the data would have to be padded out by NA values for states containing fewer than Q Economic Areas.

It is assumed that the pairs of intercepts β_{j1} and β_{j2}, $j = 1, \ldots 51$, are bivariate Normal with regression means ν_{jh} determined by state incomes, RINC$_j$ expressed relative to the US average. Then

$$\nu_{j1} = \gamma_{11} + \gamma_{12}\text{RINC}_j$$
$$\nu_{j2} = \gamma_{21} + \gamma_{22}\text{RINC}_j$$

Estimates for this model (Table 4.9) are based on three chains taken to 5000 iterations, with 500 burn-in. They show a positive effect of income on male cancer levels – this might be seen as countering subject matter knowledge whereby relative economic

Table 4.9 Lung cancer deaths in state economic areas

Model A	Mean	St. devn.	2.5%	'97.5%
Correlation, male & female cancer rates	0.509	0.102	0.289	0.689
Likelihood (model for males)	−10910	5.058	−10920	−10900
Likelihood (model for females)	−8851	5.007	−8862	−8842
Intercept Males (γ_{11})	−0.559	0.182	−0.896	−0.204
Income Effect Males (γ_{12})	0.520	0.185	0.157	0.861
Intercept females (γ_{21})	0.204	0.211	−0.224	0.572
Income effect females (γ_{22})	−0.247	0.214	−0.624	0.191
Model B	Mean	St. devn.	2.5%	'97.5%
Intercept Males	−0.078	0.165	−0.377	0.233
Income Effect Males	0.095	0.170	−0.223	0.400
Persistence Effect Males	0.618	0.058	0.506	0.726
Intercept females	0.449	0.154	0.156	0.748
Income effect females	−0.443	0.155	−0.751	−0.149
Persistence Effect females	0.360	0.045	0.270	0.450
Correlation, male & female levels	0.238	0.129	−0.034	0.486
Correlation, male levels and persistence	−0.002	0.143	−0.287	0.287
Correlation, female levels and persistence	−0.034	0.140	−0.305	0.226
Likelihood (model for males)	−7481	6.135	−7495	−7470
Likelihood (model for females)	−5430	6.713	−5444	−5418
Model C	Mean	St. devn.	'2.5%	'97.5%
Region (Males)				
East	−0.042	0.108	−0.226	0.126
G. Lakes	−0.042	0.109	−0.225	0.139
C. North	−0.064	0.119	−0.264	0.127
S. East	0.036	0.100	−0.126	0.199
S. West	−0.068	0.112	−0.252	0.113
Rocky Mts	−0.143	0.114	−0.329	0.044
Far West	0.072	0.110	−0.118	0.246
Region (females)				
East	−0.040	0.110	−0.219	0.136
G. Lakes	0.004	0.115	−0.185	0.191
C. North	−0.084	0.119	−0.285	0.104
S. East	0.116	0.102	−0.045	0.295
S. West	−0.035	0.119	−0.228	0.155
Rocky Mts	−0.144	0.112	−0.323	0.043
Far West	−0.049	0.112	−0.234	0.130
Other parameters				
Intercept Males	−0.162	0.242	−0.538	0.250
Income Effect Males	0.203	0.233	−0.186	0.575
Persistence Effect Males	0.612	0.060	0.517	0.710

(*continues*)

Table 4.9 (*continued*)

Intercept females	0.216	0.249	−0.239	0.580
Income effect females	−0.191	0.238	−0.532	0.231
Persistence Effect females	0.357	0.047	0.278	0.436
Correlation, male & female levels	0.171	0.151	−0.082	0.410
Correlation, male levels and persistence	0.065	0.182	−0.240	0.362
Correlation, female levels and persistence	0.127	0.152	−0.127	0.370
Likelihood (model for males)	−7481	6.995	−7493	−7470
Likelihood (model for females)	−5430	6.888	−5442	−5419

hardship is associated with worse health and mortality. The two outcomes have a correlation of around 0.5 in this model.

Two extended models (Models B and C) allow for persistence of mortality differences. Model B takes the form

$$Y_{ih} \sim \text{Poi}(\mu_{ih})$$

$$\log(\mu_{ih}) = \log(E_{ih}) + \beta_{S_i,\,h,\,1} + \beta_{S_i,\,h,\,2} M_{ih}$$

where M_{ih} is the maximum likelihood SMR for male and female lung cancer ($h = 1, 2$) in SEA i in 1950–69 in the earlier period. The parameters $\beta_{j,\,h,\,2}$ therefore express continuity.

For simplicity of notation, this model is written as

$$\begin{aligned} \log(\mu_{i1}) &= \log(E_{i1}) + b_{S_i,\,1} + b_{S_i,\,2} M_{1i} \\ \log(\mu_{i2}) &= \log(E_{i2}) + b_{S_i,\,3} + b_{S_i,\,4} M_{2i} \end{aligned} \tag{4.16}$$

The state-specific regression effects are modelled as multivariate Normal of order 4 (there are $H = 2$ outcomes and $K = 2$ predictors with cluster specific effects), in which parameters $b_{S_i,\,1}$ and $b_{S_i,\,3}$ (the mortality level parameters) are related to state incomes RINC_{S_i}. So for states $j = 1, 51$, the regression means are modelled as

$$\begin{aligned} \nu_{j,\,1} &= \gamma_{11} + \gamma_{12}\text{RINC}_j \\ \nu_{j,\,2} &= \gamma_{21} \\ \nu_{j,\,3} &= \gamma_{31} + \gamma_{32}\text{RINC}_j \\ \nu_{j,\,4} &= \gamma_{41} \end{aligned} \tag{4.17}$$

The coefficients γ_{21} and γ_{41} for $b_{S_i,\,2}$ and $b_{S_i,\,4}$ represent persistence effects. We evaluate this model, where the intercepts $b_{S_i,\,1}$ and $b_{S_i,\,3}$ depend only upon state incomes RINC_j in 1970–94, against an alternative which adds region in Equation (4.17); these are Models B and C, respectively. Note that examination of the three chain trace plots and Gelman–Rubin statistics for the log-likelihood and parameters shows convergence only after about 500–750 iterations for these models.

By contrast to Model A, Model B shows a significant effect, in the parameter γ_{32} in Equation (4.17), of state incomes on female but not male cancer. The highest female rates at SEA level are in the lowest income states, more in line with subject matter knowledge. The average slopes γ_{21} and γ_{41} relating later to earlier SMRs in Equation (4.16) are 0.62 for males and 0.36 for females, showing that male differences are more enduring. The two outcomes themselves are moderately correlated in this model (median 0.24 with credible interval from −0.03 to 0.49), but there is no correlation between the degree of cancer persistence and cancer levels.

Adding region to Model B to explain differences in SEA mortality (in Model C with New England as base region) attenuates the income effects, expressed by γ_{12} and γ_{32}, while clear region effects are mostly absent. The average likelihoods show no improvement over Model B and penalising for the extra 14 parameters would therefore show a clear deterioration in fit. There is a slight effect for the Rocky Mountain region (lower rates for both female and male mortality), but the credible interval is not confined to negative values. A weakly positive effect for the South East region is apparent on female mortality.

4.6 SMALL DOMAIN ESTIMATION

Survey data are often cost effective ways of obtaining information on a wide variety of topics and at frequent intervals in time. However, they may become sparse in terms of deriving social or health indicators for sub-populations defined by combinations of characteristics, generically termed 'small areas' or 'domains'. For example, a domain might be defined by stratifying on demographic survey variables such as age, locality, sex and ethnicity.

One may seek to make inferences about the average outcome in a domain based on a survey or some other sample, where this survey includes few units, perhaps none, from that domain itself. The direct survey estimates, even if design weighted, are likely to have low precision because sample sizes are not large enough at the domain level. Small area estimation describes a set of empirical and fully Bayes hierarchical regression procedures to combine survey information over similar small areas and to make inferences for the total domain populations (Ghosh and Rao, 1994). These methods may include ancillary information on the small areas from other sources (e.g. Census data) to improve the pooling.

Thus, Folsom *et al.* (2000) describe how small area estimation is used in designing health promotion interventions by linking national and state survey outcome data (on disease and health behaviours) with local area predictors, such as non-survey indicators of social and age structure, in order to make local estimates of prevalence. Random effect small area estimation models have also been applied to discrete outcomes from a Bayesian pespective. For example, Nandram *et al.* (1999) report small area estimates of mortality and Malec *et al.* (1997) consider estimation of the probability of visiting a doctor in the last year, using data from the United States National Health Interview Survey.

Let y_{ijk} be a univariate response for subject i in class or domain k and cluster j, where $j = 1, .. J$, $k = 1, .. K$ and $i = 1, .. n_{jk}$. The class might be defined by demographic attributes (e.g. sex, age band, or ethnic group) while the cluster might be a geographic area. Let x_k denote a set of p categorical predictors assumed to be the same for all subjects in class k, regardless of cluster. Thus, if $p = 2$ and x_1 were sex (1 = male, 0 = female) and x_2 ethnicity (1 = non-white, 0 = white), then a white male would have $x_1 = 1$ and $x_2 = 0$. Also, let z_j be q cluster level variables (e.g. average county or state incomes). Let $\boldsymbol{\beta}_j = \{\beta_{j1}, \dots \beta_{jp}\}$ denote varying slopes over clusters on the categoric variables. Then for a Gaussian outcome with random intercepts and slopes over all predictors,

$$y_{ijk} = \alpha_j + \boldsymbol{\beta}_j x_k + u_{ijk}$$

where $u_{ijk} \sim \mathrm{N}(0, \sigma^2)$. Setting $\boldsymbol{\delta}_j = \{\alpha_j, \boldsymbol{\beta}_j\}$, then random variation in intercepts and slopes may be related to cluster variables z_j via a multivariate normal model

$$\boldsymbol{\delta}_j = \gamma z_j + \boldsymbol{\epsilon}_j \quad \boldsymbol{\epsilon}_j \sim N_{p+1}(0, \Sigma)$$

or equivalently,

$$\boldsymbol{\delta}_j \sim N_{p+1}(\gamma z_j, \Sigma) \tag{4.18}$$

with γ of dimension $(p+1)$ by q.

Other possibilities include the random intercepts model (Moura and Holt, 1999; Battese *et al.*, 1988):

$$y_{ijk} = \alpha_j + \beta x_k + u_{ijk}$$

and models where domain and cluster are conflated into a single index j, and each survey unit is characterised by continuous predictors x_{ij} in a model such as (Hulting and Harville, 1991)

$$y_{ij} = \alpha_j + \boldsymbol{\beta}_j x_{ij} + u_{ij} \tag{4.19}$$

For a binary outcome (e.g. long term ill or not) defined by cluster and domain, we might have

$$y_{ijk} \sim \mathrm{Bern}(\pi_{ijk}) \tag{4.20}$$

$$\mathrm{logit}(\pi_{ijk}) = \alpha_j + \boldsymbol{\beta}_j x_k \tag{4.21}$$

with the same model as in Equation (4.18) for pooling strength over clusters. The random intercepts model in this case (Farrell *et al.*, 1997) is then

$$\mathrm{logit}(\pi_{ijk}) = \alpha_j + \beta x_k$$

The goal of small domain estimation is to make predictions of characteristics Y or proportions P for the entire population of a domain, namely for populations generically denoted N_{jk} for the total population of cluster j and domain k (e.g. white females in California). One might also wish to make predictions for aggregates of domains and/or clusters. Thus, if the binary outcome y_{ijk} denoted long term ill status, the domains were defined by age band, ethnicity and marital status, and the clusters were counties, one might be interested in estimating the proportion of all males who were long term ill in a particular set of counties. So if K' denotes relevant domains, J' denotes relevant clusters, then the estimate of the numerator of P combines

(a) the known survey total in the relevant domains and clusters,

$$\sum_{j \varepsilon J'} \sum_{k \varepsilon K'} \sum_{i=1}^{n_{jk}} y_{ijk}$$

(b) a regression prediction for the population parts $R_{jk} = N_{jk} - n_{jk}$ not included in the survey, obtained from a model such as Equations (4.20)–(4.21) with appropriately defined predictors.

So if $\hat{y}_{ijk}$ denotes a prediction of the binary outcome for the non-surveyed unit, the prediction of P is

$$\sum_{j\varepsilon J'}\sum_{k\varepsilon K'}\sum_{i=1}^{n_{jk}} y_{ijk} / \sum_{j\varepsilon J'}\sum_{k\varepsilon K'} n_{jk} + \sum_{j\varepsilon J'}\sum_{k\varepsilon K'}\sum_{i=1}^{R_{jk}} \hat{y}_{ijk}$$

In practice, the prediction for the non-surveyed population is likely to be at an aggregated level. An example of this approach, from a Bayesian prediction standpoint, is provided by Malec *et al.* (1997), who consider a binary outcome based on recent medical consultation or not.

As an example of appropriately defined predictors for the non-survey prediction, suppose the survey model (4.19) involved a single continuous individual level regressor x_{ij} such as income, and the goal was to predict the population wide cluster or domain mean $\bar{Y}_j$. Suppose $\bar{X}_j$ were a cluster or domain wide average on the regression variable, obtained probably from other sources. Then the non-survey population prediction would be

$$\begin{aligned}\bar{Y}_j &= \beta_{j1} + \beta_{j2}\bar{X}_j \\ &= \gamma z_j \bar{X}_j + \varepsilon_j \bar{X}\end{aligned}$$

where

$$\boldsymbol{\beta}_j = \gamma z_j + \boldsymbol{\epsilon}_j \quad \boldsymbol{\epsilon}_j \sim N_2(0, \Sigma)$$

as above. Domain wide predictions for univariate continuous outcomes (perhaps using the simpler random intercepts model) are exemplified by small area models for per capita income (Fay and Herriott, 1979), and for county crop areas (Battese *et al.*, 1988).

Example 4.8 Economic participation This example takes a sub-sample of the data presented by Farrell *et al.* (1997) relating to economic participation among working age women. There were $T_N = 72521$ women in the 1990 Census sample used by Farrell *et al.*, located in a set of $J = 33$ US states. A sub-sample of 10% of these data, amounting to $T_n = 7184$ women, is taken. As in Farrell *et al.*, a logit model for economic activity, namely working or being available to work, is proposed. The first two categorical predictors $\{x_1, x_2\}$ in this model are marital status, namely married vs. otherwise, and presence of children under 18 vs. otherwise. We allow for state varying impacts of these factors and for homogenous effects of a third factor, x_3, namely the woman's age group $(1 = 15-24, 2 = 25-34, 3 = 35-44, 4 = 45-54, 5 = 55+)$. The three factors may be taken to define classes, $k = 1, .. K$ (where $K = 20$), as above. Thus, in a multilevel pooling model (Model A in Program 4.7), the model for states j and classes k with the form

$$\begin{aligned} y_{ijk} &\sim \text{Bern}(\pi_{ijk}) \\ \text{logit}(\pi_{ijk}) &= \alpha_j + \beta_j x_k + \lambda w_k \end{aligned} \tag{4.22}$$

as in Equations (4.20)–(4.21) above, with β_j of dimension 2, and allowing the age group predictors to have constant effects λ. Writing $\delta_j = \{\alpha_j, \beta_j\}$ we then have $\delta_j \sim N_3(\Delta, V)$. The fixed effects Δ and λ are assigned N(0, 1000) priors. (Note that in Program 4.8, a single string vector is used for the sample subjects, so that the actual program does not take this nested structure.)

Using the estimated parameters $\{\delta_j, \lambda\}$ from applying the pooling model to the sub-sample, one may seek to predict the full population state economic activity rates P_j – here the 'population' is all $T_N = 72\,521$ women. The population wide estimate combines actual sample numbers active (among the 7184) with regression predictions for the remaining 65 337 women. Specifically the model parameters $\{\delta_j, \lambda\}$ are applied to state wide proportions married, with dependent children and in the five age bands. The predictions for the non-sampled population are combined with the actual sample numbers active by state. The overall activity predictions for states may be compared with estimates based on the same logit model but without any pooling of strength over states. The latter option is a fixed effects or 'no pooling' model, and coded as in model B in Program 4.7. This keeps the state varying effects of marital and dependent children status as in Equation (4.22) but as fixed effects with N(0, 1000) priors.

Convergence was rapid (under 500 iterations) for estimating Model A under three parallel chains, and summaries are based on iterations 500–2500. Under this pooling strength model the highest state level intercepts are in Maryland, Indiana and Massachusetts, while the least deterrence of married status on activity is in South Carolina, and the least deterrence of children in the household is in Florida and Washington State. The correlations r_{ab} between the random effects show that deterrence from these two factors tends to be lower where activity rates themselves are higher (Table 4.10).

Table 4.10 Random effect and parameter estimates (Model A)

State	Intercepts		Effects of being married		Effects of children	
	Mean	St. devn.	Mean	St. devn.	Mean	St. devn.
Alabama	0.52	0.28	−1.33	0.29	−0.90	0.23
Arkansas	0.67	0.33	−1.48	0.33	−1.20	0.29
California	0.64	0.19	−1.52	0.19	−1.02	0.19
Connecticut	1.17	0.32	−1.83	0.32	−1.00	0.26
Florida	0.29	0.34	−1.32	0.31	−0.63	0.28
Georgia	0.59	0.23	−1.17	0.26	−0.87	0.22
Illinois	0.77	0.20	−1.53	0.20	−1.14	0.20
Indiana	1.31	0.36	−1.68	0.34	−1.09	0.26
Iowa	0.47	0.32	−1.55	0.30	−0.73	0.27
Kansas	0.73	0.33	−1.60	0.33	−0.83	0.27
Kentucky	0.52	0.29	−1.83	0.31	−0.96	0.28
Louisiana	0.42	0.30	−1.52	0.30	−0.70	0.27
Maryland	1.41	0.32	−1.67	0.30	−1.27	0.27
Massachusetts	1.32	0.24	−2.09	0.24	−1.09	0.25
Michigan	0.69	0.29	−1.62	0.27	−1.03	0.23
Minnesota	1.02	0.29	−1.92	0.31	−0.88	0.26
Mississippi	0.41	0.34	−1.37	0.30	−0.82	0.28
Missouri	0.77	0.31	−1.60	0.26	−0.83	0.24
Nebraska	0.77	0.33	−1.82	0.35	−0.79	0.26
N. Jersey	1.20	0.24	−2.03	0.25	−0.94	0.23
N. York	1.06	0.18	−1.88	0.17	−1.20	0.16
N Carolina	0.90	0.24	−1.43	0.26	−0.81	0.22
Ohio	0.90	0.21	−1.82	0.21	−0.89	0.18

(*continues*)

Table 4.10 (*continued*)

Oklahoma	0.92	0.29	−1.63	0.28	−0.80	0.25
Oregon	0.91	0.40	−1.56	0.38	−0.92	0.30
Pennsylvania	0.74	0.17	−1.97	0.19	−0.88	0.18
S Carolina	0.44	0.29	−1.03	0.29	−0.81	0.28
Tennessee	0.63	0.25	−1.28	0.24	−1.03	0.24
Texas	0.73	0.19	−1.66	0.21	−1.11	0.17
Virginia	0.82	0.26	−1.55	0.28	−1.16	0.23
Washington	0.43	0.35	−1.54	0.31	−0.62	0.29
West Virginia	0.31	0.32	−1.80	0.33	−1.02	0.29
Wisconsin	0.98	0.25	−1.68	0.26	−0.89	0.24

	Mean	St. devn.	2.5%	97.5%
Mean Effects				
Intercept (1)	0.78	0.10	0.57	0.98
Married (2)	−1.63	0.10	−1.84	−1.43
Children (3)	−0.92	0.09	−1.09	−0.74
Correlations between Effects				
r_{12}	−0.41	0.21	−0.75	0.03
r_{13}	−0.34	0.23	−0.74	0.17
r_{23}	0.01	0.27	−0.49	0.51

Let P_j denote the actual state wide activity rate among all $T_N = 72521$ women and $\hat{P}_j$ the prediction. Then Model A has average likelihood −3739 and sum ΣD_j of absolute predicted relative deviations

$$D_j = |(P_i - \hat{P}_j)/\hat{P}_j|$$

averaging 5.3. The effective parameter procedure (Chapter 2) shows there to be 64 estimated parameters in Model A. This estimate is obtained from comparing the mean likelihood and the likelihood at the mean (the likelihood at the posterior mean is −3707). By contrast, the no pooling model involves 103 fixed effects; it has likelihood averaging −3742 and ΣD_j averaging 7.2. So both in terms of predictive fit beyond the sample and penalised fit to actual sample data, the pooling model appears to be preferred.

The prediction of full state population economic activity is only one of a wide range of predictions that could be made. One could also make predictions over combinations of the cluster categories and/or over combinations of the states (e.g. economic activity rates among women with children aged 25–44 in states grouped by region).

4.7 REVIEW

The motivations for multi-level analysis often combine both the need (cf. Chapter 2) to pool strength on estimates of unit level effects at level 2 or above (e.g. school exam success rates) and an interest, as in Chapter 3, in regression slopes for predictors at various levels. The most popularly used packages for multi-level analysis adopt an empirical Bayes methodology which may lead to understatement of the uncertainty in these various parameters. The fully Bayes approach has been succesfully applied to multilevel discrete outcomes (e.g. Carlin *et al.*, 2001), to multivariate multilevel outcomes (Thum, 1997) and to small area estimation (Ghosh and Rao, 1994). Other applications of Bayesian modelling extend to multilevel survival data and to structural equation models for multi-level data (Jedidi and Ansari, 2001). Whereas non-parametric analysis has figured prominently in certain areas of Bayesian application, applications of such approaches to multi-level data are less common. Another area for development is in analysis of multi-level survey data where the sampling process is informative, for example with inclusion probabilities proportional to the size of a school (Pfefferman *et al.*, 2002).

REFERENCES

Battese, G., Harter, R. and Fuller, W. (1988) An error-components model for prediction of county crop areas using survey and satellite data. *J. Am. Stat. Assoc.* **83**, 28–36.

Blakely, T. and Woodward, A. (2000) Ecological effects in multi-level Studies. *J. Epidemiol Commun. Health* **54**(5); 367–374.

Brodsky, A. E., O'Campo, P. J. and Aronson, R. (1999) PSOC in community context: multi-level correlates of a measure of psychological sense of community in low-income, urban neighborhoods. *J. Community Psychol.* **27**(6), 659–679.

Brooks, S. and Gelman, A. (1998) General methods for monitoring convergence of iterative simulations. *J. Computational and Graphical Stat.* **7**, 434–455.

Browne, W. and Draper, D. (2000) Implementation and performance issues in the Bayesian and likelihood fitting of multilevel models. *Computational Stat.* **15**, 391–420.

Browne, W. J., Draper, D., Goldstein, H. and Rasbash, J. (2000) Bayesian and likelihood methods for fitting multilevel models with complex level-1 variation. Institute of Education, University of London.

Carlin, J., Wolfe, R., Brown, C. and Gelman, A. (2001) A case study on the choice, interpretation and checking of multilevel models for longitudinal binary outcomes. *Biostatistics*, **2**(4), 397–416.

Clayton, D. (1996) Generalized linear mixed models. In: Gilks, W. R. *et al.* (eds.), *Markov Chain Monte Carlo in Practice*. London: Chapman & Hall.

Courgeau, D. and Baccaini, B. (1997) Multi-level analysis in the social sciences. *Population* **52**(4), 831–863.

Daniels, M. (1999) A prior for the variance in hierarchical models. *Can. J. Stat.* **27**, 567–578.

Daniels, M. and Kass, R. (1999) Nonconjugate Bayesian estimation of covariance matrices and its use in hierarchical models. *J. Am. Stat. Assoc.* **94**, 1254–1263.

Deely, J. and Smith, A. (1998) Quantitative refinements for comparisons of institutional performance. *J. Roy. Stat. Soc. A* **161**, 5–12.

Duncan, C., Jones, K. and Moon, G. (1999) Smoking and deprivation: are there neighbourhood effects? *Social Sci. & Med.* **48**(4), 497–505.

Everson, P. and Morris, C. (2000) Inference for multivariate normal hierarchical models. *J. Roy. Stat. Soc., Ser. B (Statistical Methodology)* **62**(2), 399–412.

Farrell, P., MacGibbon, B. and Tomberlin, T. (1997) Empirical Bayes small-area estimation using logistic regression models and summary statistics. *J. Bus. Econ. Stat.* **15**(1), 101–108.

Fay, R. and Herriott, R. (1979) Estimates of income for small places: an application of James-Stein procedures to census data. *J. Am. Stat. Assoc.* **74**, 269–277.

Folsom, R., Shah, B., Singh, A., Vaish, A., Sathe, N. and Truman, L. (2000) Small area estimation to target high-risk populations for health intervention. Center for Health Statistics Research, University of North Carolina.

Ghosh, M. and Rao, J. (1994) Small-area estimation – an appraisal. *Stat. Sci.* **9**(1), 55–76.

Goldstein, H. (1987) *Multilevel Models in Education and Social Research*. London: Arnold.

Goldstein, H. (1991) Nonlinear multilevel models, with an application to discrete response data. *Biometrika* **78**(1), 45–51.

Heagerty, P. and Zeger, S. (2000) Marginalized multilevel models and likelihood inference. *Stat. Sci.* **15**(1), 1–19.

Hulting, F. and Harville, D. (1991) Bayesian analysis of comparative experiments and small-area estimation. *J. Am. Stat. Assoc.* **86**, 557–568.

Humphreys, K. and Carr-Hill, R. (1991) Area variations in health outcomes – artifact or ecology. *Int. J. Epidemiology* **20**(1), 251–258.

Ishwaran, H. and Zarepour, M. (2000). Markov chain Monte Carlo in approximate Dirichlet and beta two-parameter process hierarchical models. *Biometrika*, **87**, 371–390.

Jedidi, K. and Ansari, A. (2001) Bayesian structural equation models for multilevel data. In: Marcoulides, G. and Schumacker, R. (eds.), *Structural Equation Modeling and Factor Analysis: New Developments and Techniques in Structural Equation Modeling*. New York: Laurence Erlbaum.

Johnson, V. and Albert, J. (1999) *Ordinal Data Modeling. Statistics for Social Science and Public Policy*. New York, NY: Springer.

Langford, I. and Lewis, T. (1998) Outliers in multilevel data. *J. Roy. Stat. Soc. Ser. A – Statistics in Society* **161**(2), 153–160.

McCullagh, P. and Nelder, J. (1989) *Generalized Linear Models*. London: Chapman & Hall/CRC.

Malec, D., Sedransk, J., Moriarity, C. and LeClere, F. (1997) Small area inference for binary variables in the National Health Interview Survey. *J. Am. Stat. Assoc.* **92**, 815–826.

Martinius, J. (1993) The developmental approach to psychopathology in childhood and adolescence. *Early Human Develop.* **34**(1–2), 163–168.

Mortimore, P., Sammons, P., Stoll, L., Lewis, D. and Ecob, R. (1988) *School Matters, the Junior Years*. Wells: Open Books.

Moura, F. and Holt, D. (1999) Small area estimation using multilevel models. *Survey Methodology* **25**, 73–80.

Nandram, B., Sedransk, J. and Pickle, L. (1999) Bayesian analysis of mortality rates for US health service areas. *Sankhya* **61B**, 145–165.

Pfefferman, D., Moura, F. and Silva, P. (2002) Fitting multilevel models under informative probability sampling. *Multilevel Modelling Newsletter* **14**(1), 8–17.

Rabe-Hesketh, S. and Pickles, A. (1999). Generalised, linear, latent and mixed models. In: Friedl, H., Bughold, A. and Kauermann, G. (eds.), *Proceedings of the 14th International Workshop on Statistical Modelling*. pp. 332–339.

Rice, N., Jones, A. and Goldstein, H. (2001) Multilevel models where the random effects are correlated with the fixed parameters. Department of Economics, University of York.

Seltzer, M. (1993) Sensitivity analysis for fixed effects in the hierarchical model – a Gibbs sampling approach. *J. Educational Stat.* **18**(3), 207–235.

Shouls, S., Congdon, P. and Curtis, S. (1996) Modelling inequality in reported long term illness in the UK: combining individual and area characteristics. *J. Epidemiology and Community Health* **50**(3), 366–376.

Snijders, T. A. B. and Bosker, R. (1999) *Multilevel Analysis. An Introduction to Basic and Advanced Multilevel Modeling*. London: SAGE Publication.

Thum, Y. (1997) Hierarchical linear models for multivariate outcomes. *J. Educational and Behavioural Res.* **22**, 77–108.

EXERCISES

1. In Example 4.2, fit Model A (as described in the text), namely random intercepts at area level but with Student t form. First try the direct Student t model with unknown degrees of freedom (df) using an appropriate prior (e.g. exponential) on df. Then try the scale mixture approach to the Student t, and assess whether any districts are clear 'outliers' (if the degrees of freedom is unknown in the scale mixture method then a discrete prior on it may be used in BUGS).

2. In Example 4.2, try a bivariate cluster effects model with random intercepts and slopes on social class IV and V. Is the correlation between varying intercepts and slopes clearly negative?

3. In Example 4.3 consider assessing heteroscedasticity at level 1 in terms of (a) the IQ categories of Table 4.3, and (b) both IQ category and gender. Is there a more economical IQ categorisation (by which level 1 variances are differentiated) which achieves comparable fit.

4. In Example 4.4, assess the improvement in fit from Model A to model B after correcting for the extra parameters – for example, via the harmonic mean of likelihoods or via the DIC approach, after estimating the effective parameters in Model B. Also try a model where the District percent Protestant acts as a potentially contextual variable, by making the impacts of individual Catholic or Protestant affiliation variable over Districts and dependent on this District variable (this is probably more simply done with homogenous level 2 variance).

5. In Example 4.6, use the full measurement error covariance at stage 1 as as supplied by Everson and Morris, namely

$$V = \begin{pmatrix} 148.9 & 140.4 \\ 140.4 & 490.6 \end{pmatrix}$$

   and re-estimate the model parameters. Set up a procedure to test whether any hospital has above average 'true' problem rates θ_{jh} on both outcomes, relative to the average true rates $\bar{\theta}_h$.

6. In Example 4.7, try a bivariate Normal model for state intercepts (as in Model A) but including region effects in the linear predictor. Similarly try adding the 1960 income ratio to the contemporary income ratio. Are these models well identified, as assessed from posterior summaries, from signs of coefficients, and from the effective parameters criterion (the DIC method discussed in Chapter 2)

7. In Example 4.8, try fitting a fixed effects model (as in Model B), but with constant effects of marital and dependent children status, as opposed to state varying fixed effects. How does this compare to Model A in predicting the population wide state activity rates.

CHAPTER 5

Models for Time Series

5.1 INTRODUCTION

In a variety of scientific disciplines we are faced with choosing appropriate models for representing series of observations generated in time, and for predicting the future evolution of the series. Often the series, although varying continuously in time, is observed at discrete intervals, $t = 1, .., T$. Many series are in fact averages over discrete intervals, or assessed at one point during them; thus many econometric series are monthly or quarterly, and stock price series are daily.

The goals of time series models include smoothing an irregular series, forecasting series into the medium or long-term future, and causal modelling of variables moving in parallel through time. Time series analysis exploits the temporal dependencies both in the deterministic (regression) and stochastic (error) components of the model. In fact, dynamic regression models are defined when model components are indexed by time, and a lag appears on one or more of them in the model specification (Bauwens *et al.*, 1999). For instance, a dynamic structure on the exogenous variables leads to a distributed lag model, and a dynamic structure may also be specified for the endogenous variables, the error terms, the variances of the error process, or the coefficients of the exogenous variables.

While simple curve fitting (e.g. in terms of polynomials in time) may produce a good fit it does not facilitate prediction outside the sample and may be relatively heavily parameterised. By contrast, models accounting for the dependence of a quantity (or error) on its previous values may be both parsimonious and effective in prediction. The autoregressive models developed by Box and Jenkins (1976) and Zellner (1971) are often effective for forecasting purposes (see Section 5.2), but dynamic linear and varying coefficient models (Section 5.5) have perhaps greater flexibility in modelling non-stationary series.

Bayesian methods have been widely applied in time series contexts and have played a significant role in recent developments in error correction and stochastic volatility models (Sections 5.4 and 5.6). They have advantages in simplified estimation in situations where non-standard distributions or nonlinear regression are more realistic. For example, in state space modelling the standard Kalman filtering methods rely on linear state transitions and Gaussian errors, whereas Bayes methods provide a simple approach to include both continuous and discrete outcomes. In ARMA models a Bayesian perspective may facilitate approaches not limited to stationarity, so that stationarity

Applied Bayesian Modelling P. Congdon
 ISBN: 0-471-48695-7

and non-stationarity are assessed as alternative models for the data series. A Bayes approach may also assist in analysis of shifts in time series, where likelihood methods may either be complex or inapplicable. Examples include the analysis of a permanent or temporary break point in a series (see Section 5.7), where a prior for the break point or points might be taken as uniform over a range (T_1, T_2) within $(1,T)$.

5.2 AUTOREGRESSIVE AND MOVING AVERAGE MODELS UNDER STATIONARITY AND NON-STATIONARITY

A starting point in dynamic regression models is often provided by considering dynamic structures in the outcomes. Autoregressive process models describe data driven dependence in an outcome over successive time points. For continuous data y_t, observed at times $t = 1, .. T$ the simplest autoregressive dependence in the outcomes is of order 1, meaning values of y at time t depend upon their immediate predecessor. Thus an, AR(1) model typically has the form

$$y_t = \mu + \rho_1 y_{t-1} + u_t \quad t = 2, .. T \tag{5.1}$$

where μ represents the level of the outcome, and ρ models the autocorrelation between successive observations. After accounting for such observation driven serial dependence, the errors may (at least initially) be taken as exchangeable white noise and to follow, for example, a Normal density, $u_t \sim N(0, \sigma^2)$ with constant variance and precision $\tau = 1/\sigma^2$ across all time points t, and cov $(u_s, u_t) = 0$.

If the data are centred, then a simpler model may be estimated

$$y_t = \rho y_{t-1} + u_t \tag{5.2}$$

Additional dependence on lagged observations $y_{t-2}, y_{t-3}, . y_{t-p}$ leads to AR(2), AR(3), .. AR(p) processes. It may be noted that (5.1), if taken to apply to $t = 1$ also, implies reference to unobserved or latent data y_0. If a prior on y_0 is included in the model specification, this leads to what is known as a full likelihood model. For an AR(p) model with times $t = 1, .. p$ included there are p implicit latent values, $y_0, y_{-1}, .. y_{1-p}$ in the full likelihood model.

Classical estimation and forecasting with the AR(p) model rest on stationarity, which essentially means that the process generating the series is the same whenever observation starts: so the vectors $(y_1, .. y_k)$ and $(y_t, ... y_{t+k})$ have the same distribution for all t and k. Specifically, the expectations $E(y_t)$ and covariances $C(y_t, y_{t+k})$ are independent of t. For the stationary AR(p) model to be applicable, an observed data series may require initial transformation and differencing to eliminate trend. Non-stationary time series can often be transformed to stationarity by differencing of order d, typically first or second differencing at most. This may be combined with a scale transformation, e.g. $Y_t = \log(y_t)$.

Using the B operator to denote a backward movement in time, a first difference $(d = 1)$ in y_t

$$z_t = y_t - y_{t-1}$$

may equivalently be written

$$z_t = y_t - By_t = (1 - B)y_t$$

Then with z_t now the outcome, the AR(1) model (5.2) becomes

$$z_t - \rho z_{t-1} = u_t$$

or

$$1 - \rho^B$$
$$z_t(1 - B) = u_t$$

An AR(p) process in z_t leads to a pth order polynomial in B, so that

$$z_t(1 - \rho_1 B - \rho_2 B^2 - \ldots \rho_p B^p) = u_t$$

for which an alternative notation is

$$\rho(B)z_t = u_t$$

The process is stationary if the roots of $\rho(B)$ lie outside the unit circle. For instance if $p = 1$, the series is stationary if $|\rho_1| < 1$.

In the AR(p) model, an outcome depends upon its past values and a random error or innovation term u_t. If the impact of u_t is in fact not fully absorbed in period t, there may be moving average dependence in the error term also. Thus, for centred data, the model

$$z_t - \rho_1 z_{t-1} = u_t - \theta_1 u_{t-1} \tag{5.3}$$

defines a first order moving average $MA(1)$ process in u_t combined with AR(1) dependence in the data themselves. In BUGS, moving average effects in the u_t for metric (e.g. Normal or Student t) data may require an additional measurement error term to be introduced, because the centred density adopted in BUGS for Normal and Student t data assumes unstructured errors. In general an ARIMA(p, d, q) model is defined by dependence up to lag p in the observations, by q lags in the error moving average, and by differencing the original observation (y_t) d times. An ARMA(p, q) model in y_t, therefore, retains the original data without differencing. In the general ARMA(p, q) representation,

$$\rho(B)y_t = \theta(B)u_t \tag{5.4}$$

the process is stationary if the roots of $\rho(B)$ lie outside the unit circle, and invertible if the roots of $\theta(B)$ lie outside the unit circle. For instance, if $p = q = 1$, the series is stationary and invertible if $|\rho_1| < 1$ and $|\theta_1| < 1$.

In a distributed lag regression predictors x_t, and their lagged values, are introduced in addition to the lagged observations y_{t-1}, y_{t-2}, etc. A distributed lag model for centred data has the form

$$y_t = \sum_{m=0} \beta_m x_{t-m} + u_t \tag{5.5}$$

while a model with lags in both y and x may be called an Autoregressive Distributed Lag (ADL or ARDL) model (see Bauwens *et al.*, 1999; Greene, 2000):

$$\rho(B)y_t = \beta(B)x_t + u_t$$

The latter form leads into recent model developments in terms of error correction models.

Dependent Errors

In the specifications above, the errors u_t are assumed temporally uncorrelated with diagonal covariance matrix and autocorrelation is confined to the observations themselves. However, if correlation exists between the errors then the covariance matrix is no longer diagonal. Let ε_t be correlated errors with

$$y_t = \alpha + \beta x_t + \varepsilon_t \tag{5.6a}$$

and suppose that an AR(p) transformation of the ε_t is required

$$\gamma(B)\varepsilon_t = u_t$$

in order that u_t is unstructured with constant variance, where

$$\gamma(B) = 1 - \gamma_1 B - \gamma_2 B^2 - \ldots \gamma_p B^p$$

A frequently occurring model is one with AR(1) errors ε_t such as

$$y_t = \alpha + \beta x_t + \varepsilon_t$$
$$\varepsilon_t = \gamma\varepsilon_{t-1} + u_t$$

More generally, regression models may be defined with ARMA(p, q) errors

$$\varepsilon_t - \gamma_1\varepsilon_{t-1} - \gamma_2\varepsilon_{t-2}\ldots - \gamma_p\varepsilon_{t-p} = u_t - \theta_1 u_{t-1} - \theta_2 u_{t-2}\ldots - \theta_q u_{t-q} \tag{5.6b}$$

To facilitate estimation, the AR(1) error model may be re-expressed in non-linear autoregressive form, for observations $t > 1$ subsequent to the first, and with homogenous errors u_t,

$$\begin{aligned} y_t &= \gamma y_{t-1} + \alpha - \alpha\gamma + \beta x_t - \gamma\beta x_{t-1} + u_t \\ &= \gamma(y_{t-1} - \beta x_{t-1}) + \alpha(1-\gamma) + \beta x_t + u_t \end{aligned} \tag{5.7}$$

So the intercept in the original model is obtained by dividing the intercept in the transformed data model by $1 - \gamma$.

Multivariate series

The above range of models may be extended to modelling multivariate dependence through time, with each series depending both on its own past and the past values of the other series. One advantage of simultaneously modelling several series is the possibility of pooling information to improve precision and out-of-sample forecasts. Vector autoregressive models have been used especially in economic forecasts for related units of observation, for example of employment in industry sectors or across regions, and of jointly dependent series (unemployment and production).

For example, a time series of K centred metrical variables $Y_t = (y_{1t}, y_{2t}, \ldots y_{Kt})'$ is a multivariate Normal autoregression of order p, denoted VAR(p), if it follows the relations

$$\begin{aligned} Y_t &= \Phi_{p1} Y_{t-1} + \ldots .\Phi_{pp} Y_{t-p} + U_t \\ U_t &\sim N_K(0, V) \end{aligned} \tag{5.8}$$

where the matrices $\Phi_{p1}, \ldots \Phi_{pp}$ are each $K \times K$, and the covariance matrix is for exchangeable errors $u_{1t}, u_{2t}, \ldots, u_{Kt}$. Then if $K = 2$, Φ_{p1} would consist of own-lag coefficients relating Y_{1t} and Y_{2t} to the lagged values $Y_{1,\,t-1}$ and $Y_{2,\,t-1}$ and cross-lag coefficients relating Y_{1t} to $Y_{2,\,t-1}$ and Y_{2t} to $Y_{1,\,t-1}$.

5.2.1 Specifying priors

Bayesian time series applications with autoregressive and moving average components have included all the above modelling approaches, and have also been applied in models combining state-space and classical time series concepts (Huerta and West, 1999). Among the questions that are involved in specifying priors for ARMA type model parameters are whether stationarity and invertibility constraints are taken, whether a full or conditional likelihood approach is used, and assumptions made about the innovation errors. As discussed in Chapter 1, prior elicitation consists in incorporating relevant background knowledge into the formulation of priors on parameters[1]. Often, relevant knowledge is limited, and diffuse or 'just proper' priors are called on. This raises questions of sensitivity to prior specifications, for instance on variances (Daniels, 1999) or on time series assumptions (e.g. on stationarity or otherwise or on initial conditions), and the reader is encouraged to experiment with alternative priors in the worked examples of the chapter.

Autoregressive and ARMA models without stationarity

Consider first the autoregressive AR(p) model in the endogenous variable,

$$\rho(B)y_t = u_t$$

Unlike classical approaches, a Bayesian analysis of the AR(p) model is not confined to stationary processes. As emphasized by Zellner (1971) a prior assumption of stationarity in the AR(p) process may be regarded as a modelling assumption to be assessed, rather than a necessary restriction. Hence in an iterative sampling framework, an autoregressive model may be applied to observations y_t without pre-differencing to eliminate trend, and the probability of stationarity assessed by the proportion of iterations $s = 1, .., S$ where stationarity in the coefficients $\rho^{(s)}$ at iteration s actually held in terms of roots located outside the unit circle. A significant probability of non-stationarity would then imply the need for differencing, different error assumptions, or model elaboration, for example, to a higher order AR model (Naylor and Marriott, 1996).

One approach of Zellner (1971) without a stationarity constraint is to use a non-informative reference prior, such as Jeffrey's prior, with

$$\pi(\rho_1, \ldots \rho_p, \tau) \propto \tau^{-1}$$

where $\tau = 1/\sigma^2$. In BUGS implementation of this approach would require direct sampling from the full conditionals. As with any non-informative prior, potential problems of identifiability may be increased, whereas identifiability generally improves as just proper or informative proper priors are adopted.

The reference prior approach may be generalised to include AR(p) processes with normal-gamma conjugate priors (Broemeling and Cook, 1993). For example, with a gamma $G(a, b)$ prior for τ, $\rho|\tau$ is taken to be multivariate Normal $N(r, \tau\Sigma_0)$, where r is the prior mean on the lag coefficients, and Σ_0 is a $p \times p$ positive definite matrix. A straightforward analysis is defined by conditioning the likelihood on the first p observations $Y_1 = \{y_1, y_2, \ldots, y_p\}$, so avoiding the specification of a prior on the latent pre-series value. The likelihood then only relates to observations $Y_2 = \{y_{p+1}, y_{p+2}, \ldots, y_n\}$. The conditional likelihood is then

[1] In Bayesian econometrics, a distinction is sometimes made between two types of formal elicitation procedures, structural or predictive – structural methods involve assessment of the quantiles of the prior distribution of parameters, drawing on theoretical models or past experience (Bauwens *et al.*, 1999; Kadane, 1980).

$$f(Y_2|Y_1, \rho, \tau) \propto \tau^{0.5(n-p)} \exp\left(-0.5\tau \sum_{t=p+1}^{n} [\rho(B)y_t]^2\right) \tag{5.9}$$

The posterior is proportional to the product of the two priors and the conditional likelihood.

Naylor and Marriott (1996) discuss a full likelihood analysis of the ARMA model without stationarity constraints by using proper but relatively 'weak' priors on the latent pre-series values $Y_0 = (y_0, y_{-1}, \ldots, y_{1-p})$ and $E_0 = (u_0, u_{-1}, \ldots, u_{1-q})$. For instance, if the observed series is assumed Normal with mean μ and conditional variance σ^2, Naylor and Marriott suggest the pre-series values Y_0 be taken as Student t with low degrees of freedom, having the same mean as the main series but a variance larger by a factor $\kappa \geq 1$, namely $\kappa\sigma^2$. (This is equivalent to dividing the precision τ by κ.) If there are several pre-series values (when $p > 1$), a multivariate Student t might be used. Note that Zellner (1971, p. 87) suggests a prior for y_0 that does not involve any of the parameters of the main model.

Priors on error terms

The assumption of white noise errors in the AR(p) model may need to be assessed with more general priors allowing for outlier measurements. However, interpretations of outlying points in time series are made cautiously. Outliers at time t may be clearly aberrant, and either excluded or replaced by interpolated values taking account of surrounding values (Diggle, 1990). On the other hand, especially in economic time series, they may reflect aspects of economic behaviour which should be included in the specification (Thomas, 1997).

Some fairly conventional approaches are for a Normal mixture distribution or Student t errors to replace the usual Normal error assumption (Hoek *et al.*, 1995; West, 1996). Thus, let Δ be the small probability of an outlier (e.g. $\Delta = 0.05$), and let the binary indicator

$$J_t \sim \text{Bernoulli}(\Delta)$$

govern whether the observation t is an outlier. Then one alternative (West , 1996) to the Normal errors AR(1) model in (5.1) is

$$y_t = \mu + \rho y_{t-1} + u_t$$

where $u_t \sim N(0, K_t\sigma^2)$, and where random or fixed effect parameters $K_t > 1$ inflate the variance when $J_t = 1$. This is known as an innovation outlier model and may be written as

$$u_t \sim (1 - \Delta)N(0, \sigma^2) + \Delta N(0, K\sigma^2) \tag{5.10}$$

If the Student t is used as an outlier model for the innovations then the most appropriate option is the scale mixture form – this includes weights w_t averaging 1 which scale the precision – so low weights (e.g. under 0.5) indicate possible outliers. In models with autocorrelated errors, such as

$$y_t = \alpha + \beta x_t + \varepsilon_t$$

$$\varepsilon_t - \gamma_1 \varepsilon_{t-1} = u_t$$

the innovation outlier model would apply to the u_t.

One may also define additive outliers corresponding to shifts in the observation series that may not occur for all time points. So, following Barnett *et al.* (1996) one might define a model

$$y_t = \alpha + \beta x_t + o_t + \varepsilon_t$$
$$\varepsilon_t - \gamma_1 \varepsilon_{t-1} = u_t$$

with $o_t \sim N(0, K_{1t}\sigma^2)$, where K_{1t} is either 0 or positive (corresponding to times when an additive outlier does or does not occur), and $u_t \sim N(0, K_{2t}\sigma^2)$, where K_{2t} is either 1 or greater than 1. One might then model the two outliers jointly, for instance via a discrete set of possible values for $K_t = \{K_{1t}, K_{2t}\}$. Barnett *et al.* illustrate this with a prior for K_t consisting of (0, 1), (3, 1), (10, 1), (0, 3), (0, 10) with selection among them based on a multinomial rather than binary indicator J_t, but with prior probabilities possibly biased towards the null option (0, 1). For example, prior probabilities on the just named options might be (0.9, 0.025, 0.025, 0.025, 0.025).

Another approach to additive outliers is developed by McCulloch and Tsay (1994). They consider first a random level-shift autoregressive (RLAR) model

$$y_t = \mu_t + \varepsilon_t$$
$$\mu_t = \mu_{t-1} + \delta_t \eta_t$$
$$\varepsilon_t = \gamma_1 \varepsilon_{t-1} + \gamma_2 \varepsilon_{t-2} + \ldots + u_t$$

where δ_t is Bernoulli with probability Δ and governs the chance of a level shift at time t, the terms $\eta_t \sim N(0, \xi^2)$ describe the shifts, the ε_t are autoregressive errors and the white noise errors u_t are $N(0, \sigma^2)$. The shift variance ξ^2 is taken as a large multiple (e.g. 10, or 100) times the white noise variance σ^2. The probability of a shift Δ is beta with parameters favouring low probabilities, for instance $\Delta \sim \text{Beta}(5, 95)$. The above model may be re-expressed as

$$y_t = \mu_t + \gamma_1(y_{t-1} - \mu_{t-1}) + \gamma_2(y_{t-2} - \mu_{t-2}) + .. + u_t$$
$$\mu_t = \mu_{t-1} + \delta_t \eta_t$$

McCulloch and Tsay also propose a specialised additive outlier model, namely

$$y_t = o_t + \varepsilon_t$$
$$o_t = \delta_t \eta_t$$
$$\varepsilon_t = \gamma_1 \varepsilon_{t-1} + \gamma_2 \varepsilon_{t-2} + \ldots + u_t$$

which is the same as the RLAR model except that the level $\mu_t(= o_t)$ no longer depends upon its previous value. This model can be re-expressed as

$$y_t = o_t + \gamma_1(y_{t-1} - o_{t-1}) + \gamma_2(y_{t-2} - o_{t-2}) + .. + u_t$$
$$o_t = \delta_t \eta_t$$

Another version of this model (Martin and Yohai, 1986) has

$$y_t = (1 - \delta_t)\varepsilon_t + \delta_t \eta_t$$

with ε_t again autoregressive.

Priors consistent with stationarity and invertibility

Prior assumptions regarding stationarity or non-stationarity (and invertibility or non-invertibility) may interrelate to other aspects of time series model specification. Thus, specifying priors for a full likelihood involving all observations may be more straightforward for a stationary model. Consider the AR(1) model

$$y_t = \mu + \rho y_{t-1} + u_t$$

Then for a stationary process with $\rho \in [-1, 1]$, and with exchangeable errors u_t with mean 0 and variance σ^2, the first observation y_1 has mean μ and conditional variance

$$\sigma^2/(1-\rho^2) \tag{5.11}$$

rather than σ^2. This analytic form corresponds to assuming an infinite history for the process. For $p > 2$, a matrix generalisation of (5.11) is involved. For instance for $p = 2$ with lag coefficients $\{\rho_1, \rho_2\}$, the equivalent of (5.11) is a bivariate Normal for $Y_1 = \{y_1, y_2\}$ with covariance matrix $\sigma^2\Sigma$, where

$$\Sigma = R\Sigma R' + K_1(2)K_1(2)'$$

where $K_1(2) = (1, 0)'$ is a (2×1) vector, and

$$R = \begin{bmatrix} \rho_1 & \rho_2 \\ 1 & 0 \end{bmatrix}$$

A prior constrained to stationarity might alternatively involve ensuring lag parameters in the acceptable region combined with a prior density on the pre-series values. Thus, for $p = 1$ the prior for the AR(1) model (5.2), the prior would be $\pi(y_0, \rho, \tau)$, rather than $\pi(\rho, \tau)$, while for $p > 1$ it would be necessary to specify priors on $y_0, y_{-1}, \ldots y_{1-p}$. Specifying priors on the pre-series values under stationarity may be avoided by backcasting (Ravinshanker and Pay, 1997, p. 182), a procedure which takes advantage of the fact that the model for a stationary time series taken forward in time also applies with time reversed (see Box and Jenkins, 1970).

Some authors suggest simple rejection sampling, with samples of ARMA coefficients ρ or θ accepted (as a block) if they lie in the acceptable region (Chib and Greenberg, 1994). Another option on the priors on the AR coefficients $P = \{\rho_1, \ldots \rho_p\}$ consistent with stationarity, is reparameterisation of the ρ_j in terms of the partial correlations r_j of the AR(p) process (Marriott and Smith, 1992; Marriott *et al.*, 1996). In the AR(p) model let

$$\rho^{(p)} = (\rho_1^{(p)}, \rho_2^{(p)}, \ldots \rho_p^{(p)})$$

with $\rho_j^{(p)}$ the jth coefficient. Then the stationarity condition is equivalent to restrictions that $|r_k| < 1$ for $k = 1, 2, \ldots p$. The transformations linking priors on r_j to the implied ρ_j are for ($k = 1, \ldots p$ and $i = 1, \ldots k-1$)

$$\begin{aligned} \rho_k^{(k)} &= r_k \\ \rho_i^{(k)} &= \rho_i^{(k-1)} - r_k \rho_{k-i}^{(k-1)} \end{aligned} \tag{5.12}$$

So for $p = 3$ the transformations would be

$$\begin{aligned} \rho_3^{(3)} &= r_3 \\ \rho_1^{(3)} &= \rho_1^{(2)} - r_3\rho_2^{(2)} = \rho_1^{(2)} - r_3 r_2 \quad (\text{for } k = 3, i = 1) \\ \rho_2^{(3)} &= \rho_2^{(2)} - r_3\rho_1^{(2)} = \mathrm{r}_2 - r_3\rho_1^{(2)} \quad (\text{for } k = 3, i = 2) \\ \rho_1^{(2)} &= \rho_1^{(1)} - r_2\rho_1^{(1)} = \mathrm{r}_1 - r_2 r_1 \quad (\text{for } k = 2, i = 1) \end{aligned}$$

One may use $U(-1, 1)$ priors on the r_j but alternative transformations have be proposed. Thus, let $r_j{}^*$ be a normal or uniform draw on the real line, and then r_j are given by solving $r_j^* = \log([1 + r_j]/[1 - r_j])$. Jones (1987) proposes beta variables $r_1^*, r_2^*, r_3^*, .. r_k^*$, and then transforming to the interval [−1,1] via $r_1 = 2r_1^* - 1$, $r_2 = 2r_2^* - 1$, etc. Priors on the θ coefficients in Equation (5.4) consistent with invertibility may be obtained by a parallel reparameterisation, for $k = 1, .. q$ and $i = 1, .. k - 1$ (Monahan, 1984; Marriott *et al.*, 1996).

Chib and Greenberg (1994) discuss priors for the ARMA(p, q) errors model (5.6) that are constrained to stationarity and invertibility. They emphasise that nonstationarity in the data themselves may still be modelled by unrestricted coefficients ρ on lagged y_t.

Further aspects of priors on regression coefficients

If a stationarity constraint is not imposed then a flat prior may be chosen for the $P = \{\rho_1, .. \rho_p\}$ but proper priors, such as $\rho_j \sim N(0, 1), j = 1, .., p$ provide a relatively vague but proper alternative. To take priors on the ρ_j that had larger variances would neglect the typical pattern of autoregressive coefficients on the endogenous variable, with values exceeding 1 being uncommon, except in short term explosive series. More specialised priors on the coefficients on lagged endogenous or exogenous variables may be applied in particular forms of model. In vector autoregressions, Litterman (1986) specifies a prior mean of unity for the first own-lag coefficient (relating y_{kt} to $y_{k,\, t-1}$), but zero prior means for subsequent own-lag coefficients, and for all cross variable lags. The form of the own first lag prior follows the observed trend behaviour of many undifferenced economic variables in approximating a random walk, namely

$$y_{kt} = y_{k,\, t-1} + u_{kt} \quad k = 1, .. K;\ t = 2, \ldots, T$$

In distributed lag models, Akaike (1986) discusses smoothness priors, namely Normal or Student t priors for differences of order d in the distributed lag coefficients β_m. For example, a first order smoothness prior would specify β_m as $N(\beta_{m-1}, \sigma^2_\beta)$ or equivalently $\Delta_m = \beta_m - \beta_{m-1} \sim N(0, \sigma^2_\beta)$.

However, the prior on autoregressive, moving average or regression parameters is expressed, model selection may be developed by procedures such as those of George and McCulloch (1993) or Kuo and Mallick (1994), and discussed in Chapter 3. In this way, alternative models defined by inclusion or exclusion of coefficients may be evaluated. For instance, if $J_k = 1$ or $J_k = 0$ according as the kth autoregressive coefficient ρ_k is included or excluded, then one might set prior probabilities that $J_k = 1$ which decline as k increases (Barnett *et al.*, 1996).

5.2.2 Further types of time dependence

Other forms of time dependence in the y_t may be combined with autoregression of y_t on previous values of the series, while still assuming the errors are uncorrelated. This rather catholic approach to time series modelling is more likely when there is no necessary restriction to stationarity and the autoregressive dependence in the y_t is just one among several modelling options. Thus one might consider modelling trend without necessarily differencing beforehand. However, as well as simple trends in t or $\log(t)$, there are often periodic fluctuations in time series such that a series of length T contains K cycles (timed from peak to peak or trough to trough). So the frequency or number of cycles per unit of time, is

$$f = K/T$$

An appropriate model for a series with a single cycle is then

$$y_t = A\cos(2\pi ft + P) + u_t$$

where A is the amplitude and P the phase of the cycle, and period $1/f$, namely the number of time units from peak to peak. To allow for several (r) frequencies operating simultaneously in the same data, the preceding may be generalised to

$$y_t = \sum_{j=1}^{r} A_j \cos(2\pi f_j t + P_j) + u_t$$

For stationarity to apply, the A_j may be taken as uncorrelated with mean 0 and the P_j as uniform on $(0, 2\pi)$. Because

$$\cos(2\pi f_j t + P_j) = \cos(2\pi f_j t)\cos(P_j) - \sin(2\pi f_j t)\sin(P_j)$$

the model is equivalently written

$$y_t = \sum_{j=1}^{r} \{\alpha_j \cos(2\pi f_j t) + \beta_j \sin(2\pi f_j t)\} + \mathrm{u_t} \tag{5.13}$$

where $\alpha_j = A_j \cos P_j$, $\beta_j = -A_j \sin P_j$.

5.2.3 Formal tests of stationarity in the AR(1) model

There is a wide literature on the question of trend stationarity of the outcome y_t in the AR(1) model for, with non-correlated Normal errors u_t:

$$y_t = \mu + \rho y_{t-1} + u_t \tag{5.14a}$$

where $u_t \sim N(0, \sigma^2)$. If $|\rho| < 1$ in this model, then the process is stationary with a variance $\sigma^2/(1-\rho^2)$ and long run mean

$$\mu_e = \mu/(1-\rho) \tag{5.14b}$$

as can be seen by taking expectations in Equation (5.14a). If $|\rho| < 1$ the series will tend to revert to its mean level after undergoing a shock. If, however, $\rho = 1$, the process is a nonstationary random walk with its mean and variance undefined by parameters in Equation (5.14).

Tests for nonstationarity may therefore compare the simple null hypothesis $H_0: \rho = 1$ with the composite alternative $H_1: |\rho| < 1$ and classical tests of nonstationarity revolve around this type of unit root test. If the hypothesis $\rho = 1$ is not rejected, then this implies that the differences $\Delta y_t = y_t - y_{t-1}$ are stationary (this is known as difference stationarity as opposed to trend stationarity in the undifferenced outcome).

The simple model (5.14) may be extended (Schotman and van Dijk, 1991) to suit the observed series being considered by adding trends in t (e.g. linear growth) and lags in Δy_t. These modifications are intended to improve specification and ensure that the assumption that the errors u_t are uncorrelated in fact pertains. As an example, Hoek *et al.* (1995, Equation (16)) consider the series of Nelson and Plosser (1982) in terms of an extended model

$$y_t = \mu + \rho y_{t-1} + \lambda t + \phi_1 \Delta y_{t-1} + \phi_2 \Delta y_{t-2} + u_t \tag{5.15}$$

where λt models a linear trend. Bauwens *et al.* (1999, p. 166) consider an alternative non-linear form of the AR model, also involving a trend in time λt. For an AR(1) model this is expressed as

$$(1 - \rho B)(y_t - \mu - \lambda t) = u_t$$

which can be rewritten as

$$y_t = \rho y_{t-1} + \rho\lambda + (1 - \rho)(\mu + \lambda t) + u_t$$

When lags in Δy_t are introduced, an extended version of the nonlinear form would be specified by the model

$$y_t = \rho y_{t-1} + \rho\lambda + (1 - \rho)(\mu + \lambda t) + \phi_1 \Delta y_{t-1} + \phi_2 \Delta y_{t-2} + u_t \tag{5.16}$$

Bauwens *et al.* report differences in the behaviour of nonlinear and linear versions of the AR model under nonstationarity or unit root situations.

Bayesian tests of nonstationarity in the original AR(1) model (5.14) or in these extended versions may follow the classical procedure in testing explicitly for the simple hypothesis H_0: $\rho = 1$ versus the composite alternative $|\rho| < 1$. Thus Hoek *et al.* (1995) considers a prior for ρ confined to non-explosive values, but putting a mass of 0.5 on the unit root $\rho = 1$. Other values of ρ are uniformly distributed with mass 1/(2A) between [1-A, 1) where $1 \geq A > 0$. For instance, taking $A = 1$ gives the prior

$$\pi(\rho) = 0.5 \quad \rho = 1$$
$$\pi(\rho) = 0.5 \quad \rho \varepsilon [0, 1)$$

Bayesian tests of non-stationarity may also (Lubrano, 1995a) compare the composite alternatives

$$H_0: \rho \geq 1 \text{ as against } H_1: \rho < 1$$

If there is genuinely explosive behaviour in the series, then artificially constraining the prior to exclude values of ρ over 1 may be inconsistent with other aspects of appropriate specification. The posterior probability that $\rho \geq 1$ is then a test for nonstationarity.

In either approach, other considerations may be important. The first is the impact of outliers or nonconstant variance. Hoek *et al.* (1995) demonstrate that assuming Student *t* errors for the innovations provides robustness against outliers that cause a flatter estimate of ρ than the true value, and so lead to over-frequent rejection of non-stationarity. The second consideration is the impact of initial values – for example, the unobserved value y_0 in the AR(1) model (1). If $\rho = 1$, then this particular model may be expressed as

$$y_t = \sum_{i=0}^{t-1} \rho^i u_{t-i} + \rho^t y_0$$
$$= \sum_{i=0}^{t-1} u_{t-i} + y_0$$

namely as an accumulation of innovations or shocks plus the initial value. Hence, conditioning on initial values (e.g. taking y_1 as fixed in the AR(1) model and so not referring to the unknown y_0) may distort tests of nonstationarity.

5.2.4 Model assessment

Time series fit measures often adopt cross-validatory principles as well as standard measures of fit to all the data points. Within sample fit measures include the DIC or BIC criteria, marginal likelihood approximations (Gelfand and Dey, 1994), or the minimum predictive loss criteria of Gelfand and Ghosh (1998) and Laud and Ibrahim (1995). Out-of-sample cross validation is demonstrated by Nandram and Petrucelli (1997), who use a predictive density $f(y_{t+1}|y_1, y_2, .. y_t)$ for assessing one step ahead forecasts within the span $t = 1, .., T-1$ of the observed series.

If model comparisons involve selection of subsets of coefficients (e.g. by one or more binary inclusion indicators), then posterior vs. prior probabilities of inclusion are obtainable from standard MCMC output. As discussed above, model choice may be influenced by (or be sensitive to) features of the data or model specification such as outliers and assumptions about priors. Model checking (e.g. in assessing possible outliers by CPO statistics) may interact with procedures assessing sensitivity to priors – if for instance, robust error assumptions are adopted to lessen the effect of outliers.

Example 5.1 Oxygen inhalation The first example is of an AR(p) model for the outcome y_t without a prior stationary constraint, and with uncorrelated errors. Specifically, we consider undifferenced data on a burns patient reported by Broemeling and Cook (1993). The series consists of 30 readings y_t of the volume of oxygen inhaled at two minute intervals (Figure 5.1)[2]. Broemeling and Cook adopt the AR(1) model in Equation (5.14) with the simplification provided by a conditional likelihood, taking the first observation y_1 as a known constant and avoiding reference to the initial condition, or latent value y_0. They investigate the question of stationarity of the process, by sampling directly from the analytic form of the marginal posterior for ρ, and assessing the proportion of samples where $|\rho| < 1$. If stationarity is confirmed then it makes sense to derive the long run average μ_e in the AR(1) model of Equation (5.14). Taking $\tau = 1/\sigma^2$, Broemeling and Cook adopt a flat prior on $\theta = (\mu, \rho, \tau)$, namely

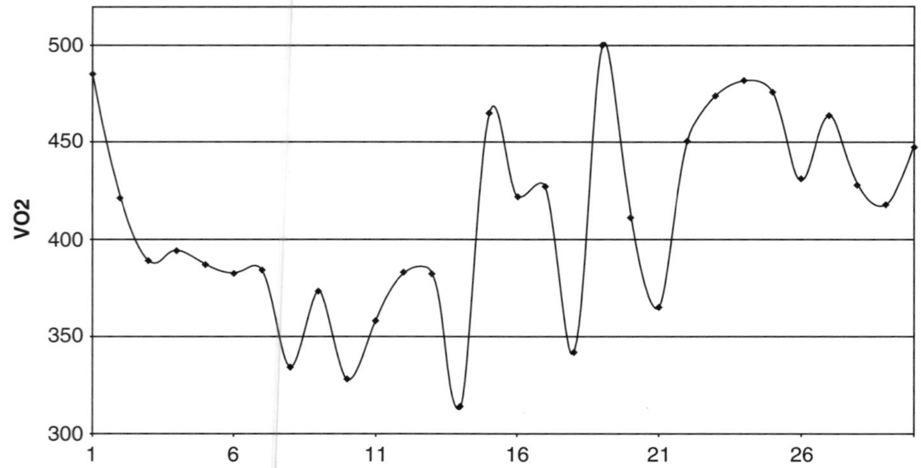

Figure 5.1 VO2 Readings

[2] Data kindly provided by Peyton Cook.

$$\pi(\theta, \tau) \propto \tau^{-1} \tag{5.17}$$

Here two forms of just proper gamma prior for τ are adopted, one $\tau \sim G(0.001, 0.001)$ which approximates Equation (5.17), whereas the prior $\tau \sim G(1, 0.001)$ approximates a uniform prior for the precision. The prior for μ is taken as $N(300, 10^8)$, weakly reflecting the average level of the observations, and ρ is taken as N(0, 1), consistent with a model not assuming a stationarity constraint. Finally, the fact that the outcome is positive means the Normal sampling for y is truncated below at zero.

To implement this conditional likelihood method, three chains with dispersed[3] starting values are run for 20 000 iterations, with convergence apparent from around iteration 1000. With the prior $\tau \sim G(0.001, 0.001)$, this gives posterior means for μ and ρ of 287 (standard deviation 79) and 0.30 (s.d. 0.19), similar to those obtained by Broemeling and Cook. The probability of non-stationarity is negligible at 7E-4. With the alternative prior $\tau \sim G(1, 0.001)$, ρ is estimated at around 0.295 (s.d. 0.18) and posterior summaries are very similar for other parameters.

The Durbin–Watson statistic (DW in Program 5.1) is used to assess auto-correlation in the residuals of the fitted model and is approximately 2 when there is no correlation. One might assess this (and hence the potential relevance of a model with correlated errors) by a step function in BUGS comparing DW with 2. Here it is clear that a conditional likelihood approach leaves no correlation in the errors, and the Durbin-Watson statistic has posterior average of 2.07, very close to 2. Another option to assess the validity of the white noise assumption for u_t is an analysis of the posterior residual estimates, $\hat{u}_t = y_t - \hat{y}_t$. For example, an effectively zero correlation parameter between $\hat{u}_t$ and $\hat{u}_{t-1}$ would be consistent with white noise.

In terms of fit criteria to compare against succeeding models, the conditional likelihood AR(1) approach yields, with $\tau \sim G(1, 0.001)$, a CPO based marginal likelihood estimate of -155.8 (Gelfand and Dey, 1994), a Predictive Loss Criterion (PLC) of 134 200 (Gelfand and Ghosh, 1998), and a mean square prediction error (one step ahead) of 4680. The individual CPOs, when scaled to have maximum 1, show the lowest probability of belonging to the data for y_{19}, with a scaled CPO of 0.027.

For a full likelihood approach without a stationary constraint (Model B in Program 5.1), it is necessary to specify priors on the latent pre-series data or errors. Thus, take $y_0 \sim t(300, \kappa\sigma^2, 2)$, where κ is assigned a G(0.01, 0.01) prior, which has mean 1 but allows values of κ much in excess of 1. The posterior mean (and s.d.) of ρ is then 0.24 (0.19), and the posterior median of κ around 10. The one step predictive error as compared to the conditional likelihood model (and defined over the same times as in that model) is worsened to 4785, and the predictive loss criterion is also higher at 137 100. On the other hand the marginal likelihood estimate (over cases 2–30) is virtually unchanged at -155.9. The pre-series value is estimated at 506 compared to $y_1 = 485$.

A further elaboration (Model C in Program 5.1) allows for outliers, with a Normal mixture replacing the standard error assumption. Thus,

$$u_t \sim (1 - \Delta)N(0, \sigma^2) + \Delta N(0, \kappa\sigma^2)$$

where $\kappa > 1$ inflates the variance for outliers. Here, Δ is set to 0.05, and a discrete prior with five values {5, 10, 15, 20, 25} is set on κ. This prior is to ensure an outlier mixture as opposed to a more general mixture, where Δ might for instance be around 0.5, and κ not

[3] These are null start values, and the values provided by the 2.5th and 97.5th percentiles of a trial run.

markedly greater than 1. The estimate of ρ is similar to Model B, with mean around 0.27 (and s.d. 0.20). The highest outlier probability, around 20%, is for observation 19 (501). However, fit does not improve; for instance, the one step ahead predictive error is 5410.

In a final analysis, Model D, the full likelihood model (Model B) is modified to include a coefficient selection indicator, consistent with the approach of Kuo and Mallick (1994). Thus

$$y_t = \mu + J\rho y_{t-1} + u_t$$

where $J\rho$ is binary. The prior probability π_J that $J = 1$ is taken as 0.9, so posterior values of π_J (based on counting samples where $J = 1$) clearly lower than 0.9 would tend to cast doubt on the value of including an AR(1) coefficient. In fact, π_J is estimated around 0.85, so there does not seem clear evidence for or against including this parameter.

Example 5.2 Nelson–Plosser series As an illustration of models for analysing trend as in Section 5.2.3, one of the 14 Nelson–Plosser series on the US economy (Nelson and Plosser, 1982) is considered. This is the real GNP series, taken up to 1988. Both linear and nonlinear AR models, as in Equations (5.15) and (5.16) respectively, are considered, and full likelihoods referring to latent pre-series values are assumed. The original series is shifted by subtracting the first observation from all observations, i.e.

$$z_t = y_t - y_1$$

and this provides justification – when combined with low precisions – for assuming a prior mean of zero for the pre-series values, which are here z_0, z_{-1}, z_{-2}.

Student t innovation errors via a scale mixture with ν degrees of freedom and weights w_t are adopted, since Hoek *et al.* (1995) report some of these series to be clearly heavy tailed. A gamma prior for the degrees of freedom is taken with sampling constrained to [1, 100], so encompassing both the Cauchy density ($\nu = 1$) and an effectively Normal density ($\nu = 100$). Hoek *et al.* (1995, p. 43) suggest varying the upper limit for the degrees of freedom as a sensitivity analysis. As in Example 5.1, a G(0.01, 0.01) prior is taken for the factor κ, which scales up the main data variance in a way appropriate for the Student t distributed pre-series latent data.

Summaries are based on three chains with 20 000 iterations and 3000 burn-in (to ensure convergence of ν). The real GNP series is stationary according to the linear model, with zero probability that $\rho > 1$ (Table 5.1). However, GNP is marginally non-stationary (with the same probability standing at 0.09) under the nonlinear model (5.16). By comparison, Bauwens *et al.* (1999, p. 189) find probabilities that $\rho > 1$ of 0.002 and 0.033 under the linear and nonlinear models. The linear model does, however, have slightly better one step-ahead prediction error, error under predictive loss, and marginal likelihood than the non-linear model. Figure 5.2 plots z_t against the weights of the scale mixture: lower weights apply to observations at odds with the remainder of the data. The lowest weight is for the depression year 1932.

As a third modelling approach, an additive outlier model (Section 5.2.1) is applied. This modifies the linear model in Equation (5.15) to the form

$$z_t = \mu + \gamma(z_{t-1} - o_{t-1}) + \lambda t + \phi_1 \Delta z_{t-1} + \phi_2 \Delta z_{t-2} + u_t$$
$$o_t = \delta_t \eta_t$$

Following McCulloch and Tsay (1994), o_t for years preceding the series are taken as zero. This model finds evidence of outliers in the years 1921, 1929, 1938 and, to a lesser

Table 5.1 GNP series (second halves of runs of 25 000, iterations)

	Mean	St. devn.	2.5%	Median	97.5%
Linear AR					
Prob of Nonstationarity	0				
μ	−0.035	0.020	−0.076	−0.035	0.001
δ	0.006	0.001	0.003	0.005	0.009
γ	0.824	0.047	0.729	0.827	0.909
ϕ_1	0.374	0.106	0.163	0.375	0.579
ϕ_2	−0.007	0.109	−0.219	−0.008	0.208
ν	18.9	24.0	2.3	7.2	89.5
Non-Linear					
Prob of Nonstationarity	0.029				
μ	−0.374	0.307	−0.991	−0.365	0.273
δ	0.032	0.004	0.022	0.032	0.038
γ	0.876	0.063	0.759	0.872	1.002
ϕ_1	0.370	0.104	0.162	0.372	0.570
ϕ_2	−0.052	0.118	−0.286	−0.051	0.175
κ	9.2	19.8	1.0	3.2	59.2
ν	19.6	24.5	2.3	7.3	89.6
Linear AR, Additive Outlier Model					
Prob of Nonstationarity	0.0				
μ	−0.031	0.020	−0.076	−0.029	0.004
δ	0.005	0.002	0.002	0.005	0.009
γ	0.838	0.053	0.721	0.842	0.925
ϕ_1	0.426	0.106	0.214	0.426	0.632
ϕ_2	−0.046	0.110	−0.254	−0.048	0.178
ν	13.6	19.2	2.0	5.6	77.1

degree, 1954. The outlier probability Δ is estimated at 0.051. The one step and predictive loss prediction errors of this model are better than either the linear or non-linear AR(1) models.

Example 5.3 US coal production Moving average modelling for innovations combined with autoregression in a metric outcome, as in Equation (5.4), may be illustrated with data from Christensen (1989). He considers a time series y_t of US coal production from 1920 to 1987 ($T = 68$), and its analysis for forecasting purposes by various ARIMA models. Because a broadly upward trend is apparent in the observed series he advocates first differencing, so that the analysis is of

$$z_t = y_{t+1} - y_t \quad t = 1, ..67$$

It may be noted that a linear trend (e.g. the λt term in Equation (5.15)) corresponds to a nonzero mean in the series of first differences. Figure 5.3 shows the original undifferenced series and the steady growth in output after 1960. Christensen concludes that $(p, d, q) = (1, 1, 2)$ models with and without intercepts are the best fit, and that the version

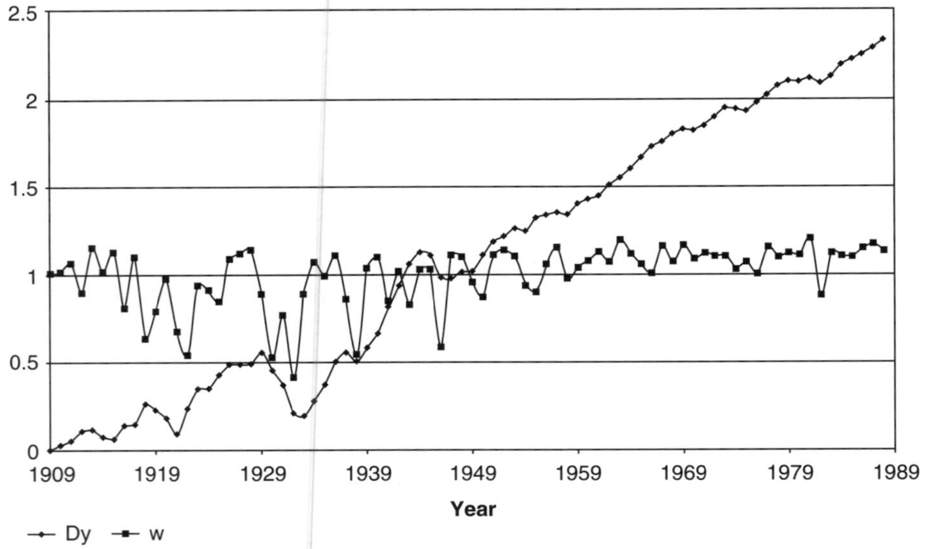

Figure 5.2 Plot of differenced GNP series and scale mixture weights (w)

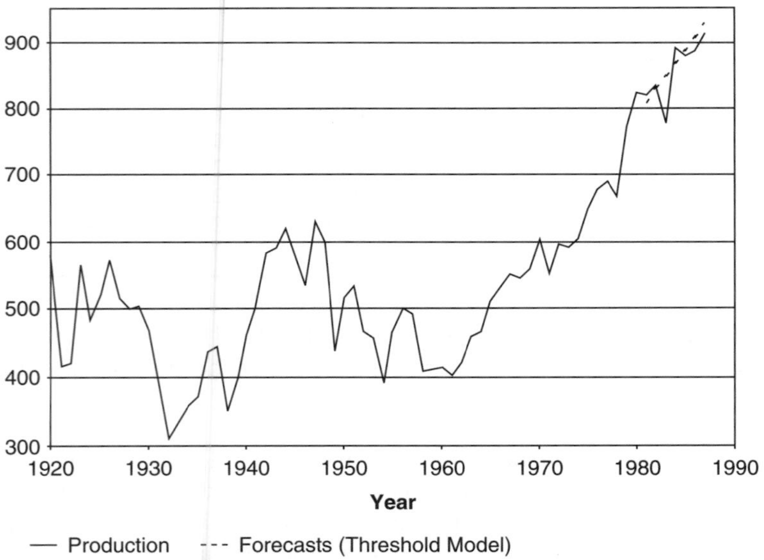

Figure 5.3 US coal

with an intercept is more plausible as it better reflects the actual upward trend in the data when applied to making forecasts.

However, to allow for the possibility of a model with a zero intercept, the first model adopted here uses a binary indicator $J_\lambda \sim \text{Bern}(0.5)$ to select a zero or non-zero intercept across all periods. Hence, the model is

$$z_t = J_\lambda \lambda + \rho z_{t-1} + u_t - \theta_1 u_{t-1} - \theta_2 u_{t-2} \tag{5.18}$$

The second model here proposes a threshold τ within the estimation period such that $I(t \geq \tau) = 1$ and $I(t < \tau) = 0$. So after τ there is a trend, but not before. Then

$$z_t = I(t \geq \tau)\lambda + \rho z_{t-1} + u_t - \theta_1 u_{t-1} - \theta_2 u_{t-2}$$

Christensen undertakes estimation with the series up to 1980 ($T_1 = 61$) and 'extended' forecasting beyond then, for the seven years up to 1987. Within the estimation series, one step ahead forecasts are obtained as

$$y_{\text{new},\, t+1} = z_{\text{new},\, t} + y_t$$

where $z_{\text{new},\, t}$ is sampled from the model for the z_t. Extended forecasts assume u_t sampled from the same density as the innovations during the estimation period. Both sets of forecasts are compared with actual data via a squared error criterion.

Because BUGS uses the centred Normal form, it is necessary to assume an additional measurement error e_t to model the moving average terms such that, for the first model,

$$z_t = J_\lambda \lambda + \rho z_{t-1} + u_t - \theta_1 u_{t-1} - \theta_2 u_{t-2} + e_t$$

The variance of e_t is taken to be less than that of the innovations u_t and use beta prior is adopted for $\lambda = \text{var}(e_t)/\text{var}(u_t)$. An extra parameter $B \sim U(1,\ 1000)$ is introduced to guide the decomposition of variance so that

$$\lambda \sim \text{Beta}(1,\ B)$$

On the AR and MA parameters, priors in line with stationarity and invertibility are assumed. An unconditional estimation, including all observations in the likelihood, is used, so that a latent (differenced) observation z_0, and latent innovations u_0 and u_{-1} are referred to. The latent innovations are assumed a priori to follow a Student t density with $\nu = 2$ degrees of freedom and with variance five times greater than the main innovation series. z_0 is taken as a distinct fixed effect.

Christensen obtains maximum likelihood estimates (with SEs) of

$$\lambda = 6.3(7.6),\ \rho = -0.42(0.23),\ \theta_1 = -0.47(0.22),\ \theta_2 = 0.38(0.16)$$

and forecasts under ARIMA(2, 1, 2) intercept model as shown in Table 5.2. Forecasts under the no-intercept model fluctuate between 780 and 785 million short tons right through from 1981 to 1987.

For the first model in Equation (5.18) a three chain run of 50 000 iterations (5000 burn-in) gives an estimate for J_λ around 0.0273. Hence the Bayes factor against an intercept is approximately 35. This model yields similar forecasts to those cited by

Table 5.2 Threshold model forecasts, US coal production

Year	Mean	St. devn.	2.5%	Median	97.5%	Actual	Christensen (1989, p. 243) forecasts
1981	808	54	701	807	915	818	787
1982	827	80	671	827	986	834	796
1983	844	92	667	843	1029	778	798
1984	863	105	661	861	1078	892	804
1985	881	117	657	879	1122	879	808
1986	900	130	651	896	1170	886	812
1987	918	142	647	914	1216	913	817

Christensen for the no intercept model. The prediction for 1987 is 778 with 95% credible interval (560,1010).

The threshold model for the intercept (Model B in Program 5.3) has a considerably better predictive loss error within the series, and improved out of sample forecasts. It finds the threshold τ to average 40, corresponding to the year 1959. The intercept λ then averages 22, which may be compared to the average increment z_t from 1960–61 to 1979–80 of around 20.5 million tons. Under this model the lag parameter ρ and first MA lag θ_1 are smaller absolutely than obtained by Christensen, with posterior means and SDs

$$\rho = -0.20(0.22),\ \theta_1 = -0.21(0.27),\ \theta_2 = 0.45(0.21)$$

This model predicts the 1987 production to be 918 million short tons (with 95% interval 647 to 1216) against an actual 913. One might envisage other options for this series, such as a dynamic intercept for this model (see Section 5.5).

Example 5.4 Investment levels by firms Maddala (1979) compares several procedures, including maximum likelihood, in the estimation of investment levels by a set of US firms. This analysis illustrates autocorrelated errors as in Equation (5.6). Maddala's investigation derives from an earlier study by Grunfeld and Griliches (1960) on the validity of using aggregate data to draw inferences about micro-level economic functions (e.g. consumption and investment functions). The aggregate is defined by ten firms, while the autoregressive errors model is applied to the yearly investment time series (1935–54) of a particular firm, General Motors. The predictor series (C_t, V_t) is available from 1934.

The model relates General Motors' gross investment y_t in year t to lagged levels of the firm's value V_{t-1} and capital stock C_{t-1}; thus the generic x_t in Equation (5.6a) is here defined by $x_t = \{V_{t-1}, C_{t-1}\}$. Maddala assumes AR(1) dependence in the errors leading to a specification for years 1936–54:

$$\begin{aligned} y_t &= \beta_0 + \beta_1 V_{t-1} + \beta_2 C_{t-1} + \varepsilon_t \\ \varepsilon_t - \gamma\varepsilon_{t-1} &= u_t \end{aligned}$$

with $u_t \sim N(0, \tau^{-1})$ being unstructured white noise. This model can be expressed in the form of (5.7), giving the model

$$y_t = \gamma y_{t-1} + \beta_0(1-\gamma) + \beta_1(V_{t-1} - \gamma V_{t-2}) + \beta_2(C_{t-1} - \gamma C_{t-2}) + u_t \tag{5.19}$$

The first model (Model A) assumes stationary errors ε, and a uniform prior on the AR parameter is assumed, namely $\rho \sim U(-1, 1)$. The model for the year 1935 ($t = 1$) can then be written

$$\begin{aligned} y_1 &= \beta_0 + \beta_1 V_{t-1} + \beta_2 C_{t-1} + \varepsilon_1 \\ \varepsilon_1 &\sim N(0, 1/\tau_1) \\ \tau_1 &= (1-\gamma^2)\tau \end{aligned}$$

The prior for the intercept β_0 is set to be appropriate for a series with values exceeding 1000. One step ahead forecasts in Equation (5.6a) require knowledge of x_{t+1}, and use current (years t) values of y, V and C.

The posterior estimates of the parameters β_1 and β_2 (from a three chain run of 20 000 iterations with 1000 burn-in) are close to the maximum likelihood estimates cited by Maddala. They show both coefficients to be above zero, with β_1 and β_2 having means and 95% credible interval 0.086 (0.045, 0.12) and 0.42 (0.28, 0.57), respectively. The autoregressive coefficient γ is estimated to have mean 0.74 with 95% credible interval (0.34, 0.98).

A second model (Model B) avoids assuming stationarity in the error process at the outset (Zellner and Tiao, 1964). Accordingly, the prior on γ is not confined to absolute values under 1, and a Normal density with mean 0 and variance 1 is adopted instead. The model for the first observation (for the year 1935) now treats it as separate effect

$$y_1 \sim N(\mu_1, V_1)$$

with V_1 set large and mean, following Equation (5.19), of

$$\mu_1 = \beta_0(1 - \gamma) + \beta_1 V_{t-1} + \beta_2 C_{t-1} + M$$

where M is a composite parameter represents the missing term $\gamma(y_0 - \beta x_0)$. This shows the posterior of β_1 unchanged, but the mean effect of C_{t-1} is reduced, with 95% interval (−0.11, 0.68) now straddling zero. There is a 35% chance of γ exceeding 1, and its mean is 0.93 with 97.5% point 1.28. The within series forecasts under this model are shown in Figure 5.4, and trace the observed series reasonably well. The value for 1953 is under-predicted and this feature is worse under the non-stationary model; the result is that the marginal likelihood based on the CPO statistics is lower by −1.7, and the pseudo Bayes factor on Model A as against B is 5.3.

Example 5.5 Swedish unemployment and production: bivariate forecasts To illustrate VAR models involving autoregression in the endogenous variables, consider the bivariate series analysed by Kadiyala and Karlsson (1997). This consists of quarterly data on unemployment and industrial production in Sweden between 1964.1 and 1992.4, with the production index in logged form. The analysis follows Kadiyala and Karlsson in only considering the latter part of the series (here from 1978.1 onwards). A lag four model in each component of the bivariate outcome is adopted. It may be anticipated

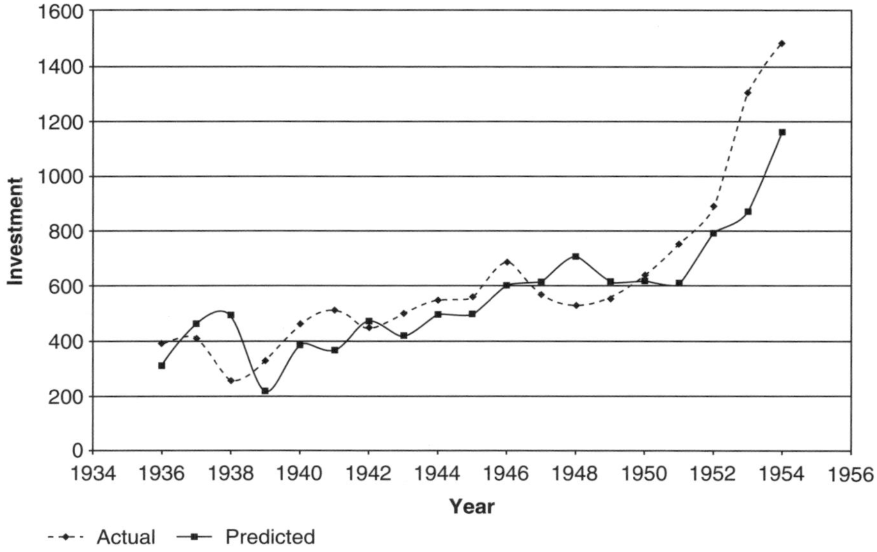

Figure 5.4 Within series forecats

that some of the cross-variable lags are negative, because unemployment increases as production falls. This model is in fact likely to be most effective in short term forecasting of unemployment, since influences on production (e.g. measures of consumer demand) are not included in the model. There are also seasonal influences on both variables and seasonal dummies are therefore included as exogenous predictors. Here the 52 quarters of the series from 1979.1 to 1991.4 are analysed (conditioning on 1978.1 to 1978.4) and then predictions made for the remaining year.

Kadiyala and Karlsson take the undifferenced series with N(1, 1) priors on the own first lags, and N(0, 1) priors on the other own-lag coefficients, and for all the cross variable lags. Experimentation here with an undifferenced series analysis showed poor identifiability and convergence on the cross-variable lag effects. Identifiability was improved by modelling aspects of trend, such as by polynomials in time to take account of the upward shift in unemployment in the early 1990s (Figure 5.5), following a decline during the 1980s. However, convergence was still problematic.

By contrast, analysis of differenced series was less subject to such problems. For this analysis N(0, 1) priors are taken on all lags, including first own lags. There are no stationarity constraints. Priors on seasonal effects take the first quarter as reference (with coefficient zero).

In Model A, all cross-variable lags (unemployment on production and vice versa) are included, and the u_{kt} are taken as multivariate Normal. With a two chain run taken to 10 000 iterations (and burn-in of 1000), the clearest defined lags are for unemployment on its own third and fourth lags, with posterior means (standard deviations) of 0.34 (0.15) and 0.28 (0.16). All cross lags of unemployment on production are negative, with the mean of the second cross lag being around -1.20 (s.d. $=0.78$). Lags of production on unemployment are insignificant (close to zero). The cross-variable correlation in the errors u_{kt} is estimated at around -0.15 (with 95% credible interval (-0.44, 0.18). Forecasts of unemployment in the short term match the rise during 1992 (Figure 5.6).

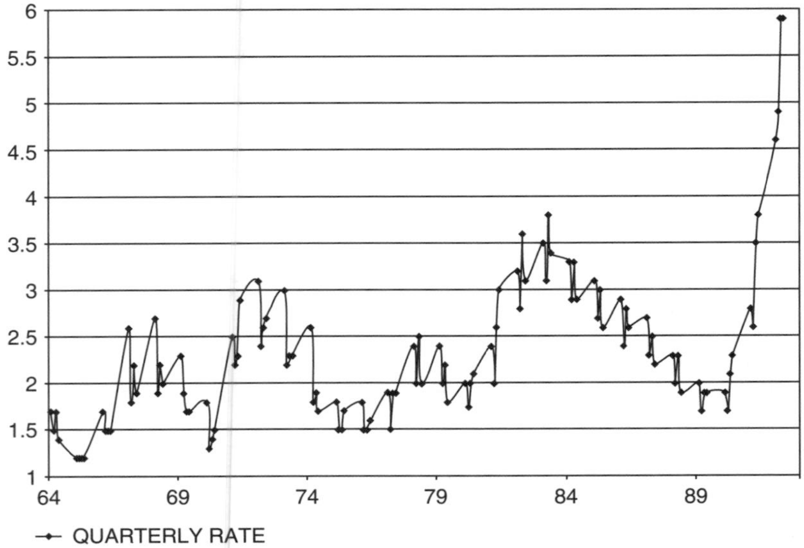

Figure 5.5 Undifferenced unemployment series

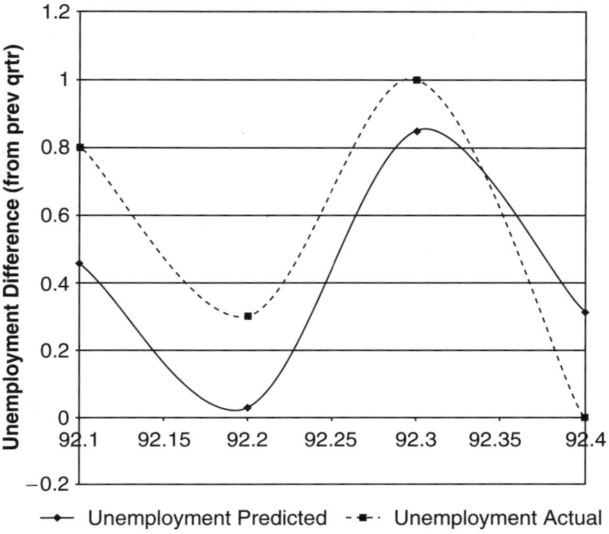

Figure 5.6 Unemployment in 1992

In a reduced model (Model B), the lags of production on unemployment are set to zero. This produces an improvement in marginal likelihood (and in other fit criteria) and no great change in the pattern of remaining coefficients or the forecasts for 1992. In both models the clearest outlier is at 1982.4, with unemployment difference $3.1 - 3.6 = -0.5$.

5.3 DISCRETE OUTCOMES

For discrete outcomes, dependence on past observations and predictors may be handled by adapting metric variable methods within the appropriate regression link. Thus for Poisson outcomes

$$y_t \sim \text{Poi}(\mu_t)$$

an AR(1) dependence on previous values in the series could be specified

$$\log(\mu_t) = \rho y_{t-1} + \beta x_t \tag{5.20}$$

Here, non-stationarity or 'explosive' behaviour would be implied by $\rho > 0$ (Fahrmeir and Tutz, 2001, p. 244), and in an MCMC framework stationarity would be assessed by the proportion of iterations for which ρ was positive. Autoregressive errors lead to specification such as

$$\log(\mu_t) = \beta x_t + \varepsilon_t$$

with

$$\varepsilon_t = \gamma \varepsilon_{t-1} + u_t$$

for $t > 1$, and u_t being white noise.

For binary data a similar framework would involve lags in the binary outcomes. Thus, if $y_t \sim \text{Bern}(\pi_t)$, then one might have a regression model with a single lag in y_t such as

$$\text{logit}(\pi_t) = \beta x_t + \rho y_{t-1}$$

For multi-category data with K categories there are $K-1$ free category probabilities, and these might be related to lagged values on up to $K-1$ dummy variables. This leads to models similar to VAR(p) models for multivariate metric outcomes, in that there are 'own' and 'cross' lags. Thus, Pruscha (1993) outlines suggests the scheme

$$y_t \sim \text{Categorical}(P_t)$$

where both y_t and P_t are of dimension K, and the probability P_{tk} that the tth value of the series is in category $k(k > 1)$ is a function of dummy indicators $D_{t-1,\,k} = 1$ if $Y_{t-1} = k$. The probabilities may be made functions of the previous response via a model such as

$$P_{tk} = \exp(\beta_k x_t + \rho_{k2} D_{t-1,\,2} + \ldots \rho_{kK} D_{t-1,\,K}) / \left[1 + \sum_k \exp(\beta_k x_t + \rho_{k2} D_{t-1,\,2} + \ldots \rho_{k,\,K} D_{t-1,\,K})\right]$$

and the lag coefficients model transitions between states.

While the canonical links (e.g. log for Poisson data and logit for binomial data) are the most common choices, one may regard the link as a modelling area in its own right. For binomial data, the modelling of a choice between skewed alternatives to the logistic model, as in Lang (1999) was considered above (Section 3.3.4). Another methodology (Mallick and Gelfand, 1994) is applicable to modelling all kinds of link; for a Poisson with 'base' log link, it involves a transformation from the R^1 scale of $\eta = \log(\mu)$ to a (0, 1) scale in which beta mixture modelling can be carried out – for instance, $J(\eta) = \exp(\eta)/[1 + \exp(\eta)]$.

Modelling the link for count data is also possible in models involving multiplicative lags in a transformed version of y_t, with mean specified as

$$\mu_t = \exp(x_t \beta)(y'_t)^\rho$$

with possible transformations being, as in Cameron and Trivedi (1999),

$$y'_{t-1} = y_{t-1} + c \quad (c > 0) \tag{5.21a}$$

or

$$y'_{t-1} = \max(c, y_{t-1}) \quad (0 < c < 1) \tag{5.21b}$$

Either c would be an additional parameter or taken as a default value such as $c = 0.5$ or $c = 1$. A generalisation is to models such as

$$(\mu_t)^\kappa = \exp(x_t \beta)(y'_t)^\rho$$

Another approach, especially for binary and categorical time series, invokes an underlying metric variable. This approach has been suggested for Poisson count data (van Ophem, 1999), but the most frequent application is for binary data using the method of Albert and Chib (1993) and Carlin and Polson (1992). Thus, for binary y_t, a positive value of the latent series Y^*_t corresponds to $y_t = 1$ and negative values of the latent series to $y_t = 0$. One might then assume an underlying true series x_t, largely free of measurement error, such that

$$Y_t^* = x_t + v_t$$
$$x_t = \rho x_{t-1} + u_t$$

with $|\rho| < 1$ corresponding to stationarity.

5.3.1 Autoregression on transformed outcome

For Poisson or binomial data, it might be sensible that the lagged value of the outcome is in the same form as the transformed mean of the current outcome value. Thus a symmetry with the log link for Poisson counts would involve a lag in the log of a transformed version of y_{t-1},

$$\log(\mu_t) = \beta x_t + \rho \log(y'_{t-1})$$

where y'_t is defined as above.

If one wished to consider extended lags or moving average effects for frequent binomial events or counts, then unmodified ARMA methods – applied as if the outcomes were effectively metric, and using Normal approximations for the binomial or Poisson – may be appropriate. However, there are potential problems in applying standard ARMA models to count data since the assumption of Normality (or of any symmetric density) may not be appropriate, especially for rare events.

If the autocorrelation is postulated in the regression errors, then a full model might take the form, for a Poisson outcome

$$y_t \sim \text{Poi}(\mu_t)$$
$$\log(\mu_t) = \beta x_t + \varepsilon_t$$
$$\varepsilon_t = \gamma \varepsilon_{t-1} + u_t$$

However, a reduced parameterisation, excluding errors u, may still reproduce the essential aspects of the alternative form (5.7). Following Zeger and Qaqish (1988), one might propose, for a Poisson outcome, the model

$$\log(\mu_t) = \beta x_t + \gamma(\log y'_{t-1} - \beta x_{t-1})$$

with y'_{t-1} as in Equations (5.21a)–(5.21b). A lag two model would then be

$$\log(\mu_t) = \beta x_t + \gamma_1(\log y'_{t-1} - \beta x_{t-1}) + \gamma_2(\log y'_{t-2} - \beta x_{t-2})$$

and 'moving average' terms would compare $\log y'_{t-j}$ with $\log \mu_{t-j}$ so that an ARMA(1, 1) type model would be

$$\log(\mu_t) = \beta x_t + \gamma(\log y'_{t-1} - \beta x_{t-1}) + \theta(\log y'_{t-2} - \log \mu_{t-2})$$

5.3.2 INAR models for counts

Integer valued autoregressive (INAR) schemes are oriented to discrete outcomes, and have a close affinity with ARMA models for metric outcomes (McKenzie, 1986). In particular, their specification often includes devices to ensure stationarity (e.g. of the underlying mean count through time), though a Bayesian approach may make this constraint less necessary. INAR schemes introduce dependence of the current count y_t on previous counts y_{t-1}, $y_{t-2}, \ldots$, and also allow an integer valued innovation series w_t (a form of random shock). The autoregressive component of the model can be seen as a survival model to time t for each particle in the previous overall counts y_{t-1}, y_{t-2}, etc.

Thus, for an INAR(1) model, one considers the chance ρ that each of the y_{t-1} particles survives through to the next period. If, say, $y_{t-1} = 4$, this amounts to conceiving a Bernoulli model Bern(ρ) for continuation of each of the four particles. The autoregressive component of the INAR(1) model for y_t is

$$C_t = \sum_{k=1}^{y_{t-1}} \text{Bern}(\rho)$$

Equivalently, C_t is binomial with y_{t-1} subjects, and ρ the probability of success. This approach to autoregression for non-negative integers is known as a binomial thinning operation, and is denoted as

$$\rho^{\circ} y_{t-1}$$

An INAR(2) process would refer to two preceding counts, y_{t-1} and y_{t-2}, and involve two survival probabilities, ρ_1 and ρ_2. Note that for an INAR(p) process, stationarity is defined by

$$\sum_{k=1}^{p} \rho_k < 1$$

(Cardinal *et al.*, 1999). For overdispersed data, McKenzie (1986) suggested that the 'survival probabilities', such as ρ_{1t} in an INAR(1) model, be time varying, and one might then envisage autoregressive priors on these probabilities.

If $y_{t-1} = 0$ then there is no first order lag autoregressive component in the model for y_t. In BUGS the binomial can still be used when $y_t = 0$, so one can code an INAR(1) model using the binomial, rather than program the full thinning operation. Note that an INAR model requires an initialising prior for the first value y_1 of the series, and McKenzie (1986) proposes

$$y_1 \sim \text{Poi}(\theta)$$

As well as survival of existing particles (which might have time-varying covariates attached to them so that ρ_{1t} is modelled via logit regression), there is an 'immigration' process, analogous to the innovations of an ARMA model for a metric outcome. Thus, new cases w_t are added to the 'surviving' cases from previous periods. McKenzie and others have envisaged y_t as then being the summation of two separate Poisson processes with different means. Thus, McKenzie's INAR(1) model has

$$y_t = \rho^{\circ} y_{t-1} + w_t \tag{5.22}$$

where the mean of w_t is Poisson with mean $\theta(1 - \rho)$ to ensure stationarity in the mean for y.

One might also consider unconstrained Poisson densities for w_t (not tied to ρ in an INAR(1) model), especially if there is overdispersion. Thus Franke and Seligmann (1993) propose a mixed Poisson with two possible means λ_1 and λ_2 for w_t in an analysis of epileptic seizure counts. Switching in the innovation process at time t is determined by binary variables Q_t (which may in turn be drawn from an overall beta density). Another option is to allow the mean of the w_t to be time dependent.

One might also envisage (in terms of its compatibility with the BUGS computing environment) having a single mean μ_t for y_t, but composed of the survival term $\rho_1{}^\circ y_{t-1}$, and an additional series ω_t, that follows a positive density (e.g. a gamma). For a stationary INAR(1) model, this leads to

$$\begin{aligned} y_t &\sim \text{Poi}(\mu_t) \\ \mu_t &= \rho^\circ Y_{t-1} + \omega_t \\ \omega_t &\sim G(b\theta(1-\rho), b) \end{aligned} \tag{5.23}$$

5.3.3 Continuity parameter models

Harvey (1989) and Ord *et al.* (1993) also propose a model for count series combining two sources of randomness. One concerns changes in the underlying level (as does the survival term of an INAR model) and the other refers to the distribution of observations around that level. Thus, with Poisson sampling, the mean is itself gamma distributed with time evolving parameters (a_t, b_t),

$$\begin{aligned} y_t &\sim \text{Poi}(\mu_t) \\ \mu_t &\sim G(a_t, b_t) \end{aligned} \tag{5.24}$$

The gamma parameters are related to previous parameters (a_{t-1}, b_{t-1}) via a common continuity parameter ϕ. This takes values between 0 and 1 that applies to both scale and index of the gamma. Thus

$$\begin{aligned} a_t &= \phi a_{t-1} \\ b_t &= \phi b_{t-1} \end{aligned} \tag{5.25}$$

The initial values a_0, b_0 are assigned a prior ensuring positive values (e.g. log-normal or gamma). To avoid improper priors for later time periods, one may modify Equation (5.25) by adding a small constant – indicating the minimum prior scale and index in the prior for the μ_t. This might be taken as an extra parameter or preset. Thus

$$\begin{aligned} a_t &= \phi a_{t-1} + c \\ b_t &= \phi b_{t-1} + c \end{aligned}$$

It is also possible to drop the constraint $\phi < 1$ and assess the probability that ϕ is in fact consistent with information loss (i.e. with accumulated discounting of past observations as $\phi < 1$ implies).

5.3.4 Multiple discrete outcomes

The approach of Equations (5.24)–(5.25) may be extended to multivariate count series,

$$y_{kt}, k = 1, .. K, t = 1, .. T$$

by modelling the total count at time t

$$Y_t = \sum_k y_{kt}$$

in the same way as a univariate count with parameter ϕ_1. The disaggregation to the individual series is modelled via a multinomial-Dirichlet model, with the evolution of the Dirichlet parameters governed by a second parameter ϕ_2.

Jorgensen *et al.* (1999) also consider a method for multiple count series $y_{1t}, y_{2t}, .. y_{Kt}$ observed through time, but in terms of an underlying univariate latent factor θ_t that accounts for much of their interdependency. The components of the multivariate series are conditionally independent given θ_t. Thus

$$y_{kt} \sim \text{Poi}(\nu_{kt}\theta_t) \tag{5.26}$$

where ν_{kt} is a term dependent on regressors X_k relevant to the kth outcome. The regression is typically applied via a log link:

$$\log(\nu_{kt}) = \beta_k X_k$$

and β_k differs by outcome.

The series are envisaged as related (e.g. counts of diseases caused by similar risk factors), and the latent process θ_t evolves as a gamma Markov process

$$\theta_t \sim G(c_t, d_t) \tag{5.27}$$

The gamma density has the benefit of conjugacy with the Poisson as in the continuity parameter models above, and the evolution of $\{c_t, d_t\}$ might follow the stationary scheme of Section 5.3.3. However, Jorgensen *et al.* (1999) propose a nonstationary scheme, whereby

$$c_t = \theta_{t-1}/\sigma^2 \tag{5.28a}$$

$$d_t = 1/[b_t\sigma^2] \tag{5.28b}$$

where σ^2 is a variance parameter, and the mean of θ_t given preceding values $\theta_0, .. \theta_{t-1}$ is $b_t\theta_{t-1}$. So, one may modify the impact of the preceding latent value by a regression of b_t on additional covariates z_t. To ensure b_t positive, one may take

$$\log(b_t) = \alpha z_t \tag{5.28c}$$

Jorgensen suggests z_t be defined in terms of first order differences $z_t = \Delta Z_t = Z_t - Z_{t-1}$, where Z_t are viewed as long term influences, and the X_t as short term.

Example 5.6 Meningococcal infection As an example of the INAR(p) process, consider a series of 104 counts in 28 day periods of meningococcal infections in Quebec. These span the period 1986–1993. The maximum observed count in any single period is six. As Cardinal *et al.* (CRL, 1999) note, the autocorrelation and partial autocorrelation functions of the (undifferenced) count series are virtually indistinguishable from a series of random shocks, except for a slight lag at five months. Using the model of Equation (5.22), but with w_t as a general Poisson process (not linked to the ρ coefficients), CRL confirm the significant effect at lag 5. They find a Ljung–Box portmanteau statistic of fit with a non-significant vale ($p = 0.63$ at the maximum likelihood estimate).

Here the specification in Equation (5.23) is adopted, and following CLR an INAR(5) model is estimated for the first seven years' data ($n = 91$). The 'autoregressive coefficients' ρ_k are assigned a beta, $B(1, 1)$, prior. For additional flexibility (Model A), the innovation error ω_t is modelled as an exponential density with changing parameter Λ_t, where $\log\Lambda_t$ is modelled as a first order Normally distributed random walk. An alternative assumption on the innovation errors is that they are independently gamma with parameters α_1 and α_2 (Model B). Forecasts are made for the 13 periods of 1993 for each model, though the coding in BUGS is not very elegant because stochastic quantities are not allowed as indices in sums or do-loops.

As one check on fit, the Ljung–Box criterion is derived for $L = 1, 2, \ldots 22$ lags. A posterior predictive check (Gelman *et al.*, 1995) is incorporated based on sampling new counts from the model mean and deriving a 'new' data Ljung–Box criterion. This is compared to the real data criterion, and the predictive check is based on the proportion of iterations where the new criterion improves on the real one.

Posterior summaries from the second half of a three chain run of 5000 iterations show a similar impression of a lag at five months, but also more apparent lower lag effects (i.e. with lower standard deviations) than in the analysis of CRL. The posterior means (and s.d.) of the lag coefficients are $\rho_1 = 0.15(0.09)$, $\rho_2 = 0.09(0.07)$, $\rho_3 = 0.12(0.09)$, $\rho_4 = 0.15(0.10)$, and $\rho_5 = 0.24(0.12)$. The minimum value of the Ljung–Box statistic is 7.3, roughly approximating the maximum likelihood value. The Ljung–Box statistic is approximately chi-square with $L - p(=17)$ degrees of freedom if the model is appropriate, so there is no evidence of lack of fit. The predictive criterion is close to 0.5 (around 0.58), and indicates the same.

Model B produces similar estimates of the coefficients, and similar short term forecasts (Figure 5.7). Its marginal likelihood, namely -151.2, is slightly better than Model A at -152.4. The most suspect observation under both models is the seventh, with five cases, compared to noughts and ones in adjacent periods.

Example 5.7 England vs. Scotland football series To illustrate the continuity parameter model, consider the goals scored by England in the England vs. Scotland international, which has been running since 1872. Excluding war years, there were a total of $T = 104$ matches up to 1987. A subset of this series (England away games) is analysed using the Poisson-gamma model of (24)–(25) in Harvey and Fernandes (1989). A uniform prior on the continuity parameter ϕ is assumed. Flat gamma priors are taken

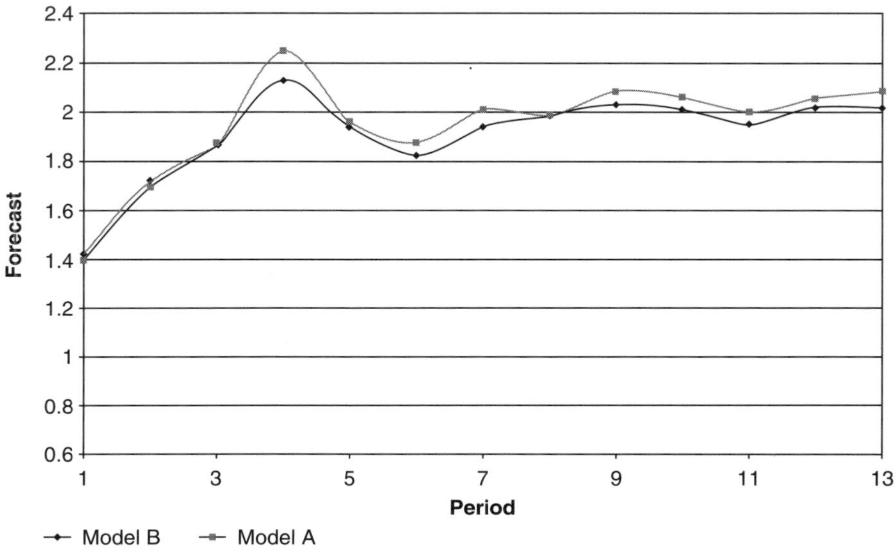

Figure 5.7 Forecast cases in 1993

on the initial conditions $\{a_0, b_0\}$. This defines Model A. As an alternative model for these data, an INAR(1) model is also considered.

A three chain run of 5000 iterations on model A leads to a 95% interval for ϕ from 0.91 to 0.99 with median 0.93. By contrast, Harvey and Fernandes obtain a value of 0.844 for the 53 matches played away by England at Hampden Park. The forecast value for England's score in the next match following the observed series (for year $T + 1 = 105$) is 1.73. The predictive probability distribution (obtained by extracting the iterations of y.new in Program 5.7) has 36% of the distribution being for zero goals.

An illustration of the fit to the observed series is provide by matches 71–80, where 7 and 9 goals were scored in the 72nd and 78th matches. These are the two clearest outliers using a CPO criterion. The 95% intervals do not include these exceptional scores, though the raised means of 2.3 and 3.1 reflect them (Figure 5.8).

The INAR model has a slightly lower pseudo marginal likelihood for this example (−182.2 vs. −180.9 under Model A), and provides a lower (posterior mean) forecast for the next score, namely 1.56. The lag 'coefficient' ρ is estimated at 0.097 (s.d. 0.057).

Example 5.8 Polio infections in the USA Fahrmeier and Tutz (1994) and others have analysed all or part of a time series y_t of new polio infections per month in the USA between January 1970 and December 1987. A question often raised with these data is the existence or otherwise of a linear trend in time, after accounting for seasonal variations. These are represented by sine and cosine terms as in Equation (5.13), namely

(a) cosine of annual periodicity, beginning with 1 in January 1970 (i.e. frequency 1/12);
(b) sine of annual periodicity, beginning with 0 in January 1970;
(c) cosine of semi-annual periodicity, beginning with 1 January 1970 (frequency 1/6); and
(d) sine of semi-annual periodicity, beginning with 0 in January 1970.

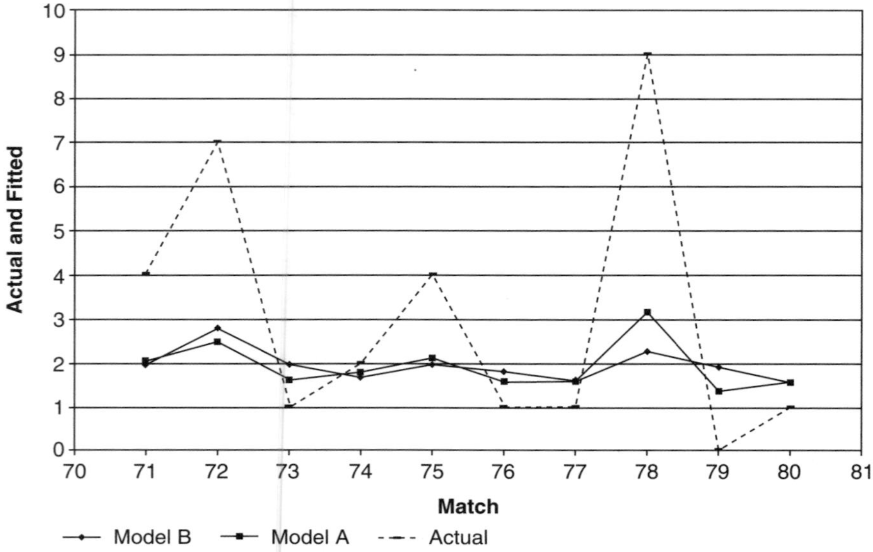

Figure 5.8 England vs. Scotland

We consider a linear trend model including these periodic impacts, and follow Fahrmeier and Tutz in assuming lags up to order 5 in the count itself, as in Equation (5.20). This model therefore has 11 parameters (intercept, trend, seasonals and lags) and 163 time points.

To illustrate forecasts with this approach, the model is fitted till the end of 1983 and forecasts made beyond then. Estimation is conditional on the first five observations, so that no model is required for the five latent data values ($y_0, y_{-1}, \ldots y_{-4}$). From a two chain run of 5000 iterations, with

$$y_t \sim \text{Poi}(\mu_t) \quad t = 6, 168$$

the model coefficients (posterior means and standard deviations) are

$$\begin{aligned}\log(\mu_t) = & \underset{(0.19)}{-0.14} \underset{(0.0015)}{- 0.0032t} \underset{(0.11)}{- 0.22\cos(2\pi t/12)} \underset{(0.11)}{- 0.47\sin(2\pi t/12)} \\ & \underset{(0.11)}{+0.13\cos(2\pi t/6)} \underset{(0.11)}{- 0.37\sin(2\pi t/6)} \underset{(0.03)}{+ 0.084y_{t-1}} \underset{(0.04)}{+ 0.038y_{t-2}} \\ & \underset{(0.04)}{-0.044y_{t-3}} \underset{(0.04)}{+ 0.028y_{t-4}} \underset{(0.03)}{+ 0.079y_{t-5}}\end{aligned}$$

The 95% credible interval for the time effect is confined to negative values, namely (−0.0062, −0.0003), and so supports a downward trend. By contrast, Fahrmeier and Tutz (2001, p. 254) find only a 9.5% probability that the maximum likelihood estimate of this parameter is significant.

Some of the coefficients in the full model above are not well defined. A second option in Model A allows for coefficient selection or exclusion using the Kuo–Mallick method. Inclusion rates are found to be lowest for lags 3 and 4, though in fact are below 5% for all coefficients except the time term, the sine effects and the first lag in y.

In the forecasting model (Model B in Program 5.8), the lag 3 and 4 coefficients are accordingly set to zero, though further model reductions might be made. The actual counts of cases beyond 1983 are generally 0 or 1, exceeding 1 in only 3 of the 48 months, and the forecast counts are all between 0.3 and 1.5 in line with the trend to lower incidence.

As an illustration of the latent variable approach of Section 5.3.4 to these data, consider again the series from 1970 to 1983 (there is only one series so $K = 1$). Jorgensen *et al.* take $Z_t = t$ to model the trend demonstrated by the above analysis so that the 'long term' model reduces to using a constant $\Delta Z_t = Z_{t+1} - Z_t = 1$, the coefficient of which, as in Equation (5.28c), represents the trend coefficient. The standard deviation σ in Equation (5.28a) is assigned a uniform (0.001, 100) prior – see Model C in Program 5.8. The initial parameter θ_0 is assigned a diffuse gamma prior.

Convergence with a two chain run is achieved after 12 000 iterations and from the subsequent 4000, the posterior mean of σ averages around 0.25. The short-term effects are modelled as the coefficients of $\cos(2\pi t/12)$, $\sin(2\pi t/12)$, $\cos(2\pi t/6)$ and $\sin(2\pi t/6)$, and estimates for these are similar to those reported in another analysis of these data by Chan and Ledolter (1995).

The trend coefficient does not appear significant under the Poisson-gamma model of Section 5.3.4. Jorgensen *et al.* argue that lack of evidence for a trend is accounted for by the nonstationary nature of the latent process θ_t, whereas Chan and Ledolter use a stationary AR(1) process for the latent series. Specifically, the model of Chan and Ledolter (see Model D in Program 5.8) is

$$Y_t \sim \text{Poi}(\mu_t)$$
$$\log(\mu_t) = \beta X_t + \varepsilon_t$$
$$\varepsilon_t = \gamma\varepsilon_{t-1} + u_t$$
$$u_t \sim N(0, \sigma_u^2)$$

In BUGS this alternative AR(1) model may be implemented using the centred Normal form for ε_t, namely $\varepsilon_t \sim N(\gamma\varepsilon_{t-1}, \sigma_u^2)$, and we obtain an estimate for γ around 0.72. The trend coefficient, namely the coefficient on time within βX_t, is mostly confined with this model to negative values (consistent with a downward tend in polio cases), with 95% interval $\{-0.010, 0.002\}$. Comparison of Model D with Model C in terms of pseudo-marginal likelihoods gives a Bayes factor in favour of *D* of around 5.

5.4 ERROR CORRECTION MODELS

Time series regressions in economics and elsewhere are defined in statistical terms by questions of identifiability and parsimonious parameter choice, but also by their being sensible in terms of the substantive application. It is in this sense that apparently significant regression relationships involving non-stationary outcomes y_t and predictors $x_{1t}, x_{2t}, ..$ may in reality be subject to a spurious correlation problem. If one or more explanatory variable shows a distinct trend (i.e. is nonstationary), and so does the outcome, then an apparently significant relationship may occur even if y and the putative predictors are independent. Granger and Newbold (1974) considered spurious regression relationships between two nonstationary series y_t and x_t, separately generated by random walk processes with changing variances, while Nelson and Plosser (1982) show that many economic time series do appear to follow random walk error processes. If a series is a first order random walk (or effectively so with autocorrelation ρ_1 in Equation (5.1) of, say, 0.99) then its first difference is stationary.

The analysis of Granger and Newbold was of 'integrated' series, namely those not differenced in an attempt to gain stationarity. Consider a regression for such series

$$y_t = \beta_0 + \beta_1 x_t + u_t \tag{5.29}$$

where successive u_t are taken to be non-correlated. Then taking first differences in y_t or $Y_t = \log(y_t)$, and in the predictors, may reduce the chance of spurious regression findings, and is often the recommended method for modelling non-stationary variables. Under differencing one obtains the model

$$\Delta y_t = \beta_1 \Delta x_t + u_t - u_{t-1} \tag{5.30}$$

which in fact defines a moving average error.

Suppose, however, that Equation (5.29) represents a long-term equilibrium relationship (e.g. between consumption and income), with the errors representing temporary disequilibria. In this case differencing on both sides will be considering only the short run dynamics of the underlying relationship. If a long run association holds, then it is likely that changes in y will depend not only on changes in x, as assumed in Equation (5.30), but also on the long term relationship (5.29) between y and x, and on the disequilibrium between y and x at lags $t-1$, $t-2$, etc. Moreover, estimates of transformed data models such as Equation (5.30) may in fact show significant intercept terms, implying that y continues to change, even if $\Delta x_t = 0$.

In practice, models such as Equation (5.29) may be inappropriate for representing long-term relationships, because of delays in the adjustment of y to changes in x. A model allowing lagged impacts of both y and x on the outcome is more likely to be observed, with form such as

$$y_t = \mu + b_1 x_t + b_2 x_{t-1} + \rho y_{t-1} + u_t \tag{5.31}$$

This is known as an Autoregressive Distributed Lag model, denoted ADL or ARDL, of order (1, 1) (Bauwens *et al.*, 1999, p 136; Greene, 2000, p. 724). To avoid problems raised by y_t and x_t being non-stationary, Equation (5.31) may be rearranged by first substracting y_{t-1} from both sides and then subtracting $b_1 x_{t-1}$ (from $b_1 x_t$ and adding it to $b_2 x_{t-1}$) to yield

$$\Delta y_t = \mu + b_1 \Delta x_t + (b_1 + b_2) x_{t-1} - (1 - \rho) y_{t-1} + u_t$$

Setting $\delta_1 = (b_1 + b_2)/(1 - \rho)$, and $\delta_0 = \mu/(1 - \rho)$, then gives a further restatement of Equation (5.31) as

$$\Delta y_t = b_1 \Delta x_t + (\rho - 1)(y_{t-1} - \delta_0 - \delta_1 x_{t-1}) + u_t \tag{5.32}$$

The term multiplied by $\alpha = \rho - 1$ may be regarded as modelling the negative or positive feedback of a disequilibrium error from the previous period, and so describes an error correction mechanism. So Equation (5.32) may incorporate the underlying long-term relationship more effectively than Equation (5.30), and may therefore perform better in long-term forecasting. One motivation for using an ECM version of a model is to incorporate prior information, for example on economic multiplier effects of x on y which appear explicitly in the ECM version of an ADL model, but not in its original version as in Equation (5.31) (Bauwens *et al.*, 1999).

These ideas can be extended to the case where the hypothesised equilibrium relationship, analogous to Equation (5.29), involves two predictors:

$$y_t = \beta_0 + \beta_1 x_t + \beta_2 z_t + u_t \tag{5.33}$$

The typical observed short-run (disequilibrium) relationship, parallel to Equation (5.31), might then be

$$y_t = b_0 + b_1 x_t + b_2 x_{t-1} + c_1 z_t + c_2 z_{t-1} + \rho y_{t-1} + u_t$$

Re-expressing in ECM form then leads to

$$\Delta y_t = b_1 \Delta x_t + c_1 \Delta z_t + (\rho - 1)(y_{t-1} - \delta_0 - \delta_1 x_{t-1} - \delta_2 z_{t-1}) + u_t$$

Cointegrated series

One situation where error correction specifications are implied is when the outcome y_t and predictor(s) are co-integrated. Co-integration is the statistical expression of a stable long run relationship between y_t and a predictor x_t, and may be defined in terms of the stationarity or otherwise of y, x and the error in models such as Equation (5.29) and (5.33). Thus, if a series y_t follows a first order random walk, then its first difference is stationary, and y_t may be termed an integrated process of order 1, denoted $I(1)$; however, if y_t is stationary without differencing, it is considered to be integrated of order zero, $I(0)$.

Two nonstationary series y_t and x_t are cointegrated if they are both $I(1)$ but the errors u_t, as in Equation (5.29),

$$u_t = y_t - \beta_0 - \beta_1 x_t$$

are stationary, that is $I(0)$. This is because if y and x are governed by a long-term relation then any disequilibrium error should hover around zero (Engle and Granger, 1987). If, by contrast, y_t and x_t are not cointegrated, then the u_t are $I(1)$, and nonstationary. Whether or not the u_t are stationary may be assessed by regressing u_t on u_{t-1}:

$$u_t = \lambda u_{t-1} + \eta_t$$

and if $|\lambda| < 1$, then u is stationary.

Example 5.9 Voting intentions Although economic relations form the major application for ECM models, they have also been applied to political affiliations and attitudes. Clarke *et al.* (1998) investigate changes in voting intentions for a future General Election in Britain, specifically the proportion intending to vote for the Labour Party in the five years preceding the May 1997 election victory for Labour. Thus for 64 months from January 1992 to April 1997, the proportion intending to vote Labour was related to factors such as:

(a) underlying Labour party identification (iden[ ] in Program 5.9);
(b) the proportion of respondents seeing the Labour leader as the best potential prime minister (bpm[ ] in Program 5.9);
(c) expectations of personal economic welfare (the 'feel good' factor denoted fg[ ]);
(d) economic indicators such as unemployment and interest rates (denoted intr[ ]);
(e) perceptions of the most important problem (inflation, unemployment, health, etc.). The proportions seeing unemployment and inflation as most important are denoted mipu[ ] and mipi[ ].

Also used to explain voter intentions are short-term temporary issues or 'events', such as political scandals and interest rate crises.

Clarke *et al.* argue that a long-term cointegrating relationship exists between Labour voting intentions, and factors (a) and (b) above, namely Labour party identification and seeing the Labour leader as best Prime Minister. Therefore it is necessary to introduce an error correction

$$\mathrm{ECM}_t = \mathrm{LAB}_t - \delta_1 \mathrm{LBPM}_t - \delta_2 \mathrm{LPID}_t \tag{5.34}$$

into the voting intentions model. Their preferred model (Clarke *et al.*, Table 3, Personal Prospective Model) selects from a fuller set of predictors and has the form (for two temporary events)

$$\begin{aligned}\Delta\mathrm{LAB}_t = \beta_0 &+ \beta_1 \Delta\mathrm{LBPM}_t + \beta_2 \mathrm{LPID}_{t-2} + \alpha \mathrm{ECM}_{t-1} + \beta_3 \Delta\mathrm{FG}_t \\ &+ \beta_4 \Delta\mathrm{INTR}_t + \beta_5 \Delta\mathrm{MIPI}_t + \beta_6 \Delta\mathrm{MIPU}_t \\ &+ \beta_7 \mathrm{EVENT1}_t + \beta_8 \mathrm{EVENT2}_t + u_t\end{aligned} \tag{5.35}$$

The events are handled as short-term effects, and since the model is for differenced variables, a dummy is coded +1 in the month when the event occurs (or is assigned to have occurred), −1 in the next month, and 0 otherwise. Two events assumed likely to boost Labour general election voting intentions are included, namely the Conservative election scandals of March 1997, and the European elections of June 1994 with a large Labour victory. (The model of Clarke *et al.* included several other events of this type.)

The results obtained with this model (Table 5.3) are similar to those of Clarke *et al.* (1998) in terms of the parameters in the ECM itself, namely δ_1 and δ_2 in Equation (5.34), and the negative sign on the ECM coefficient α in the main model Equation (5.35).

The relatively small size of α (Clarke *et al.* obtain $\alpha = -0.15$) translates into a long 'half-life' for the delayed negative feedback of factors included in the generic term x_{t-1} in Equation (5.32). For example, a negative 'event' which reduced the Labour voting intention by three points in month t would be carried over to a reduction of $(3 - 3 \times 0.197) = 2.4$ points in the next month, $(2.4 - 0.197 \times 2.4) = 1.9$ points the next month, and so on.

5.5 DYNAMIC LINEAR MODELS AND TIME VARYING COEFFICIENTS

Classical time series methods assume fixed relationships between y_t and x_t and stationarity (absence of upward or downward trend) in y_t, or a transformation of y_t, so that both intercepts and regression coefficients are fixed. In practice, relationships between variables are likely to vary over time: for example, the response of the birth rate to economic conditions in successive years, or of sales to advertising over successive weeks, is unlikely to be constant whatever the level of birth rate or the economy. Autoregressive moving average models with fixed regression effects are often of limited use in understanding the processes which generated such relationships, and may best be used as a

Table 5.3 Labour voting intentions, parameter summary

Variable		Cointegrating Regression				Clarke *et al.*	
		Mean	St. devn.	2.5%	97.5%	MLE	s.e. (MLE)
Labour Leader Best Prime Minister	δ_1	0.637	0.074	0.493	0.777	0.635	0.083
Labour Party Identification	δ_2	0.583	0.069	0.448	0.717	0.585	0.078
		Error Correction Model					
Intercept	β_0	0.264	0.285	−0.298	0.833	0.47	0.24
Labour Leader Best Prime Minister	β_1	0.463	0.074	0.318	0.613	0.27	0.08
Labour Party Identification	β_2	0.158	0.092	−0.020	0.340	0.19	0.09
ECM	α	−0.197	0.084	−0.366	−0.034	−0.15	0.07
Feel Good Factor	β_3	−0.047	0.049	−0.142	0.054	−0.08	0.04
Interest Rates	β_4	0.866	0.930	−0.955	2.713	1.4	0.86
Inflation Most Important Problem	β_5	0.343	0.150	0.044	0.629	0.55	0.14
Unemployment Most Important Problem	β_6	0.071	0.054	−0.040	0.175	0.15	0.05
Euro Election	β_7	3.416	1.400	0.712	6.099	4.73	1.33
Conservation campaign scandals	β_8	3.551	1.342	0.846	6.238	5.2	1.31

first approximation to summarise the data in a parsimonious way for forecasting purposes (Diggle, 1990). They may also have limitations in forecasting itself: a model with good fit in the sample period may perform poorly after that if the underlying parameters are in fact evolving through time (West and Harrison, 1989).

To model stochastic shifts in regression parameters one may call upon random effects models which specify time dependence between successive parameter values in the form of smoothness priors. For example, whereas a fixed coefficient time series regression for univariate y might be

$$y_t = \alpha + \beta x_t + \rho y_{t-1} + u_t$$

a model with nonstationary levels and β coefficient might be specified as

$$y_t = \alpha_t + \beta_t x_t + \rho_t y_{t-1} + u_t$$

with the parameters evolving according to

$$\alpha_t = \alpha_{t-1} + \omega_{1t} \tag{5.36a}$$

$$\beta_t = \beta_{t-1} + \omega_{2t} \tag{5.36b}$$

$$\rho_t = \rho_{t-1} + \omega_{3t} \tag{5.36c}$$

with the 'state' equations in α_t, β_t and ρ_t defined for $t = 2, .. T$. The most general dynamic linear model has errors u_t and ω_{kt} with time dependent variances V_t and $W_t = (W_{1t}, W_{2t}, W_{3t})$, respectively.

In practice, the variances V_t and W_{kt} may be taken to be constant over all time points $t = 1, .., T$, so that there is a single level of volatility describing evolution in levels or regression effects. In this example, then one might take

$$\omega_{1t} \sim N(0, W_1)$$
$$\omega_{2t} \sim N(0, W_2)$$
$$\omega_{3t} \sim N(0, W_3)$$

where W_1, W_2 and W_3 are hyperparameters to be estimated. It is also possible to assume that the ω_{kt} follow a multivariate form, with a constant dispersion matrix.

The system is set in motion by vague priors on the levels, β coefficients, and the autoregressive parameters ρ in the first period. For example, one might assume the first period regression parameter has a prior

$$\beta_1 \sim N(b_1, C_1)$$

where b_1 and C_1 are both known (typical values might be $b_1 = 0$, $C_1 = 1000$).

The priors on α_t and β_t in Equations (5.36a) and (5.36b) are first-order random walk priors and are equivalent to taking the differences $\alpha_t - \alpha_{t-1}$ and $\beta_t - \beta_{t-1}$ to be random with variances W_1, W_2 and W_3. One might instead consider higher order differences to be random, e.g. take $\Delta^2\alpha_t = \Delta(\alpha_t - \alpha_{t-1}) = \alpha_t - 2\alpha_{t-1} + \alpha_{t-2}$ as having zero mean and variance W_1. In this case, the prior on the levels is a second order random walk and can be written (for $t > 3$) as

$$\alpha_t = 2\alpha_{t-1} - \alpha_{t-2} + \omega_{1t}$$

To accommodate changing volatility without excess parameterisation, Ameen and Harrison (1985) suggest a discounting prior for the variances, which specifies a prior for the precisions at time 1, and then discounts later precisions to reflect a time decay factor. So for the observation error variance and the variance of the levels at time 1 one might take

$$V_1^{-1} \sim G(s_1, t_1), \; W_{11}^{-1} \sim G(s_2, t_2)$$

Subsequent precisions are downweighted by a factor δ, where $0 < \delta < 1$, and for $t > 1$

$$V_t^{-1} = \delta_1 V_{t-1}^{-1}$$
$$W_{1t}^{-1} = \delta_2 W_{1t-1}^{-1}$$

Other approaches to stochastic variances involve ARCH-GARCH and structural shift models, and are discussed below.

5.5.1 State space smoothing

A major use for state-space models is in semi-parametric additive regression, where the object is to construct a smooth 'signal' $f(t)$ from noisy data $y(t)$. In this case, there are no covariates, and $f(t)$ describes the smoothly changing level assumed to underlie the observations on $y(t)$ which are subject to measurement error. Another common model task is to achieve a smooth representation of the changing nonlinear effect of a covariate. There is a close link (e.g. Fahrmeir and Lang, 2000) between state space models for dynamic general linear models and general additive models involving smooth functions of metric predictors, as discussed in Section 3.6. The orientation in this chapter is to smoothing of time series *per se*, together with non-linear regression analysis achieved via time varying regression coefficients.

To illustrate techniques for smoothing time series, consider a univariate series $y(t)$ observed at equidistant points, $t = 1, 2, 3, \ldots T$. Then the model has the form

$$y(t) = f(t) + e(t)$$

The $e(t)$ are typically taken as exchangeable errors such as $e(t) \sim N(0, \sigma^2)$, but the true series $f(t)$ follows a random walk of order k. For example, if $k = 2$,

$$f(t) = 2f(t-1) - f(t-2) + u(t) \tag{5.37}$$

with $u(t) \sim N(0, \tau^2)$. One may expect the conditional variance τ^2 of the true series to be less than that of the noisy series $y(t)$, with the noise to signal ratio $\lambda^2 = \sigma^2/\tau^2$ then being greater than 1, and $1/\lambda^2$ being under 1. So a prior (e.g. gamma) on $1/\lambda^2$ might be taken that favours small positive values. Alternatively, one might take a uniform $U(0, 1)$ prior on the ratio $\sigma^2/[\sigma^2 + \tau^2]$. Higher values of λ^2 correspond to greater smoothing (as the variance τ^2 of the smooth function becomes progressively smaller).

With $k = 2$ and diffuse priors on $f(1)$ and $f(2)$ in Equation (5.37), and with Normal metrical data $y(t)$, the posterior means $\hat{f}(1), \hat{f}(2) \ldots \hat{f}(T)$ provide values which minimise the penalised fit function

$$\sum_{t=1}^{T} \{y(t) - f(t)\}^2 + \lambda^2 \sum_{t=3}^{T} \{\Delta^2 f(t)\}^2$$
$$= \sum_{t=1}^{T} \{y(t) - f(t)\}^2 + \lambda^2 \sum_{t=3}^{T} \{u(t)\}^2$$
$$= \sum_{t=1}^{T} \{y(t) - f(t)\}^2 + \lambda^2 \sum_{t=3}^{T} \{f(t) - 2f(t-1) - f(t-2)\}^2$$

Higher order difference priors may be used for seasonal effects. For example, for quarterly data, a possible smoothness prior is

$$h(t) = s(t) + s(t-1) + s(t-2) + s(t-3) \sim \mathrm{N}(0, \tau.s)$$

For monthly data, the analogous scheme is

$$h(t) = s(t) + s(t-1) + s(t-2) + \ldots. + s(t-11) \sim \mathrm{N}(0, \tau.s)$$

Instead of simple random walk priors, autoregressive priors involving lag coefficients $\phi_1, \ldots, \phi_k$ may be specified as smoothness priors. For example, an AR(2) prior in the true series would be

$$f(t) \sim \mathrm{N}(\phi_1 f(t-1) + \phi_2 f(t-2), \tau^2)$$

Kitagawa and Gersch (1996) illustrate the use of such priors (with high order k) to estimate the spectral distribution of a stationary time series.

Example 5.10 Asymmetric series This example illustrates the detection of a signal in noisy data when the form of the signal is exactly known. Thus, Kitagawa and Gersch (1996, Chapter 4) simulate a time series according to the truncated and asymmetric form

$$y(t) = f(t) + e(t) \quad t = 1, 200$$

where the true series or signal is

$$f(t) = (12/\pi)\exp(-\{t-130\}^2/2000)$$

and $e(t) \sim \mathrm{N}(0, \sigma^2)$, where $\sigma^2 = 1$. The maximum value of the true series is just under 4 at $t = 130$, with the true series being effectively zero for $t < 50$. Kitagawa and Gersch (1996, p.109) contrast the AIC fit values obtained with different orders k in the random walk smoothness prior, and select $k = 2$ on the basis of the AIC criterion, so that

$$f(t) = 2f(t-1) - f(t-2) + u(t)$$

with $u(t) \sim \mathrm{N}(0, \tau^2)$. The $k = 1$ model is found by Kitagawa and Gersch to be too ragged, while the smoothing obtained with values $k = 2, 3, 4$ is visually indistinguishable.

Here two alternative priors are set on the variance of τ^2 conditional on σ^2, one a uniform prior $U(0, 1)$ on $\sigma^2/[\sigma^2 + \tau^2]$, the other a $G(0.1, 0.2)$ prior on τ/σ. The latter prior favours values under 1 in line with variability about the signal being expected to be less than that around the observations. A G(1, 0.001) prior on $1/\sigma^2$ is adopted.

The median value of τ^2 obtained under the first prior, from the second half of a two chain run to 20 000 iterations, stands at 1.03E-4, as compared to the value of 0.79E-4 cited by Kitagawa and Gersch using a series generated by the same process. The median observational variance is estimated at 1.11. The true series is reproduced satisfactorily (Figure 5.9). This prior leads to convergence in under 5000 iterations. Other priors, whether gamma or uniform on the ratios τ^2/σ^2 or τ/σ tend to converge much slower. The U(0, 1) prior on τ/σ takes 100 000 iterations to obtain σ^2 around 1.1 and a median on τ^2 of 0.6E-4, and provides a slightly better fit to the high values of the series.

Convergence may be facilitated by direct sampling from one or perhaps both the full conditionals on $P_1 = 1/\sigma^2$ and $P_2 = 1/\tau^2$, namely

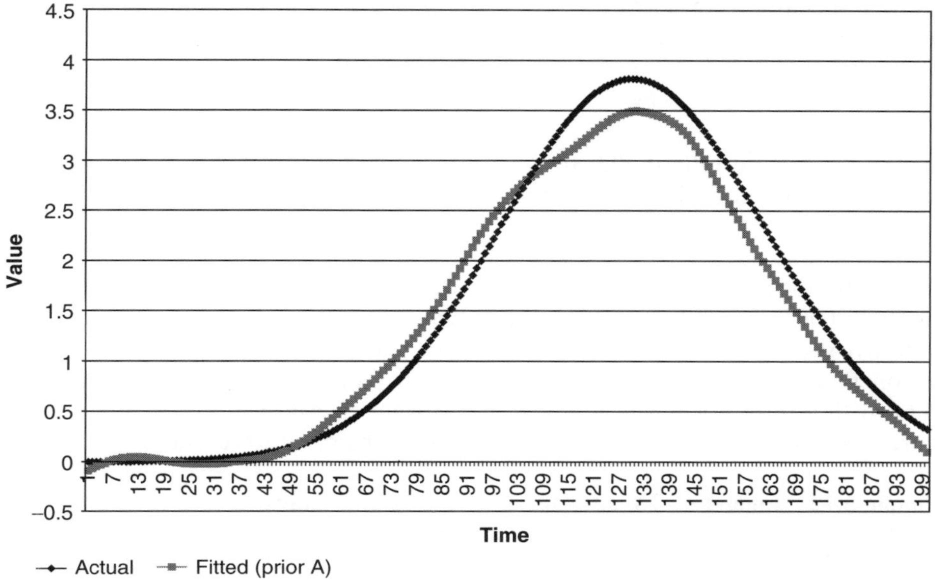

Figure 5.9 Actual and fitted smooth curve (asymmetric Gaussian)

$$P_1 \sim \mathrm{G}(a_1 + 0.5T,\ b_1 + 0.5\sum_{t=1}^{T} e^2(t))$$

$$P_2 \sim G(a_2 + 0.5(T-k),\ b_1 + 0.5\sum_{t=\mathrm{k}}^{T} u^2(t))$$

where one might take $a_j = b_j = 0.001$, $j = 1, 2$. An exercise is to compare the fit of an RW(3) model with the RW(2) model using the pseudo-marginal likelihood or other model assessment approach.

Example 5.11 UK coal consumption As an illustration of a smoothing model for a time series with seasonal effects, as well as a secular trend (decline) over time, we follow Harvey (1989) and Fruhwirth-Schatner (1994) in considering a UK coal consumption series C_t of 108 quarterly observations from 1960–1986. The Basic Structural Model (BSM) of Harvey (1989) may then be applied, with a seasonal smoothing prior appropriate to quarterly data. The outcome y_t is the log of the original consumption series divided by 1000, namely

$$y_t = \log(C_t/1000)$$

Assuming Normality in this outcome, the model has the following components

$y_t \sim \mathrm{N}(x_t, \theta_4)$	(Observation model)
$x_t = \mu_t + s_t$	(Underlying trend after allowing for seasonal effect)
$\mu_t \sim \mathrm{N}(\mu_{t-1} + a_{t-1}, \theta_1)$	(Evolution in mean)
$a_{t-1} \sim \mathrm{N}(a_{t-2}, \theta_2)$	(Increments to mean)
$s_t \sim \mathrm{N}(-s_{t-1} - s_{t-2} - s_{t-3}, \theta_3)$	(Seasonality)

Fruhwirth-Schatnor considers the series for 1960–1982 and the utility of Gibbs sampling as against the Kalman forward filtering-backward sampling algorithm. Here, a uniform prior $U(0, 1)$ is set on the ratio θ_1/θ_4, and with the precisions

$$\theta_4^{-1}, \theta_2^{-1}, \theta_3^{-1}$$

assigned the diffuse G(0.01, 0.00001) priors adopted by Fruhwirth-Schatner. As an alternative model, consider a lag 1 autoregression in a_{t-1}, namely

$$a_{t-1} \sim \mathrm{N}(\phi a_{t-2}, \theta_2)$$

To assess fit, the predictive loss criterion of Laud and Ibrahim (1995) is used. Two chain runs of 75 000 iterations on both models (with summaries based on the last 50 000 iterations) show a lower predictive loss on the autoregressive model, with variance component estimates as in Table 5.4. The density of ϕ is concentrated on negative values, with a posterior mean of −0.29, but is right skewed with 97.5% point of 0.65. Figure 5.10 shows the actual and fitted series with the basic structural model.

Example 5.12 Impact of TV advertising This example illustrates evolving regression impacts as in Equation (5.36b), and also involves a binomial outcome. Thus, a study on the impact TV advertising involved asking a set weekly total of 66 individuals a 'yes or no' question about an advert for a chocolate bar, and was continued over 171 weeks (Migon and Harrison, 1985). The number of positive answers r_t in week t is modelled as binomial with logit link to a single covariate, $x_t =$ weekly expenditure on advertisements. Thus

$$r_t \sim \mathrm{Bin}(66, \pi_t)$$
$$\mathrm{logit}(\pi_t) = \alpha_t + \beta_t x_t$$

with the parameters evolving according to

$$\alpha_t = \alpha_{t-1} + \omega_a$$
$$\beta_t = \beta_{t-1} + \omega_b$$

RW(1) priors for α_t and β_t are adopted with a diagonal dispersion matrix with variance terms $\mathrm{var}(\omega_a) = \sigma^2_\alpha$ and $\mathrm{var}(\omega_b) = \sigma^2_\beta$. Priors for the initial conditions α_1 and β_1 are relatively diffuse, namely N(0, 1000).

Table 5.4 Components of variance, coal consumption

Basic Structural Model	Mean	St. devn.	2.5%	Median	97.5%
Variance of smooth series	0.00156	0.00138	0.00005	0.00118	0.00528
Variance of trend	0.000018	0.000019	3E-6	0.000013	0.000070
Seasonal Variance	0.00009	0.00014	0.00000	0.00004	0.00045
Residual Variance	0.01544	0.00276	0.01049	0.01526	0.02135
Autoregressive Model					
Variance of smooth series	0.00339	0.00191	0.00023	0.00316	0.00785
Variance of trend	0.00110	0.00204	0.00001	0.00030	0.00640
Seasonal Variance	0.00009	0.00016	0.00000	0.00004	0.00049
Residual Variance	0.01336	0.00280	0.00842	0.01321	0.01922

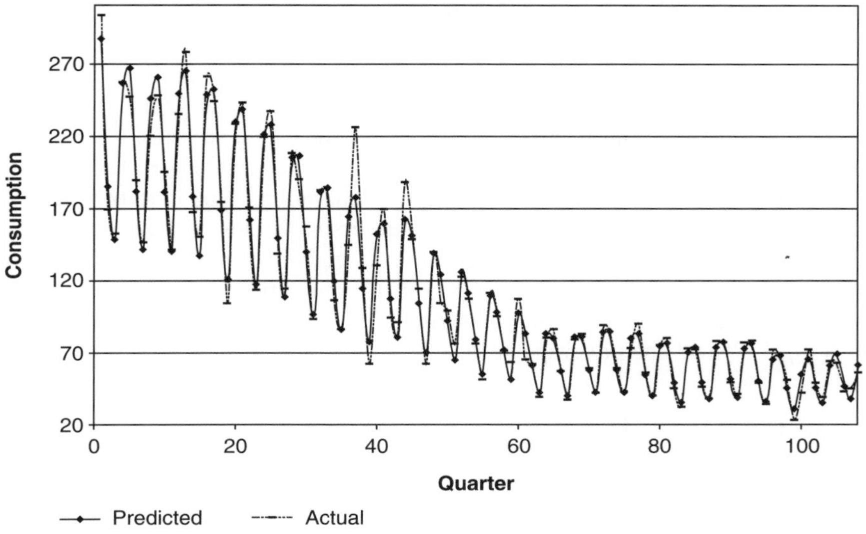

Figure 5.10 Actual and predicted coal consumption

With three chain runs of 15 000 iterations, posterior mean estimates of σ^2_α and σ^2_β are close to those cited by Fahrmeir and Tutz, though their variance for the slopes at 0.0002 is smaller than the estimate here of around 0.00045. The plot of the parameters themselves (Figure 5.11) shows a decrease in the mean (i.e. awareness level) though the positive impact of advertising is more or less stable. The parameters in Figure 5.11 are based on a run of 7500 iterations (2500 burn in) with every tenth iterate retained.

An alternative model (Model B) with a multivariate Normal prior on α_t and β_t jointly yields very similar estimates for σ^2_β, and on the two sets of evolving parameters themselves.

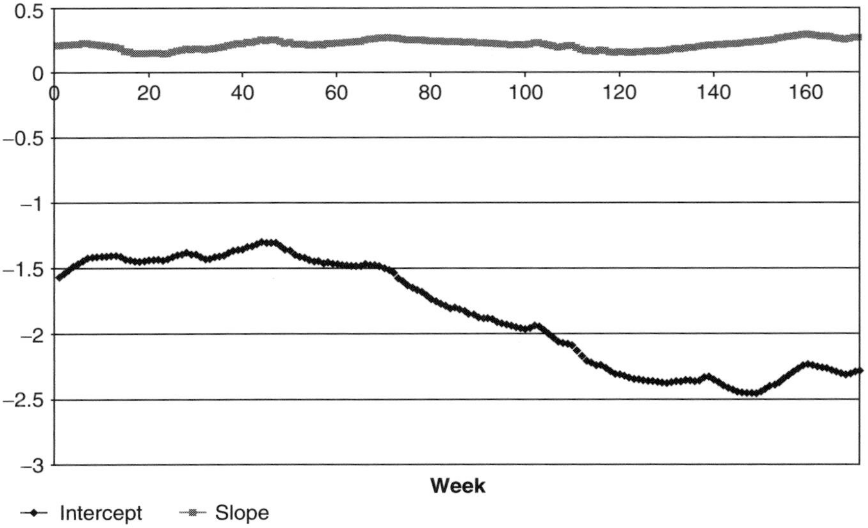

Figure 5.11 TV advertising, intercepts and slopes

5.6 STOCHASTIC VARIANCES AND STOCHASTIC VOLATILITY

There are many instances, including the dynamic coefficient models just discussed, where it may be necessary to model observed time series y_t or make forecasts, when the variance is not fixed, but itself stochastic over time. Such situations are exemplified by stock price and exchange rate series where large forecast errors tend to occur in clusters, when the series are unsettled or rapidly changing. This is known as volatility clustering, and suggests dependence between successive values of the squared errors. In many applications of such models, the series has effectively a zero mean; for example, in many financial time series (e.g. exchange rates or price indices) the ratio of successive values E_t/E_{t-1} averages 1, and a series defined by the log of these ratios $y_t = \log(E_t/E_{t-1})$ will then average zero.

5.6.1 ARCH and GARCH models

Following Engle (1982), consider a time series regression

$$y_t = \beta x_t + \varepsilon_t$$

in which the error variances for subsets of the full period are different. One way to accommodate heteroscedasticity through time (i.e. changes in scale) is to introduce an exogenous variable z (or lagged values of the outcome), with the model now being perhaps

$$y_t = \beta x_t + z_t^{\gamma} \varepsilon_t$$

Forecasts of y now depend upon z_t, as well as the preceding values of the outcome itself. Another option involves an autoregressive conditional heteroscedastic or ARCH model (Greene, 2000), such that

$$y_t = \beta x_t + \varepsilon_t = \beta x_t + u_t \sqrt{h_t} \tag{5.38}$$

where the u_t have mean zero and variance 1, and the h_t depend upon previous squared errors

$$h_t = \alpha_0 + \alpha_1 \varepsilon_{t-1}^2 \tag{5.39}$$

with both α_0 and α_1 positive. While the most usual assumption is $u_t \sim \mathrm{N}(0, 1)$, Bauwens and Lubrano (1998) consider $u_t \sim \mathrm{Student}(0, 1, \nu)$.

Then $E(\varepsilon_t|\varepsilon_{t-1}) = 0$, and the conditional variance

$$\begin{aligned} V_t = \mathrm{var}(\varepsilon_t|\varepsilon_{t-1}) = E(\varepsilon_t^2|\varepsilon_{t-1}) &= E(u_t^2)[\alpha_0 + \alpha_1 \varepsilon_{t-1}^2] \\ &= \alpha_0 + \alpha_1 \varepsilon_{t-1}^2 \end{aligned}$$

is heteroscedastic with respect to ε_{t-1}. Specifically, Equation (5.39) defines an ARCH(1) model, whereas an ARCH(2) model would involve dependence on ε_{t-2}^2. Thus

$$h_t = \alpha_0 + \alpha_1 \varepsilon_{t-1}^2 + \alpha_2 \varepsilon_{t-2}^2$$

If $|\alpha_1| < 1$ in Equation (5.39), then the unconditional variance has mean zero and variance $\alpha_0/(1 - \alpha_1)$. The log likelihood for the tth observation under Equation (5.38) is

$$\log L_t = -0.5 \log(2\pi) - 0.5 \log h_t - 0.5 \varepsilon_t^2 / h_t \tag{5.40}$$

where $\varepsilon_t = y_t - \beta x_t$. If y_t has an effectively zero mean and the regression model does not involve predictors x_t, then one may follow Engle (1982) and write

$$y_t = u_t\sqrt{h_t} \quad (5.41)$$

where for an ARCH(1) model

$$h_t = \alpha_0 + \alpha_1 y_{t-1}^2$$

The log likelihood for the tth observation is then

$$\log L_t = -0.5\log(2\pi) - 0.5\log h_t - 0.5y_t^2/h_t \quad (5.42)$$

In the GARCH model the conditional variance V_t depends upon previous values of V_t (or of h_t), as well as on lags in ε_t^2. Whereas lags in ε_t^2 are analogous to moving average errors in an ordinary ARMA time series, lags in V_t are parallel to the autoregressive component (Greene, 2000). A GARCH(p, q) model involves a lag of order p in V_t and one of order q in ε_t^2. Thus, a GARCH(1, 1) model would be either

$$h_t = \alpha_0 + \gamma V_{t-1} + \alpha_1\varepsilon_{t-1}^2$$

or

$$h_t = \alpha_0 + \gamma h_{t-1} + \alpha_1\varepsilon_{t-1}^2$$

where $\alpha_0 > 0$. The likelihood is as in Equation (5.40). The stationarity conditions here are that $\alpha_1 + \gamma < 1$, together with $\alpha_1 \geq 0$, $\gamma \geq 0$. The specification in Equation (5.41) leads to a GARCH(1, 1) model in which

$$h_t = \alpha_0 + \gamma V_{t-1} + \alpha_1 y_{t-1}^2 \quad (5.43)$$

with likelihood as in Equation (5.42).

5.6.2 Stochastic volatility models

Another option for modelling changing variances is known as stochastic volatility, and includes models within the state-space framework (Kitagawa and Gersch, 1996). Thus in

$$y_t = \beta x_t + \varepsilon_t$$

with $\varepsilon_t \sim \mathrm{N}(0, V_t)$, one may assume the evolution of $\Delta^k \log V_t$ follows a random walk process. For example, taking $k = 1$, and setting $g_t = \log V_t$, gives a first order random walk which may follow a Normal or Student form:

$$g_t \sim \mathrm{N}(g_{t-1}, \sigma_g^2)$$

An alternative stochastic volatility formulation (Pitt and Shepherd, 1998) involves autoregressive dependence in latent variables κ_t, which represent the evolving log variances. Thus for a series with zero mean and no regressors, one might specify first order dependence in the latent log variances (Harvey *et al.*, 1994), with

$$y_t = u_t \exp(\kappa_t/2) \quad (5.44)$$

and

$$\kappa_t = \phi\kappa_{t-1} + \eta_t$$

where the u_t are N(0, 1) and $\eta_t \sim N(0, \sigma^2_\eta)$. If $|\phi| < 1$, then the κ_t are stationary with variance $\sigma^2_\eta/(1-\phi^2)$.

Another option is denoted the unobserved ARCH model (Shephard, 1996), in which an ARCH model still holds but is observed with error. This is generally classified as a stochastic volatility approach. For a zero mean observation series y and no covariates, a measurement error model combined with an ARCH model leads to

$$\begin{aligned} y_t &\sim N(\lambda_t, \sigma^2) \\ \lambda_t &\sim N(0, h_t) \\ h_t &= \alpha_0 + \alpha_1 \lambda^2_{t-1} \end{aligned} \tag{5.45}$$

To ensure h_t is positive, α_0 and α_1 are constrained to be positive, and the further restriction $0 \leq \alpha_1 \leq 1$ ensures that the ARCH series is covariance stationary. If there were covariates or lags in the model for y, then the λ_t would be distributed as

$$\lambda_t \sim N(\mu_t, h_t)$$

where, for instance, $\mu_t = \beta x_t$.

For multivariate series (e.g. of several exchange rates) subject to volatility clustering, factor analysis type models have been proposed to model the interrelated volatility (Pitt and Shephard, 1999; Harvey *et al.*, 1994). For instance, for two series y_{tk}, $k = 1, 2$ and one factor f_t, one might have

$$\begin{aligned} y_{t1} &= \beta_1 f_t + \omega_{t1} \\ y_{t2} &= \beta_2 f_t + \omega_{t2} \end{aligned}$$

with f_t and the ω_{tk} evolving in line with stochastic volatility. Thus $f_t \sim N(0, \exp(\kappa^f_t))$, $\omega_{t1} \sim N(0, \exp(\kappa^{\omega 1}_t))$, $\omega_{t2} \sim N(0, \exp(\kappa^{\omega 2}_t))$. Then first order autoregressive dependence in the latent log variances would imply

$$\begin{aligned} \kappa^f_t &= \rho^f \kappa^f_{t-1} + \eta^f_t \\ \kappa^{\omega 1}_t &= \rho^{\omega 1} \kappa^{\omega 1}_{t-1} + \eta^{\omega 1}_t \\ \kappa^{\omega 2}_t &= \rho^{\omega 2} \kappa^{\omega 2}_{t-1} + \eta^{\omega 2}_t \end{aligned}$$

Example 5.13 Spot market index To illustrate in a comparative fashion both ARCH models and stochastic volatility models, consider the weekly spot market index series E_t of Bauwens and Lubrano (1998) relating to the shares of Belgian firms at the Brussels stock exchange for 3-1-86 to 26-1-96, namely 508 observations. The transformed outcome is the index return, given by the first difference of the logarithm of the index (times 1000), namely $y_t = 1000[\log(E_t) - \log(E_{t-1})]$.

We first consider a variant of the ARCH model, with

$$y_t = \mu + \varepsilon_t = \mu + u_t\sqrt{h_t}$$

where $u_t \sim N(0, 1)$ and

$$h_t = \alpha_0 + \alpha_1 \varepsilon^2_{t-1} = \alpha_0 + \alpha_1 (y_{t-1} - \mu)^2 \tag{5.46}$$

The coefficient α_1 is constrained to be between 0 and 1, and α_0 to be positive. Fit is assessed by the CPO estimate based on the average of the inverse likelihoods, and by the predictive loss criterion of Gelfand and Ghosh (1998). The initial variance h_1 is taken as a separate fixed effect. The transformed outcome used by BL leads to large variances, and to accommodate this a uniform prior U(0, 1000) for α_0 is adopted.

A two chain run to 5000 iterations (and 1000 burn-in) shows the lag coefficient α_1 in Equation (5.46) to have 95% interval (0.28, 0.66), and α_0 one of (182, 261). The predictive loss criterion is 380 200.

A second approach is provided by a GARCH(1, 1) model applied to the series $y'_t = y_t - 2.246$, where $\bar{y} = 2.246$. This avoids introducing a parameter for the mean of y_t, and h_t is then as in Equation (5.43). The posterior means for α_1 and γ are, respectively, 0.26 (s.e. 0.06) and 0.63 (s.e. 0.07). The predictive loss falls to 372 500. It may be noted that initial iterations with this model are very slow.

A third model (Model C) for these data is provided by the unobserved ARCH model of Equation (5.44), with a non-zero mean μ assumed for the λ_t, so that

$$y_t \sim \text{N}(\lambda_t, \sigma^2)$$
$$\lambda_t \sim \text{N}(\mu, h_t)$$
$$\varepsilon_t = \lambda_t - \mu$$
$$h_t = \alpha_0 + \alpha_1 \varepsilon_{t-1}^2$$

This also yields an improvement in fit over the ARCH model of Equation (5.46), with predictive error of 373 200. This is assessed from the second half of a two chain run over 6000 iterations. Convergence in the parameters α_0 and α_1 is only obtained after about 3000 iterations using the over-relaxation option. The lag coefficient α_1 has a median of 0.33, but is not precisely identified, having 95% credible interval {0.02, 0.96}. The most distinct outlier (with lowest CPO) is associated with the sharp drop between weeks 90 and 91 from 4516 to 3955.

The final model (Model D) is provided by a random walk model for the stochastic log of the variance V_t, with

$$y_t = \mu + \varepsilon_t$$

where

$$\varepsilon_t \sim \text{N}(0, V_t)$$

Because BUGS is parameterised in terms of precisions, one may set $P_t = 1/V_t$ and then take $H_t = \log(P_t)$ as an RW(1) process:

$$H_t \sim \text{N}(H_{t-1}, \sigma_H^2)$$

It is worthwhile experimenting with alternative priors on the smoothing variance, since this parameter is crucial to the performance of the model (see Exercises). Taking the prior

$$1/\sigma_H^2 \sim \text{G}(1, 0.0001)$$

leads to a predictive loss criterion of 365 900. The plot of the precisions P_t in Figure 5.12 shows the highest volatility (i.e lowest precision) at weeks 90–100 and 225–235.

Example 5.14 Exchange rate Durbin and Koopman (2000) and Harvey et al. (1994) consider stochastic volatility models of the form

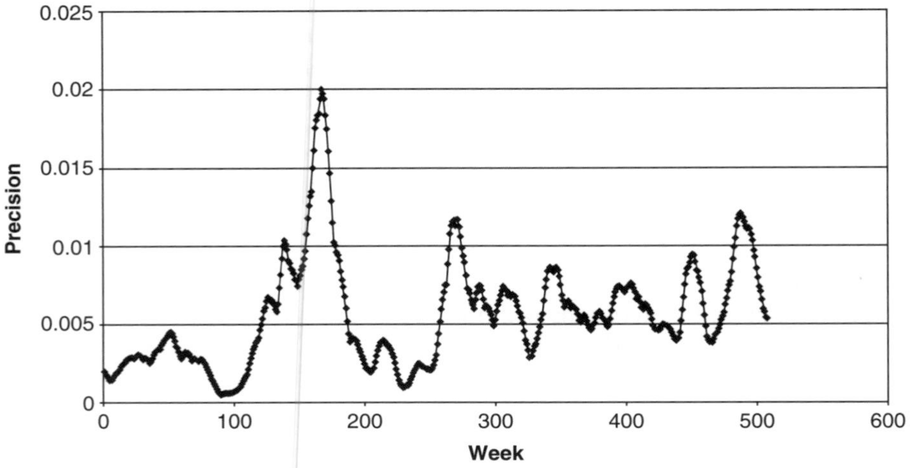

Figure 5.12 Spot market index volatility

$$y_t = \beta x_t + u_t\sqrt{h_t}$$
$$u_t \sim N(0,\ 1)$$
$$h_t = \sigma^2 \exp(\theta_t)$$
$$\theta_t = \phi\theta_{t-1} + \eta_t$$

where the η_t are normal with variance σ^2_η. Durbin and Koopman consider a series on the pound-dollar exchange rate between October 1st 1981 and June 28th 1985, and define a model with no predictor term βx_t or constant β_0. It is valid to assume that $\beta_0 = 0$, since the observations consist of differences in logged exchange rates E_t, with $y_t = \Delta \log(E_t)$.

In BUGS, the time varying dispersions are parameterised via precisions

$$P_t = 1/h_t$$

and the first model follows Durbin and Koopman in assuming stationarity in θ, with

$$0 < \phi < 1$$

We find – from the second half of a two chain run of 5000 iterations – median estimates for σ and σ_η of 0.50 and 0.16, whereas Durbin and Koopman cite values of 0.63 and 0.17. They estimate a lag coefficient ϕ of 0.973, whereas the value obtained here, using a U(0, 1) prior, has mean 0.982 and a 95% interval {0.959, 0.997}.The variances are below 0.5 for most of the period but increase to over 1 in the spring of 1985 (see Figure 5.13), exceeding 2.5 for some days.

An alternative model drops the stationarity assumption, since ϕ seems to approach 1 and hence non-stationarity might be implied. A strategy allowing non-stationarity was proposed by Gamerman and Moreira (1999), though it is theoretically implausible in the long run. Despite setting a prior $\phi \sim N(0,\ 1)$, the posterior 95% interval is still entirely below 1. The pseudo marginal likelihood is slightly lower (−911.6) than the stationary option (−909.6).

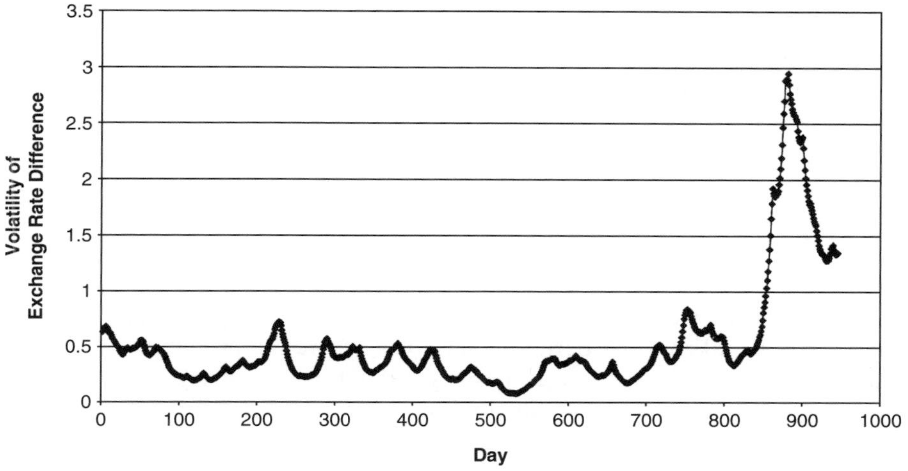

Figure 5.13 Volatility of exchange rate series

5.7 MODELLING STRUCTURAL SHIFTS

State space models are designed to accommodate gradual or smooth shifts in time series parameters. Often, however, there are temporary or permanent shifts in time series parameters that occur more abruptly, and a more appropriate model allows for changes in regression regimes and other shifts in structure. Section 5.2 considered innovation and additive outliers. Here we consider models that allow for repeated switching between distinct regimes according to a latent Markov series, and models for shifts in both the mean and variance of autoregressive series. A further category includes switching regression models (Maddala and Kim, 1996).

5.7.1 Binary indicators for mean and variance shifts

McCulloch and Tsay (1994) consider autoregressive models allowing for shifts in mean and/or variance. By allowing for variance shifts as well as changes in level, nonstationary trends that might otherwise have been attributed to changes in level may more appropriately be seen as due to heteroscedasticity. McCulloch and Tsay choose to focus explicitly on autoregressive models, rather than introduce moving average effects because given a sufficiently large AR model, one may achieve similar results to a stipulated ARMA model. Thus let

$$y_t = \mu_t + \varepsilon_t$$

where a change in level is accommodated by letting

$$\mu_t = \mu_{t-1} + \delta_{1t}\nu_t \tag{5.47}$$

The δ_{1t} are binary variables for each time point which equal 1 if a shift in mean occurs and ν_t models the shift that occurs, conditional on $\delta_{1t} = 1$. The ν_t are usually modelled as normal with mean zero and low precision τ_ν. The autoregressive component of the series is the p-lag model for ε_t, namely

$$\varepsilon_t = \gamma_1\varepsilon_{t-1} + \gamma_2\varepsilon_{t-2} + \ldots + \gamma_{t-p}\varepsilon_{t-p} + u_t \qquad (5.48)$$

where shifts in the variance of u_t are allowed. Thus, let $u_t \sim N(0, V_t)$, and let δ_{2t} be an additional binary series such that

$$\begin{aligned} V_t &= V_{t-1} \quad (\delta_{2t} = 0) \\ &= V_{t-1}\omega_t \quad (\delta_{2t} = 1) \end{aligned} \qquad (5.49)$$

where ω_t models the proportional change in the variance. Alternatively

$$V_t = V_{t-1}(1 + \omega_t\delta_{2t})$$

The ω_t are positive variables with, for example, a gamma prior. The binary shift indicators may be taken to be Bernoulli variables, where the probabilities that δ_{2t} and δ_{2t} equal 1 are known and small (e.g. $\eta_1 = \eta_2 = 0.05$). Alternatively, these probabilities may be assigned beta priors that favour low values. The relative importance of the mean and variance shift components (as reflected in the sizes of η_1 and η_2 if they are free parameters) will be affected by prior specifications on the ω_t and the variance of the v_t.

5.7.2 Markov mixtures

A different approach to changes in regime involves the Markov mixture model of Chib (1996), Leroux and Puterman (1992), and others. Thus suppose for each time point the process is in one of m states $\{s_t\}(t > 1)$, as determined by an $m \times m$ stationary Markov chain $P = \{p_{ij}\}$, where

$$p_{ij} = Pr[s_t = j | s_{t-1} = i] \qquad (5.50)$$

The first state (namely s_1) is determined by drawing from a multinomial with m categories. Given the underlying state $s_t = k$, the observation itself has the kth of the m possible components of the mixture, and these components might differ in means, variances or other summary shape parameters.

A model with both mean-variance shifts and reference to a latent Markov series is suggested by Albert and Chib (1993). Their model has $m = 2$ and values $s_t = 0$ or $s_t = 1$. Thus, an order p autoregression in the regression errors allows for shifts in mean and variance of y_t within the specification

$$\begin{aligned} y_t|s_t = \beta x_t + \psi s_t + \gamma_1(y_{t-1} - \beta x_{t-1} - \psi s_{t-1}) + \gamma_2(y_{t-2} - \beta x_{t-2} - \psi s_{t-2}) + \\ \ldots\ldots + \gamma_p(y_{t-p} - \beta x_{t-p} - \psi s_{t-p}) + u_t \end{aligned} \qquad (5.51)$$

where $u_t \sim N(0, V_t)$. The variance shifts are produced according to the model

$$V_t = \sigma^2(1 + \omega s_t) \qquad (5.52)$$

where ω is the proportionate shift in variance when $s_t = 1$. This model involves possible correlations between β and the mean shift parameter ψ, especially if x_t just consists of an intercept. Improper or weakly informative priors may reduce identifiability.

5.7.3 Switching regressions

Switching regression models originate in classical statistics with Quandt (1958), and have received attention in Bayesian terms in works by Geweke and Terui (1993),

Lubrano (1995b), among others. In such models, the conditional expectation of the endogenous variable may follow two or more regimes. Suppose, for example, that the error variance does not also switch and that some regression effects (applying to exogenous variables z_t) are not included in the switching.

The choice between regimes is determined by a threshold function K_t that drives either abrupt switching by a step function or a smooth transition function. The latter is typically a cumulative distribution function between 0 and 1, such as the logit (Bauwens *et al.*, 1999).

For instance, a step function Δ_t might be defined as one if a trend in time exceeds a threshold τ, and zero otherwise. If the trend were simply measured by the linear term t, then

$$\begin{aligned} K_t &= t - \tau < 0 \Rightarrow \Delta_t = 0 \\ K_t &= t - \tau > 0 \Rightarrow \Delta_t = 1 \end{aligned} \tag{5.53}$$

The simplest model then allows for two regimes:

$$y_t = \gamma z_t + (1 - \Delta_t)\beta_1 x_t + \Delta_t \beta_2 x_t + u_t \tag{5.54}$$

where, for example, $u_t \sim N(0, \sigma^2)$. The threshold function might also be defined by lags on the outcome, such as in the step function scheme

$$\begin{aligned} K_t &= y_{t-1} - d < 0 \Rightarrow \Delta_t = 0 \\ K_t &= y_{t-1} - d > 0 \Rightarrow \Delta_t = 1 \end{aligned}$$

More generally, the appropriate lag r in y_t, such that

$$\Delta_t = 1 \quad \text{if } y_{t-r} > d$$

is an additional unknown. Geweke and Terui (1993) consider joint prior specification for $\{r, d\}$ in models where the alternative regimes are different order lags in y; such as an AR(p_1) model if $\Delta_t = 1$, and an AR(p_2) model (with different coefficients throughout) if $\Delta_t = 0$.

A smooth transition function in these cases might take the form

$$\Delta_t = \exp(\varphi\{y_{t-1} - d\})/[1 + \exp(\varphi\{y_{t-1} - d\})]$$

or

$$\Delta_t = \exp(\varphi\{t - \tau\})/[1 + \exp(\varphi\{t - \tau\})]$$

where $\varphi > 0$ governs the smoothness of the transition.

Example 5.15 Fetal lamb movements An example of the Markov mixture model is provide by a time series of lamb fetal movement counts y_t from Leroux and Puterman (1992), where the presence in the mixture of more than one component leads to Poisson overdispersion. One might model such over dispersion by a gamma mixture (leading to a marginal negative binomial model).

Alternatively, suppose, following Leroux and Puterman, that a two class Markov mixture applies, with shifts between two Poisson means determined by a Markov chain (i.e. $m = 2$). Relatively diffuse Dirichlet priors for each row of P are adopted, such that for the elements in Equation (5.50) one has

$$p_{i,\, 1:m} \sim D(1, 1, \ldots 1)$$

The same prior is used for the first period state choice model. For the two Poisson means $G(1, 1)$ priors are stipulated, with an identifiability constraint that one is larger – an initial run justified such a constraint, showing the means to be widely separated.

With this model, a two chain run of 5000 iterations (1000 burn-in) shows a state occupied most of the periods (about 220 from 240), which has a low average fetal movement rate, and a minority state with a much higher rate, around 2.2–2.3. The majority state has a high retention rate (reflected in the transition parameter p_{22} around 0.96) while movement out of the minority state is much more frequent (Table 5.5).

The actual number of movements is predicted closely, though Leroux and Puterman show that using $m = 3$ components leads to even more accurate prediction of actual counts. The model with $m = 2$ shows relatively small CPOs for the movements at times 85 and 193 (counts of 7 and 4, respectively).

For comparison, and since the outcome is a count, Model B consists of an INAR(1) model. The 'innovation' process is governed by Bernoulli switching between means λ_1 and λ_2 (with $\lambda_2 > \lambda_1$ to guarantee identifiability). Thus,

$$Y_t \sim \text{Poi}(\mu_t)$$
$$\mu_t = \pi^\circ Y_{t-1} + \lambda_1\delta_t + \lambda_2(1 - \delta_t) \quad t > 1$$
$$\mu_1 = \lambda_1\delta_1 + \lambda_2(1 - \delta_1)$$

with $\delta_t \sim \text{Bern}(\eta)$ and η assigned a beta prior. This model also identifies a sub-population of periods with a much higher movement rate, around 4.5, than the main set of periods. It has a very similar pseudo-marginal likelihood to the two-state Markov switching model (−180 vs. −179).

Example 5.16 US unemployment As an illustration of models allowing both mean and variance shifts, consider the US unemployment time series analysed by Rosenberg

Table 5.5 Lamb movements, Markov mixture model parameters and predictions

	Mean	St. devn.	2.5%	Median	97.5%
$p_{1,1}$	0.66	0.15	0.35	0.67	0.93
$p_{1,2}$	0.34	0.15	0.07	0.33	0.65
$p_{2,1}$	0.04	0.03	0.01	0.03	0.12
$p_{2,2}$	0.96	0.03	0.88	0.97	0.99
Periods with $s_t = 1$	17.8	9.4	6.0	17.0	40.0
Periods with $s_t = 2$	222.2	9.4	200.0	223.0	234.0
λ_1	2.28	0.74	1.22	2.15	4.01
λ_2	0.23	0.05	0.14	0.23	0.32

Number of movements, actual and predicted	Actual Events	Predicted Events
0	182	180.2
1	41	43.7
2	12	9.9
3	2	3.6
4	2	1.5
5	0	0.6
6	0	0.2
7	1	0.1

and Young (1995) with the original six month average percent rates U_t transformed according to

$$y_t = 100 \times \ln(1 + U_{t+1}/100) - 100 \times \ln(1 + U_t/100)$$

Here the full data set consists of monthly data from 1954–1992 inclusive, providing 78 six monthly averages. As in Section 5.7.1, assume the binary indicators δ_{1t} and δ_{2t} for shifts in the means μ_t, and in the variances of the x_t series are Bernoulli with probabilities η_1 and η_2, respectively. These are taken as extra parameters, and the priors for η_1 and η_2 determine the level of shifting and the results of the analysis may depend on how they are set. The relative size of η_1 and η_2 is also affected by the variances assumed for the variance of the ν_t and the informativeness of the prior assumed for the ω_t.

Following Rosenberg and Young, a relatively low prior chance of shifts in either mean or variance is assumed, with $\eta_1 \sim B(1, 19)$ and $\eta_2 \sim B(1, 19)$. The autoregressive series is taken as order $p = 1$. Rosenberg and Young experimented with a lag 2 model, but found it to produce no improvement in fit. The proportional shifts, as in Equation (5.49), are taken to have a gamma prior favouring a concentration around an average of 1, namely

$$\omega_t \sim G(5, 5)$$

One might experiment with different prior assumptions on the degree of concentration around the average of 1, e.g. by alternatively taking G(1, 1) and G(10, 10) priors for the ω_t. As to the variance of the ν_t, this can be preset, and Rosenberg and Young suggest using a large multiple (e.g. 10 times) the residual variance from a standard ARMA model. Fitting ARMA(1, 1), ARMA(2, 2) and similar models showed a residual variance around 0.25, and so taking $\nu_t \sim N(0, 2.5)$, is one option. It might be sensible to assess alternatives involving other multiples, such as $\nu_t \sim N(0, 5)$ or $\nu_t \sim N(0, 1)$. One might also take the variance of the ν_t as an additional parameter.

Here a two-chain run to 2500 iterations and with $\nu_t \sim N(0, 1)$ shows probabilities of 0.014 for level shift and 0.124 for variance shift. Identifiability is improved by directly sampling the shift probabilities η_1 and η_2 from their full conditional densities. The lag 1 parameter γ_1 in the autoregressive series is estimated as 0.48. Rosenberg and Young, in an analysis of quarterly rather than six monthly series, also found a higher probability η_2 of a variance shift than a mean shift, but with the excess of η_2 over η_1 (0.086 vs 0.015) smaller than that estimated under the model here. A very close fit to the observed differenced and transformed series is achieved, with no discrepant CPOs (the smallest, around 2% of the maximum CPO, is for observation 35 in the series of 77).

For comparison, a relatively simple state space model in the undifferenced and untransformed six-monthly unemployment rate is also fitted (Model B in Program 5.16). This involves a first order random walk model in the 'true' series. The sum of squared prediction errors of actual vs. predicted unemployment rates (E.s in Program 5.14), shows this relatively simple model to have broadly comparable fit to the mean and variance shift model.

Example 5.17 Consumption function for France To illustrate regression switching, we follow Bauwens, Lubrano and Richards (1999, p. 248) in considering a stochastic consumption function for France. This is applied to logged consumption and income for 116 quarterly points from 1963Q1 to 1991Q4. The outcome is the first difference $\Delta \log C_t$ in logged consumption, and is related to the comparable income variable

$\Delta \log Y_t$, and to the lag 4 difference in log consumption. Conditioning on the first five observations gives for $t = 6$, 116

$$\Delta \log C_t = \delta + \eta D.1969Q2 + \gamma\Delta \log C_{t-4} + \beta\Delta \log Y_t + (\rho - 1)[\log C_{t-1} - \nu \log Y_{t-1}] + u_t \tag{5.55}$$

where $u_t \sim N(0, \sigma^2)$. The term $D.1969Q2$ reflects a short-run distortion due to the wage increases following the Matignon negotiations of 1969. As to the error correction term in square brackets this reflects a long-term equilibrium between consumption and income

$$\log C^* = \log K + \nu \log Y^*$$

Providing ν, the long run elasticity, is 1, then $C^* = KY^*$ and the propensity to consume K is constant. In a dynamic (first differences) framework, the long-term equilibrium is expressed as

$$\Delta \log C^* = \nu\Delta \log Y^*$$

This implies that a shift in the propensity to consume (i.e. a change in K) only changes the constant in Equation (5.55), with all other regression effects unaffected by the switching.

The observed propensity to consume K in this period in France is distorted not just by the 1969 negotiations, but by a longer run upward movement from around 0.77 in 1978 to 0.89 ten years later. The question is whether, despite this apparent trend, an underlying constant propensity can be obtained by suitable parameterisation of the consumption function (5.55), and by switching of the regression constant δ in this function.

Bauwens *et al.* first fit Equation (5.55) without any switching mechanism, and find that no equilibrium is defined. They then allow a single permanent shift in the propensity to consume via a step function as in Equation (5.53). They find this leads to ν under 1, a situation incompatible with a constant K. They then investigate whether a single, but non-permanent, shift in the mean propensity to consume restores a stable consumption function as defined by $\nu = 1$ (or by a 95% credible interval for ν including 1). This implies a return to a previous equilibrium after the temporary transition from equilibrium.

They therefore introduce a double parameter transition function such that $I(\tau_1, \tau_2) = 1$ for quarters t between τ_1 and τ_2 (both set within the period spanned by the observations) and $I(\tau_1, \tau_2) = 0$ otherwise, so that

$$\Delta \log C_t = \delta + \kappa I(\tau_1, \tau_2) + \eta D69.2 + \gamma\Delta \log C_{t-4} + \beta\Delta \log Y_t + (\rho - 1)[\log C_{t-1} - \nu \log Y_{t-1}] + u_t \tag{5.56}$$

The unit elasticity model with $\nu = 1$ is accepted with this transition function, and Bauwens *et al.* obtain a final equation (where ν is set to 1) as follows (posterior means and SDs):

$$\begin{aligned}\Delta \log C_t = &- \underset{[0.0041]}{0.0088} - \underset{[0.0017]}{0.0071} I(\tau_1, \tau_2) + \underset{[0.0067]}{0.019} D69.2 - \underset{[0.078]}{0.26}\Delta \log C_{t-4} \\ &+ \underset{[0.069]}{0.23}\Delta \log Y_t - \underset{[0.022]}{0.11}[\log (C_{t-1}/Y_{t-1})]\end{aligned}$$

with the mean of τ_1 estimated as 1973.3 and of τ_2 as 1984.1. The prior ranges (within $t = 6, .. 116$) for these two threshold parameters τ_1 and τ_2 are $U(29, 61)$ and $U(62, 98)$. These ranges are separated for identifiability and chosen by trial and error (Bauwens *et al.*, 1999, p. 250).

Estimates with this model may be sensitive to prior specifications on the single or double break points, τ_1 and τ_2. Thus, uniform priors over the full range of times (6–116) may well give different estimates to priors restricted to an interior subinterval (e.g. 20–100). Similarly, a gamma prior such as G(0.6, 0.01) with average 60, approximately half way through the periods, but with large variance – and with sampling constrained to the range (6, 116)–might be used, combined with a constraint $\tau_2 > \tau_1$. This may lead to different results than a uniform prior.

There are also possible identification and convergence problems entailed in the non-linear effects of ρ and ν in Equation (5.56), when ν is a free parameter. Here ρ is allowed to be outside the interval $[-1, 1]$. One way to deal with the identifiability problem is to introduce a conditional prior for ν given ρ, or vice versa (see Bauwens *et al.*, 1999, p.142), and so ρ is taken to be a linear function of ν, namely $a_1 + a_2\nu$. Then N(0, 1) priors are adopted on a_1 and a_2, and all regression coefficients with the exception of ν, which is assigned an N(1, 1) prior, are constrained to positive values.

We first fit Equation (5.56) with a single break point (i.e. a permanent shift in the propensity to consume), ν a free parameter, and a G(0.6, 0.01) prior on the breakpoint τ_1. Convergence in all parameters in a three chain run occurs after 15 000 iterations and from iterations 15 000–20 000, a 95% credible interval for ρ of {1.02, 1.19} is obtained, and a pseudo-marginal likelihood of 390. The density for ν is concentrated below unity, with 95% interval {0.86, 0.95}. The density for τ_1 is negatively skewed and has some minor modes; however, there is a major mode at around $t = 85$ to $t = 90$, with the posterior median at 87 (i.e. 1984.3).

To fit Equation (5.56) with ν still a free parameter, and two breakpoints (i.e. a temporary shift in the propensity to consume), the intervals (29, 61) and (62, 98) of Bauwens *et al.* are used in conjunction with gamma priors (Model C in Program 5.17). Taking wider intervals within which sampled values may lie, such as (7, 61) for τ_1, causes convergence problems. Even with the same intervals for the breakpoints as adopted by Bauwens *et al.*, convergence on ν is slow. The Gelman–Rubin scale reduction factor on ν remains at around 1.2 after iteration 8500 in a three chain run, and the 95% interval from 5000 iterations thereafter is {0.88, 1.02}, including the equilibrium value of 1. On this basis the same model (5.56), but with ν set to 1, may be fitted (see exercises).

It may be noted that an alternative methodology uses a smooth rather than abrupt transition function, such as

$$\Delta_t = 1/[1 + \exp(-\varphi(t - \tau_1)(t - \tau_2))]$$

Other variations might include a stationarity constraint with $|\rho| < 1$.

5.8 REVIEW

Bayesian time series analysis offers flexibility in several areas, and is now a major theme in new time series developments. Among major modelling areas illustrating the benefits of a Bayesian approach may be mentioned:

- the lesser restriction to stationarity in autoregressive time series as compared to classical ARIMA analysis (e.g. Nandram *et al.*, 1997);
- the representation of nonlinear time series and time-varying parameters through dynamic linear models (West and Harrison, 1997); a recent survey of state-space time series modelling (including non-Gaussian cases) is provided by Tanizaki and Mariano (1998);
- the reduction of information in a large set of variables by using a factor structure in stochastic volatility applications (Aguilar and West, 2000; Meyer and Yu, 2000); time-varying volatility is also tackled by ARCH models with recent Bayesian studies including Vrontos *et al.* (2000);
- the ability to model structural changes in different aspects of the time series such as level trend and variance (see Section 5.7 and Wang and Zivot, 2000);
- recent developments in dynamic linear and semiparametric modelling of discrete outcomes (Cargnoni *et al.*, 1997; Fahrmeir and Lang, 2001).

REFERENCES

Aguilar, O. and West, M. (2000) Bayesian dynamic factor models and portfolio allocation. *J. Bus. Economic Stat.* **18**, 338–357.

Akaike, H. (1986) The selection of smoothness priors for distributed lag estimation. In: Goel, P. and Zellner, A. (eds.), *Bayesian Inference and Decision Techniques*. Amsterdam: Elsevier, pp. 109–118.

Albert, J. and Chib, S. (1993) Bayes inference via Gibbs sampling of autoregressive time series subject to Markov mean and variance shifts. *J. Bus. Economic Stat.* **11**, 1–15.

Ameen, J. R. M. and Harrison, P. (1985) Normal discount Bayesian models. In: Bernardo, J., De Groot, M., Lindley, D. and Smith, A. (eds.), *Bayesian Statistics* 2. Amsterdam: North-Holland, pp. 271–298.

Barnett, G., Kohn, R. and Sheather, S. (1996) Bayesian estimation of an autoregressive model using Markov Chain Monte Carlo. *J. Econometrics* **74**(2); 237–254.

Bauwens, L. and Lubrano, M. (1998) Bayesian inference on GARCH models using the Gibbs sampler. *The Econometrics J.* **1**, C23–C46.

Bauwens, L., Lubrano, M. and Richard, J. (2000) *Bayesian Inference in Dynamic Econometric Models*. Oxford: Oxford University Press.

Box, G. E. P. and Jenkins, G. (1976) *Time Series Analysis: Forecasting and Control (Rev Ed)*. Holden-Day.

Broemeling, L. and Cook, P. (1993) Bayesian estimation of the mean of an autoregressive process. *J. Appl. Stat.* **20**, 25–39.

Cameron, A. and Trivedi, P. (1998) *Regression Analysis of Count Data*. Oxford: Oxford University Press.

Cardinal M., Roy, R. and Lambert, J. (1999) On the application of integer-valued time series models for the analysis of disease incidence. *Stat. in Med.* **18**, 2025–2039.

Cargnoni, C., Müller, P. and West, M. (1997) Bayesian forecasting of multinomial time series through conditionally Gaussian dynamic models. *J. Am. Stat. Assoc.* **92**, 640–647.

Carlin, B. and Polson, N. (1992) Monte Carlo Bayesian methods for discrete regression models and categorical time series. In: Bernardo, J. *et al.* (eds.), *Bayesian Statistics 4*. Oxford: Clarendon Press.

Chan, K. and Ledolter, J. (1995) Monte-Carlo EM estimation for time series models involving counts. *J. Am. Stat. Assoc.* **90**(429), 242–252.

Chib, S. (1996) Calculating posterior distributions and modal estimates in Markov mixture models. *J. Econometrics* **75**(1), 79–97.

Chib, S. and Greenberg, E. (1994) Bayes inference in regression models with ARMA(p, q) errors. *J. Econometrics* **64**, 183–206.

Christensen, R. (1989) *Log-Linear Models and Logistic Regression, 1st ed.* New York, NY: Springer.

Clarke, H., Stewart, M. and Whiteley, P. (1998) New models for New Labour: The political economy of Labour Party support, January 1992 to April 1997. *Am. Polit. Sci. Rev.* **92**(3), 559–575.

Daniels, M. (1999) A prior for the variance in hierarchical models. *Can. J. Stat.*, **27**, 567–578.

Diggle, P. (1990) *Time Series; A Biostatistical Introduction.* Oxford: Oxford University Press.

Durbin, J. and Koopman, S. (2000) Time series analysis of non-Gaussian observations based on state space models from both classical and Bayesian perspectives. *J. Roy. Stat. Soc. B* **62**, 3–29.

Engle, R. (1982) Autoregressive conditional heteroscedasticity with estimates of the variance of United Kingdom inflation. *Econometrica* **50**, 987–1007.

Engle, R. and Granger, C. (1987) Co-integration and error correction: representation, estimation and testing. *Econometrica* **55**, 251–276.

Fahrmeir, L. and Lang, S. (2001) Bayesian inference for generalized additive mixed models based on Markov random field priors. *J. Roy. Stat. Soc., Ser. C* **50**, 201.

Fahrmeir, L. and Tutz, G. (2001) *Multivariate Statistical Modelling based on Generalized Linear Models.* New York, NY: Springer.

Franke, J. and Seligmann, T. (1993) Conditional maximum likelihood estimates for INAR1 processes and their application to modelling epileptic seizure counts. In: Subba Rao, T. (ed.), *Developments in Time Series Analysis.* London: Chapman & Hall, pp. 310–330.

Gamerman, D. and Moreira, A. (1999) Comments on Pitt, M and Shephard, N (op cit). In: Bernardo, J. M. *et al.*, (eds.), *Bayesian Statistics 6.* Oxford: Clarendon Press.

Gelman, A., Carlin, J. B., Stern, H. S. and Rubin, D. B. (1998) *Bayesian Data Analysis.* London: Chapman & Hall.

George, E. and McCulloch, R (1993) Variable selection via Gibbs sampling. *J. Am. Stat. Assoc.* **88**, 881–889.

Geweke, J. and Terui, N. (1993). Bayesian threshold auto-regressive models for nonlinear time series. *J. Time Series Anal.* **14**(5), 441.

Granger, C. and Newbold, P. (1974) Spurious regressions in econometrics. *J. Econometrics* **2**, 111–120.

Greene, W. (2000) *Econometric Analysis, 4th ed.* Englewood Cliffs, NJ: Prentice Hall.

Grunfeld, Y. and Griliches, Z. (1960) Is aggregation necessarily bad? *Rev. Economics and Stat.* **42**, 1–13.

Hamilton, J. (1989) A new approach to the economic-analysis of nonstationary time-series and the business-cycle. *Econometrica* **57**, 357–384.

Harvey, A. (1989) *The Econometric Analysis of Time Series, 2nd ed.* New York, NY: Philip Allan.

Harvey, A. C. and Fernandes, C. (1989) Time series models for count or qualitative observations. *J. Bus. Economic Stat.* **7**, 407–417.

Harvey, A., Ruiz, E. and Shepherd, N. (1994) Multivariate stochastic variance models. *Rev. Economic Stud.* **61**, 247–264.

Hoek, H., Lucas, A. and Vandijk, H. (1995) Classical and Bayesian aspects of robust unit-root inference. *J. Econometrics* **69**, 27–59.

Huerta, G. and West, M. (1999) Priors and component structures in autoregressive time series models. *J. Roy. Stat. Soc., Ser. B* **61**, 881–899.

Jones, M. (1987) Randomly choosing parameters from the stationarity and invertibility region of autoregressive-moving average models. *Appl. Stat.* **36**, 134–138.

Jørgensen, B., Lundbye-Christensen, S., Song, P. and Sun, L. (1999) A state space model for multivariate longitudinal count data. *Biometrika* **86**, 169–181.

Kadane, J. (1980) Predictive and structural methods for elicitating prior distributions. In: Zellner, A. (ed.), *Bayesian Analysis in Econometrics and Statistics.* Amsterdam: North-Holland, pp. 89–93.

Kadiyala, K. R. and Karlsson, S. (1997) Numerical methods for estimation and inference in Bayesian VAR-models. *J. Appl. Economics* **12**, 99–132.

Kuo, L. and Mallick, B. (1998) Variable selection for regression models. *Sankhya* **60B**, 65–81.

Lang, J. (1999) Bayesian ordinal and binary regression models with a parametric family of mixture links. *Computational Stat. and Data Anal.* **31**(1), 59–87.

Laud, P. and Ibrahim, J. (1995) Predictive model selection. *J. Roy. Stat. Soc., Ser. B* **57**, 247–262.

Leroux, B. and Puterman, M. (1992) Maximum penalized likelihood estimation for independent and Markov-dependent mixture models. *Biometrics* **48**, 545–558.

Lesage, J. and Magura, M. (1991) Using interindustry input-output relations as a Bayesian prior in employment forecasting models. *Int. J. Forecasting* **7**, 231–238.

Litterman, R. (1986) Forecasting with Bayesian vector autoregressions: five years of experience. *J. Bus. Economic Stat.* **4**, 25–38.

Lubrano, M. (1995a) Testing for unit roots in a Bayesian framework. *J. Economics* **69**(1), 81–109.

Lubrano, M. (1995b) Bayesian tests for cointegration in the case of structural breaks. *Recherches Economiques de Louvain* **61**, 479–507.

McCulloch, R. and Tsay, R. (1994) Bayesian inference of trend and difference stationarity. *Econometric Theory* **10**, 596–608.

McKenzie, E. (1986) Autoregressive moving-average processes with negative-binomial and geometric marginal distributions. *Adv. in Appl. Probability* **18**, 679–705.

Maddala, G. (1979) *Econometrics*. New York, NY: McGraw-Hill.

Maddala, G. and Kim, I. (1996) Structural change and unit roots. *J. Stat. Planning and Inference* **49**(1), 73–103.

Mallick, B. and Gelfand, A. (1994) Generalized linear models with unknown link functions. *Biometrika* **81**, 237–245.

Marriott, J. and Smith, A. (1992) Reparameterisation aspects of numerical Bayesian methodology for ARMA models. *J. Time Series Anal.* **13**, 327–343.

Marriott, J., Ravishanker, N., Gelfand, A. and Pai, J. (1996) Bayesian analysis of ARMA processes: complete sampling-based inference under full likelihoods. In: Berry, D. *et al.* (eds.), *Bayesian Analysis in Statistics and Econometrics*. New York: Wiley, pp. 243–256.

Marriott, J. and Smith, A. (1992) Reparametrization aspects of numerical Bayesian methodology for autoregressive moving-average models. *J. Time Series Anal.* **13**, 327–343.

Martin, R. and Yohai, V. (1986) Influence functionals for time series. *Ann. Stat.* **14**, 781–818.

Meyer, R., and Yu, J. (2000): BUGS for a Bayesian analysis of stochastic volatility models. *The Econometrics Journal*, **3**(2), 198–215.

Migon, H. S. and Harrison, P. J. (1985) An application of non-linear Bayesian forecasting to television advertising. In: Bernardo, J., De Groot, M., Lindley, D. and Smith, A. (eds.), *Bayesian Statistics, 2*. Amsterdam: North-Holland, pp. 271–294.

Monahan, J. (1984) A note on enforcing stationarity in autoregressive-moving average models. *Biometrika* **71**, 403–404.

Nandram, B. and Petrucelli, J. (1997) A Bayesian analysis of autoregressive time series panel data. *J. Bus. Economic Stat.* **15**, 328–334.

Naylor, J. and Marriott, J. (1996) A Bayesian analysis of non-stationary AR series. In: Bernardo, J. *et al.* (eds.), *Bayesian Statistics 5*. Oxford: Oxford University Press.

Nelson, C. and Plosser, C. (1982) Trends and random-walks in macroeconomic time-series – some evidence and implications. *J. Monetary Economics* **10**(2), 139–162.

Newbold, P (1974) The exact likelihood function for a mixed autoregressive-moving average process. *Biometrika* **61**, 423–426.

Ord, K., Fernandes, C. and Harvey, A. (1993) Time series models for multivariate series of count data. In: Subba Rao, T. (ed.), *Developments in Time Series Analysis*. London: Chapman & Hall, pp. 295–309.

Partridge, M. and Rickman, D. S. (1998) Generalizing the Bayesian vector autoregression approach for regional interindustry employment forecasting. *J. Bus. and Economic Stat.* **16**, 62–72.

Pitt, M. and Shephard, N. (1998) Time-varying covariances: A factor stochastic volatility approach. In: Bernardo, J. M. *et al.* (eds.), *Bayesian Statistics 6*. Oxford: Clarendon Press.

Puterman, M. (1988) Leverage and influence in autocorrelated regression models. *Appl. Stat.* **37**, 76–86.

Quandt, R. (1958) The estimation of parameters of a linear regression system obeying two separate regimes. *J. Am. Stat. Assoc.* **53**, 873–880.

Ravishanker, N. and Ray, B. K. (1997) Bayesian analysis of vector ARMA models using Gibbs sampling. *J. Forecasting* **16**, 177–194.

Rosenberg, M. and Young, V. (1999) A Bayesian approach to understanding time series data. *North Am. Actuarial J.* **3**, 130–144.

Schotman, P. and van Dijk, H. (1991) On Bayesian Routes to Unit Roots, *Journal of Applied Econometrics*, **6**, 387–401.

Tanizaki, H. and Mariano, R. (1998) Nonlinear and non-gaussian state-space modeling with Monte Carlo simulations. *J. Econometrics* **83**(1, 2), 263–290.
Thomas, R. (1997) *Modern Econometrics: An Introduction*. NJ Addison-Wesley.
Van Ophem, H. (1999) A general method to estimate correlated discrete random variables. *Econometric Theory* **15**, 228–237.
Vrontos, I., Dellaportas, P. and Politis, D. (2000) Full Bayesian inference for GARCH and EGARCH models. *J. Bus. and Economic Stat.* **18**, 187–198.
Wang, J. and Zivot, E. (2000) A Bayesian time series model of multiple structural changes in level, trend, and variance. *J. Bus. and Economic Stat.* **18**, 374–386.
West, M. (1996) Bayesian time series: models and computations for the analysis of time series in the physical sciences. In: Hanson, K. (ed.), *Maximum Entropy and Bayesian Methods*. Dordrecht: Kluwer.
West, M. and Harrison, J. (1989) *Bayesian Forecasting and Dynamic Models, 1st ed.* Berlin: Springer-Verlag.
Zeger, S. and Qaqish, B. (1988) Markov regression models for time series: A quasi-likelihood approach. *Biometrics* **44**(4), 1019–1031.
Zellner, A. (1971) *An Introduction to Bayesian Inference in Econometrics*. Wiley, New York.
Zellner, A. and Tiao, G. (1964) Bayesian analysis of the regression model with autocorrelated errors. *J. Am. Stat. Assoc.* **59**, 763–778.

EXERCISES

1. In Example 5.1, assess the evidence for a non-zero ρ by applying the binary coefficient selection method to the outlier innovation model (Model C in Program 5.1). Is this inference affected by taking a smaller value of Δ (e.g. 0.01)?

2. In Example 5.2, try the same modelling approach for the velocity series (1869–1988). This series is identified as nonstationary (with significant probability that $\rho > 1$) by Bauwens *et al.* (1999). The data for this series are included in Example 5.2.

3. In Example 5.2, assess the inferences regarding stationarity when the innovations are assumed Normal rather than Student t.

4. In Example 5.10, fit a true series evolving according to an RW(3) model such that

$$f(t) = 3f(t-1) - 3f(t-2) + f(t-3) + u(t)$$

   and assess its fit against the RW(2) model.

5. In Example 5.13, try the ARCH Model A but introduce a lag in y_{t-1} into the model for the y series. Does this improve the predictive loss criterion?

6. In Example 5.13, try alternative priors on the random walk variance in Model D (such as a uniform prior on $V_H^{0.5}$ or alternative gamma priors on $1/V_H$), and assess effects on model fit.

7. In Example 5.15, fit a Markov switching model with a three state transition matrix, and similarly the INAR(1) variant model with an innovation process mixing three different means. How do these compare in terms of the DIC criterion and pseudo-marginal likelihood?

8. Fit the French consumption model of Example 5.17, as in Equation (5.55), with two breakpoints and ν set to 1. Compare inferences obtained with gamma priors on τ_1 and τ_2 constrained to the intervals (29, 61) and (62, 98) with those obtained using uniform priors on the same intervals. How does the fit obtained compare to a model with one breakpoint only?

CHAPTER 6

Analysis of Panel Data

6.1 INTRODUCTION

Panel or longitudinal data sets occur when the continuous or discrete response Y_{it} of each subject $i(i = 1, .., N)$ is observed on several occasions $t = 1, .. T_i$. Occasion totals T_i may differ between subjects, as may times of observations v_{it}, and so spacings $\Delta_{it} = v_{it} - v_{i,\,t-1}$ between observations. Such a panel or serial data set may pertain to individual patients in clinical trials or subjects of follow-up surveys, or to aggregated units such as population age groups or geographical areas. The analysis of change in serial measurements over individuals or groups plays a major role in social and biomedical research, and is fundamental in understanding causal mechanisms of disease or social pathology, in assessing the impact of policy or treatment interventions, and in the analysis of developmental and growth processes. For instance, in economic applications, the panel may be at individual, household or firm level and relate to questions such as economic participation or consumption (for households or individuals) or patent activity and investment levels at firm level.

Major methodological questions in panel data include the modelling, via random effects, of permanent subject effects, of growth curve parameters, and autocorrelated errors. Also important are the extension to categorical outcomes (binary, multinomial or count data) of methods originally developed for continuous outcomes (Chib and Carlin, 1999). Missing data often occur in panel studies (see Section 6.5), especially permanent loss or 'attrition' of subjects, where T_i is less than the maximum span of the study in clinical trials or panel studies of economic interventions (Hausman and Wiseman, 1979); Bayesian perspectives on this issue include those of Little and Rubin (1987).

The accumulation of information over both times and subjects increases the power of statistical methods to identify effects (e.g. treatment effects in medical applications), and permits the estimation of parameters (e.g. permanent effects or 'frailties' for subjects i) that are not identifiable from cross-sectional analysis or from repeated cross-sections on different subjects. While cross-section data can be used to estimate age or cohort related change, these estimates rely on differences between groups rather than individual change profiles (Ware, 1985). Diggle *et al.* (1994) provide an example of a cross-sectional relation between reading ability Y_i and age X_i

$$Y_i = \alpha + \beta X_i + \varepsilon_i$$

Applied Bayesian Modelling P. Congdon
 ISBN: 0-471-48695-7

where β is the average difference in ability for groups differing by a single year of age. With T repeated yearly observations on N individuals one may identify a more substantively informative model

$$Y_{it} = \alpha + \beta X_{i1} + \delta(X_{it} - X_{i1}) + \varepsilon_{it} \tag{6.1}$$

where β is the same as in Equation (6.1), and δ is the growth in ability for an extra year of age. Growth curve models generalise Equation (6.1) by using random effects to model variability in baseline ability or growth over subjects.

In economic applications, a cross-sectional model into, say, patent activity by firms is limited to considering the impact of observed predictors on the outcome. A longitudinal model gives better scope to assess the role of more or less constant unobserved heterogeneity between firms, in terms of entrepreneurial and technical skills which affect patent applications, and may be difficult to capture with observable variables (Winkelmann, 2000).

Longitudinal designs provide information to describe patterns of change and development, enabling predictions of individual growth or change beyond the observed path that take account of not only the impact of age or time, but of additional subject variables. For example, Lee and Hwang (2000) consider the best choice of prior for the purposes of extended prediction, namely prediction beyond the observed time range of the sample. Predictive applications also occur in demographic and actuarial contexts, where observations of mortality or other vital events are recorded for several periods in succession, and classified by age, sex or other demographic characteristics (Hickman and Miller, 1981). Here one may be interested in forecasts of vital events in these categories in future years.

6.1.1 Two stage models

The modelling of subject effects via univariate or multivariate random effects leads into a wide class of two stage models for both growth data and other types of longitudinal observations. Thus, in a simple linear growth curve

$$Y_{it} = \alpha_i + \delta_i t + \varepsilon_{it} \tag{6.2}$$

the random effects describe differences in baseline levels of the outcome (α_i), such as the underlying average attainment for subject i, or in the linear growth rates of subjects (δ_i), such as differences in attainment growth. In more general models of this type, also including covariate effects varying over subjects, the distribution of the random effects, whether parametric or non-parametric, and if parametric, whether Normal or otherwise (Butler and Louis, 1992), constitutes the first stage of the prior density specification. The hyperparameters on the density of $\theta_i = \{\alpha_i, \delta_i \dots\}$ form the second stage of the prior specification.

In many studies, the interest may especially be in identifying subject level effects from panel data with greater reliability than is possible with cross-sectional data (Horrace and Schmidt, 2000), with a typical specification for continuous data taking the form

$$Y_{it} = \alpha_i + \beta X_{it} + \varepsilon_{it} \tag{6.3}$$

with α and ε both Normal, and with the α_i modelled as random effects independent of other information. Heterogeneity between subjects i in their levels on α_i may be interpreted as unobserved differences that impact on the outcome, and reflect stable unmeasured characteristics of individuals. Control for unobserved heterogeneity is then the basis for obtaining consistent estimates of the systematic (regression) part of the model, involving observable predictors (Hamerle and Ronning, 1995). An alternative 'factor analytic' perspective on permanent effects is provided by Dagne (1999), in which the α_i may themselves be related to covariates; one might have loadings λ on the α_i varying either by time

$$Y_{it} = \lambda_t \alpha_i + \beta X_{it} + \varepsilon_{it}$$

or by subject

$$Y_{it} = \lambda_i \alpha_i + \beta X_{it} + \varepsilon_{it}$$

with the variance of the α_i pre-defined (e.g. var $(\alpha) = 1$) for identifiability. However, one does not necessarily need to adopt this approach to relate the α_i to fixed covariates W_i.

Certain prior specifications on the α_i in the conventional random effects model (6.3) may improve identifiability if the ε_{it} are autocorrelated. For continuous outcomes, suppose the errors follow a first order autoregression

$$\begin{aligned} Y_{it} &= \alpha_i + \beta X_{it} + e_{it} \\ e_{it} &= \gamma e_{i,\, t-1} + u_{it} \quad t > 1 \\ u_{it} &\sim \mathrm{N}(0,\, \sigma_u^2) \\ e_{i1} &\sim \mathrm{N}(0,\, \sigma_1^2) \end{aligned}$$

Then, following Chamberlain and Hirano (1999), rather than the prior

$$\alpha_i \sim \mathrm{N}(0,\, \sigma_\alpha^2)$$

for the permanent effects, one might link the initial conditions and the α_i via the prior

$$\alpha_i \sim \mathrm{N}(\psi e_{i1},\, \sigma_\alpha^2)$$

where ψ can be positive or negative. This amounts to assuming a bivariate density for α_i and e_{i1} with independence between them corresponding to ψ being effectively zero. Note that if γ is close to 1, it may be more difficult to identify α_i and e_{i1} separately.

Enduring differences between individuals represented by fixed subject level errors may be associated with a form of structural or spurious state dependence in panel data (Heckman, 1981). This is particularly so in panel studies of binomial events (e.g. unemployment, accidents, labour participation) or of multinomial choices (e.g. choices between product brands). The individual effect α_i here may be interpreted as a propensity to experience the event, or as the utility of a certain choice, and such variation of itself induces correlation over time. Thus, if subject level effects have variance σ_α^2 and the (uncorrelated) observation errors have variance σ_ε^2, and denoting the combined error

$$\eta_{it} = \varepsilon_{it} + \alpha_i \tag{6.4a}$$

the correlation between η_{it} at periods s and t is

$$\tau = \sigma_\alpha^2 / (\sigma_\alpha^2 + \sigma_\varepsilon^2) \tag{6.4b}$$

Unmeasured differences between individuals may also operate through an autocorrelated structure. Suppose that the observations Y_{it} were binary such that on a continuous latent scale $Z_{it} \geq 0$ if and only if $Y_{it} = 1$, and $Y_{it} = 0$ otherwise. Then

$$
\begin{aligned}
Z_{it} &= \beta X_{it} + \alpha_{it} \\
\alpha_{it} &= \gamma \alpha_{i,\,t-1} + \varepsilon_{it}
\end{aligned}
\tag{6.5}
$$

so that a subject who has a high propensity α_{it} at time t will have a higher propensity than average at $t+1$ if γ is positive.

Autoregression in the observations By contrast, true state dependence in a model for a binary panel outcome would mean model (6.5) would be extended to include autoregression on Y_{it} itself, for instance as follows:

$$
\begin{aligned}
Z_{it} &= \beta X_{it} + \rho Y_{i,\,t-1} + \alpha_{it} \\
\alpha_{it} &= \gamma \alpha_{i,\,t-1} + \varepsilon_{it}
\end{aligned}
\tag{6.6}
$$

Here $\rho Y_{i,\,t-1}$ may be seen as measuring the association between the event in the preceding period and the utility or propensity in the current period. With appropriate parameterisation, a binary model involving a lag in observed outcome $Y_{i,\,t-1}$ may be cast as a Markov chain model (Hamerle and Ronning, 1995). In practice, models such as Equation (6.6) may be difficult to identify for binary outcomes, but may be feasible for count or continuous outcomes.

Autoregressive panel models for continuous outcomes may involve several lags, for example in an AR(2) model

$$Y_{it} = \beta X_{it} + \rho_1 Y_{i,\,t-1} + \rho_2 Y_{i,\,t-2} + \varepsilon_{it}$$

and may be extended to included moving average terms in the ε_{it}. Although autoregressive (or ARMA) models and growth curve models are often seen as competing alternatives, there is scope for combining them. Curran and Bollen (2001) consider models such as

$$Y_{it} = \alpha_i + \delta_i t + \rho y_{i,\,t-1} + \varepsilon_{it}$$

that they term Autoregressive Latent Trait (ALT) models.

6.1.2 Fixed vs. random effects

While random effects models of subject level variability are increasingly used as standard, there are possible caveats against random effects models in observational (non-experimental) panel studies. Suppose a model for count outcomes Y_{it}, with means μ_{it} has a log link regression of form

$$\log \mu_{it} = \alpha_i + \beta X_{it} + \varepsilon_{it}$$

Such a random effects model may assume subject effects α_i to be independent of observed characteristics, so that individuals with different levels of W_i or X_{it} have the same expected value of α_i. This assumption may be realised under randomisation (e.g. in medical trials) but may be less likely in observational settings, where selectivity effects operate. Allison (1994) cites the relation between depressive reaction to abortion Y and religion W_{1i}: Catholic women are much less likely to have abortions, and so unless religion is included as a predictor, the effect of the event abortion ($X_{2it} = 1$ or 0) on the depressive reaction outcome will be confounded with differences in religion. Therefore, fixed effects models may be less restrictive in terms of their underlying assumptions. As well as the benefit of not assuming the independence of α_i and $\{W_i, X_{it}\}$, there is the improvement in robustness in not needing to specify the density of the α_i. On the other hand, estimation and identifiability are problematic for large N and small T.

In a Bayesian approach, there may be less of a problem when α_i is correlated with W_i, since that correlation may be modelled in the random effects distribution, by linking the permanent effects to known covariates in a regression format – as is commonly done in multi-level growth curve models, but not confined to that setting. Thus, instead of regarding individual effects as nuisance factors, the latent frailty, severity, or attitude may be modelled as functions of covariates. Dagne (1999) contrasts random effects models with a slope of unity on the terms α_i, to latent variable models which include both regression of the α_i on covariates and allow a varying slope over subjects on the latent variable.

6.1.3 Time dependent effects

As well as using panel data to sharpen inferences about individual differences, in some circumstances a two stage model with random effects over time (rather than over individuals) at the first stage may be relevant. As compared to time series data, fewer points are needed to model evolving regression coefficients. Thus let counts of a health event Y_{it} be assumed Poisson

$$Y_{it} \sim \text{Poi}(E_{it}\mu_{it})$$

where E_{it} are expected events using demographic standardisation. One may be interested in the changing impacts of covariates, fixed W_i, or varying, X_{it}, as follows:

$$\log(\mu_{it}) = \beta_{0t} + \beta_{1t}X_{it}$$

where β_{0t} and β_{1t} are random over time. For example, in a study of changing levels of suicide mortality over the $N = 33$ boroughs of London, Congdon (2001) discusses models for the changing impact of socio-economic variables (area deprivation and community stability) on the health outcome. A further modelling choice is thus between a focus on variation in change between subjects or on temporal change in the impact of fixed or fluctuating covariates defined for subjects.

One may also consider evolution in other parameters, such as lag coefficients on previous values of the outcome. In a model for continuous Y_{it} with lag on $Y_{i,\,t-1}$, one might specify, following Curran and Bollen, 2001), an ALT model with time varying AR(1) parameter:

$$Y_{it} = \alpha_i + \delta_i t + \rho_t Y_{i,\,t-1} + \varepsilon_{it}$$

6.2 NORMAL LINEAR PANEL MODELS AND GROWTH CURVES FOR METRIC OUTCOMES

This section considers in more detail the specification of the linear regression model for panel data on continuous outcomes, which sets the basis for panel models for discrete outcomes. Suppose the observations $\underset{\sim}{Y}_i = (Y_{i1}, Y_{i2}, \ldots.\ Y_{iT})$ on subjects $i = 1, .., N$ are of equal length T, and their means μ_{it} depend upon a vector (of length p) of fixed covariates X_i. Then

$$Y_{it} = X_i\beta + \varepsilon_{it} \tag{6.7}$$

where $\varepsilon_{i,\,1:T}$ is multivariate normal with mean zero and $T \times T$ dispersion matrix Σ. Extension to time varying covariates X_{it} is straightforward. Time varying predictors in growth curve models might well functions of the times $t = 1, .. T$.

Assume $\psi = \Sigma^{-1}$ has a Wishart prior density with scale matrix $\mathbf{R}$ and degrees of freedom r, and β has a multivariate normal prior with mean β_0 and dispersion matrix $\mathbf{B_0}$. Then the full conditional distribution of β given Σ^{-1} is multivariate normal

$$N_q(\beta^*, \mathbf{B}^*) \tag{6.8a}$$

where

$$\beta^* = \mathbf{B}^*(\mathbf{B}_0^{-1}\beta_0 + \sum_{i=1,\,N} X_i \psi \underset{\sim}{Y_i}) \tag{6.8b}$$

and

$$(\mathbf{B}^*)^{-1} = \mathbf{B}_0^{-1} + \sum_{i=1,\,N} X_i \psi X_i \tag{6.8c}$$

The full conditional of ψ is Wishart with $r + N$ degrees of freedom and scale matrix $\mathbf{R}^*$, where

$$(\mathbf{R}^*)^{-1} = \mathbf{R}^{-1} + \sum_{i=1,\,N} \underset{\sim}{\varepsilon_i} \underset{\sim}{\varepsilon_i'} \tag{6.8d}$$

In a growth curve analysis, the design matrix X_{it} would typically be time specific, but with equal values over subjects i. It might consist of an intercept $X_{it1} = 1$ for all subjects and times, with succeeding covariates being powers or other functions (e.g. orthogonal polynomials) of time or age t. Thus for a linear growth model $X_{it2} = t$, while a quadratic growth model would involve a further column in X, namely $X_{it3} = t^2$. If common coefficients β_1, β_2, β_3, etc. are assumed across subjects, they represent the relationship between the mean outcome and time or age t. For example, studies of mean marital quality over time suggest a more or less homogenous linear decline over the course of marriage (Karney and Bradbury, 1995).

6.2.1 Growth curve variability

However, average growth curves will often conceal substantial variability in development that longitudinal research is designed to address. Such variability in growth (e.g. in the linear growth effects of $X_{it2} = t$) may be correlated with variability in the individual levels on the outcome, leading to growth curves with multivariate random effects. For instance, a commonly observed effect in panel and growth curve models is regression to the mean, whereby higher growth occurs from lower base levels (so that growth and level are inversely related). An alternative to the notation $\{\alpha_i, \delta_i\}$ in Equation (6.2) is the multivariate one b_{ik}, $k = 1, 2$. For a linear growth curve with random intercepts and growth rates, a frequently used formulation is

$$Y_{it} = b_{i1} + b_{i2}t + \varepsilon_{it} \tag{6.9a}$$

with

$$\varepsilon_{i,\,1:T} \sim N_T(0, \Sigma)$$

and the random effects $b_{i,\ 1:2}$ following their own density, such as

$$b_{i,\ 1:2} \sim N_2(\mu_b, \Sigma_b)$$

The mean values of the b_{ik} in this case would be the intercept μ_{b1} and average linear growth rate μ_{b2}.

Given the role of $\underset{\sim}{b}_i$ in representing individual variations, including correlations between the growth paths and the levels of each subject, it may become more reasonable after introducing the $\underset{\sim}{b}_i$ to assume that the ε_{it} are independent. Hence, a simplifying assumption

$$\varepsilon_{it} \sim \mathrm{N}(0, \sigma^2 \mathrm{I}) \tag{6.9b}$$

may be made and assessed against other schemes involving a general unstructured dispersion matrix Σ or some specified time dependence (e.g. AR(1) dependence). Other questions of interest might include establishing whether variations in linear growth rates b_{i2} could be explained by other fixed attributes W_i of individuals: for example, whether differential declines in marital quality are related to initial spouse age, or to spouse education.

If individuals i have different observation times, or are nested hierarchically within groups j, then more complex growth curve models have been suggested. Diggle (1988) proposes a model for panel data in which observation times v_{it} may vary between subjects. Then the series for individual i may be modelled as

$$Y_i(v_{it}) = \mu_i(v_{it}) + W_i(v_{it}) + \varepsilon_{it} + \alpha_i \tag{6.10}$$

This representation contains a simple measurement error or white noise term ε_{it}, as well as autoregressive errors $W_i(v)$. The prior for the latter would incorporate a model for correlation $\rho(\Delta)$ between successive observations according to the time difference $\Delta_t = v_{t+1} - v_t$ between readings. The error association typically decreases in Δ, since measurements closer in time tend to be more strongly associated. The model includes constant subject level errors α_i which may depend upon covariates. These stable effects may also pre-multiply covariates, including the times v_{it} themselves, in which case they become variable growth rates.

Suppose individuals i are classified by group $j = 1, .. J$, as well as by individuals i within groups. Assume for simplicity equally spaced observation times for all subjects. Then the corresponding model to Equation (6.10) contains measurement error, as well as autoregressive dependence, at observation level, constant effects α_{ij} specific to subject i and group j, and growth curve parameters varying over group or over individuals. For example, a group varying linear growth model might take the form

$$Y_{ijt} = A_j + B_j t + \alpha_{ij} + e_{ijt} + \varepsilon_{ijt} \tag{6.11}$$

where the e_{ijt} are autoregressive with

$$e_{ijt} = \gamma_1 e_{ij,\ t-1} + u_{ijt}$$

and both the ε_{ijt} and u_{ijt} are exchangeable measurement errors. The group effects (varying intercepts and linear growth rates) might be taken to have bivariate dependence.

6.2.2 The linear mixed model

The growth curve with random variation in trajectory parameters over subjects, or subjects within groups, is a special case of the linear normal mixed model for subjects $i = 1, .. N$ and times $t = 1, .. T$

$$Y_{it} = X_i\beta + W_i b_i + \varepsilon_{it} \tag{6.12}$$

In this specification, the ε_{it} are commonly taken to be distributed according to the homoscedastic assumption (6.9b), W_i is usually but not necessarily a q-dimensional subset of regression vector X_i, and b_i a $q \times 1$ random effect with mean zero and covariance matrix Σ_b. If the density of b_i is multivariate normal, then the mean and variance of $\underset{\sim}{Y}_i$ unconditionally on b_i, are respectively

$$\boldsymbol{E}(\underset{\sim}{Y}_i) = \boldsymbol{X}_i\boldsymbol{\beta}$$

and

$$\boldsymbol{V}(\underset{\sim}{Y}_i) = \boldsymbol{\sigma}^2\boldsymbol{I}_T + \boldsymbol{W}_i\Sigma_b\boldsymbol{W}_i$$

There may be additional grouping variables is 6.12, e.g. exam results over time for pupils i within schools j, or clinical measures over time for patients i within hospitals j. In this case, the multi-level random effects b_{ij} may be assumed, or b_i may be related to fixed characteristics of the higher level grouping.

Suppose a Wishart prior is assumed on Σ_b^{-1} with degrees of freedom r_b and scale matrix $\boldsymbol{R}_b$, and further that $\tau = \sigma^{-2}$ is Gamma with parameters ν_1 and ν_2 and that β has a multivariate normal prior with mean β_0 and dispersion matrix $\mathbf{B}_0$. Then the full conditional densities of β, Σ_b and τ can be obtained (see Chib, 1995) by rewriting Equation (6.12) as in Equation (6.7):

$$Y_{it} - W_i b_i = X_i\beta + \varepsilon_{it}$$

The conditional density of β then has the same form as Equations (6.8a)–(6.8c) with dispersion matrix $\Sigma = \sigma^2 I_T$. The conditional density of τ is Gamma with parameters

$$\nu_1 + NT/2$$

and

$$\nu_2 + 0.5 \sum_{i=1,\ N} \ \sum_{t=1,\ T} \varepsilon_{it}^2$$

while that of Σ_b^{-1} is Wishart with degrees of freedom $r_b + N$, and scale matrix $\mathbf{R}_b{}^*$ where

$$(\mathbf{R}_b^*)^{-1} = \mathbf{R}_b^{-1} + \sum_{i=1,\ N} b_i b_i'$$

analogous to Equation (6.8d). Similarly, the full conditional of the b_i conditions on β, and (6.12) can be rewritten as

$$H_{it} = Y_{it} - X_i\beta = W_i b_i + \varepsilon_{it}$$

The b_i then have variances V_i, given by

$$(V_i)^{-1} = \Sigma_b^{-1} + \tau W_i W_i'$$

and means

$$V_i \tau W_i H_i$$

With suitable adaptations, the linear mixed model may be applied with discrete outcomes. An alternative parameterisation in models such as (6.12) is considered by Chib *et al.* (1998) in a discussion of Poisson outcomes. When W_i is a subset of X_i, e.g. $W_i = X_{ik}$ in the case $q = 1$, it may be preferable to merge the 'fixed effect' term $X_{ik}\beta_k$ with the random effect term $X_{ik}b_i$, such that b_i has a non zero mean, for example $b_i \sim N(\mu_b, \sigma_b^2)$, where $\mu_b = \beta_k$. There is then no overlap between the fixed effect predictors X and the random effects predictors W. This may provide improved identifiability in Markov Chain Monte Carlo sampling

6.2.3 Variable autoregressive parameters

While random coefficients attached to powers of time may represent diversity in growth curves, another option is to incorporate elements of an autoregressive approach. One possibility is the autoregressive latent trait model mentioned above. Another is to allow AR coefficients varying over subjects. Thus, for homogenous impacts of previous values in a trajectory on current values, and with exogenous attributes X_i (e.g. treatments, gender, etc.) as additional influences, a growth model in p lags might take the form

$$Y_{it} = \rho_{0i} + \rho_1 Y_{i,\,t-1} + \rho_2 Y_{i,\,t-2} + \ldots\ldots + \beta_t X_i + \varepsilon_{it} \tag{6.13}$$

where ρ_{0i} represent different levels of the outcome for individuals. To allow for differing shapes in the growth trajectories of subjects, one or more of the lag coefficients $\rho_1, \rho_2, ..$ may be allowed to be random as well as the intercepts ρ_{0i}. Then $\{\rho_{0i}, \rho_{1i}, ..\rho_{pi}\}$ are taken to be multivariate Normal or Student t with means $\{\rho_0, \rho_1, ..\rho_p\}$. This may allow greater flexibility than differences in linear or polynomial growth rates, in time t or log time. One may also introduce autoregressive variability in the model for the error structure. For example, an AR(1) error structure in the errors of the means of a Poisson outcome might be taken to vary over subjects:

$$\log \mu_{it} = \beta X_{it} + \alpha_i + e_{it}$$
$$e_{it} = \gamma_i e_{it-1} + \varepsilon_{it}$$

In a study of suicide trends, Congdon (2001) compares this 'differential persistence' model with a model allowing variable growth rates.

Example 6.1 Growth curve analysis: plasma citrate readings An example of growth curve variability with continuous outcomes as in Equation (6.8a), and possibly autocorrelated errors, is considered by Hand and Crowder (1996). The measurements at $t = 1, ..5$ equally spaced times and on $i = 1, ..10$ subjects are of plasma citrate concentrations in micromoles per litre. Hand and Crowder assume a two stage growth model with varying intercepts and growth rates, and also allow for AR(1) errors in the observation errors e_{it}, so that

$$Y_{it} = b_{i1} + b_{i2}t + e_{it} \tag{6.14a}$$

for $t > 1$, with

$$e_{it} = \gamma_1 e_{i,\,t-1} + u_{it} \tag{6.14b}$$

where the u_{it} are exchangeable with mean 0 and variance σ_u^2. Instead of the direct form (6.14), one may accommodate this form of structured errors by transforming the regression term for times $t > 1$. This entails subtracting $\gamma_1 Y_{i,\,t-1} = \gamma_1 b_{i1} + \gamma_1 b_{i2}(t-1) + \gamma_1 e_{i,\,t-1}$ from Equation (6.14a), to give

$$Y_{it} = \gamma_1 Y_{i,\,t-1} + (1-\gamma_1)b_{i1} + b_{i2}(t - \gamma_1 t + \gamma_1) + u_{it}$$

For the first observation ($t = 1$), one option is to take $e_{i1} = \gamma_1 e_{i0} + u_{i1}$ as a distinct error term, with its own variance σ_1^2.

Here the intercepts b_{i1} and growth rates b_{i2} are initially assumed to be drawn from a bivariate Normal density, before considering whether extensions such as a mixture of Normals for the b_{ik} are supported by the data. To assess fit and aid model choice we consider, as earlier, a predictive loss criterion, and a pseudo marginal likelihood measure[1]. In subsequent examples the DIC is also used. The mean intercept and slope μ_{b1} and μ_{b2}, are assigned N (0, 10^8) and N(0, 10^2) priors respectively. The correlation γ_1 is taken to be N(0, 1), and so is not constrained to stationarity. For the precision matrix Σ_b^{-1} of the b_{ik}, a Wishart prior with an identity scale matrix and 2 d.f. is assumed.

For the standard multivariate Normal, three chains are run for 20 000 iterations with dispersed starting values[2]; convergence was obtained by around 1000 iterations. The resulting estimates (from iterations 1000–20 000) show intercepts to be negatively related to slopes, as can be seen from the correlation C_{12} in Table 6.1. The clearest reductions over time in subject concentrations (i.e. the most negative b_{i2}) are in those subjects (4, 9, 10) with initially high readings, illustrating a regression to the mean. However, the average coefficient for linear growth μ_{b2} straddles zero, and there is only a 65% probability that μ_{b2} is negative (see 'test' in Program 6.1). There is a weakly positive auto-correlation in the residuals e_{it}; this might suggest that a simple uncorrelated error structure be adopted instead, but there is a clear worsening of fit when this is actually applied.

The second model involves a discrete mixture of two sub- groups of patients, since some subjects have fairly clear evidence of decline, while others have static or even slightly rising readings. An identifiability constraint is applied on the intercepts of the two groups. Convergence with this model is much slower (only by 25 000 iterations in a three chain run of 50 000 iterations). This analysis shows a low intercept group with positive (but insignificant) trend in time, and a high intercept group with trend coefficient biased to negative values (95% interval from −8.1 to 2.1). This model has a similar pseudo-marginal likelihood to the one group multivariate Normal, but a clearly worse predictive loss: the precision of samples of new data is worsened.

Example 6.2 Hypertension trial As an example of a hierarchical data set, Brown and Prescott (1999) present data on $i = 1, .. 288$ patients randomised to receive one of three

[1] For the predictive loss criterion (see Gelfand and Ghosh (1998), Sahu *et al.* (1997) and Ibrahim *et al.* (2001), 'new data' Z_{it} are sampled from $f(Z|\theta)$, where f is the same density assumed for Y_{it} and θ are samples from the posterior density $p(\theta|Y)$. Let ζ_{it} and φ_{it} be the posterior mean and variance of the Z_{it}. Then for w positive, one possible criterion has the form

$$\sum_i \sum_t \{\varphi_{it} + \left(\frac{w}{w+1}\right)(Y_{it} - \zeta_{it})^2\}$$

This criterion would be compared between models at selected values of w, typical values being $w = 1$, $w = 10$ and $w = 10\,000$. If the Y_{it} are missing or censored, then the second term in the criterion is omitted and only the precision of prediction is relevant. Another useful diagnostic tool is the Monte Carlo estimate of the conditional predictive ordinate, obtained as the harmonic mean of the likelihood for the $\{i, t\}$th observation (Gelfand and Dey, 1994). The product of the CPOs yields a pseudo-marginal likelihood.

[2] Null parameter values and the upper and lower 2.5th points of a trial run.

Table 6.1 Plasma citrate, AR(1) model with variable patient effects

	Mean	St. devn.	2.50%	Median	97.50%
Overall intercept and growth rate					
μ_{b1}	118.8	8.3	102.3	118.7	135.5
μ_{b2}	−0.56	1.74	−4.07	−0.54	2.94
Dispersion matrix for subject effects					
Σ_{b11}	698.6	408.3	264.6	592.8	1776.0
Σ_{b12}	−115.5	78.5	−320.1	−95.3	−30.8
Σ_{b22}	20.9	16.7	3.6	16.6	64.1
Correlation between intercepts and slopes					
C_{12}	−0.97	0.06	−1.00	−0.99	−0.85
Intercepts					
Subject 1	88.2	1.9	84.5	88.2	91.9
Subject 2	115.9	1.4	113.1	116.0	118.5
Subject 3	127.2	1.5	124.4	127.2	130.1
Subject 4	149.4	1.8	145.7	149.4	153.0
Subject 5	102.7	1.5	99.6	102.7	105.6
Subject 6	108.3	1.6	105.3	108.3	111.6
Subject 7	83.9	2.0	80.0	84.0	87.8
Subject 8	116.1	1.4	113.4	116.1	118.7
Subject 9	157.7	2.1	153.5	157.8	161.8
Subject 10	141.6	1.7	138.2	141.6	144.9
Growth Rates					
Subject 1	4.79	1.88	1.15	4.78	8.53
Subject 2	0.05	1.36	−2.56	0.02	2.87
Subject 3	−2.22	1.45	−5.14	−2.20	0.57
Subject 4	−5.40	1.84	−8.98	−5.42	−1.65
Subject 5	2.28	1.53	−0.64	2.25	5.42
Subject 6	0.67	1.58	−2.64	0.72	3.64
Subject 7	5.06	1.96	1.31	5.03	8.92
Subject 8	−0.07	1.35	−2.68	−0.09	2.63
Subject 9	−6.74	2.08	−10.79	−6.76	−2.53
Subject 10	−4.57	1.69	−7.92	−4.56	−1.23
Correlation of residuals					
γ_1	0.16	0.19	−0.22	0.16	0.55
Residual variance					
σ^2	219.8	54.5	136.2	212.1	348.8

drug treatments for hypertension (A=Carvedilol, B=Nifedipine, C=Atenolol), with drug A being the new drug and the other two being existing standard treatments. Patients are allocated to one of $j = 1, .. 29$ clinics. The analysis here considers a pre-treatment (week 1) baseline reading B_i of Diastolic Blood Pressure (DBP), and four post-treatment readings Y_{it} at two weekly intervals (weeks 3, 5, 7 and 9 after treatment). Treatment success is judged in terms of reducing blood pressure.

A first analysis of these data (Model A) is a fixed effects model without random effects over patients or patients × clinics. It involves just baseline β and treatment η effects, with the new treatment A as reference category

$$Y_{it} = \mu + \beta B_i + \eta_B + \eta_C + \varepsilon_{it}$$

Because a single baseline measure is missing, an imputation is made based on random missingness (see Section 6.5) such that the response mechanism itself does not need to be modelled. Convergence with three chains was achieved at under 500 iterations, and the summary is based on iterations 500–5000. The results of this analysis (Table 6.2) show the lowest DBP readings for drug C, once baseline morbidity is controlled for; centred treatment effects may be obtained by subtracting $\eta_A(=0)$, η_B and η_C from their mean. A low CPO is apparent for the third visit of patient 249, with DBP reading of 140, compared to previous readings of 120 and 118. The predictive loss criterion, the pseudo-marginal likelihood and DIC[3] are, respectively, 126 353 (for $w = 1$), −3925 and 9569.

Introducing a subject level random intercept leads to the model (Model B in Program 6.2A)

$$Y_{it} = \mu + b_i + \beta B_i + \eta_B + \eta_C + \varepsilon_{it}$$

Define $\eta_{it} = \varepsilon_{it} + b_i$ as in Equation (6.4a). Then the variance of the constant subject effects is determined by the correlation τ as in Equation (6.4b), which is assigned a

Table 6.2 Hypertension trial, alternative models (without clinic effects)

No random effects	Mean	St. devn.	2.5%	97.5%
Nifedipine (η_B)	−1.23	0.65	−2.52	0.05
Atenolol (η_C)	−2.95	0.64	−4.22	−1.70
Baseline (β)	0.51	0.06	0.39	0.62
μ	40.4	5.8	29.1	52.7
Normal subject random effect				
Var(b)	39.9	4.4	32.0	48.9
Nifedipine (η_B)	−1.34	1.02	−3.31	0.68
Atenolol (η_C)	−3.16	1.01	−5.22	−1.19
Baseline (β)	0.49	0.09	0.32	0.66
μ	42.2	9.1	24.6	60.3
τ	0.51	0.03	0.44	0.57
Dirichlet process				
Nifedipine (η_B)	−1.03	1.01	−2.97	0.98
Atenolol (η_C)	−2.73	1.04	−4.72	−0.66
Baseline (β)	0.44	0.09	0.27	0.61
μ	46.6	9.4	28.0	65.1

[3] This DIC is based on minus twice the likelihood.

uniform prior over (0, 1). Estimates for this model show slightly enhanced treatment effects in absolute terms, but also reduced precision, so that a more substantial portion of the density of η_B is above zero. In particular, its effect is indistinguishable from that of drug A. The model assessment criteria all agree on the gain in introducing permanent patient random effects; their distribution is shown in Figure 6.1. The predictive loss criterion falls to 67 400 (with w=1), the log pseudo-marginal likelihood improves from −3925 to −3671, and the DIC falls to 9030.

Following Butler and Louis (1992), one possible model extension (Model C in Program 6.2) is a non-parametric mixture of sub-populations on the effects b_i. For example, one option they suggest is sub-populations having different means and variances, but with the sub-population means summing to zero; they also apply the non-parametric maximum likelihood model of Laird (1978). Parametric assumptions about the permanent subject effects are also avoided by a Dirichlet process mixture (Escobar and West, 1998). So it is assumed that

$$b_i \sim N(\nu_i, \phi_i)$$

where

$$(\nu_i, \phi_i) \sim G$$

$$G \sim D(G_0, \alpha)$$

In practice, there will be clustering of the ν_i and ϕ_i values, and a maximum of $J = 20$ clusters is assumed for the 288 patients The baseline prior G_0 has the form

$$\begin{aligned} \nu_j &\sim N(0, f_j\phi_j) \\ \phi_j &\sim G(1, 0.001); j = 1, .., J \end{aligned} \tag{6.15}$$

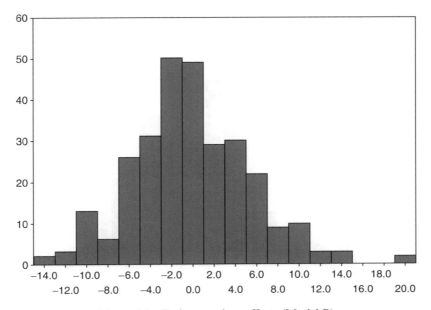

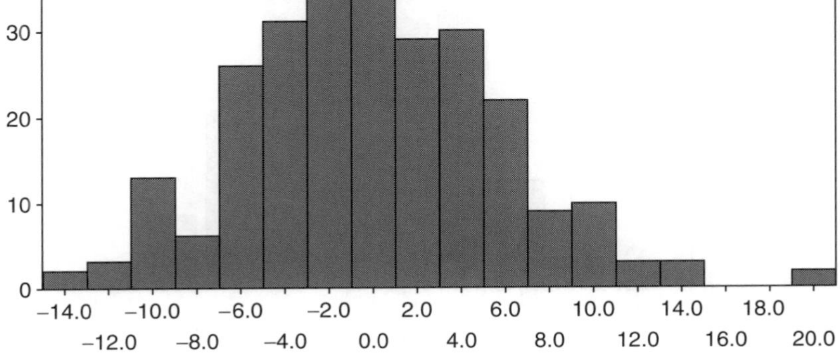

Figure 6.1 Patient random effects (Model B)

where the f_j determine the relative spread of cluster means and patient specific random effects, and are assigned a G(1, 1) prior. The category C_i to which patient i is assigned has prior

$$C_i \sim \text{Categorical}(p_1, p_2, \ldots p_J)$$

Only J^* clusters (between 1 and J) are selected at any iteration as appropriate for one or more observations. The mixture weights $p_1, p_2, \ldots p_J$ are determined by the stick-breaking method (see Sethuraman 1994, and Sethuraman and Tiwari, 1981). Thus, let $r_1, r_2, \ldots r_{J-1}$ be a sequence of Beta (1, α) random variables (and $r_J = 1$) , and set

$$p_1 = r_1$$
$$p_2 = r_2(1 - r_1)$$
$$p_3 = r_3(1 - r_2)(1 - r_1)$$
$$\ldots$$

The precision parameter α is preset here to 1, though it may be updated using the algorithm described by Escobar and West, 1998, p. 9). Sensitivity to alternative values of α may be assessed by setting alternative values; larger values of α imply more clusters J^* and greater differentiation between the b_i.

A two chain run shows convergence after 2500 iterations, and the summary is based on iterations 2500–5000. The posterior averages of J^* is 5.9, and Figure 6.2 shows most of the density for J^* is contained under 10 clusters. The predictive loss criterion falls to 67 140 (with $w = 1$), and the log pseudo-marginal likelihood and DIC also improve over Model B (respectively, −3648 and 9026). The treatment effects are absolutely smaller than in Model B, but have similar precision, while the baseline effect is smaller than in the first two models. Figure 6.3 shows a more peaked density of patient effects than Figure 6.1.

To introduce the information on clinics into the analysis, one may adopt a form of the multi-level growth curve model, as in Equation (6.11). See Program 6.2, Model D, which has data input in a different form. Corresponding to the broad decline over time

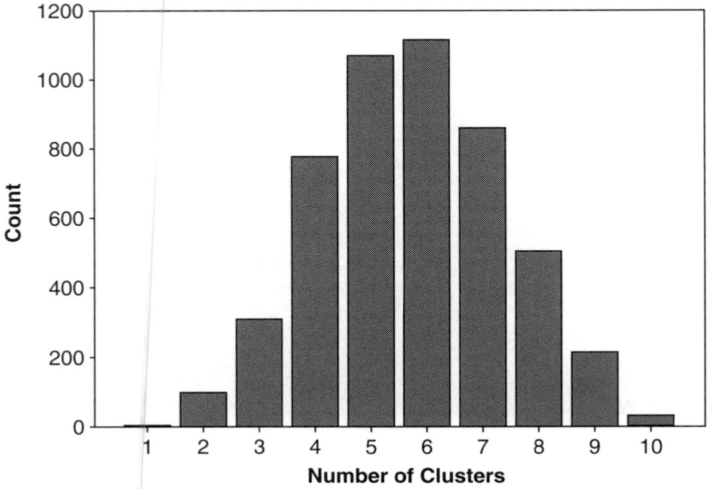

Figure 6.2 Number of clusters (two chaings)

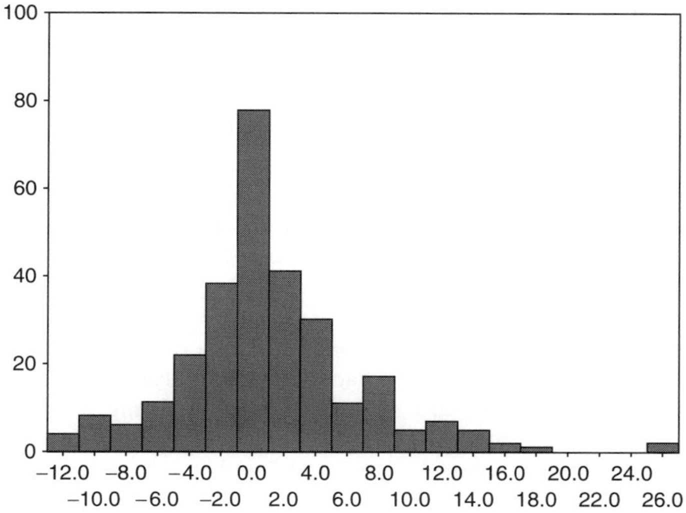

Figure 6.3 Patient random effects, Model C

in DBP readings, this model introduces a linear growth effect at clinic level. Intercepts and baseline effects also vary at clinic level, and there is an error term additionally at patient-clinic level. Thus with j denoting clinic and patient indices $i = 1, .. n_j$ now regarded as being nested within j (so $\Sigma_j n_j = 288$), the model has the form

$$Y_{ijt} = \beta_{1j} + \beta_{2j}t + \beta_{3j}B_{ij} + \eta_B + \eta_C + \tau_{ij} + e_{ijt}$$

with

$$e_{ijt} = \gamma e_{ij,\ t-1} + u_{ijt} \tag{6.16}$$

and with u_{ijt} an unstructured measurement error. The initial conditions e_{ij1} have a distinct variance term. The clinic effects $\beta_{kj}(k = 1, 3)$ have means and variances $\{\mu.\beta_k, \phi.\beta_k\}$.

Convergence with three chains is achieved after 5000 iterations with the aid of the over-relaxation option (there being a delayed convergence in $\mu.\beta_1$ and $\mu.\beta_3$). This model confirms a significant linear decline in the blood pressure readings with 95% interval for $\mu.\beta_2$ between −1.48 and −0.62. This model confirms the apparently beneficial effect of drug C (Table 6.3). The baseline parameter remains important (averaging 0.40 over

Table 6.3 Hypertension trial: multi-level model, parameter Summaries (iterations 5000–7500)

	Mean	St. devn.	2.5%	97.5%
η_B	−1.32	0.98	−3.33	0.68
η_C	−3.08	0.98	−4.95	−1.21
$\mu.\beta_1$	54.9	5.9	45.7	66.5
$\mu.\beta_2$	−1.03	0.22	−1.48	−0.62
$\mu.\beta_3$	0.40	0.06	0.29	0.50
γ	0.89	0.05	0.79	0.98

clinics), though smaller than in the first three models. This form of dependence on the initial observation coexists with autoregressive dependence in the errors, with γ in Equation (6.16) averaging 0.89.

Example 6.3 Protein content in milk Rahiala (1999) analyses data on the percentage protein content y_{it} of cow milk according to diet regime – see also Diggle *et al.* (1994). 52 cows were observed for a maximum of 19 weeks, though some were observed for less than this – the numbers of weeks vary from 14–19. There are a few missing cases in intervening weeks, before the end of the observation period on particular cows; these are assumed to be missing at random (Table 6.4).

One might transform the percentage variables to ensure they are in the appropriate range when predictions are transformed back, but the analysis here follows Rahiala in treating the data as metric. We also follow Rahiala in fitting a lag 5 model which conditions on the first five observations – with non-zero lags assumed at 1, 2, 3 and 5 weeks, but varying coefficients only at lag 2 and 5 and the mean lag at 5 taken as zero. Then with notation as in Equation (6.13)

$$Y_{it} \sim \mathrm{N}(\nu_{it}, \sigma_1^2)$$
$$\nu_{it} = \rho_{0i} + \rho_1 Y_{i,\,t-1} + \rho_{2i} Y_{i,\,t-2} + \rho_3 Y_{i,\,t-3} + \rho_{5i} Y_{i,\,t-5} + \beta X_i$$
$$\{\rho_{0i}, \rho_{2i}, \rho_{5i}\} \sim \mathrm{N}_3\left[\psi, \Sigma_\rho\right]$$

where $\psi = (\rho_0, \rho_2, 0)$. The model also includes a single covariate X_i for diet type (barley=1, 0=lupins) with its coefficient treated as a fixed effect. Note that in BUGS the data is provided in a 52×19 format even though some cows are only observed for 14 weeks.

Because not all the first five values in each animal's series is observed (e.g. $y_{20,\,2}$ is missing), the first five points must still be modelled in some way under the conditional approach, even though they are not included in the differential autoregression scheme. One might assume a random effects model for the first five periods, with the observations Y_{it} for animal i drawn from a Normal density with variance σ_2^2 and means $m.y_i$, and the means themselves drawn randomly from a normal density with mean M and variance σ_3^2.

Table 6.4 Variable lag effects, growth curve model

	Mean	St. devn.	2.5%	97.5%
Correlations between random effects				
Between Intercept and lag 2	−0.03	0.62	−0.96	0.97
Between Intercept and lag 5	−0.18	0.60	−0.98	0.92
Between Lag 2 and Lag 5	−0.69	0.33	−0.99	0.18
Coefficients (ρ_0 and ρ_2 are mean random effects, others are fixed effects)				
Intercept (ρ_0)	0.81	0.10	0.58	1.01
Lag 1 (ρ_1)	0.48	0.04	0.41	0.57
Lag 2 (ρ_2)	0.08	0.04	0.00	0.16
Lag 3 (ρ_3)	0.19	0.04	0.12	0.27
Diet (β)	0.06	0.02	0.01	0.11

With a three chain run, convergence is apparent from iteration 10 000 and the summary is based on iterations 10 000–40 000. There are clear positive lags at 1 and 3 weeks, and also a negative correlation between lag 2 and lag 5 growth coefficients. The effect of the lupin diet is positive.

A second model allows the mean coefficient ρ_5 to be non-zero. This option shows a clear improvement in the predictive loss criterion of footnote 1 (from 183 to 171 when $w = 1$). This model raises ρ_0 to around 1, while ρ_5 has mean −0.09 and 95% credible interval (−0.15, −0.03).

6.3 LONGITUDINAL DISCRETE DATA: BINARY, ORDINAL AND MULTINOMIAL AND POISSON PANEL DATA

As discussed in Chapter 3 and elsewhere, discrete outcomes may often be modelled in terms of latent continuous variables. Thus, underlying a panel of binary observations Y_{it}, we may posit a continuous latent variable or underlying propensity Z_{it} such that $Y_{it} = 1$ if Z_{it} is positive, and $Y_{it} = 0$ if Z_{it} is negative. Following Heckman (1981), one may formulate the model for Z_{it} in terms of measured or endogenous effects V_{it} (for example, $V_{it} = \beta X_{it}$, where X_{it} are known predictors) and a stochastic error ε_{it} so that

$$Z_{it} = V_{it} + \varepsilon_{it} \tag{6.17}$$

Then $Z_{it} \geq 0$ if and only if $Y_{it} = 1$, while $Z_{it} < 0$ if and only $Y_{it} = 0$. In a frequently used Bayesian approach developed by Albert and Chib (1993), the latent scale may be sampled via truncated sampling, with the truncation ranges determined by the observed Y_{it}. It is assumed that the probability of success is expressed as $\pi_{it} = F()$, where $F(.)$ is a distribution function, and so lies between 0 and 1. So a success occurs according to

$$\Pr(Y_{it} = 1) = \Pr(Z_{it} > 0) = \Pr(\varepsilon_{it} > -V_{it}) = 1 - F(-V_{it})$$

For forms of F that are symmetric about zero, such as the cumulative distribution function of a standard normal variable, the last element of this expression equals $F(V_{it})$.

If the chosen distribution function is the cumulative Normal, then Z may be sampled from a truncated normal: truncation is to the right (with ceiling zero) if the observation is $Y_{it} = 0$, and to the left by zero if $Y_{it} = 1$. To approximate a logit link, Z_{it} can be sampled from a Student t density with eight degrees of freedom, since, following Albert and Chib (1993), a $t(8)$ variable is approximately 0.634 times a logistic variable. This sampling based approach to the logit link additionally allows for outlier detection if the scale mixture version of the Student t density is used, rather than the direct Student t form. The scale mixture option retains truncated Normal sampling but adds a mixture variables λ_i, such that

$$Z_{it} \sim \mathrm{N}(\beta x_i, \lambda_i^{-1}) \quad \mathrm{I}(\mathrm{L}, U)$$

with λ_i sampled from a Gamma density $G(4, 4)$. The resulting regression coefficients need to be scaled from the $t(8)$ to the logistic. $L = 0$ when $Y_{it} = 1$ and $U = 0$ when $Y_{it} = 0$.

With Y_{it} still a binary outcome, the Z_{it} in Equation (6.17) may be expressed as

$$Z_{it} = \beta X_{it} + \alpha_i + \kappa_t + \varepsilon_{it} \tag{6.18}$$

where the κ_t are interpreted as period effects invariant over individuals (e.g. reflecting moves in national economic conditions on firm level outcomes), and the α_i are permanent effects attached to the subjects. For binary outcomes, restrictions are needed for identifiability. Thus, if a constant variance ϕ of the ε_{it} is assumed, then it is necessary that $\phi = 1$ (or possibly some other preset value). Note, however, that time varying variances ϕ_t may be identifiable provided one of them (e.g. ϕ_1) is set to a pre-specified value. Heckman extended this error structure to include coefficients on the α_i. This is known as the one factor model

$$Z_{it} = \beta X_{it} + \lambda_t \alpha_i + \kappa_t + \varepsilon_{it} \tag{6.19}$$

where, if the ε_{it} are taken as uncorrelated, the λ_t may be used to describe the correlation between time points t and s (see Example 6.6). For count outcomes, Dagne (1999) proposes a similar model with the loadings either over subjects or times. Thus, $Y_{it} \sim \text{Poi}(\mu_{it})$, and for loadings varying by subject

$$\mu_{it} = \lambda_i \alpha_i + \beta X_{it} + \varepsilon_{it}$$

Ordinal panel data Models for ordinal responses over time are important because in many settings involving human subjects, classifications are on a graded scale, with precise quantification not being possible. Examples include pre- and post-treatment observations on rankings of illness symptoms (e.g. no symptoms, mild, definite) or changed illness states, as well as survey questions on changing views on controversial topics. As discussed in Chapter 3, a continuous scale may often be envisaged to underlie the grading, with a series of thresholds $\tau_1, \tau_2, .. \tau_{C-1}$ defining which of C categories a subject lies in. Then with Z_{it} taking one of the forms as above, for example

$$Z_{it} = \beta_0 + \beta X_{it} + \varepsilon_{it}$$

we have

$$Y_{it} = \begin{cases} 1 & \text{if } Z_{it} < \tau_1 \\ 2 & \text{if } \tau_1 < Z_{it} < \tau_2 \\ \ldots & \\ C-1 & \text{if } \tau_{C-2} < Z_{it} < \tau_{C-1} \\ C & \text{if } Z_{it} > \tau_{C-1} \end{cases}$$

Alternative parameterisations are possible to ensure identifiability: either taking $\beta_0 = 0$, or $\tau_1 = 0$, or $\tau_C = C$ ensure that the mean of the Z is identified (Long, 1997, p. 122). The variance of Z may be identified by taking ε to be N(0, 1). This leads to the ordinal probit model, whereas taking ε to be Student $t(8)$ have variance 1 leads to the ordinal logit model.

In panel setting one might consider shifts in the location of thresholds by making the cut-points time specific – for example, if the analysis was intended to assess whether there had been a shift in attitudes. Random variations in intercepts between subjects or variation in trends across time (random time slopes) may also be modelled in panel settings because of the repetition of observations over subjects.

6.3.1 Beta-binomial mixture for panel data

Many repeated observations of social choice processes are available as binary series at times $1, 2, \ldots T$ and aggregate to binomial series for total subject populations or sub-populations of subjects. For example, the events migration, job change or divorce are binary though $Y_{it} = 1$ may sometimes include more than one event. As Davies *et al.* (1982) point out, such time series may show three kinds of systematic variation: heterogeneity in event probabilities across individuals or sub-populations, non-stationarity in the event rates at individual or aggregate level or both, and event history effects. The latter are exemplified by first order effects, when the state occupied at time $t-1$ influences the state occupied at time t. The factors underlying these choice sequences may be heterogeneous over sub-populations, as well as varying over time.

Suppose the binary series are of length T, with 2^T possible series being potentially observable, For instance, if $T = 3$, one may observe the sequences $\{1,1,1\}$, $\{1,1,0\}$, $\{1,0,1\}$, $\{0,1,1\}$, $\{1,0,0\}$, $\{0,1,0\}$, $\{0,0,1\}$, $\{0,0,0\}$. One may observe population sub-groups classified by demographic and social attributes which differ in terms of their distribution over these patterns. For example, if the event were migration or purchasing trips then groups with high event probabilities will be concentrated in sequences with higher occurrences of ones, e.g. higher proportions of sequences such as $\{1,1,1\}$ or $\{1,0,1\}$. This is the heterogeneity issue. There may also be non-stationarity, if for example the sequence $\{1,0,1\}$ is more frequent than the sequence $\{1,1,0\}$ because the event rate is higher at time 3 than at time 2.

Let p_{kt} denote the aggregate event rate at time t for subgroup k. Often heterogeneity is modelled within a logit or other link, so that if y_{kt} denotes the number undergoing the event at time t from n_{kt} at risk, then one might stipulate

$$y_{kt} \sim \text{Bin}(p_{kt}, n_{kt})$$
$$\text{logit}(p_{kt}) = a_t + b_k + e_{kt}$$

where the e_{kt} are Normal. However, heterogeneity in discrete outcomes over time might also be modelled in terms of conjugate mixture densities. Thus, heterogeneity in event probabilities p over individuals or sub-populations may be represented by a beta mixture $g(p)$. In the case of variations at individual level, the beta parameters may be linked to selected characteristics of individuals.

Suppose we have a stationary series with the rates p_t at different times t being constant, $p_t = p$. Then the probability of different sequences such as $\{1,1,0\}$ or $\{1,1,1\}$ may be modelled in terms of the moments

$$\mu'_j = E[p^j] = \int_0^1 p^j g(p)dp \quad j = 1, 2, \ldots \tag{6.20}$$

For a series of length T, μ'_j is equivalent to the probability of exactly j events occurring – regardless of their sequencing within the T points. For example, if $T = 3$, the probability of an event at every point in the series may be denoted

$$\pi_{111} = \int_0^1 (p.p.p)g(p)dp = E[p^3] = \mu'_3$$

and the probability of the sequences $\{1,1,0\}$, $\{0,1,1\}$, and $\{1,0,1\}$ are equal since

$$\pi_{110} = \int_0^1 (p \cdot p \cdot (1-p))g(p)dp = E[p^2 - p^3] = \mu'_2 - \mu'_3$$

$$\pi_{011} = \int_0^1 ((1-p) \cdot p \cdot p)g(p)dp = E[p^2 - p^3] = \mu'_2 - \mu'_3$$

$$\pi_{101} = \int_0^1 (p \cdot (1-p) \cdot p)g(p)dp = E[p^2 - p^3] = \mu'_2 - \mu'_3$$

If, however, the chance of the event is varying over time (i.e. there is non-stationarity), with $p_1, p_2, .. p_T$ possibly different, this model has to be generalised. Suppose the form of heterogeneity is fixed and, taking the first period as reference, may be denoted $g(p_1)$. A simple option to account for non-stationarity is linear scaling of event probabilities

$$p_t = \delta_t p_1$$

This 'scaling' is somewhat analogous to using a piecewise exponential in hazard regression, with the δ_t modelling the fluctuations in the p_t without necessarily presupposing a parametric form. Then the chance of the series {1,1,1} would be

$$\pi_{111} = \int_0^1 (p_1 \cdot \delta_1 p_1 \cdot \delta_2 p_1) g(p_1) dp$$

The likelihood for $T = 3$ consists of N_{111} persons with pattern {1,1,1}, N_{110} persons with pattern {1,1,0}, etc. If the chance of the event is low, then linear scaling may be appropriate even though in general linear scaling may lead to rates exceeding one – unless the maximum rate (rather than the first rate) is taken as a reference, and all δ_t are then under one.

However, one may also specify logistic scaling such that there is appropriate {0,1} bounding of probabilities. Thus, if successive odds ratios are linked according to

$$[p_t/(1 - p_t)] = \kappa_t[p_1/(1 - p_1)]$$

then successive logits are linked as

$$\text{logit}(p_t) = \log(\kappa_t)\text{logit}(p_1)$$

Davies *et al.* adopt the parameterisation

$$p_t/(1 - p_t) = (1/\kappa_t)[p_1/(1 - p_1)] \tag{6.21}$$

so that

$$p_t = \theta_t p_1 \quad t > 1$$

where

$$\theta_t = 1/(\kappa_t - \kappa_t p_1 + p_1) \tag{6.22}$$

Similar models may be developed for Poisson series.

Example 6.4 Migration histories As an example of a conjugate mixture approach, consider data from Crouchley *et al.* (1982) on ten year migration histories for 10 000 residents of Wisconsin, USA, disaggregated into $k = 1, .. 4$ sub-populations defined by two age bands and two household tenures (Table 6.5). Sequences of urban migrations are defined by five two year periods ($T = 5$), with MMMMM the same as {1,1,1,1,1} above while MMMMS denotes four migrations followed by 'staying' or non-migration.

Considerations of the housing market and lifecycle migration processes indicate that the parameters θ_{kt} in Equation (6.22) are likely to vary both over sub-populations (e.g. younger people are more likely to migrate) and over time.

There are $i = 1, .., 32$ possible migration sequences of five outcomes (migrate or stay), so that the likelihood is defined over $32 \times 5 \times 4$ points. Let $d_{it} = 1$ if a migration occurs at period t (e.g. $d_{i1} = 1$, $d_{i2} = 1$, $d_{i3} = 0$, $d_{i4} = 1$, $d_{i5} = 1$ for the sequence MMSMM) and

Table 6.5 Inter-urban moves over five two-year periods in Milwaukee, Wisconsin, USA

Renters		Owners		Move Pattern
25–44	46–64	25–44	46–64	(M=Migrate, S=Not Migrate)
511	573	739	2385	SSSSS
222	125	308	222	SSSSM
146	103	294	232	SSSMS
89	30	87	17	SSSMM
90	77	317	343	SSMSS
43	24	51	22	SSMSM
27	16	62	19	SSMMS
28	6	38	5	SSMMM
52	65	250	250	SMSSS
17	20	48	14	SMSSM
26	19	60	25	SMSMS
8	4	10	3	SMSMM
8	9	54	21	SMMSS
11	3	18	1	SMMSM
10	3	21	1	SMMMS
4	1	8	2	SMMMM
41	29	134	229	MSSSS
16	15	23	10	MSSSM
19	13	36	25	MSSMS
2	4	1	0	MSSMM
11	10	69	24	MSMSS
11	2	15	3	MSMSM
1	9	13	2	MSMMS
2	2	2	0	MSMMM
7	5	40	18	MMSSS
4	2	9	2	MMSSM
8	1	15	3	MMSMS
1	0	5	0	MMSMM
8	1	22	7	MMMSS
3	2	7	2	MMMSM
5	0	9	2	MMMMS
6	3	5	0	MMMMM

n_{ik} be the number of persons of type k undergoing that sequence. Then the likelihood for sequence i, time t and sub-population k is defined as

$$y_{ikt} = d_{it} n_{ik}$$
$$y_{ikt} \sim \text{Bin}(p_{kt}, n_{ik})$$

As Table 6.5 shows, the event rate is low. However, one may apply the odds ratio scaling model in Equations (6.21)–(6.22), so that for the kth group the model for p_{kt} combines the updating parameters $\theta_{kt}(t > 1)$ with the initial period baseline migration rates p_{k1} which are assigned a mixing density $g(p_{k1})$. Thus a beta prior might be assumed for p_{k1}, and then

$$\text{logit}(p_{kt}) = \text{logit}(p_{k1}) - \log(\kappa_t)$$

where κ_t is as in Equation (6.21). Gamma $G(1, 0.001)$ priors are adopted for the κ_t.

The estimates[4] of θ_{kt} (for $k = 1, 4$ and $t > 1$) for the four sub-populations are as in Table 6.6. These show greatest non-stationarity for the young renters sub-population.

The elements needed to obtain the variances of migration rates in the groups, namely μ_{2k} and μ_{1k} may be monitored, as may the chances of different sequences (allowing for non-stationarity). The variances are then obtained from the posterior means of μ_{2k} and μ_{1k} as $\mu_{2k} - (\mu_{1k})^2$. The variances are higher in the renter groups, reflecting a well known feature of migration behaviour. As an example of a sequence probability Program 6.4 includes the probability of the sequence MMMMS by group.

Example 6.5 Ratings of schizophrenia To illustrate ordinal outcomes y_{it} over time, we follow Hedecker and Gibbons (1994) in considering data on the impacts of alternative drug treatments on symptom severity rankings in 324 schizophrenic patients, collected as part of the NIMH Schizophrenic Collaborative Study. There were originally four

Table 6.6 Panel migrations, parameter summary

$\theta_{k,t}$ Parameters (by subpopulation & time)	Mean	St. devn.	2.5%	97.5%
θ_{12}	1.18	0.13	0.96	1.44
θ_{13}	1.79	0.17	1.47	2.15
θ_{14}	2.54	0.23	2.12	3.03
θ_{15}	3.11	0.27	2.62	3.68
θ_{22}	1.34	0.17	1.04	1.72
θ_{23}	1.63	0.19	1.30	2.05
θ_{24}	2.08	0.24	1.67	2.61
θ_{25}	2.37	0.26	1.89	2.93
θ_{32}	1.42	0.08	1.26	1.59
θ_{33}	1.73	0.10	1.56	1.93
θ_{34}	1.62	0.09	1.46	1.82
θ_{35}	1.55	0.09	1.38	1.72
θ_{42}	1.06	0.08	0.91	1.22
θ_{43}	1.37	0.09	1.18	1.56
θ_{44}	1.01	0.07	0.87	1.16
θ_{45}	0.91	0.07	0.78	1.05
$\mu_{1,k}$				
$\mu_{1.1}$	0.10480	0.00815	0.08928	0.12170
$\mu_{1.2}$	0.08775	0.00813	0.07202	0.10390
$\mu_{1.3}$	0.14820	0.00666	0.13530	0.16130
$\mu_{1.4}$	0.08537	0.00457	0.07698	0.09466
$\mu_{2.k}$				
$\mu_{2.1}$	0.01104	0.00173	0.00797	0.01480
$\mu_{2.2}$	0.00777	0.00143	0.00519	0.01079
$\mu_{2.3}$	0.02201	0.00198	0.01831	0.02602
$\mu_{3.4}$	0.00731	0.00078	0.00593	0.00896

[4] A two chain run shows convergence at under 500 iterations.

treatments: chloropromazine, fluphenazine, thioridazine, and placebo. Since previous analysis revealed similar effects of the three anti-psychotic drugs, the treatment is reduced to a binary comparison of any drug vs. placebo. The severity score is derived form item 79 of the Inpatient Multidimensional Psychiatric Scale.

Hedecker and Gibbons collapse the seven points of that item into four: (1) normal or borderline mental illness; (2) mild illness; (3) marked illness; and (4) severely ill. Here we retain the full scale with 1=normal, 2=borderline, 3=mildly ill, 4=moderately ill, 5=markedly ill, 6=severely ill and 7=extremely ill. More pronounced gaps might be anticipated in thresholds between some of these categories (e.g. 5 vs. 4) than others. There are three repetitions after the first (week 0) reading, which is coincident with treatment, namely at weeks 1, 3 and 6.

The likelihood model is then, for subjects $i = 1, .. N$ and times $t = 1, ... T$ and $C = 7$ levels on the outcome

$$\begin{aligned}
y_{i,t} &\sim \text{Categorical}(p_{i,t,\ 1:C}) \\
\text{logit}Q_{i,t,\ j} &= \tau_j - \mu_{it} \\
p_{i,t,\ 1} &= Q_{i,t,\ 1} \\
p_{i,t,\ j} &= Q_{i,t,\ j} - Q_{i,t,\ j-1} \quad j = 2, \ldots .\ C-1 \\
p_{i,t,\ C} &= 1 - Q_{i,t,\ C-1} \\
\mu_{i,t} &= \beta X_{it}
\end{aligned}$$

The predictors X_{it}, some time varying, include time itself (specifically square root of weeks ν_{it}), the main treatment effect, a treatment by time interaction and the sex of the patient. Two models are considered (Models A and B in Program 6.5), one with a fixed intercept and impact of time; the other involves a bivariate model for varying intercepts and slopes on time

$$\mu_{i,\ t} = b_{i1} + b_{i2}\nu_{it}^{0.5} + \beta_1 X_{1i} + \beta_2 X_{2i} + \beta_3 X_{1i}\nu_{it}^{0.5}$$

where

$$b_i \sim N_2(\mu_b, \Sigma_b)$$

the means of b_{ik} are the intercept and average growth rate, X_1 is treatment and X_2 is gender. Identification of the $C - 1 = 6$ thresholds involves setting the parameter τ_1 (governing the transition from normality to borderline illness) to zero and estimating the remaining five threshold parameters subject to monotonicity constraints.

Table 6.7 shows for the fixed effects model[5] a clear time effect, an inconclusive main drug effect, but a greater improvement over time for the drug group. The same effects show under random effects, but all coefficients are amplified. The DIC under the two models are 4121 and 3444, so there is clear evidence of heterogeneity. Under both models there seems to be a major change in threshold values for severely vs. extremely ill, but the thresholds are larger under random effects – the latter feature is also apparent in Table 3 of Hedeker and Gibbons (1994).

As usual, one might assess sensitivity by assuming alternate forms of random effects, as in Example 6.2. The Exercises include fitting a scale mixture version of the random

[5] With three chains convergence is apparent at 500 and 1500 iterations under the fixed effects and random effects models respectively; summaries are based on iterations 500–3000 and 1500–3000, respectively.

Table 6.7 Schizophrenia severity ratings (posterior summaries) intercept and time effects fixed

	Mean	St. devn.	2.5%	97.5%
Constant	4.69	0.24	4.21	5.16
Time	−0.54	0.12	−0.77	−0.31
Drug	0.175	0.207	−0.216	0.582
Male	0.173	0.098	−0.018	0.365
Drug by Time	−0.72	0.13	−0.98	−0.46
τ_2	1.27	0.12	1.06	1.51
τ_3	2.26	0.13	2.02	2.53
τ_4	3.40	0.14	3.13	3.69
τ_5	4.94	0.16	4.65	5.26
τ_6	7.52	0.22	7.10	7.97
Intercept and time effects random				
Constant (Av.)	7.66	0.40	7.02	8.43
Time (Av.)	−0.85	0.16	−1.20	−0.56
Drug	0.35	0.31	−0.33	0.93
Male	0.18	0.24	−0.26	0.67
Drug by Time	−1.19	0.19	−1.53	−0.77
τ_2	2.05	0.18	1.70	2.40
τ_3	3.72	0.21	3.30	4.11
τ_4	5.64	0.24	5.12	6.07
τ_5	8.16	0.29	7.54	8.68
τ_6	11.90	0.41	11.04	12.65
Dispersion matrix				
Σ_{b11}	3.68	0.73	2.52	5.23
Σ_{b12}	−0.62	0.30	−1.29	−0.10
Σ_{b22}	1.18	0.25	0.73	1.69

effects model (equivalent to a multivariate t, and with degrees of freedom an unknown); the DIC suggests a slight gain in fit for this model compared to the simple multivariate Normal.

Example 6.6 Rheumatoid arthritis Additional issues arising from unequal spacing and missing data are illustrated by binary series Y_{it} from Fitzmaurice and Lipsitz (1995) following a randomised trial of therapies for $i = 1, .., 51$ patients with classic or definite rheumatoid arthritis. The treatments were auranofin and a placebo, and the binary response measured at baseline ($t = 1$) and four times thereafter ($t = 2, 3, 4, 5$) is a patient self-assessment of arthritis status (1=good, 0=poor). Some self-assessments (17 from 255) are missing, and Fitzmaurice and Lipsitz adopt an MCAR assumption to these missing responses (i.e missingness is unrelated both to any observed responses and covariates or to the missing outcomes themselves). The gap between the baseline and first post-treatment assessment is one week, with succeeding gaps being four weeks (i.e. the five readings are at 0, 1, 5, 9 and 13 weeks). The predictors of status are time, treatment, age at baseline and gender.

Fitzmaurice and Lipsitz adopt a logit link

$$Y_{it} \sim \text{Bern}(\pi_{it})$$
$$\text{logit}(\pi_{it}) = \theta_{it}$$

and consider models for the correlation between Y_{is} and Y_{it} $(t \neq s)$, defined by

$$\vartheta_{ist} = (\pi_{ist} - \pi_{is}\pi_{it})/[\pi_{is}(1 - \pi_{is})\pi_{it}(1 - \pi_{it}))^{0.5}$$

where π_{ist} is the joint probability[6] of a good status at times s and t. An exponential model for the correlation takes the form

$$\rho_{ist} = \vartheta^{|t-s|}$$

where $0 < \vartheta < 1$. They in fact adopt a reformulation in terms of the odds ratios of good status at times s and t:

$$\psi_{ist} = \pi_{ist}(1 - \pi_{is} - \pi_{it} + \pi_{ist})/[(\pi_{is} - \pi_{ist})(\pi_{it} - \pi_{ist})]$$

modelled as

$$\psi_{ist} = \alpha^{1/|t-s|}$$

with $1 < \alpha < \infty$.

The first model adopted here allows for autocorrelation via lag one autoregressive errors in the logit scale, with

$$Y_{it} \sim \text{Bern}(\pi_{it}) \quad (6.23a)$$

$$\text{logit}(\pi_{it}) = \theta_{it} = \beta X_{it} + e_{it} \quad (6.23b)$$

Allowing for the differential spacing (in terms of time in weeks) between observations $t-1$ and t gives

$$e_{it} = \gamma^{|v_t - v_{t-1}|} e_{i,\, t-1} + u_{it} \quad t > 1 \quad (6.23c)$$

where $\text{var}(u) = \sigma^2$. Since one may write

$$e_{i2} = \gamma e_{i1} + u_{i1}$$

and $\text{var}(e_{i2}) = \text{var}(e_{i1})$, one may specify

$$e_{i1} \sim \text{N}(0, \sigma^2/(1 - \gamma^2))$$

Fitzmaurice and Lipsitz assume γ is between 0 and 1, and an informative prior constraining γ to be positive is adopted here, namely a uniform prior

$$\gamma \sim \text{U}(0, 1)$$

For identifiability, the variance of u is set at $\sigma^2 = 1$. The missing responses for certain subjects are taken as missing at random, that is possibly depending on observed responses and covariates for these subjects, but not on the missing responses themselves.

[6] One might obtain an estimate of this correlation by defining a vector of length 4 for each of the $T_i(T_i - 1)/2$ pairs of binary observations for the ith subject, with the first item of the vector being 1 if both y_{is} and y_{it} are 1, the second being 1 if y_{is} alone is 1, the third being 1 if y_{it} alone is 1, and the fourth being 1 if y_{it} and y_{is} are both 0. Then multinomial sampling would be used, with N series of T_i observations being reformed as $NT_i(T_i - 1)/2$ multinomial observations of length 4.

We obtain[7] the estimates shown in Table 6.8 for time, female gender, age and the therapy. These compare to the Fitzmaurice and Lipsitz estimates for the effects of these covariates of −0.013 (s.e. 0.022), −0.61 (0.42), −0.015 (0.018) and 1.45 (0.45). The predictive loss criterion (with $w = 1$) is 46. The high correlation confirm replicates the work of Fitzmaurice and Lipsitz, who found a strong odds ratio form of dependence between successive patient self-ratings.

An alternative approach is provided by the 'one factor model' of Heckman (1981, p. 130) namely

$$\theta_{it} = \beta X_{it} + \lambda_t \alpha_i + \varepsilon_{it}$$

where the ε_{it} are unstructured errors with variances ϕ_t, the α_i are latent propensities for good health status with variance σ^2_α, and the λ_t are time-varying loadings. For identifiability, it is assumed that $\lambda_1 = 1$, and $\phi_t = \phi = 1$. The variance σ^2_α is defined by a $U(0, 1)$ prior on

$$\tau = \sigma^2_\alpha/[\sigma^2_\alpha + \phi]$$

Defining the ratios

$$R_t = \lambda^2_t\sigma^2_\alpha/[\lambda^2_t\sigma^2_\alpha + \phi_t]$$

of 'permanent' to total variance[8], the correlation between disturbances at times t and s is then $C_{ts} = A_t A_s$, where

$$A_t = R_t^{0.5}$$

N(1, 1) priors are assumed on the $\lambda_t (t > 1)$, and constrained to being positive.

Fitting this model[9] enhances the mean therapy effect to 2.2, with 95% interval {0.75, 4.1}, but shows weaker effects than in Table 6.8 for gender or age. The correlations range between 0.47 (between times 3 and 4) and 0.74 (between times 1 and 2). The individual propensities α_i range from 3.3 for patient 4, who records good health status at all time points, despite being on the placebo treatment, to patient 32 with a score of −3.8 and classed as poor status at all points. The predictive loss criterion stands at 45, showing slight gain to using this model. Subject to identifiability more complex one factor models might improve fit, for example taking ϕ_t variable by time.

Finally, the method of Albert and Chib (1993) offers a direct approach to sampling the underlying latent variables in Equation (6.17). It is applied to the model of

Table 6.8 Arthritis status, differential spacing model, parameter summary

	Mean	St. devn.	2.50%	97.50%
Intercept	1.86	1.68	−1.5	5.17
Time	0.022	0.05	−0.075	0.12
Female	−1.14	0.84	−2.82	0.49
Age	−0.017	0.034	−0.085	0.052
Therapy	1.82	0.79	0.39	3.48
γ	0.93	0.03	0.86	0.97

[7] A three chain run shows convergence at 1000 iterations and summaries are based on iterations 1000–5000.
[8] The composite error $\lambda_t\alpha_i$ multiplies the permanent subject error by a time varying factor.
[9] A two chain run is taken to 5000 iterations with 1000 burn-in.

Equation (6.23) but might equally be applied to the one factor model. In the presence of missing observations, one option is to use imputations of the missing binary responses Y_{it}, so permitting one to define the sampling ranges for the latent continuous indicators Z_{it}. For illustration a single drawing of the missing Y_{it} is used (at 5000 iterations in the run for Model A). A more complete multiple imputation procedure would pool over several analyses with different imputations. An alternative approach if there are missing data points is to use unconstrained sampling of the Z for those points. One might then estimate probabilities such as $\Pr(Z_{it} > 0) \equiv \Pr(Y_{it} = 1)$ for the missing observations.

Here the latent sampling approach is applied with Student $t(8)$ errors, so approximating the logit link. This entails defining gamma variables φ_{it} with density $G(4, 4)$ and then sampling (within the relevant constraints) the Z_{it} with means $\theta_{it} = \beta X_{it} + e_{it}$ and precisions φ_{it}. One might also take the precisions at subject level φ_i, and not observation level. Model fit may be assessed by sampling new Z values without upper and lower constraints, and then classifying new Y values according to whether the new Z exceed 0; on this basis, the predictive criteria of Gelfand and Ghosh (1998) and others may then be derived, comparing actual and new Y.

Under the imputation approach[10], the correlation under differential spacing is 0.91 (see Table 6.9). This approach shows a lower precision on the therapy effect (though it is still judged beneficial), and the gender effect is amplified as compared with the usual logit model in Equation (6.23). The predictive loss criterion under this model is 39.4. Under unrestricted sampling[11] of Z_{it} when there are missing Y_{it} (Model D in Program 6.6), the posterior mean for γ is 0.89 and the treatment effect averages 2.1 with 95% interval (0.4, 4). For predicting missing values under this approach the model for Z_{it} could well be improved, for instance using autoregression on previous values of Z_{it} or of Y_{it}, particularly since the first occurrence of missing Y is at $t = 3$.

Example 6.7 Patent applications As an illustration of generalised Poisson regression models for panel data on counts y_{it}, we follow Hausman, Hall and Griliches (1984) and Cameron and Trivedi (1999) in considering patent applications by technology firms. Trends in patent activity may be partly explained in terms of levels of current and past research inputs, type of firm, and time t itself. However, unobserved variation is likely to remain between firms in terms of factors such as entrepreneurial and technical skills – suggesting the need for a permanent firm effect. Autocorrelation in residuals is reported by Hausman *et al.* in a simpler model without such unobserved factors included, and by

Table 6.9 Latent student t model, posterior summary

	Mean	St. devn.	2.5%	97.5%
Intercept	3.61	2.15	−0.17	8.00
Time	−0.013	0.059	−0.126	0.104
Female	−1.48	1.12	−4.00	0.58
Age	−0.04	0.04	−0.12	0.03
Therapy	2.38	1.00	0.43	4.39
γ	0.91	0.03	0.85	0.95

[10] Convergence with three chains is apparent at 3000 iterations (with delay in the age effect converging); summaries are based on iterations 3000–10 000.

[11] The sampling intervals for Z are (−20, 0) when Y is 0, (0, 20) when Y is 1, and (− 20, 20) when Y is missing.

modelling this type of heterogeneity the need for a serial correlation model for the errors may be avoided.

Hausman *et al.* adopt a multiplicative conjugate gamma form for the heterogeneity, such that

$$y_{it} \sim \text{Poi}(\nu_i \mu_{it}) \tag{6.24a}$$

$$\nu_i \sim \text{G}(\delta, \delta) \tag{6.24b}$$

$$\log(\mu_{it}) = \beta x_{it} \tag{6.24c}$$

For this specification, the marginal means and variances are of negative binomial form with $E(y_{it}) = \mu_{it}$ and $V(y_{it}) = \mu_{it} + \text{Var}(\nu)\mu_{it}^2 = \mu_{it} + \mu_{it}^2/\delta$.

The analysis here follows Hausman *et al.* in adopting a random effects specification for firms, in addition to a regression model including time t and logs of current and lagged research expenditures $R_0, R_{-1}, \ldots R_{-5}$. Note that the original inputs have been centred in Program 6.7 to improve convergence. As noted by Cameron and Trivedi (1998, p. 286), the coefficients on $\log(R_{-j})$ are elasticities, so that the coefficients on $\log R_0$, $\log R_{-1}, \ldots \log R_{-5}$, would sum to unity if a 50% increase in patents filed follows a 50% increase in research spending (i.e. a proportional effect).

The first model used here has the form has an additive (log-normal) form for the extra-variation between firms.

$$\log(\mu_{it}) = \beta x_{it} + \alpha_i$$
$$y_{it} \sim \text{Poi}(\mu_{it})$$
$$\alpha_i \sim \text{N}(0, \tau)$$

where α_i are interpretable as permanent firm effects. The covariates are the research lags and time itself. The estimated[12] sum of the research elasticities (Table 6.10) comes close to Table III of Hausman *et al.* The return to research spending in patent output is not proportional, though higher than reported by Cameron and Trivedi (1998, p. 286).

The second model follows Hausman *et al.* in taking a conjugate gamma-Poisson model (Model B in Program 6.7), as in Equation (6.24), with

$$\nu_i \sim \text{G}(\delta, \delta)$$
$$\delta \sim \text{G}(1, 1)$$

where sensitivity to the prior for δ might involve gamma priors $G(g, g)$ with alternative values g. The $\log(\nu_i)$ are analogous to the α_i of the Poisson log-normal model. Note that a diffuse prior on δ (e.g. with $g = 0.001$) may lead to numerical problems. This approach gives broadly similar results to the preceding model, and also shows a less than proportional research input gain. The DIC and predictive loss criterion are both slightly improved under the gamma-Poisson model; the predictive loss (with $w = 1$) falls from 204 500 to 202 400, and the DIC from 9760 to 9650.

To generalise to bivariate effects, at firm level the Poisson log-normal model may be preferable. Following Chib *et al.* (1998), the slope on $\log R_0$ is taken to be variable over firms, together with the intercept. Rather than assuming zero means for these parameters (denoted $\alpha_{i,\, 1::2}$), and retaining separate 'fixed effects' for the intercept and the

[12] In the first and second models, three chain runs show convergence at around 3000 iterations and summaries are based on iterations 3000–5000.

Table 6.10 Models for firm heterogeneity: regression parameters and estimated firm effects (Firms 1–4)

	Mean	St. devn.	2.5%	Median	97.5%
Log-Normal Heterogeneity					
Intercept	1.02	0.07	0.89	1.02	1.16
Sum of Coefficients on R_{-j}	0.80	0.02	0.75	0.79	0.85
Coefficient on					
R_0	0.49	0.04	0.41	0.49	0.57
R_{-1}	−0.03	0.05	−0.12	−0.03	0.06
R_{-2}	0.10	0.04	0.02	0.10	0.18
R_{-3}	0.12	0.04	0.04	0.12	0.20
R_{-4}	0.04	0.04	−0.03	0.04	0.11
R_{-5}	0.08	0.03	0.02	0.08	0.14
Time	−0.065	0.003	−0.071	−0.065	−0.059
Effects for firms 1–4					
α_1	2.34	0.08	2.18	2.34	2.50
α_2	0.17	0.35	−0.55	0.19	0.81
α_3	0.36	0.10	0.16	0.36	0.54
α_4	−1.86	0.53	−2.96	−1.82	−0.91
Gamma-Poisson heterogeneity					
δ_1	1.10	0.11	0.92	1.10	1.33
Sum of Coefficients on R_{-j}	0.76	0.06	0.67	0.75	0.84
Intercept	2.46	0.06	2.35	2.46	2.58
Coefficient on					
R_0	0.48	0.14	0.28	0.50	0.68
R_{-1}	−0.07	0.10	−0.23	−0.08	0.12
R_{-2}	0.11	0.05	0.03	0.08	0.20
R_{-3}	0.12	0.05	0.05	0.11	0.22
R_{-4}	0.03	0.05	−0.04	0.01	0.15
R_{-5}	0.08	0.06	−0.02	0.08	0.17
Time	−0.064	0.005	−0.072	−0.064	−0.054
Effects for firms 1–4 (logs of gamma effects)					
$\log(\nu_1)$	1.90	0.07	1.77	1.90	2.04
$\log(\nu_2)$	−0.19	0.37	−0.96	−0.17	0.50
$\log(\nu_3)$	−0.03	0.13	−0.25	−0.04	0.22
$\log(\nu_4)$	−2.75	0.83	−4.68	−2.62	−1.48

coefficient on $\log R_0$, it may be preferable for MCMC identifiability and convergence to take $\alpha_{i,\,1:2}$ to be bivariate with mean η, where η_1 is equivalent to the fixed effect intercept β_0 and η_2 to the fixed effect β_1. So

$$\log(\mu_{it}) = \alpha_{i1} + \alpha_{i2} \log R_{i,\,t} + \beta_2 \log(R_{i,\,t-1}) + \ldots \beta_6 \log(R_{i,\,t-5})$$

To further improve identifiability an N(1, 1) prior on η_2 is adopted, and N(0, 1) priors on β_2, β_3, β_4, β_5 and β_6. This model yields[13] an improved fit with DIC at 9105 and predictive loss criterion at 147 000. Its improvement occurs especially for firms with low (e.g. runs of zeroes) patent counts such as 13 and 21, and firms with high patent activity, such as firms 45 and 242. Under this model, the coefficient on $\log R_0$ increases to around 0.64, with the sum of elasticities around 0.89. It is interesting to note a correlation of −0.76 between the firm specific slopes and intercepts.

None of the above models directly approaches the issue of serial correlation in the errors, though this is modest in the bivariate firm effects model[14]. A fourth model (Model D in Program 6.7) therefore assesses autocorrelation under the Poisson log-normal model, specifically a lag 1 dependence in the errors. There are various options here, with a common choice being based on stationarity

$$\log(\mu_{it}) = \beta x_{it} + \alpha_i + e_{it} \quad (6.25a)$$

$$e_{it} \sim N(\gamma e_{i,\,t-1}, \sigma^2) \quad t > 1 \quad (6.25b)$$

$$e_{i1} \sim N(0, \sigma^2/(1-\gamma^2)) \quad (6.25c)$$

However, in the short runs (small T) typical of panel data, a non-stationary model is feasible, with the variance of the first error a separate parameter not linked to σ^2, and γ following a prior such as N(0, 1). A transformation within the link for μ_{it} can also be made, so that for $t > 1$

$$\log(\mu_{it}) = \beta x_{it} + \alpha_i + \gamma[\log(\mu_{i,\,t-1}) - \beta x_{i,\,t-1} - \alpha_i] + u_{it}$$

where $u_{it} \sim N(0, \sigma^2)$. One might also assume a lag 1 correlation in an estimate of the regression errors based on comparing actual and predicted counts. This approach uses the transformation (Cameron and Trivedi, 1998)

$$y^*_{it} = y_{it} + h, \; h > 0$$

so that

$$\log(\mu_{it}) = \beta x_{it} + \alpha_i + \gamma(\log y^*_{i,\,t-1} - \beta x_{i,\,t-1} - \alpha_i)$$

with h either set to a small constant or forming an extra parameter.

Another issue with including observation level heterogeneity is that the e_{it} may effectively model the observations, making the regression terms in effect redundant. Unlike for binary panel data, it is not necessary to set σ^2 to 1. However, it may be advisable to set a prior favouring relatively high precision in the e_{it} (for a discussion of similar questions in modelling binomial heterogeneity, see Valen and Albert, 1998). Also, a high lag 1 correlation may make separate identification of e_{i1} and α_i difficult. One may also adopt the strategy of linking these two effects, as discussed above.

Exploratory analysis with model (6.25) suggested that there was a high lag coefficient close to unity, whether non-stationary or stationary priors on γ were used. To improve identifiability, the permanent firm effects α_i are therefore excluded. Also, a relatively

[13] A two chain run shows convergence at around 5000 iterations and inferences are based on iterations 5000–7500.

[14] Defining errors by firm and period as $(y_{it} - \mu_{it})/\mu_{it}$ gives a correlation between errors at lag 1 and at lag 2 in this model of 0.12, compared to 0.29 in the Poisson-gamma model.

informative G(5, 1) prior on $1/\sigma^2$ was taken together with N(1, 1) priors on β_1 (the coefficient on $\log R_0$) and N(0, 1) priors on β_2, β_3, β_4, β_5 and β_6.

Under a stationary prior for γ in the model (6.25), a two chain run (with convergence from iteration 5000 in a total of 7500) shows a sum of elasticities more in line with proportional returns, in contrast to the models without temporal error correlation (Table 6.11). The posterior mean for the sum of elasticities is 0.9, though the 95% upper point does not exceed 1. The posterior mean for $1/\sigma^2$ is 21.5 compared to the prior mean of 5 so that the prior does not seem incompatible with the data. The 95% credible interval (0.97, 0.98) for γ comes close to a random walk. The DIC and predictive loss fit measures both show a gain from using this model; the DIC is 8590 compared to 9650 under the Poisson-gamma model for permanent firm effects.

A variety of other modelling approaches may be applied to these data, including variable selection methods (e.g. Kuo and Mallick, 1998), since some of the lag coefficients may be unnecessary, or priors specifically adapted to distributed lags (Shiller, 1974).

6.4 PANELS FOR FORECASTING

Panel data in econometrics, demography and biometrics reinforce information on trend by repetition over subjects. Such subjects vary by setting, and might be patients or pupils (growth curves), regions or firms (differences in productivity growth), or age groups (demographic schedules through time). While evaluation of time series models usually involves cross-validatory assessment in time (e.g. via one step ahead forecasts), a broader set of checks are possible in panel data analysis. Suppose there are N subjects observed for T times. Granger and Huang (1997) consider out-of-sample prediction in econometric panel models, and provide a useful framework distinguishing between out-of-sample predictions (for subjects or firms $N+1$, $N+2, .. N+M$ not in the sample), post-sample predictions (for times $T+1$, $T+2, \ldots$ beyond the sample) and predictions both post and out-of-sample (for both times $t=1$, $T+1, ..$ and subjects $i=N+1$, $N+2, ..$).

Table 6.11 Temporal correlation model, parameter summary

	Mean	St. devn.	2.5%	97.5%
Sum of Coefficients on R_j	0.90	0.03	0.84	0.95
Coefficient on				
R_0	0.44	0.03	0.36	0.51
R_{-1}	0.001	0.07	−0.10	0.13
R_{-2}	0.11	0.05	0.00	0.21
R_{-3}	0.12	0.05	0.04	0.22
R_{-4}	0.14	0.04	0.06	0.22
R_{-5}	0.09	0.04	0.02	0.16
Time	−0.06	0.01	−0.07	−0.05
Error correlation				
γ	0.98	0.003	0.97	0.98

As a particular frequent application, growth curves raise similar issues of prediction both beyond the observed time points, and also in terms of incomplete trajectories for subsets of subjects. We might also have observed cohort c for T time points and want to use their growth profile parameters to predict the future growth of a younger cohort $c+1$ observed for fewer than T times. In the balanced growth curve case where equal numbers of times are observed for each subject, there are observations for N subjects at T times points $v_1, v_2, v_3, \ldots v_T$, forming a $T \times N$ matrix Y. Let X be a $T \times m$ design matrix, where m is the degree of the growth curve polynomial. Thus for linear growth, $m = 2$, and

$$X = \begin{bmatrix} 1 & v_1 \\ 1 & v_2 \\ \cdots & \cdots \\ 1 & v_T \end{bmatrix}$$

In the random effects case,

$$Y_{it} = \beta_{i1} + \beta_{i2} v_t + \varepsilon_{it}$$

the intercepts β_{i1} and linear growth terms β_{i2} vary over subjects i, for example via a bivariate density, and the ε may be unstructured or correlated.

Other common options involve the growth parameters β being taken as homogenous over G subgroups of the subjects, for example defined by gender or assigned treatment. Let **A** be an assignment matrix of order $G \times N$, with $A_{ji} = 1$ if subject i is in group j. For instance, if there were gender subgroups of size N_1 and N_2, then **A** contains N_1 columns with entries

$$\begin{matrix} 1 \\ 0 \end{matrix}$$

and N_2 columns of entries

$$\begin{matrix} 0 \\ 1. \end{matrix}$$

Then β will be of dimension $m \times G$, and for $m = 2$ would contain group specific intercepts and linear growth parameters. The growth curve model in this case may then be specified as

$$Y = X\beta A + \eta \qquad (6.26)$$

where η is a $T \times N$ error matrix.

The specification of η may be important if the goal is extended prediction for all N subjects to new times v_{T+1}, v_{T+2}, etc. outside the current growth curve. Other types of prediction involve projecting incomplete trajectories for certain subjects observed only for $S < T$ time points.

Thus, an unstructured error matrix with

$$\text{cov}(\eta_{i,\ 1:T}) = \sum$$

might be adequate to describe a growth pattern (e.g. a clinical trial of known duration), where extrapolation beyond the trial is not an issue. However, structured modelling of the η_{it} may be required if post-sample prediction is the goal (or autoregression in the Y themselves). Thus one might adopt an AR(1) structure with

$$\eta_{it} = \gamma\eta_{i,\ t-1} + u_{it} \quad j > 1$$

where u_{it} are white noise errors with mean 0 and variance σ^2. For a growth curve with $m = 2$ and homogenous effects within subgroups, the model for the first time point in the AR(1) model has mean defined by the group index $a_i \varepsilon \{1, .. G\}$

$$\mu_{i1} = \beta_{1[a_i]} + \beta_{2[a_i]}\nu_1 + \eta_{i1}$$

where η_{i1} has variance $\sigma^2/(1 - \gamma^2)$.

The question of optimal prediction then occurs. Standard discrepancy measures comparing predicted and actual trajectories may be used if actual values are known. Thus, Lee and Geisser (1996) consider extending the sample re-use or predictive cross-validation method which for cross-sectional data involves omitting each case at a time and predicting it on the basis of the remaining $N - 1$ cases. The 're-use' occurs because each observation is used $N - 1$ times in separate models. In the panel data context, with T repeated observations for the N subjects, a sample re-use procedure might be used to predict the final (i.e Tth) observation in the series for a single subject i on the basis of that subject's prior series data, and the full series for the remaining $N - 1$ subjects. This procedure would be repeated N times.

Example 6.8 Dental development data We consider the dental data from Potthof and Roy (1964) relating to 27 subjects, 11 girls ($a_i = 1$) and 16 boys ($a_i = 2$). In the full data set there are $T = 4$ observations on each subject, these being distances in mm from the centre of the pituitary to the pteryomaxillary fissure at two yearly intervals (so that centred times are $v_1 = -3$, $v_2 = -1$, $v_3 = 1$, $v_4 = 3$). Following Lee and Geisser (1996), observation 20 is omitted as an outlier. Further the original data are modified so that the last female vector has its fourth observation, namely 28, missing: so $T_{11} = 3$, and the first three observations for that subject are (24.5, 25, 28). Also the first male vector (with first three observations 26, 25 and 29) is regarded as having the fourth observation, namely 31, missing.

Prediction of the missing data is carried out under three models: first, a linear growth curve model with gender specific intercepts and slopes (cf. Equation (6.26)) and uncorrelated errors. Thus,

$$Y_{it} = \beta_{1,\, a_i} + \beta_{2,\, a_i} v_t + \varepsilon_{it}$$

with ε multivariate Normal with mean 0 and dispersion matrix Σ. The second is a growth curve model with both unstructured univariate Normal errors ε and autocorrelated errors e_{it} (lag of order 1). Thus, for $t > 1$

$$Y_{it} = \beta_{1,\, a_i} + \beta_{2,\, a_i} v_t + e_{it} + \varepsilon_{it}$$

where $e_{it} = \gamma e_{i,\, t-1} + u_{it}$, where the u are unstructured. An N(0,1) prior is assumed for γ. The third model involves a first order lag autoregression in the observations themselves. The means for the third model are of the form (for $t > 1$)

$$\mu_{it} = \beta_{a_i} + \rho_{a_i} y_{i,\, t-1}$$

For the first period alternative means are assumed, with

$$\mu_{i1} = \beta^*_{a_i}$$

Another model for $t = 1$ might retain the same intercept and model the latent y_{i0} in

$$\mu_{i1} = \beta_{a_i} + \rho_{a_i} y_{i0}$$

Table 6.12 Dental development data, alternative structures for data and errors, model parameters and predictions of missing data points

1. Unstructured errors				
Correlations for times $(t+1, t)$	Mean	St. devn.	2.5%	97.5%
(2, 1)	0.61	0.12	0.33	0.81
(3, 1)	0.71	0.10	0.47	0.87
(3, 2)	0.83	0.06	0.68	0.92
(4, 1)	0.49	0.15	0.16	0.74
(4, 2)	0.71	0.10	0.47	0.87
(4, 3)	0.84	0.07	0.67	0.93
Intercepts				
β_{11}	22.69	0.53	21.65	23.75
β_{12}	25.31	0.51	24.31	26.31
Growth Rates				
β_{21}	0.47	0.10	0.27	0.67
β_{22}	0.78	0.10	0.60	0.97
Predictions				
Predicted $y_{11,4}$	27.97	1.26	25.51	30.48
Predicted $y_{12,4}$	29.71	1.23	27.24	32.13
MAD	1.24	0.65	0.23	2.70
2. Structured & unstructured errors				
Intercepts	Mean	St. devn.	2.5%	97.5%
β_{11}	22.67	0.58	21.44	23.76
β_{12}	25.27	0.51	24.32	26.39
Growth Rates				
β_{21}	0.46	0.12	0.19	0.68
β_{22}	0.77	0.11	0.57	0.97
Predictions				
Predicted $y_{11,4}$	27.77	1.43	24.78	30.44
Predicted $y_{12,4}$	29.91	1.41	27.09	32.69
MAD	1.29	0.69	0.29	2.97
Autocorrelation, γ	0.77	0.07	0.63	0.91
3. Autoregression in observations (gender specific)				
Intercepts	Mean	St. devn.	2.5%	97.5%
β_1	2.96	1.65	0.93	7.18
β_2	6.55	2.20	2.57	11.25
β_1^*	21.20	0.72	19.77	22.64
β_2^*	22.87	0.61	21.69	24.10

(continues)

Table 6.12 (*continued*)

Autoregressive parameters				
ρ_1	0.91	0.07	0.72	1.00
ρ_2	0.80	0.09	0.60	0.97
Predictions				
Predicted $y_{11,4}$	28.42	1.32	25.71	30.88
Predicted $y_{12,4}$	29.72	1.32	27.10	32.22
MAD	1.30	0.69	0.23	2.87

as being from a more robust density than the observed y_{it} (e.g. multivariate t with low degrees of freedom, but the same mean and dispersion as in the multivariate Normal for the observed y_{it}).

The mean absolute deviation is used as a discrepancy measure and obtained by averaging over the two missing data points. There are relatively slight differences in predictive success under these different models (Table 6.12). The growth curve with unstructured errors[15] gives a posterior mean of 28 for the fourth observation of the last female and 29.7 for the fourth observation of the first male, with average MAD of 1.24. Allowing autocorrelated errors changes forecasts little, namely 27.8 and 29.9 for the missing female and male data points respectively. The growth rate β_{2k} is higher for males ($k = 2$) than females ($k = 1$) under both growth curve models, and the autoregressive parameter in the third model is correspondingly higher for males. The third model yields a higher autoregressive parameter for females and slightly higher prediction of $y_{11,4}$.

6.4.1 Demographic data by age and time period

Profiles of demographic outcomes by age, marital status, ethnicity, etc. may be observed repeatedly over time, and this forms the basis for projecting to future times. The analysis of such data often also focuses on smoothing observed schedules of mortality, marriage rates, fertility, over the demographic categories, for example to take account of regularities in the relationship between adjacent rates, especially age specific rates. Hickman and Miller (1981) consider in particular the projection of exit rates for populations of insured lives or properties, and of annuitants and pensioners. The goal is estimation of probabilities of remaining in or leaving these populations because of death, withdrawal, migration, etc. These probabilities in turn figure in cash flow projections of insurance companies. Similar issues arise in demographic projections, for example in projecting population mortality rates.

Hickman and Miller consider models allowing both for correlations between mortality at different ages and for correlations over calendar time (i.e. the usual temporal autocorrelation). They present totals of female annuitant deaths D_{kt} by ages k, namely 51–55, 56–60, . . 86–90, and 91–95 and for years t, namely 1953, 1958, 1963 and 1968 (where $k = 1, \ldots K$ and $t = 1, \ldots T$, with $K = 9$ and $T = 4$). These deaths occur among total exposures (total time exposed to risk in the relevant population) E_{kt}. They consider a Normal likelihood approximation based on square root transforms of the observed

[15] Convergence in model1 is obtained at under 500 iterations in two chain runs of 20 000, and summaries are based on iterations 500–20 000.

crude death rates $m_{kt} = D_{kt}/E_{kt}$. The square root transforms $v_{kt} = (m_{kt})^{0.5}$ are taken to be normal around means $\nu_{kt} = (\mu_{kt})^{0.5}$, with variances $1/(4E_{kt})$.

The model means ν_{kt} are in turn[16] assumed multinormal with means ϕ_{kt}, and with covariation among errors determined by correlations across both ages and times. The correlations between ages $i, j = 1, .. K$ are described by a matrix

$$A_{ij} = \rho_1^{|i-j|}/(4\{L_i\ L_j\}^{0.5})$$

where L_i are prior exposures, in effect prior sample sizes based on accumulated actuarial experience with such populations, and assumed known. The correlations between times $s, t = 1, .. T$ are described by a matrix

$$C_{st} = \rho_2^{|s-t|}$$

Then the prior covariance matrix of the vector μ_{kt} (of length $K.T = 36$) is the Kronecker product $G = C^*A$. Hickman and Miller assume both ρ_1 and ρ_2 to be known, though they could be assigned priors, e.g. on the (0, 1) interval.

Example 6.9 Annuitant deaths A similar structure to that used by Hickman and Miller is illustrated here except in assuming a Poisson likelihood with

$$D_{kt} \sim \text{Poi}(\mu_{kt}E_{kt})$$
$$\log(\mu_{kt}) = \alpha + \eta_{kt}$$

where the η_{kt} have the above described multinormal structure. Autocorrelation in age and time separately is also considered, in which

$$\log(\mu_{kt}) = \alpha + \varepsilon_{1k} + \varepsilon_{2t} \tag{6.27}$$

and both errors are governed by lag 1 dependence.

Hickman and Miller define a structured covariance for the η_{kt} determined by only three parameters, the overall variance and the two correlations. To illustrate the fully parameterised multinormal model, Model A includes a 36×36 dispersion matrix for the interdependence of age and time errors. This is assigned a Wishart prior with 36 degrees of freedom. This model has 24.5 effective parameters and a DIC of 58.5, and closely reproduces the data with a deviance at the mode of 10. Model B follows Hickman and Miller (1981) in taking a reduced parameterisation for the multinormal with ρ_1 and ρ_2 assumed known (with values 0.9 and 0.5, respectively). These values are based on extensive sensitivity analysis by them with the annuitant data. To allow for remaining variation in this model a simple unstructured error $u_{kt} \sim N(0, \sigma^2)$ is included, so that

$$\log(\mu_{kt}) = \alpha + \eta_{kt} + u_{kt}$$

where $1/\sigma^2 \sim G(1, 0.001)$. This model gives a slightly improved fit compared to the unstructured dispersion model, with DIC of 56. One may set priors for ρ_1 and ρ_2 (rather than assuming them preset), but direct sampling in WINBUGS is very slow. One possibility is to precalculate covariance matrices G over a grid of pairs of values of ρ_1 and ρ_2, input them as data, and then use a discrete prior to find the combination supported by the data.

A conditional error model is easier to implement in WINBUGS, as in Equation (6.27), with

[16] Hickman and Miller actually consider the transformed outcome $(1000\mu_{kt})^{0.5}$.

$$\varepsilon_{1k} \sim N(\rho_1 \varepsilon_{1k-1}, \tau_1^2)$$
$$\varepsilon_{2t} \sim N(\rho_2 \varepsilon_{2t-1}, \tau_2^2)$$

The posterior means of ρ_1 and ρ_2 (and 95% credible intervals) are, respectively, 0.88 (0.83, 0.93) and 0.77 (0.05, 0.99), with the density for ρ_2 negatively skewed. The DIC of 46.5 (with 10.5 effective parameters) improves on the multinormal models. Table 6.14 presents posterior means of the square root of death rates in relation to populations in thousands, namely

$$b_{kt} = (\mu_{kt} \times 1000)^{0.5}$$

Table 6.15 Posterior means for graduated value of square root of death rates per 1000 population (second half of run of 20 000 iterations over two chains), b_{kt} for ages k and years t

Age	Year	Deaths	Exposed lives	Empirical rate	Smoothed rate	St. devn.
51–55	1953	0	171	0.00	2.11	0.29
	1958	2	214	3.06	2.02	0.28
	1963	3	328	3.02	2.08	0.28
	1968	3	439	2.61	1.97	0.27
56–60	1953	15	1371	3.31	2.78	0.16
	1958	8	1874	2.07	2.66	0.15
	1963	20	2879	2.64	2.73	0.15
	1968	28	3597	2.79	2.59	0.14
61–65	1953	63	4899	3.59	3.49	0.11
	1958	87	7939	3.31	3.34	0.09
	1963	132	11230	3.43	3.43	0.09
	1968	174	16530	3.24	3.26	0.08
66–70	1953	111	6596	4.10	4.30	0.11
	1958	235	14463	4.03	4.12	0.09
	1963	430	22500	4.37	4.24	0.08
	1968	529	33360	3.98	4.02	0.06
71–75	1953	69	2414	5.35	5.59	0.15
	1958	180	6451	5.28	5.35	0.12
	1963	407	13668	5.46	5.50	0.10
	1968	611	22109	5.26	5.21	0.08
76–80	1953	69	925	8.64	7.89	0.22
	1958	115	2029	7.53	7.56	0.17
	1963	340	5387	7.94	7.76	0.14
	1968	611	11689	7.23	7.36	0.12
81–85	1953	35	269	11.41	10.30	0.32
	1958	59	631	9.67	9.87	0.26
	1963	158	1448	10.45	10.14	0.24
	1968	351	3941	9.44	9.62	0.20
86–90	1953	9	62	12.05	13.37	0.52
	1958	23	130	13.30	12.81	0.45
	1963	59	363	12.75	13.15	0.43
	1968	130	802	12.73	12.48	0.39
91–95	1953	2	10	14.14	16.91	1.02
	1958	7	24	17.08	16.20	0.95
	1963	13	62	14.48	16.64	0.95
	1968	48	149	17.95	15.79	0.88

These are reasonably close to those cited by Hickman and Miller (1981, Table 11). Rates under this model are smoothed towards the average for the age group, though trends are still in line with those in the crude rates. For instance, for ages 76–80, the smoothed rates (like the crude ratres) show a deterioration between 1958 and 1963, as compared to improvement between the first two years.

The final model (Model D in Program 6.9) is as just described but estimated using observations over three years, and a prediction made for 1968. For these data the posterior means of ρ_1 and ρ_2 (and 95% credible intervals) are, respectively, 0.87 (0.82, 0.91) and 0.65 (0.04, 0.99). The credible intervals of the predictions for 1968 include the actual values, except for ages 91–95, which are under-predicted.

Various generalisations of these models can be developed, either for the multinormal (simultaneous autocorrelation) or the conditional error model. For example, one might make the age correlation specific to periods, so that

$$\log(\mu_{kt}) = \alpha + \varepsilon_{1kt} + \varepsilon_{2t}$$
$$\varepsilon_{1kt} \sim N(\rho_{1t}\varepsilon_{1,\,k-1,\,t},\ \tau_1^2)$$

Random walk autoregressive priors of order 2 are investigated by Berzuini and Clayton (1994), so that one might have

$$\varepsilon_{1k} \sim N(2\varepsilon_{1,\,k-1} - \varepsilon_{1,\,k-2},\ \tau_1^2)$$

6.5 MISSING DATA IN LONGITUDINAL STUDIES

A practical feature of many surveys or trials is nonresponse, either unit nonresponse with failure to obtain any responses from certain members of the sampled population, or item nonresponse to individual questions from unit-level respondents. This raises implications of nonresponse bias, for example in connection with sub-populations with known nonresponse problems in surveys (e.g. low income minorities). Such problems occur in periodic longitudinal surveys where a separate sample of the population is chosen each time the survey is carried out. Panel surveys (following up on a cohort of subjects not open to recruitment) are additionally subject to an attrition effect as unit non-response is cumulative (Winer, 1983). Similarly, in clinical trials or other studies with a cohort design, following up treatment outcomes or quality of life measures, patients may drop out of the study because of deteriorating health, poor quality of life, or death. Such drop-out is 'informative', inducing a dependence between the variable being measured and the drop-out mechanism.

Among approaches to this problem are the development of sample weights and adjustments for non-response and survey attrition, and the use of imputation methods. Explicit models for non-response rely on observed responses: often non-response on one or more items occurs for some subjects, but complete response on certain items or survey design variables are available for all sample members. Suppose X_1 is observed for all units in a study, but X_2 is not observed for everyone. Then models or imputation methods proceed on assumptions about the mechanism relating X_1 and X_2. The assumption of ignorability is that respondents and non-respondents (on X_2) with the same X_1 differ only randomly on X_2. A non-ignorable model by contrast allows for the possibility of systematic differences in X_2 for a responder and a non-responder having the same value of X_1.

Let $R = \{R_{ijt}\}$ indicate whether response was made ($R_{ijt} = 0$) or missing ($R_{ijt} = 1$) for subjects $i = 1, .. N$ at times $t = 1, .. T$ and items $j = 1, .. k$, and that X denotes covariates fully measured for all respondents (e.g. survey design variables such as geographic area of sampling). The most general form of joint distribution of the indicators R and the outcomes, known and missing, $Y = \{Y_{\text{obs}}, Y_{\text{mis}}\}$, has the form

$$f(R, Y|\theta_R, \theta_Y, X) = f_y(Y|X, \theta_Y) f_r(R|Y, X, \theta_R)$$

where f_y is the sampling density of the data and f_r is density for the response mechanism. This form allows the response mechanism to be influenced by the outcomes Y, whether observed or missing.

A less general assumption is that of 'missingness at random' if

$$f_r(R|Y_{\text{obs}}, Y_{\text{mis}}, X, \theta_R) = f_r(R|Y_{\text{obs}}, X, \theta_R)$$

In this case, non-response can depend upon the observed outcomes on other items, or on observed responses by other subjects, but given these, it will not depend upon the missing item responses themselves. If the MAR assumption holds and models for the outcome and response are separate (i.e. θ_Y and θ_R are non-overlapping), then the missingness pattern is called ignorable (Rubin, 1976). So missingness on an item is non-ignorable if it depends upon the missing value of that outcome.

In panel studies, a particular definition of non-response occurs where drop-out (without subsequent re-entry to observation) is selective. In the selection model (Diggle and Kenward, 1994; Heckman, 1976), the joint density of Y and R is expressed as above in terms of the marginal density of Y and the conditional density of R given Y. Drop-outs at time t are then random if $f(R_t|Y) = f(R_t|Y_1, \ldots Y_{t-1})$, but if related to the current outcome, known or otherwise, are informative. In practice, assuming random drop-out means relating $\Pr(R_{ijt} = 1)$ to previous binary outcomes $Y_{ij1}, .. Y_{ij,\, t-1}$, but not to the current outcome itself. An alternative pattern mixture model, proposed by Little (1993), involves expressing the joint density of Y and R as the marginal density of R and the conditional density of Y given R, namely

$$f(R, Y|) = f_r(R) f_y(Y|R)$$

Example 6.10 Schizophrenia treatments To illustrate modelling informative as against random drop-out, this example considers patient withdrawals in a longitudinal trial comparing treatments for schizophrenia (Diggle, 1998), where the issue is permanent attrition by a large number of patients (as against intermittent missing values). 517 patients were randomly allocated either to a control (placebo) group or to treatments involving one of two drugs, with different dosages on one : haloperidol at 20mg and risperidone at 2, 6, 10 and 16mg. There are therefore six possible regimes, including the placebo. Measures were obtained at seven time points (at selection into trial, at baseline, and at weeks 1, 2, 4, 6 and 8 thereafter). Let v_t denote the number of weeks at these time points, with the baseline defined by $v_2 = 0$, and selection for the trial happening at $v_1 = -1$. The outcome measure was the PANSS scale (Positive and Negative Symptom Scale), a continuous measure of psychiatric illness, with higher scores denoting more severe illness. Let G_i denote the treatment group of patients $i = 1, .. N$. All patients on risperidone are combined into one category, so the groups are then 1 = haloperidol, 2 = placebo, and 3 = risperidone.

The cumulative level of attrition is only 0.6% at the second round of observation (when $v_2 = 0$), but reaches 1.7%, 13.5%, 23.6% and 39.7% in successive waves, and stands at 48.5% in the final wave. The drop out rate peaks therefore at the sixth week. The question is whether attrition is related to health status: if there is a positive impact of PANSS scores on the probability of drop out, then the observed time paths of PANSS scores may be distorted. The observed means (not including the missing scores due to drop out) are, respectively, 87.2, 92.5, 86.1, 80.9, 78.5, 76.1 and 73.2. It may be of interest to estimate the wave means when the informative nature of the non-response mechanism is allowed for, and wave means include estimates of the scores for the drop outs.

Suppose the R_{ijt} are defined as above, where there is just one item to consider, so we can write this indicator as R_{it}. Response at time s will be ignorable if it depends upon observed covariates or previous observed values of PANSS score Y_{it} $(t = 1, .. s - 1)$, but not if it depends on the current, possibly missing, score Y_{is}. There are two options on the definition of the likelihood. One may either consider all missing data points, so that for a drop out at the fourth wave, the response model likelihood would include the data $R_{it} = 1$ for $t = 4, .., 7$ and the outcome model would include the missing data Y_{it}, for $t \geq 4$. Alternatively, one might limit the analysis to the first instance when $R_{it} = 1$, so the likelihood extends to $T_i + 1$ for subjects who drop out at occasion T_i (when $T_i < 7$) and is $T_i = 7$ for subjects who stay under observation throughout.

The data model for patient i is taken to be normal with

$$Y_{it} \sim \mathrm{N}(\mu_{it}, 1/\tau_1) \quad t = 1, T \tag{6.28}$$

with means defined by treatment specific linear terms in time (in weeks v), by treatment effects δ_j and by random terms as follows:

$$\mu_{it} = M + \delta_{G_i} + \theta_{G_i} v_t + \gamma_{G_i} v_t^2 + U_i + e_{it} \tag{6.29}$$

with $\delta_1 = 0$ for identifiability. The permanent subject effects U_i have mean zero and variance $1/\tau_2$. A quadratic in weeks is included, since the PANSS readings tend to rise between selection (week -1) and baseline (week 0) before falling thereafter.

Apart from unstructured measurement errors, implicit in Equation (6.28), Diggle assumes an autoregressive dependence in the errors e_{it}, with covariance between e_{it} and e_{is} specified as

$$\mathrm{Cov}(e) = \sigma^2 e^{-\phi\Delta^2}$$

where $\Delta = v_s - v_t$. Here an alternative form is adopted, though with similar allowance for difference time spacings between readings. Thus, the e_{it} in the first time period are taken as Normal

$$e_{i1} \sim \mathrm{N}(0, 1/\tau_3)$$

while subsequent errors have the form

$$e_{it} \sim \mathrm{N}(E_{it}, 1/\tau_4) \quad t = 2, T$$

The means of the autoregressive errors for $t > 1$ are

$$E_{it} = \rho^{[v_t - v_{t-1}]} e_{i,\, t-1}$$

The model for response indicators is

$$R_{it} \sim \mathrm{Bern}(\pi_{it})$$

with the dependence on the outcome Y in the model for π determining whether drop-out is random or informative. For example, a non-informative model (when the likelihood includes only the first non-response wave) is

$$\begin{aligned} \text{logit}(\pi_{i,t}) &= c_1 + c_2 Y_{i,t-1} \quad t > 1 \\ \text{logit}(\pi_{i,1}) &= c_1 + c_2 Y_{i,0} \end{aligned} \tag{6.30}$$

where the unknown Bernoulli outcomes $Y_{i,0}$ are modelled in terms of an extra unknown 'response rate' at time 0, π_0. By contrast, an informative model is

$$\text{logit}(\pi_{i,t}) = c_1 + c_2 Y_{i,t} \tag{6.31}$$

since this may refer to missing scores Y. Either type of model might also include time itself as a predictor of missingness, since there is clearly a trend to increased attrition at later waves.

With the non-informative model for response (6.30), similar results to those cited by Diggle (1998, p. 221) are obtained[17] for the coefficients of the response model. Drop-out increases with PANSS score so that those remaining are increasingly 'healthier' than the true average.

The linear time terms θ_1, θ_2 and θ_3 and the treatment effects both support the effectiveness of the new drug, though its main effect δ_3 in Equation (6.29) is not as clearly defined as the decline in PANSS scores under this treatment, shown by θ_3 (Table 6.16).

Table 6.16 PANSS scores over time, non-informative drop-out, parameters for response and observation model

	Mean	St. devn.	2.5%	97.5%
Response model				
C_1	−5.32	0.27	−5.84	−4.81
C_2	0.034	0.003	0.028	0.039
Observation model				
Treatment				
δ_2	1.43	1.76	−1.95	5.02
δ_3	−1.72	1.40	−4.29	1.49
Error correlation				
ρ	0.938	0.015	0.908	0.967
Linear time				
θ_1	0.04	0.73	−1.29	1.48
θ_2	1.03	0.61	−0.11	2.25
θ_3	−1.86	0.33	−2.40	−1.16
Squared time				
γ_1	−0.066	0.099	−0.258	0.117
γ_2	−0.066	0.082	−0.232	0.095
γ_3	0.104	0.041	0.023	0.175

[17] A two chain run shows convergence at 1000 iterations and the summary is based on iterations 1000–5000. For the informative response model convergence in a 5000 iteration run is obtained at 2500 iterations.

The placebo group apparently experience a deterioration in outcome over time ($\theta_2 > 0$). A high correlation over time in errors of the regression model for the observed scores is apparent with posterior mean for ρ of 0.94.

Fit may be assessed by the predictive criterion in footnote 1, where for missing Y_{it} only the precision of prediction is relevant; the criterion comparing actual and predicted R is 328, and that comparing predicted and actual Y is 277 100.

Introducing the current PANSS score Y_{it} into the model for response R_{it} makes the drop-out model informative, as in Equation (6.31). This response model is applied with the likelihood just including the first wave of non-response for those who drop out[18]. The fit to the scores Y_{it} improves slightly (predictive criterion 270 000), but the response model fit is unchanged (Table 6.17). In terms of inferences on treatment effectiveness, both the main treatment effect and the linear time effect for risperidone treatment are now significantly negative, though the time slope is less acute than under non-informative response.

6.6 REVIEW

The analysis of panel data by fully Bayesian techniques facilitates a more flexible and robust approach, for example in terms of assumptions about:

Table 6.17 PANSS scores over time, informative drop-out, parameters for response and observation model

	Mean	St. devn.	2.5%	97.5%
Response model				
C_1	−5.58	0.38	−6.28	−4.80
C_2	0.034	0.004	0.026	0.041
Observation model				
Treatment				
δ_2	3.25	2.38	−1.16	8.07
δ_3	−4.70	2.20	−9.60	−0.15
Correlation				
ρ	0.960	0.015	0.930	0.988
Linear time				
θ_1	0.06	0.33	−0.61	0.68
θ_2	1.15	0.37	0.38	1.88
θ_3	−0.81	0.19	−1.16	−0.42
Squared time				
γ_1	−0.017	0.087	−0.186	0.158
γ_2	0.046	0.095	−0.135	0.240
γ_3	0.108	0.045	0.023	0.199

[18] One might define the likelihood over all seven waves, but this requires an informative prior on c_2 as for some subjects several waves of non-response are then involved.

- the nature of autoregressive dependence in error terms, allowing for an assessment of the stationarity assumption rather than necessarily assuming it;
- the parametric structure (or otherwise) of error terms, both time varying and fixed subject errors (e.g. Hirano, 2002)
- data augmentation to sample initial pre-series values, such as in an AR(2) model in the response referring implicitly to unobserved data at times 0 and -1 (Karlsson, 2002);
- data augmentation to impute missing data values depending on an additional panel type model for the missingness mechanism in case the latter is informative (see Section 6.5).

There are a number of techniques so far generally tackled by maximum likelihood or EM methods where the fully Bayesian method might be used to good effect: examples are INAR models for Poisson and multinomial data (Bockenholt, 1999a, 1999b) or the Poisson exponentially weighted moving average model (Brandt *et al.*, 2000). Chapter 5 illustrated how INAR models might be implemented in Bayes terms and extension to panel situations is relatively straightfoward.

REFERENCES

Albert, J. and Chib, S. (1993) Bayesian analysis of binary and polychotomous response data. *J. Am. Stat. Assoc.* **88**(422), 669–679.

Allison, P. (1994) Using panel data to estimate the effects of events. *Sociological Meth. and Res.* **23**, 174–199.

Berzuini, C. and Clayton, D. (1994) Bayesian analysis of survival on multiple time scales. *Stat. in Med.* **13**, 823–838.

Bockenholt, U. (1999a) Analyzing multiple emotions over time by autoregressive negative multinomial regression models. *J. Am. Stat. Assoc.* **94**, 757–765.

Bockenholt, U. (1999b) Mixed INAR (1) Poisson regression models: analyzing heterogeneity and serial dependencies in longitudinal count data. *J. Econometrics* **89**, 317–338.

Brandt, P., Williams, J., Fordham, B. and Pollins, P. (2000) Dynamic modelling for persistent event count time series. *Am. J. Political Sci.* **44**(4), 823–843.

Brown, H. and Prescott, R. (1999) *Applied Mixed Models in Medicine*. New York, NY: Wiley.

Butler, S. and Louis, T. (1992). Random effects models with non-parametric priors. *Stat. in Med.* **11**, 1981–2000.

Cameron, A. and Trivedi, P. (1998) *Regression Analysis of Count Data*. Econometric Society Monographs, 30. Cambridge: Cambridge University Press.

Chamberlain, G. and Hirano, K. (1999) Predictive distributions based on longitudinal earnings data. *Annales d'Economie et de Statistique* **55**, 211–242.

Chib, S. (1995) Inference in panel data models via Gibbs sampling. In: Matyas, L. and Sevestre, P. (eds.), *The Econometrics of Panel Data: A Handbook of the Theory with Applications*. Boston, MA: Kluwer.

Chib, S., Greenberg, E. and Winkelmann, R. (1998) Posterior simulation and Bayes factors in panel count data models. *J. Econometrics* **86**, 33–54.

Chib, S. and Carlin, B. (1999) On MCMC sampling in hierarchical longitudinal models. *Stat. and Comput.* **9**, 17–26.

Congdon, P. (2001) Bayesian models for suicide monitoring. *Euro. J. Population* **16**, 251–284.

Crouchley, R., Davies, R. and Pickles, A. (1982) A reexamination of Burnett's study of Markovian models of movement. *Geogr Anal* **14**(3), 260–262.

Curran, P. and Bollen, K. (2001) The best of both worlds: combining autoregressive and latent curve models. In: Collins, L. and Sayer, A. (eds.), *New Methods for the Analysis of Change*. Washington, DC: American Psychological Association, pp. 107–135.

Dagne, G. (1999) Bayesian analysis of hierarchical Poisson models with latent variables. *Comm. Stat. Theory Meth.* **28**(1), 119–136.

Davies, R., Crouchley, R. and Pickles, A. (1982) Modelling the evolution of heterogeneity in residential mobility. *Demography* **19**, 291–299.

Diggle, P. (1988) An approach to the analysis of repeated measurements. *Biometrics* **44**, 959–971.

Diggle, P. (1998) Dealing with missing values in longitudinal studies. In: Everitt, B. S. and Dunn, G. (eds.), *Statistical Analysis of Medical Data: New Developments*. London: Arnold.

Diggle, P., Liang, K. and Zeger, S. (1994) *Analysis of Longitudinal Data*. Oxford: Oxford University Press.

Diggle, P. and Kenward, M. (1994) Informative drop-out in longitudinal data analysis. *J. Roy. Stat. Soc., Ser. C* **43**(1), 49–93.

Escobar, M. and West, M. (1998) Computing nonparametric hierarchical models. In: Dey, D. et al. (eds.) *Practical Nonparametric and Semiparametric Bayesian Statistics*. New York, NY: Springer.

Fitzmaurice, G. and Lipsitz, S. (1995) A model for binary time series data with serial odds ratio patterns. *J. Roy. Stat. Soc., Ser. C* **44**(1), 51–61.

Gelfand, A. and Dey, D. (1994) Bayesian model choice: asymptotics and exact calculations. *J. Roy Stat. Soc. B*, **56**, 501–514.

Gelfand, A. and Ghosh, S. (1998) Model choice: A minimum posterior predictive loss approach. *Biometrika* **85**, 1–11.

Granger, R. and Huang, L. (1997) Evaluation of Panel Data Models: Some Suggestions from Time Series. University of California, San Diego Department of Economics, Discussion Papers DP 97–10.

Hamerle, A. and Ronning, G. (1995) Panel analysis for qualitative variables. In: Arminger, G., Clogg, C. and Sobel, M. (eds.), *Handbook of Statistical Modeling for the Social and Behavioral Sciences*. Plenum.

Hand, D. and Crowder, M. (1996) *Practical Longitudinal Data Analysis*. Texts in Statistical Science Series. London: Chapman & Hall.

Hausman, J. and Wiseman, A. (1979) Attrition bias in experimental and panel data: the Gary Income Maintenance Experiment. *Econometrica* **47**, 455–473.

Hausman, J., Hall, B. and Griliches, Z. (1984) Econometric-models for count data with an application to the patents R and D relationship. *Econometrica* **52**(4), 909–938.

Heckman, J. (1976) The common structure of statistical models of truncation, sample selection, and limited dependent variables, and a simple estimator for such models. *Ann. Economic and Social Measure*. **5**, 475–492.

Heckman, J. (1981) Statistical models for discrete panel data. In: Manski, C. and McFadden, D. (eds.), *Structural Analysis of Discrete Data with Econometric Applications*. Cambridge, MA: MIT Press, pp. 114–178.

Hickman, J. and Miller, R. (1981) Bayesian bivariate graduation and forecasting. *Scand. Actuar. J.* **1981**, 129–150.

Hirano, K. (1998) A semiparametric model for labor earnings dynamics. In: Dey, D. *et al.* (eds.), *Practical Nonparametric and Semiparametric Bayesian Statistics*. New York, NY: Springer, pp. 355–369.

Hirano, K. (2000) Semiparametric Bayesian inference in autoregressive panel data models. *Econometrica* **70**, 781–799.

Horrace, C. and Schmidt, P. (2000) Multiple comparisons with the best, with economic applications. *J. Appl. Econometrics* **15**, 1–26.

Ibrahim, J. Chen, M. and Sinha, D. (2001) Criterion-based methods for Bayesian model assessment. *Stat. Sin.* **11**(2), 419–443.

Johnson, V. and Albert, J. (1999) *Ordinal Data Modeling*. New York, NY: Springer.

Karney, B. R. and Bradbury, T. N. (1995) The longitudinal course of marital quality and stability: A review of theory, method, and research. *Psychological Bulletin*, **118**, 3–34.

Karlsson, S. (2002) Bayesian Methods in Econometrics: Panel Data. Manuscript, Department of Economic Statistics, Stockholm School of Economics.

Kuo, L. and Mallick, B. (1998) Variable selection for regression models. *Sankhya* **60B**, 65–81.

Lee, J. and Geisser, S. (1996) On the prediction of growth curves. In: Lee, J. *et al.* (eds.), *Modelling and Prediction*. Berlin: Springer.

Lee, J. and Hwang, R.(2000) On estimation and prediction for temporally correlated longitudinal data. *J. Stat. Plann. Inference* **87**(1), 87–104.

Little, R. (1993) Pattern-mixture models for multivariate incomplete data. *J. Am. Stat. Assoc.* **88**(421), 125–134.

Little, R. and Rubin, D. (1987) *Statistical Analysis with Missing Data.* New York, NY: Wiley.

Long, J. (1997) *Regression Models for Categorical and Limited Dependent Variables.* Advanced Quantitative Techniques in the Social Sciences, Volume 7. London: Sage.

Potthoff, R. F. and Roy, S. (1964) A generalized multivariate analysis of variance model useful especially for growth curve problems. *Biometrika* **51**, 313–326.

Rahiala, M. (1999) Random coefficient autoregressive models for longitudinal data. *Biometrika* **86**(3), 718–722.

Rubin, D. (1976) Inference and missing data. *Biometrika* **63**, 581–592.

Sahu, S., Dey, D., Aslanidou, H. and Sinha, D. (1997) A Weibull regression model with gamma frailties for multivariate survival data. *Lifetime Data Anal.* **3**, 123–137.

Sethuraman, J. (1994) A constructive definition of Dirichlet priors. *Stat. Sin.* **4**(2), 639–650.

Sethuraman, J. and Tiwari, R. (1982) Convergence of Dirichlet measures and the interpretation of their parameter. *Statistical Decision Theory and Related Topics III, Proc. 3rd Purdue Symposium.*

Shiller, R. (1973) A distributed lag estimator derived from smoothness priors. *Econometrica* **41**, 775–788.

Ware, J. (1985) Linear models for the analysis of longitudinal studies. *The Am. Stat.* **39**(2), 95–101.

Winer, R. S. (1983). Attrition bias in econometric models estimated with panel data. *J. Marketing Res.* 177–186.

Winkelmann, R. (2000) *Econometric Analysis of Count Data, 3rd ed.* Berlin: Springer.

EXERCISES

1. In Example 6.1, try a bivariate Student t via a scale mixture (see Program C). How does the predictive loss criterion change, and which patient has the lowest weight? Try also setting the degrees of freedom (df) in the Bivariate Student to be an extra parameter.

2. In Example 6.2, apply the DPP model, where the baseline prior has a single variance term, namely $\phi_j = \phi$ in Equation (6.15). How does fit compare with the model assuming component specific variances?

3. In Example 6.3, investigate whether an improved fit results from making all the lag coefficients random (with all mean lag coefficients also free parameters).

4. In Example 6.5, try replicating the model of Hedeker and Gibbons (1994) which reduces the ordinal outcome to four categories: I (normal or borderline, combining categories 1 and 2 of the 7 category outcome); II (mildly or moderately ill, combining categories 3 and 4 of the 7 category outcome); III (markedly ill, category 5 of the 7 category outcome); IV (severely or extremely ill, combining categories 6 and 7 of the 7 category outcome). How much does fit deteriorate through making this simplification?

5. In Example 6.5, try a scale mixture version of the random effects model, with degrees of freedom an unknown, and with the full ordinal scale. The analysis may be sensitive to the prior adopted for the degrees of freedom (e.g. a uniform between 4 and 100, a discrete prior with a grid of values over the same range). Are there any clear outliers (scale factors clearly under one)?

6. In Example 6.6, try the Albert–Chib model with Normal sampling of the Z_{it}, i.e. equivalent to a probit link. Does this affect fit or other inferences as compared to the Student $t(8)$ model? Also, try both the Student t and Normal sampling models when

the lag is in the Z_{it} themselves rather than the error. How does this compare in fit to the autocorrelated error model?

7. In Example 6.8, try prediction of the missing dental growth curve data using AR(1) error autocorrelation parameters specific to gender. Does this improve predictive accuracy?

8. In Example 6.8, try estimating the alternative autoregression model suggested above involving a multivariate t to model the latent y_{i0}.

CHAPTER 7

Models for Spatial Outcomes and Geographical Association

7.1 INTRODUCTION

The analysis of spatial data has involved recent advances in several fields, to some extent referring to a central core of knowledge, but showing many distinct features in the specialisms involved. Thus, many Bayesian applications have occurred in spatial epidemiology, with Lawson (2001), Lawson *et al.* (1999) and Elliott *et al.* (2000) providing state-of-the-art discussion. One may say that here a major element is the assessment of patterns of relative disease risk in terms of possible clustering, perhaps around environmental point sources, but also in terms of ecological regression of disease patterns in terms of known risk factors (Lawson, 2001, p. 5). A more long-standing tradition of spatial modelling has occurred in spatial econometrics, with Anselin (2001) and Anselin and Florax (2002) providing recent overviews, and with Lesage (1999) providing a review of Bayesian principles in this area. Here the major emphasis lies in describing behavioural relationships by regression models, whether the data are defined over regions and areas, or at the level of individual actors (house purchasers, firms, etc.) involved in spatially defined behaviours. A third major specialism occurs in geostatistics, where a continuous spatial framework is adopted and the goal is often to smooth or interpolate between observed readings (e.g. of mineral concentrations) at sampled locations. Providing a common thread is a central core of spatial statistics, exemplified by the works of Ripley (1981) and Cressie (1993).

One by-product of developments proceeding in sometimes disparate areas is that certain terms (e.g. heterogeneity) may be defined differently according to field of application. On the other hand, there is also often considerable benefit in applying concepts across these areas, for example using geostatistical ideas in ecological regression or in examining spatial disease patterns (Oliver *et al.*, 1992). The present chapter seeks to provide a selective introduction to some of the modelling issues involved in these diverse areas; it is inevitably selective and partial, but hopes to identify some underlying common themes.

The contrasting concerns of the major specialisms may be illustrated by a distinction sometimes drawn between models focused on spatial interactions, and those concerned with spatial disturbances, meaning regression errors (Doreian, 1982). In spatial effect or spatial interaction models, the spatial pattern or space-time pattern is the main focus of the analysis, for instance when the realised patterns reflect causal processes. Thus

Applied Bayesian Modelling P. Congdon
 ISBN: 0-471-48695-7

investigations into 'focused' clustering of excess mortality or illness around pollution point sources typically involve a hypothesis that exposure increases nearer to the source resulting in raised disease incidence (Section 7.6). Similarly, in spatial econometrics, spatial autoregressive models in the observations themselves are commonplace (e.g. see Anselin, 2001), for instance reflecting behavioural processes of spatial diffusion.

However, in many regression applications in spatial epidemiology involving aggregate disease counts in areas, spatial interaction models are relevant only for certain outcomes, such as infectious diseases (Cliff and Ord, 1981). In these applications spatial dependence[1] occurs because of omitted or unmeasured spatially correlated predictors, and so is reflected in regression errors, causing departures from the independent errors assumption of the conventional regression. In problems involving both space and time dimensions (Section 7.7), errors may be correlated in both time and space simultaneously (Lagazio *et al.*, 2001). An example of incorrect inferences when the outcomes or errors in a regression are spatially correlated is that such correlation reduces the amount of independent evidence available to model the process under investigation, and may, if not allowed for, lead to over-estimation of the significance of regression relationships (Richardson and Monfort, 2000, p. 211).

An additional issue raised clearly by writers such as Fotheringham *et al.* (2000) and Lesage (1999) is that of spatial heterogeneity[2], either in regression relationships (e.g. in terms of regression coefficients varying over space) or in terms of heteroscedasticity in a spatially unstructured error term. There may be identifiability problems in separating spatial dependence (e.g. correlation) from spatial heterogeneity (de Graaff *et al.*, 2001; Anselin, 2001).

Consider how one might allow for spatial correlation in regression errors (though heterogeneity is considered in Section 7.5). Whereas the standard linear model for metric outcomes would assume independence in the errors of e_i and e_j, with $i \neq j$ being different areas, alternatively suppose the errors of area i are not necessarily independent of the errors in other areas. Thus, let Y and ε be $n \times 1$, x be $n \times p$ and assume a linear regression

$$Y = x\beta + e \tag{7.1a}$$

with a joint prior specifying multivariate dependence between the errors, such as an MVN model with

$$e \sim N_n(0, \Sigma) \tag{7.1b}$$

where the off-diagonal elements in the $n \times n$ covariance matrix Σ reflect the spatial patterns in the data. A range of techniques (e.g. variogram analysis) explore covariation in the outcomes or in regression residuals, in relation to inter-area or inter-point distances, so as to estimate parameters in the covariance matrix (see Section 7.4). For example, Cook and Pocock (1983) use a form of variogram analysis to define the covariance matrix Σ in terms of an overall prior variance σ^2, when $d_{ij} = 0$ and a parameter γ reflecting off-diagonal spatial dependence between areas:

[1] Following Anselin (2001), spatial dependence may be considered as any departure from the independent errors model or from models excluding spatial lags in outcomes; thus in spatially dependent models, one may have $\text{cov}(Y_i, Y_j) \neq 0$ for i and j neighbours in some sense, and/or $\text{cov}(e_i, e_j) \neq 0$. The existing spatial literature is dominated by linear models and linear spatial correlation, whereas the broader concept of dependence might include non-linear spatial errors or interactions (de Graaff *et al.*, 2001).

[2] In disease mapping applications, the term excess heterogeneity is often applied to spatially unstructured errors (for Poisson overdispersion) in the log link for count outcomes.

$$\begin{aligned}\Sigma_{ij} &= \sigma^2 r_{ij}\\ r_{ij} &= \sigma^2 \exp(-\gamma d_{ij})\end{aligned} \tag{7.2}$$

where $R = [r_{ij}]$ are correlations. The distance decay parameter γ is expected to be positive so that outcomes or errors in neighbouring areas are more similar. Such 'direct modelling' of the covariance structure is also relevant for spatial interpolation in fields such as geostatistics, where readings of an outcome $Y(s_i)$ may be obtained in two or three dimensional space $s_i = (s_{1i}, s_{2i}, \ldots)$ and the model for covariation between locations s_i and s_j is used to predict Y at an unobserved location.

Alternatively, one may seek a parameterisation of Σ in terms of a known interaction scheme (Bailey and Gattrell, 1995), which may take account of contiguities between areas or distances d_{ij} between them (Section 7.2). This approach results in simplified identifiability of other aspects of the model (Anselin, 2001). The extent of spatial dependence (typically of a linear nature) in the interactions between Y or in the regression errors is modelled in terms of spatial autocorrelation between adjacent areas, usually expected to be positive. Spatial autoregressive models then include one or more unknown autocorrelation parameters ρ, applying either to spatial lags in the outcomes themselves or to the regression errors. In mapping of disease counts (Section 7.3), a particular modelling structure assumes a known interaction scheme, and also that $\rho = 1$ in the errors, and focuses on the relative variability of spatial and unstructured errors (e.g. Best *et al.*, 1999; Mollié, 2000).

7.2 SPATIAL REGRESSIONS FOR CONTINUOUS DATA WITH FIXED INTERACTION SCHEMES

Here we first consider regression models where the interaction between areas is taken as known and the focus is on making correct inferences about regression impacts, on estimating different forms of spatial correlation (e.g. in the data themselves or in the regression errors), or on allowing for spatial heterogeneity. We first focus on the observed continuous outcome case, though it may be noted that the ideas transfer to latent continuous variable models (e.g. when outcomes are binary or ordinal), using, for instance, the sampling methods of Albert and Chib (1993).

Thus, consider an $n \times n$ matrix $\mathbf{C}$ of contiguity dummies, with $c_{ij} = 1$ if areas i and j are adjacent, and $c_{ij} = 0$ otherwise (with $c_{ii} = 0$). Alternatively, a distance based interaction scheme might involve elements such as $c_{ij} = 1/d_{ij}(i \neq j)$ or $c_{ij} = 1/d_{ij}^2$, but again with $c_{ii} = 0$. Then scale the elements to sum to unity in rows, with W as the scaled matrix,

$$W = [w_{ij}] = [c_{ij}/\sum_j c_{ij}]$$

Given the known form of $\mathbf{C}$ and hence W, a model for spatial dependence in the errors for a metric outcome Y might then take the form

$$\begin{aligned}Y &= x\beta + e\\ e &= \rho We + u\end{aligned} \tag{7.3a}$$

where ρ is an unknown correlation parameter, where Y, e and u are vectors of length n, and x is of dimension $n \times p$. Here the u denote spatially unstructured errors, which are frequently taken as homoscedastic $u_i \sim N(0, \sigma^2)$. If the interactions are scaled within rows, then the permissible maximum of ρ is 1 (Anselin, 2001; Bailey and Gattrell, 1995, Chapter 7), and the permissible minimum is the smallest eigenvalue of W, which is greater than -1 but less than 0. Since spatial correlation is positive in the great majority of econometric or health applications, a prior on ρ constrained to [0, 1] is feasible in many applications.

7.2.1 Joint vs. conditional priors

Letting $Q = I - \rho W$, the precision matrix Σ^{-1} of the errors e in Equation (7.3a) may be derived as

$$\sum^{-1} = \tau Q'Q \quad (7.3b)$$

where $\tau = \sigma^{-2}$ (Richardson *et al.*, 1992). A full multinormal scheme for the e could be used, with the errors sampled simultaneously from their joint prior

$$e \sim N_n(0, \Sigma) \quad (7.3c)$$

However, a conditional scheme is possible, and may be simpler to sample from (e.g. in WINBUGS). As mentioned by Wakefield *et al.* (2000), modelling of spatially correlated errors may proceed by initially specifying either the joint multivariate distribution of the vector e, or the univariate density of each areas error, e_i, conditional on the current estimate of errors in other areas ($e_j, j \neq i$). Conditions that ensure the joint density is proper (so that the e_i are identifiable) when the model specification starts with a conditional rather than the joint prior[3] are discussed by Wakefield *et al.* (2000) and Besag and Kooperberg (1995).

One possible conditional prior (the conditional autoregressive or CAR prior) expresses e_i in Equations (7.3) in the centred univariate Normal form

$$e_i \sim N(M_i, \sigma^2) \quad (7.4a)$$

where the mean of each area's error

$$M_i = \rho \sum_j c_{ij} e_j \quad (7.4b)$$

is a weighted average of errors in contiguous areas, and ρ is bounded by the inverses of the minimum and maximum eigenvalues of C. (Note that the interaction matrix for this form of prior needs to be symmetric.) The covariance of the vector e in the corresponding joint prior is then $\Sigma = \sigma^2(I - C)^{-1}$ (Richardson, 1992; Wakefield *et al.*, 2000).

One may also have a 'spatial effects' or 'spatial autoregression' model, with spatial lags in the outcomes themselves (e.g. Ord, 1975; Anselin, 2001), with

$$Y = x\beta + \rho WY + u \quad (7.5)$$

where u is white noise. Spatial dependence in both Y and e may occur in the same model, for example:

$$Y = x\beta + \rho_1 WY + e$$
$$e = \rho_2 We + u$$

[3] The identifiability issue with the ICAR(1) model is discussed in Section 7.2.1.

where e is multinormal as in Equations (7.3). In some situations, a spatial lag in the outcome might not be substantively sensible. If it were, then a sensitivity analysis might consider both correlation in both spatial effects and spatial errors, and also consider non-constant variances in the u_i (Lesage, 2002); it might also encompass several forms of spatial interaction matrix, even if these are taken as fixed within the modelling (Richardson and Monfort, 2000; Anselin, 2001).

It may be noted that the multinormal spatial disturbances model (7.3a) may be expressed as

$$Y - \rho Wy = x\beta - \rho Wx\beta + u \tag{7.6}$$

namely as a regression of the 'filtered' outcome $Y^* = Y - \rho Wy$ on the filtered predictors $x^* = x - \rho Wx$. So the pure spatial effects model (7.5) is in fact a particular case of (7.6).

For continuous outcomes, it is in fact simpler in WINBUGS to use this filtered outcome and predictor approach[4]. Thus, in Example 7.1, the means for a metric outcome Y_i are $\mu_i = \rho Wy + x_i\beta - \rho Wx\beta$. With count or binomial outcomes, by contrast, one might have an error directly in the form (7.3) or (7.4) in the regression link. Thus, if $Y_i \sim \text{Poi}(\mu_i)$, one might take

$$\log(\mu_i) = x_i\beta + e_i$$

$$e_i \sim \text{N}(M_i, \sigma^2)$$

where M_i is as in (7.4b).Example 7.3 accordingly contains a CAR prior analysis of the Scottish lip cancer data.

Whatever the form of distribution relevant to the outcome, there are possible issues of identifiability involved[5], since for instance clustering of regression errors may be produced by a form of spatial heterogeneity, in that error variances differ between subsets of regions. Suppose, following Anselin (2001), spatial heterogeneity is taken to refer either to non-constant error variances (heteroscedasticity) or to non-constant regression coefficients across space, sometimes known as 'spatial regimes' (Anselin, 1990). Then in a single cross-section, spatial autocorrelation and spatial heteroscadisticity are observationally equivalent. Thus, a clumping of positive residuals in a set of neighbouring areas might reflect either mechanism.

Spatial heteroscedasticity may be parameterised in various ways: either one may suppose all areas to have distinct variances, or there may be groups of areas S_r with $\sigma_i^2 = \sigma_r^2$ if $i \in S_r$. Lesage (2000) proposes scale mixtures (with each area having its own variance) to robustify inferences against outlier data points. This applies even after

[4] This is because for continuous data following a Normal or Student t density, a centred form is adopted in WINBUGS which assumes unstructured errors.

[5] Highly parameterised but less well identified models (e.g. combining both spatial heterogeneity and clustering) can of course be rejected by model fit criteria which penalise complex models which do not add greatly to accurate prediction. This will be illustrated by loss of precision in fitted means or predictions of new data. One approach to such predictive fit is that of Gelfand and Ghosh (1998), Sahu *et al.* (1997) and Ibrahim *et al.* (2001). Let f_i be the observed frequencies, ϕ be the parameters, and z_i be 'new' data sampled from $f(z|\phi)$. Suppose ν_i and ς_i are the posterior mean and variance of z_i, then one possible criterion for any $w > 0$ is

$$D.z = \sum_{i=1}^{n} \varsigma_i + [w/(w+1)] \sum_{i=1}^{n} (\nu_i - f_i)^2$$

spatial autocorrelation is explicitly modelled, and so would relate to the density assumed for the u_i. Thus, a scale mixture on the σ^2 in (7.3) or (7.4) is equivalent to assuming Student t rather than normal u_i. Such a mixture is illustrated in Example 7.2 (involving the Columbus Crime dataset).

7.3 SPATIAL EFFECTS FOR DISCRETE OUTCOMES: ECOLOGICAL ANALYSIS INVOLVING COUNT DATA

Within the last decade, models for spatial dependence in discrete data (e.g. count or binary) have seen a major development. Spatial dependence figures strongly in the analysis of disease maps, where event counts rather than metric outcomes are the usual focus and where much recent conceptual and methodological development has a Bayesian orientation (e.g. Besag *et al.*, 1991). In spatial econometrics, by contrast, more attention has focused on the spatial probit model for binary outcomes (e.g. Lesage, 2000).

In epidemiological analysis of small area disease data, the main object is often to estimate the true pattern of relative risk in the face of overdispersion in the observed event counts and spatially correlated errors due to omitted predictors. Estimation of relative risks by conventional methods based on the Poisson density (e.g. by standard mortality ratios defined as ratios of observed to expected events) assumes that the disease or mortality risk is constant over areas and over individuals within areas. In fact, individual risks may vary within areas, and risks vary between areas, so that observed event counts show a greater variability than the Poisson stipulates. This variation can be modelled by expressing area relative risks in terms of one or more random effects.

Some such effects may be spatially unstructured (analogous to white noise in time series), and these are sometimes denoted as 'excess heterogeneity' (e.g. Best *et al.*, 1999, p. 132). However, overdispersion may also occur due to spatially correlated effects. As mentioned above, spatial epidemiology often focuses on clustering of adverse health events due to socio-economic or environmental factors; observations on the latter are frequently not available or at least, not at sufficiently low spatial resolation. The prior model for spatially clustered sources of overdispersion may be seen as proxying unobserved risk factors (e.g. environmental or cultural) which are themselves spatially correlated (Best, 1999).

For example, suppose a count of deaths D_i from a certain cause is observed in a set of small areas, and that expected deaths E_i (in the demographic sense) are derived from some standard schedule of death rates from the cause concerned. The outcomes may, subject to the necessity to take account of overdispersion, be taken as Poisson,

$$D_i \sim \text{Poi}(E_i\theta_i) \tag{7.7}$$

where θ_i is the relative risk of mortality in area i. Poisson sampling may be justified by considering binomial sampling of deaths by age D_{ij} in relation to populations by age P_{ij} with death rates π_{ij}, and by assuming relative risks and age rates are proportional, namely $\pi_{ij} = \theta_i\pi_j$ (Wakefield *et al.*, 2000).

Consider the Poisson sampling model as it stands, with the θ_i initially taken as fixed effects. Then with flat priors on them, the estimated θ_i will be equal to the maximum likelihood estimates of SMRs, namely $\hat{\theta}_i = D_i/E_i$. However, these may be unreliable as estimates of relative risk (e.g. see Bernardinelli and Montomoli, 1992), since maps of SMRs are subject to distortion through low event counts or populations at risk, and small changes in event totals may produce major shifts in the SMRs. In devising models to 'pool strength' and reduce such anomalies, one may envisage the total variability in the observed rates or SMRs having two components: within area sampling variation around the true underlying rate, and between area variations in the true rates, which are likely to show spatial correlation to some degree.

To allow for unstructured 'white noise' variability about the true rates, one might take the log of the risk θ_i to consist of an overall average and a white noise error u_i. The model is then

$$D_i \sim \text{Poi}(E_i\theta_i) \tag{7.8a}$$

$$\log(\theta_i) = \gamma + u_i \tag{7.8b}$$

$$u_i \sim \text{N}(0, \sigma_u^2) \tag{7.8c}$$

where γ is an intercept representing the log of the average relative risk in the areas compared to the standard population. However, there may in addition to spatially unstructured variability be spatially correlated errors e_i, so that (7.8b) becomes

$$\log(\theta_i) = \gamma + u_i + e_i \tag{7.9}$$

This is sometimes known as the 'convolution' model.

7.3.1 Alternative spatial priors in disease models

There are several alternative specifications for the e_i. One option suggested for disease mapping and ecological regression (Leyland *et al.*, 2000) assumes[6] that underlying the spatial errors e are unstructured errors v. Under this approach joint densities may then be readily specified for u and v. So

$$e_i = \sum_{j=1}^{n} w_{ij} v_j \tag{7.10}$$

where the w_{ij} are row standardised interactions as above. If the w_{ij} are based on contiguity, then $w_{ij} = 1/N_i$ if areas i and j are adjacent, with N_i being the number of neighbours of area i. In this case of contiguity interactions, $e_i = N_i^{-1}\Sigma_{j\varepsilon L_i} v_j$, with L_i denoting the neighbourhood of areas adjacent to i.

In the convolution model (7.9), one might take u_i and v_i to be bivariate Normal with covariation σ_{uv}, but this modelling approach extends readily to multivariate error forms, for outcomes $j = 1, .. J$. For instance, an MVN prior of dimension $2J$ allows correlation between outcome specific errors u_{ij} and v_{ij}, and so expresses interdependence (in regression errors) between the outcomes (Congdon, 2002a). This is illustrated below with data for Glasgow postcodes and counts from two causes of death.

As for continuous outcomes, full joint prior specifications for the spatial errors e_i may be proposed or conditional priors specified *ab initio*. One possible joint density specification (Besag *et al.*, 1991) often used in health applications is in terms of pairwise differences in errors, so that

[6] This type of error structure may also be adopted for continuous outcomes (e.g. Kelejian and Robinson, 1993), but a particular recent application is to disease counts.

$$p(e) \propto \exp\left[-0.5\kappa^{-2}\sum_{i<j}(e_i - e_j)^2\right] \quad (7.11a)$$

The corresponding conditional form (Wakefield *et al.*, 2000, Equation (7.15); Bernardinelli *et al.*, 1999, Equation (26.4)) is

$$P(e_i|e_j, j \neq i) \sim \mathrm{N}(M_i, \sigma_i^2) \quad (7.11b)$$

with

$$M_i = \sum_{j\neq i} c_{ij}e_j / \sum_{j\neq i} c_{ij} = \sum_{j\neq i} w_{ij}e_j$$

and c_{ij} being spatial interactions as above. The variances differ by area with

$$\sigma_i^2 = \kappa^2 / \sum_{j\neq i} c_{ij}$$

This scheme, known as an intrinsic conditional autoregression or ICAR(1) prior (since the correlation ρ is set by default to 1), is improper, since it specifies only differences in log relative risks, not levels.

Suppose the c_{ij} form a contiguity matrix with $c_{ij} = 1$ if areas i and j are adjacent and $c_{ij} = 0$ otherwise. In this case, the ICAR(1) model for the e_i has parameters

$$\begin{aligned} M_i &= \bar{e}_i \\ \sigma_i^2 &= \kappa^2/N_i \end{aligned} \quad (7.11c)$$

where $\bar{e}_i$ is the average of the e_j in the locality L_i of area i (with the average excluding e_i itself), and N_i is the number of neighbouring areas in L_i. Note that Student t errors e_i may be achieved by a scale mixture on κ^2 in Equation (7.11c), that is $\kappa_i^2 = \lambda_i\kappa^2$, where $\lambda_i \sim G(0.5\nu, 0.5\nu)$ and ν is the degrees of freedom in the Student t density.

Other pairwise difference priors are possible. Besag *et al.* (1991) mention a double exponential (Laplace) prior

$$p(e) \propto \chi \exp\left[-0.5\chi\sum_{i<j}|e_i - e_j|^2\right]$$

which, like the Student t, is more robust to outliers or discontinuities in the risk surface. χ is a scaling parameter, with smaller values implying smaller spatial variability.

The fact that Equation (7.11a) is improper may lead to problems in convergence and identifiability in Bayesian estimation based on repeated sampling. One way of producing identifiability is to omit the constant (such as γ in Equation (7.9)) so that the average of the e_i defines the level. Another is to constrain the e_i to sum to zero, which in practice involves centreing at each iteration in an MCMC run (Ghosh *et al.*, 1998). This identifiability option for the ICAR(1) model with e normal (i.e. κ^2 constant over areas) is implemented in WINBUGS13 (and subsequent versions) as the 'carnormal' density. The Laplace pairwise difference prior is available in WINBUGS as the 'carl1' density.

Another strategy involves a model redefinition. Thus, following Sun *et al.* (1999, 2000), propriety of the posterior is obtained by explicitly introducing a spatial correlation parameter ρ absolutely less than 1. So for contiguity interactions, $M_i = \rho\bar{e}_i$ and

$$P(e_i|e_j, j \neq i) \sim \mathrm{N}(\rho\bar{e}_i, \kappa^2/N_i) \quad (7.12)$$

(compare Sun *et al.*, 2000, Equation (2)). Whereas prior (7.11) is then an ICAR(1), the more general approach is denoted an ICAR(ρ) prior.

Consider the 'convolution' model of Equation (7.9), where e_i is based on the pairwise differences prior (7.11a) and an ICAR(1) model is assumed. To assess the relative strength of spatial and unstructured variation in this model requires estimates of marginal rather than conditional variances, and so one may form 'moment estimates' of the marginal variances, such as

$$\hat{\sigma}^2 = \sum (e_i - \bar{e})^2/(n-1)$$

This may be compared with the posterior estimate of the marginal variance of the u_i, or with a comparable moment estimator

$$\hat{\sigma}_u^2 = \sum (u_i - \bar{u})^2/(n-1)$$

There may be identifiability problems in the convolution model, especially if flat priors are taken on the overall variances (precisions) of both the u_i and e_i. The posterior results are dependent on the prior specifications regarding the balance of dispersions on the two components, especially if, say, the precision of the unstructured errors was specified to be relatively high. Based on a simulation analysis, Bernardinelli *et al.* (1995) recommend that $\text{var}(u_i) = 0.7\ \text{var}(e_i)$, where $\text{var}(e_i)$ is the conditional variance κ^2; this is known as the 'balanced dispersion' prior. More generally, priors for the variances of e_i and u_i might reflect their interdependence, e.g. by taking a prior on the ratio $\text{var}(e)/\text{var}(u)$ on the ratio $\text{var}(e)/[\text{var}(e)+\text{var}(u)]$ (see Example 1.3).

7.3.2 Models recognising discontinuities

It may be noted that doubts remain about the ability of the convolution specification to reproduce discontinuities in disease maps (e.g. a low mortality area surrounded by high mortality areas will have a smoothed rate distorted by a spatially correlated error model). Forms of discrete mixture have been proposed as more appropriate to modelling discontinuities in high disease risk (Militino *et al.*, 2001). For example, Knorr-Held and Rasser (2000) propose a scheme whereby at each iteration of an MCMC run, areas are allocated to clusters. These are defined by cluster centres and surrounding contiguous areas, and have identical risk within each of them. Clusters may be redefined at each iteration. The estimated relative risk for each area, averaged over all iterations, is then a form of non-parametric estimator, and may better reflect discontinuities.

Lawson and Clark (2002) propose a mixture of the ICAR(1) and Laplace priors, with the mixture defined by a continuous (beta) density rather than a discrete mixture. So Equation (7.9) becomes

$$\log(\theta_i) = \gamma + u_i + \eta_i e_i + (1 - \eta_i) f_i$$

where f_i follows the conditional Laplace form, and one might take the beta prior on the η_i to have fixed parameters, for instance $\eta_i \sim \text{Beta}(1, 1)$ or $\eta_i \sim \text{Beta}(0.5, 0.5)$. Options on such a scheme include setting priors on the hyperparameters $\{a, b\}$ in the beta mixture $\eta_i \sim \text{Beta}(a, b)$, or simpler mixture forms, such as just putting more emphasis on the unstructured component in some 'discontinuous' areas:

$$\log(\theta_i) = \gamma + \eta_i u_i + (1 - \eta_i) e_i$$

7.3.3 Binary outcomes

For binary data defined over areas $i = 1, .. n$

$$Y_i \sim \text{Bern}(\pi_i)$$

one might model spatial and nonspatial error effects via the logit or probit links on π_i. For binary data the variance is not identifiable, so it is preset, usually to 1. One might, for instance, specify a model with a CAR prior on a spatial error, so that

$$\text{logit}(\pi_i) = x_i\beta + e_i$$

where

$$e_i \sim N(M_i, 1)$$

and where the mean of each area's error is

$$M_i = \rho \sum_j c_{ij} e_j$$

An ICAR(1) or ICAR(ρ) model could also be specified for the e_i in this model. If both unstructured and structured errors (u_i and e_i) are envisaged, then their respective variances would be shares of an overall known quantity, for example $\text{var}(u) + \text{var}(e) = 1$.

The majority of recent Bayesian development has been on the latent variable model and on the multinormal spatial error model, as in Equation (7.3). Thus,

$$Y_i = 1 \text{ if } Y_i^* > 0$$
$$Y_i = 0 \text{ if } Y_i^* < 0$$

where

$$Y_i^* = x_i\beta + e_i$$
$$e_i = \rho \sum_j w_{ij} e_j + u_i$$
$$u_i \sim \text{N}(0, 1)$$

The vector of errors then has a multinormal form,

$$e \sim \text{MVN}(0, [(\text{I} - \rho W)'(\text{I} - \rho W)]^{-1})$$

Lesage (2000) emphasises possible heteroscedasticity in binary data, so that $u_i \sim \text{N}(0, \kappa_i)$, where the $\kappa_i \sim G(\nu, \nu)$ average 1 and ν is the Student degrees of freedom.

Example 7.1 Agricultural subsistence and road access The first worked example considers spatial dependence in the errors of a regression model for a continuous outcome, using the alternative 'filtered' form of the spatial errors model in Equation (7.6). Several studies have considered a dataset for the $i = 1, .. 26$ Irish counties relating the proportion Y_i of the county's agricultural output consumed by itself (i.e. subsistence) to a measure of its Arterial Road Accessibility (ARA); a normal approximation is generally adopted to this binomial outcome. The data is discussed and analysed in Cliff and Ord (1981). A linear model containing homoscedastic errors $u_i \sim \text{N}(0, \sigma_u^2)$ and with no allowance for spatial dependence

$$Y_i = x_i\beta + u_i \quad (7.13)$$

(with $x_{0i} = 1$) then serves as the baseline.

As one model diagnostic (though not a model choice criterion), measures of spatial interaction such as Moran's I may be monitored. Here the intention is to assess in the context of Bayesian sampling estimation how and whether such interaction measures are affected by different model spefications. Note that these are derived statistics, and there is no need to specify a prior on them; for example, in Equation (7.13) only priors on β and the variance (or in BUGS the precision) of the u_i are required. In their original applications via classical (e.g. ML) procedures, estimates of such interaction measures may have been based on regression residuals and their estimated standard error obtained by assuming particular modes of data generation. The Bayesian sampling perspective means one is not constrained in this way, and posterior standard errors are obtained by repeated sampling.

Here such an approach is exemplified using regression residuals as monitored over $t = 1, .., T$ iterations, for area i, denoted $u_i^{(t)}$. Then the sampling average of Morans I statistic is obtained as

$$I = T^{-1} \sum_t \sum_{i \neq j} w_{ij} u_i^{(t)} u_j^{(t)} / \sum_{i \neq j} [u_i^{(t)}]^2$$

where in the present application, two definitions of (row standardised) interactions w_{ij} are considered. One is based on simple contiguities, the other on standardised weights based on intercounty distances d_{ij} and length of shared boundary, so that $c_{ij} = B_{ij}/d_{ij}$, where B_{ij} is the proportion of the boundary of county i in contact with county j, and then $w_{ij} = c_{ij}/\Sigma_j c_{ij}$. The latter interactions are supplied by Cliff and Ord (1981, p. 229).

The Moran statistic typically has a small negative expectation, when applied to regression residuals (Cliff and Ord, 1981). One may be confident, however, that lower values of this statistic represent lesser autocorrelation, despite a caveat that the value of $I = 0$ does not correspond to the lowest possible autocorrelation (see Haggett *et al.*, 1977, p. 357). Note that one might also apply such measures to the unstructured errors for a discrete outcome, such as a disease count response where the model omitted spatial effects as in Equation (7.8b).

With the baseline uncorrelated errors model, three chains are run for 15 000 iterations and posterior summaries are based on the last 14 000 of these[7]. The monitoring includes Moran statistics for the regression residuals as in Table 7.1. These are similar to those cited by Cliff and Ord, for contiguity weights, namely 0.397 (s.e. = 0.12) and for standardised weights, namely 0.436 (s.e. 0.14). Although the linear regression has high predictive R^2, there is under-estimation of subsistence in the remoter counties, and over-estimation of subsistence in the less isolated eastern counties, with better road and rail links. One option would be to include measures of such transport access, e.g. whether a county is served by a direct freight link to the Irish capital, Dublin.

However, to make correct inferences about the regression estimate of subsistence on ARA, it is probably necessary to explicitly model the spatial dependence in the regression errors. Here contiguity weights are used in the spatial errors model of Equation (7.3)

[7] As elsewhere, outlier status is assessed by the CPO statistics obtained by the method of Gelfand and Dey (1994, Equation (26)); the product of these statistics (or the sum of their logged values) gives a marginal likelihood measure, leading to a pseudo-Bayes factor (Gelfand, 1996). The CPOs may be scaled as proportions of the maximum giving an impression of points with low probability of 'belonging' to the main data set.

Table 7.1 Models for subsistence rates

	Mean	St. devn.	2.5%	97.5%
Uncorrelated error model				
Moran(Distance-Boundary Weights)	0.45	0.12	0.21	0.69
Moran(Contiguity)	0.35	0.14	0.09	0.63
β_0 (Intercept)	−8.71	3.56	−15.62	−1.71
β_1 (ARA)	0.0053	0.0008	0.0038	0.0069
Spatial errors model				
Moran(Distance-Boundary Weights)	0.064	0.11	−0.101	0.324
β_0' (Intercept)	0.475	0.688	−0.797	1.926
β_1 (ARA)	0.0021	0.0007	0.0007	0.0036
ρ	0.914	0.08	0.703	0.997

and with the regression means based on the transformed model in Equation (7.6). For improved identification, the intercept parameter[8] is represented as $\beta_0' = \beta_0 - \beta_0\rho$.

A run of 20 000 iterations over three chains with 5000 burn-in ensures convergence in ρ, with its posterior density as in Figure 7.1. The median of 0.936 compares to a Bayes mode of 0.938 cited by Hepple (1995). The effect of ARA on subsistence is halved, and the Moran statistic based on the alternative spatial interaction definition (distance and common boundary) is clearly reduced in line with eliminating spatial dependence in the regression residuals (see Table 7.1).

At individual area level, the CPO estimates do not show any clear outliers; the lowest scaled CPO is 0.09 for county Mayo (area 16) which has its subsistence rate under-predicted. The overall marginal likelihood based on these CPOs is −62.7, a clear reduction compared to that for the uncorrelated error model in Equation (7.13) (namely −72.8).

Example 7.2 Columbus neighbourhood crime Anselin (1988) considers data on crime rates Y_i (burglaries and vehicle thefts per 1000 households) in 49 neighbourhoods of Columbus, Ohio, with predictors income (X_1) and housing values (X_2) in thousands of dollars. He considers first an uncorrelated errors regression which gives, with standard errors in brackets,

$$\begin{array}{cccc} Y = 68.6 & -\ 1.60X_1 & -\ 0.27X_2 \\ (4.7) & (0.33) & (0.10) \end{array} \tag{7.14a}$$

By contrast, two spatial models may be considered, namely a spatial autoregressive model[9] and a spatial disturbances model. Here the latter is estimated, according to

[8] Sampled values of the true intercept $\beta_0 = \beta_0'/(1-\rho)$ will be essentially undefined when sample values of ρ are very nearly 1, and this will affect MCMC convergence. The true intercept may be estimated using posterior means of β_0' and ρ.

[9] Thus,

$$Y = \gamma_0 + \rho_2 WY + \gamma_1 X_1 + \gamma_2 X_2 + u$$

where u might have constant or area specific variance but is spatially uncorrelated. This is estimated by Anselin to have parameters $\rho_2 = 0.43(0.12)$, $\gamma_1 = -1.03(0.31)$ and $\gamma_2 = -0.27(0.09)$.

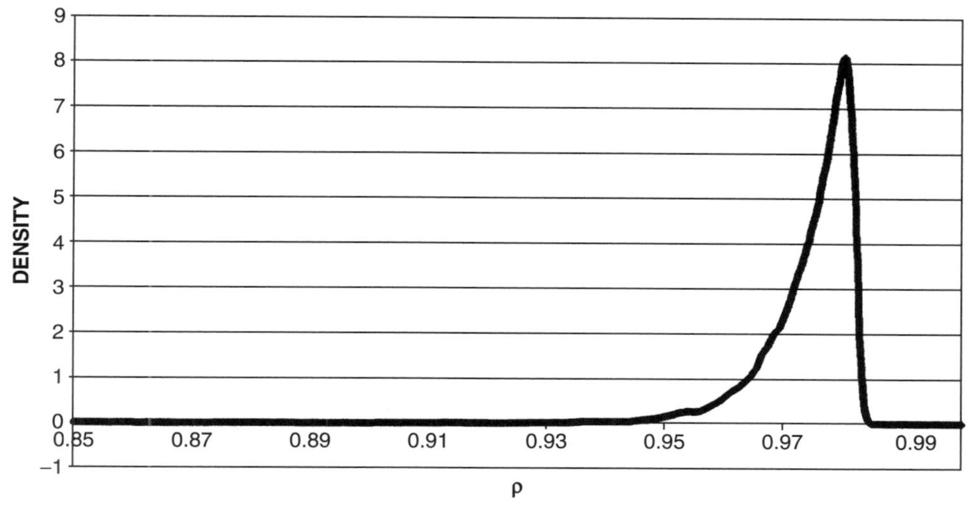

Figure 7.1 Plot of ρ

$$\begin{aligned} Y &= \beta_0 + \beta_1 X_1 + \beta_2 X_2 + e \\ e &= \rho_1 We + u \end{aligned} \tag{7.14b}$$

where u is initially taken as homoscedastic, but then a heteroscedastic alternative to account for outliers is considered. W is a row standardised weights matrix based on contiguities c_{ij}. Anselin employs maximum likelihood estimation of the spatial error model to give $\rho_1 = 0.56$ (s.e.=0.13). The coefficient of income is much reduced as compared to an uncorrelated error model (7.14a), namely to $\beta_1 = -0.94$ (s.e.=0.33), but that of housing value is enhanced, with $\beta_2 = -0.30(0.09)$.

Here the correlated error model is first estimated with u homoscedastic, and with a uniform prior over (0, 1) for ρ_1. N(0, 10) priors are adopted for β_1 and β_2 to avoid numeric overflow. The model for the regression mean is re-expressed according to the transformation (7.6), namely

$$Y = x\beta + \rho W(Y - x\beta) + u \tag{7.15}$$

with $u \sim N(0, \sigma^2)$. Early convergence in a run of 20 000 iterations over three chains is apparent, with iterations from 1001 giving posterior estimates as in Table 7.2.

These estimates are similar to those of Anselin (1988), and confirm the reduced impact of income (as compared to an uncorrelated errors model) when spatial correlation is modelled in conjunction with homogenous non-spatial errors u_i. Examination

Table 7.2 Crime rates, spatial error model, homoscedastic u

	Mean	St. devn.	2.5%	Median	97.5%
β'_0	31.3	13.0	9.3	29.7	60
β_1	−1.05	0.41	−1.87	−1.05	−0.25
β_2	--0.26	0.10	−0.47	−0.26	−0.06
ρ_1	0.55	0.20	0.13	0.56	0.90
σ^2	124.4	27.6	82.1	120.3	190.2

of the CPO statistics suggests area 4 is suspect; this area is aberrant in terms of having the lowest crime rate but the sixth lowest income. (The two are usually inversely associated.)

To allow for the possibility of non-constant variance over space (Lesage, 1999, 1997), and ensure more robust inferences about regression effects in the face of possible outliers, Student t disturbances u_i are adopted in the SAR model. This involves taking the precision of the ith observation

$$\kappa_i \tau$$

where the multiplier

$$\kappa_i \sim G(\nu/2, \nu/2)$$

This scale mixture of Normals is equivalent to a Student t with ν degrees of freedom. The degrees of freedom is also taken as a parameter. Outliers are indicated by low multipliers κ_i.

In addition to a scale mixture, an alternative prior structure on ρ_1 is also proposed as in Lesage (1997). This may be seen as achieving robustness on the spatial interaction component of the model. Hence

$$\pi(\rho_1) \propto (s+1)\rho_1^s \tag{7.16}$$

and for $s > 1$ values of ρ_1 near zero are downweighted. Lesage tries pre-determined values $s = 0$, 0.5, 1, 2 and 5, but here a discrete prior is adopted on these options, so that there is a joint prior structure

$$\pi_2(\rho_1|S)\pi_1(s)$$

The value $s = 0$ is equivalent to a uniform prior over (0, 1). The prior (7.16) is implemented in BUGS via a discrete prior on the points $\rho_1 = 0, 0.01, 0.02, \ldots 0.99, 1$, with original weights $(s+1)\rho_1^s$ scaled to sum to unity.

For the spatial errors model, this leads to a higher value of ρ_1, and ν approximately 27 – indicating some departure from normality in the crime rate outcome (see Table 7.3). The mean value of s is 1.76. This model makes the impact of income more negative, while the housing value coefficient now straddles zero. The choice between this model and the homoscedastic errors model (7.15) is based partly on its better accommodation of possible outliers, and also its better marginal likelihood (−189.7) compared to that of the homoscedastic model (−193.9). The resulting pseudo-Bayes factor clearly favours the more robust assumption on the u_i. The Student t weights κ_i still show observation 4 as a possible outlier, with $\kappa_4 = 0.48$ compared to the remaining areas, where κ_i ranges between 0.70 and 1.11.

A suggested exercise is to fit the spatial autoregression model (with a spatial lag in Y) under alternative prior assumptions (homoscedastic vs hetersoscedastic) on the errors u.

Table 7.3 Spatial errors: non-constant variance

	Mean	St. devn.	2.5%	Median	97.5%
β_0	19.09	11.78	2.00	17.22	47.98
β_1	−1.11	0.38	−1.86	−1.11	−0.35
β_2	−0.17	0.11	−0.40	−0.16	0.04
ν	27.8	27.9	2.6	14.2	93.9
ρ	0.72	0.18	0.29	0.75	0.99

Example 7.3 Poisson data: the lip cancers example Spatial models for a discrete outcome are illustrated with the Scottish lips cancer data, included in the WINBUGS examples and widely analysed elsewhere. These are often small event counts Y_i for $i = 1, .. 56$ counties, and for disease mapping purposes the goal of introducing random effects (unstructured and/or spatially dependent) is to obtain smoothed estimates of the Relative Risks (RRs) of cancer incidence by county. Inferences on regressors will be improved by taking account of possible overdispersion in the count outcome.

A single regressor is provided in this example by 0.1 times the percent of labour force in agriculture and related occupations ($x_1 = \text{AFF}/10$), so the adjusted relative risks of lip cancer will also reflect county occupational structure as well as smoothing towards the local or global average. Interactions c_{ij} between counties are in terms of dummy indicators for contiguity ($c_{ij} = 1$ if counties i and j are neighbours). Model fit and regression inferences are assessed under different specifications for the prior on the errors e_i in Equation (7.9).

First, consider the CAR prior as in Equation (7.4), with a prior for ρ defined by the eigenvalues[10] of C. Then the model is

$$\begin{aligned} Y_i &\sim \text{Poi}(E_i\theta_i) \\ \log(\theta_i) &= x_i\beta + e_i = \beta_0 + \beta_1 x_{1i} + e_i \\ e_i &\sim \text{N}(M_i, \sigma^2) \\ M_i &= \rho \sum_j c_{ij} e_j \end{aligned}$$

The posterior estimates from a three chain run of 10 000 iterations (with convergence after 500) shows a mean correlation of around 0.16, and 95% credible interval (0.12, 0.17). The AFF variable has coefficient β averaging 0.46 and 95% credible interval (0.21, 0.71). The lowest scaled CPOs are for counties 50 and 55, namely 0.03 and 0.025 (these are Dundee and Annandale). These results may be compared to those obtained by Bell and Broemeling (2000), who cite an interval (0.131, 0.170) on ρ; the paper by Stern and Cressie (1999) has a list of counties corresponding to the numbers used here.

If instead we use the ICAR framework with default correlation of $\rho = 1$ as in Equation (7.11c), a lower estimate of the coefficient on AFF is obtained, with mean 0.39 and 95% interval (0.16, 0.60). Here the model for the mean relative risk omits for the intercept for identifiability, and so is

$$\log(\theta_i) = \beta_1 x_{1i} + e_i$$

This model has an improved pseudo-marginal likelihood estimate compared to the CAR model (−156.2 vs. −162.3) giving a pseudo-Bayes factor in its favour of around 400.

However, an alternative is a data driven choice of the ICAR correlation in the ICAR(ρ) model, with the model for the means now having an intercept again. This leads to an enhancement of the AFF effect compared to the ICAR(1) model, with posterior mean 0.43 and 95% interval (0.09, 0.68). The ICAR correlation is estimated to have mean 0.92 and median 0.94. This model has a worse pseudo-marginal likelihood than the ICAR(1) model, around −159.5.

[10] The minimum and maximum eigenvalues of C for the Scottish county data are −3.07 and 5.71, so the range for ρ is between −0.326 and 0.175.

Finally, a robust mixture approach drawing on the Lawson and Clark (2002) method, as in Section 7.3.2, is fitted. This is coded directly in Model D (Program 7.3) using the double exponential density ddexp() in WINBUGS, and to produce identifiability an additional autocorrelation parameter as in Equation (7.12) is introduced. A two chain run of 5000 iterations shows convergence at around 500 iterations and gives a higher coefficient on the AFF variable than the preceding methods, with mean (and 95% interval) of 0.49 (0.22, 0.76). The autocorrelation parameters on the Normal and Laplace mixture elements have means 0.61 and 0.88. The lowest η_i (most outlying case in terms of the Normal spatial structure) is for county 49, which has an unusually low crude and smoothed relative risk compared to the surrounding counties (which both abut it and are similar in terms of urban and occupational status). Despite extra insights like this, the pseudo-marginal likelihood is worse than for the ICAR(1) model, namely−161.

Smoothed relative risk profiles are consistent across the four methods: the CAR option shows 18 out of 56 counties with RRs clearly above 1 (95% credible interval entirely above 1) and the ICAR methods both show 19 out of 56.

Example 7.4 Glasgow deaths The conditional autoregression approach has been a popular way of tackling spatial dependence in discrete outcomes (e.g. in disease mapping). However, other techniques have been proposed and may have benefits in terms of, say, allowing both multivariate outcomes and a mix of unstructured and structured errors as in the convolution model (7.9). As a case study of an alternative methodology, the analysis of this example follows Leyland *et al.* (2000) in considering a bivariate mortality outcome for a set of 143 Glasgow postcode sectors, namely cancer and circulatory deaths $\{Y_{i1}, Y_{i2}\}$. Thus defining the spatial error in terms of an underlying unstructured error as in Equation (7.10), and with two outcomes in a convolution model, there are four errors to consider. Without predictors, we might then have

$$Y_{i1} \sim \text{Poi}(E_{i1}\theta_{i1})$$
$$Y_{i2} \sim \text{Poi}(E_{i2}\theta_{i2}$$
$$\log(\theta_{i1}) = \beta_{01} + u_{i1} + e_{i1} \tag{7.17a}$$
$$\log(\theta_{i2}) = \beta_{02} + u_{i2} + e_{i2} \tag{7.17b}$$

In the first model (Model A), four errors are distributed independently of each other, and u_{i1} and u_{i2} denote independently distributed and spatially unstructured errors with variances σ_1^2 and σ_2^2. The spatial errors under Model A in Equations (7.17) are obtained as

$$e_{i1} = N_i^{-1} \sum_{j\varepsilon L_i} \nu_{j1}$$
$$e_{i2} = N_i^{-1} \sum_{j\varepsilon L_i} \nu_{j2}$$

with the ν being taken as independently Normal with variances τ_1^2 and τ_2^2,

$$\nu_{i1} \sim \text{N}(0, \tau_1^2)$$
$$\nu_{i2} \sim \text{N}(0, \tau_2^2)$$

A second option (Model B) takes the e_{i1} and e_{i2} in Equation (7.16) to be separately CAR(1), as in Equation (7.11c). Here the 'balanced dispersion prior' recommended by Bernardinelli *et al.* (1995) is used.

However, an alternative model allowing correlated errors (i.e. multivariate error dependence among all four errors) is straightforward. Denote the first two error terms are $\eta_{i1} = u_{i1}$ and $\eta_{i2} = u_{i2}$, and the third and fourth as $\eta_{i3} = v_{i1}$ and $\eta_{i4} = v_{i2}$. The η_{ij} are then taken to be multivariate Normal with mean vector (0,0,0,0) and 4×4 dispersion matrix Σ_η. A Wishart prior on Σ_η^{-1} is assumed, with 4 degrees of freedom. The standardised version of Σ_η provides correlations $R = [r_{ij}]$ between the four underlying errors.

Here interest will centre on whether the multivariate error model provides a better description of the data, and how estimated relative risks on the two types of mortality are affected by the model choice. Model summaries are based on three chain runs of 10 000 iterations with 1000 burn-in. Under model C, this leads (Table 7.4) to a correlation of only 0.08 between the two unstructured errors, but a correlation of 0.28 between the two spatial errors, v_{i1} and v_{i2}.

As noted above, spatially dependent errors may proxy unobserved covariates (or covariates not included in a regression) which are themselves spatially associated; here such variables might include area deprivation, for example. Evidence for missing spatially correlated covariates will be strengthened by high correlations between the v_{ij}. However, Table 7.4 shows that the correlations are not pronounced and none are conclusively positive.

It may be that the simpler, less parameterised model is adequate. Using the CPO statistics for areas i and outcomes j based on monitoring the inverse Poisson likelihood gives marginal likelihoods for the independent error Models A and B of −904 and −913, compared to −908 for Model C. So Model A appears to provide the best explanation for the data. In terms of inferences about risk, Model C identifies more extreme relative risks: for instance, [illegible] out of [illegible] area-outcome relative risks exceed 1.5 under Model C, but only 4 under Models A and B.

7.4 DIRECT MODELLING OF SPATIAL COVARIATION IN REGRESSION AND INTERPOLATION APPLICATIONS

The preceding discussion and examples consider continuous and discrete outcomes for zones. As described in the Introduction, one may consider outcomes for point based data in broadly the same way. Instead of a fixed interaction matrix W, defined perhaps by contiguity between areas i and j, for point data any predefined adjacency is unclear,

Table 7.4 Bivariate mortality outcome with correlated errors

Correlation	Mean	St. devn.	2.50%	97.50%
R_{12}	0.08	0.14	−0.19	0.35
R_{13}	0.25	0.15	−0.06	0.53
R_{14}	0.22	0.15	−0.09	0.51
R_{23}	0.17	0.16	−0.15	0.46
R_{24}	0.19	0.15	−0.10	0.47
R_{34}	0.28	0.19	−0.12	0.61

and the influence of proximity needs to be estimated. Consider observations $y_i\ i = 1, .. n$ observed at points $s_i = (s_{1i}, s_{2i}, \ldots..)$, which are here taken to be in two dimensional space. A starting point for estimating the effect of proximity is provided by the matrix of interpoint distances, $d_{ij} = |s_i - s_j|$.

7.4.1 Covariance modelling in regression

For instance, suppose the goal is, as before, to estimate a regression

$$Y = x\beta + e \tag{7.18a}$$

where the joint dependence between errors might be described by a multivariate Normal prior

$$e \sim N_n(0, \Sigma) \tag{7.18b}$$

such that the dispersion matrix Σ reflects the spatial interdependencies within the data. Outcomes may also be discrete, and then one might have, for binomial data, say,

$$Y_i \sim \text{Bin}(\pi_i, N_i) \quad i = 1, .. n$$

$$\text{logit}(\pi_i) = x_i\beta + e_i$$

where again, the errors may be spatially dependent. Let the $n \times n$ covariance matrix for e be

$$\Sigma = \sigma^2 R(d)$$

where σ^2 defines the overall variance (as defined along the diagonal when $i = j$ and $d_{ii} = 0$) and $R(d) = [r_{ij}(d_{ij})]$ models the correlations between the errors e_i and e_j in terms of the distances between the points[11]. The function r is defined to ensure that $r_{ii}(d_{ii}) = r_{ii}(0) = 1$ and that R is positive definite (Anselin, 2001; Fotheringham *et al.*, 2000).

Among the most commonly used functions meeting these requirements are the exponential model

$$r_{ij} = \exp(-3d_{ij}/h) \tag{7.19a}$$

where h is the range, or inter-point distance at which spatial correlation ceases to be important[12]. The Gaussian correlation function has

$$r_{ij} = \exp(-3d_{ij}^2/h^2) \tag{7.19b}$$

and the spherical (Mardia and Marshall, 1984) has $r_{ij} = 0$ for $d_{ij} > h$ and

$$r_{ij} = (1 - 3d_{ij}/2h + d_{ij}^3/2h^3) \tag{7.19c}$$

for $d_{ij} < h$; see Example 7.5 for an illustration. In each of these functions, h is analogous to the bandwidth of kernel smoothing models. If $\Sigma = \sigma^2 R(d)$, then the covariance tends

[11] Although this approach is theoretically based on point data, it can be extended to aggregated data by considering the centroid of an area to define its location in continuous space. Ideally, such centroids take account of variations in population density within areas; for instance, centroids of electoral wards in the UK may be based on aggregating over smaller sub-areas of wards (known as enumeration districts), with the ward centroid being a population weighted average of sub-area centroids.

[12] This interpretation of h is clearer in the variogram version of the exponential model, considered below.

to σ^2 as d_{ij} tends to zero. In some cases, there will further baseline variability (e.g. due to measurement error) and the covariance may be defined as

$$\Sigma = \tau^2 + \sigma^2 R(d)$$

with the limiting variance as d_{ij} tends to zero being $\tau^2 + \sigma^2$ instead of σ^2. Another formulation in this case involves a discontinuity when $d_{ii} = 0$, so that

$$\Sigma_{ij} = \sigma^2 r_{ij}(d_{ij}) \quad i \neq j$$
$$\Sigma_{ii} = \tau^2$$

Whatever the model adopted for the correlation between errors, the problem reduces to simultaneously estimating the regression coefficients and the parameters of the distance decay function.

7.4.2 Spatial interpolation

In other situations, the emphasis may be on interpolation or prediction at locations s.new, on the basis of the observations Y_i, $i = 1, \ldots, n$ made at points (taken to be in two dimensions) $s_i = (s_{1i}, s_{2i})$. An example of spatial interpolation or 'kriging' from a Bayesian perspective is provided by Handcock and Stein (1993), who consider the prediction of topological elevations Y_{new} at unobserved locations on a hillside given an observed sample of 52 elevations at two dimensional grid locations. To emphasize the definition in space, denote the observations at s_i as $Y_i = Y(s_i)$ and assume that the model for $Y(s_i)$ consists only of a mean $\mu(s_i)$ which varies over space, where space here means the two dimensional real space. The most general process is then

$$Y(s) = \mu(s) + e(s)$$

where the error vector $e(s)$ has mean zero and covariance $\sigma^2(s)R$, and the mean $\mu(s) = E[Y(s)]$ models the first order trend over space. Trend, sometimes known as large scale variation, is exemplified by the North West to South East gradient in cardiovascular disease in England (Richardson, 1992), while small scale (second order) variation represents features such as spatial correlation in neighbouring areas.

Just as for time series models, prediction to new locations may be facilitated by transformation of the data to remove trend and induce stationarity, meaning the mean and variance are independent of location, and in particular that $\text{cov}(e) = \sigma^2 R$. Methods for spatial interpolation may, for instance, be based on analysing transformed observations or residuals e_i which are free of trend. An example of such data transformation is the EDA method known as row and median polish, and illustrated by Cressie (1993) on percents coal ash. One may also use regression to account for spatial trend: this is known as trend surface analysis, where the above model takes the form

$$Y(s) = \mu(s) + e(s) = x(s)\beta + e(s)$$

and the $x(s)$ are functions of the spatial coordinates (s_{1i}, s_{2i}). One might have a trend regression model with $x_1 = s_1$ and $x_2 = s_2$ (linear terms in the grid references), x_3 and x_4 being squared terms in s_1 and s_2, respectively, and x_5 being an interaction between s_1 and s_2. Simultaneously modelling spatial effects in the mean and variance structure may, however, be problematic (Ecker and Gelfand, 1997), and often a constant mean is assumed (this is known as ordinary kriging), with

$$\mu(s) = \mu = \beta_0$$

After trend or location is allowed for in the model for $\mu(s)$, the covariation in the errors may be modelled. Direct modelling of the covariance structure of the observations Y may also be of interest provided stationarity is plausible.

7.4.3 Variogram methods

A technique often used to estimate the covariance structure focuses on functions of dissimilarity between errors or observations. Let $\Sigma(d)$ denote the covariance matrix at distance d between points. Whereas

$$\Sigma(d) = \sigma^2 R(d)$$

diminishes to zero for widely separated points and attains its maximum as d_{ij} tends to zero, the variogram function

$$\gamma(d) = \sigma^2 - \Sigma(d) = \sigma^2(I - R(d))$$

has value zero when $d_{ij} = 0$, and reaches its maximum at σ^2 as spatial covariation in $\Sigma(d)$ disappears; hence σ^2 is known as the sill in geostatistics. For instance, the variogram for the exponential model is

$$\gamma(d_{ij}) = \sigma^2(1 - e^{-3d_{ij}/h}) \tag{7.20}$$

As above, an extra variance parameter to allow for measurement error at $d = 0$ (the 'nugget' error) may be added to give

$$\gamma(d) = \nu^2 + \sigma^2[\mathrm{I} - R(d)]$$

and the sill is now $\nu^2 + \sigma^2$. An alternative version of a nugget error model is

$$\gamma(d) = \nu^2 + (\sigma^2 - \nu^2)[I - R(d)]$$

In fact, the expected value of the difference[13] between observations Y_i and Y_j (or errors e_i and e_j) at separation d_{ij} is $2\gamma(d_{ij})$. Estimation of the variogram is often carried out within narrow distance bands, such that for the $n(d)$ points within a band $\{d - \varepsilon, d + \varepsilon\}$ the estimated variogram is

$$\hat{\gamma}(d) = 0.5\frac{1}{n(d)} \sum_{d_{ij} \in \{d-\varepsilon,\ d+\varepsilon\}} (Y_i - Y_j)^2$$

[13] For points $y(s)$ and $y(t)$ at separation $d = s - t$, the variogram is

$$\begin{aligned}\gamma(s, t) &= 0.5E[y(s) - y(t)]^2 \\ &= 0.5\mathrm{Var}[y(s) - y(t)]\end{aligned}$$

For an isotropic process (where the covariation does not depend upon the direction between s and t),

$$\gamma(s, t) = \gamma(s + \Delta, t + \Delta)$$

and taking $\Delta = -t$ gives

$$\gamma(s, t) = \gamma(s - t, 0) = \gamma(d)$$

So estimation of the variogram typically considering distances between all $n(n-1)/2$ pairs $\{i, j\}$ of points and then grouping them into a relatively few (say 10 or 20) distance bands. Then the relation between $\hat{\gamma}(d)$ and d is modelled over distance bands $d = 1, \ldots, D$ via nonlinear least squares in terms of parametric forms such as Equation (7.20). If $n(n-1)/2$ is not unduly large, one might fit a nonlinear curve such as Equation (7.20) to all sets of paired differences $(Y_i - Y_j)^2$ or $(e_i - e_j)^2$.

However, recent Bayesian approaches have tended to focus on spatial interpolation consequent on direct estimation of the covariance matrix from the likelihood for $Y_i = Y(s_i)$. Thus, Diggle *et al.* (1998) consider adaptations for discrete outcomes of the cananonical model

$$Y_i = \mu + e(s_i) + u_i \tag{7.21}$$

where the errors u_i are independently $N(0, \tau^2)$, and accounts for the nugget variance, and the vector $e(s) \sim N_n(0, \sigma^2 R)$ models the spatial structure in the errors. This can be seen as a continuous version of the convolution model of Equation (7.9). Ecker and Gelfand (1997) also adopt this likelihood approach, but consider generalisations of the usual parametric covariance structures such as those of Equations (7.19a)–(7.19c).

Whatever the approach used to estimate $\Sigma(d)$, prediction of Y_{new} at a new point s.new then involves the $n \times 1$ vector g of covariances $g_i = \text{Cov}(s_{\text{new}}, s_i)$ between the new point and the sample sites $s_1, s_2 \ldots s_n$. For instance, if $\Sigma(d) = \sigma^2 e^{-3d/r}$, then the covariance vector is obtained by plugging in to this parametric form the distances $d_{1\text{new}} = |s_{\text{new}} - s_1|$, $d_{2\text{new}} = |s_{\text{new}} - s_2|$, etc. The prediction Y_{new} is a weighted combination of the existing points with weights λ_i, $i = 1, \ldots n$ determined by

$$\lambda = g\Sigma^{-1}$$

For example, the prediction Y_{new} under Equation (7.21) is obtained (Diggle *et al.*, 1998, p. 303) as

$$Y_{\text{new}} = \mu + g(\tau^2 I + \sigma^2 R)^{-1}(Y - \mu)$$

7.4.4 Conditional specification of spatial error

While joint prior specifications are the norm in such applications, one might also consider conditional specifications with the same goals (e.g. estimation of proximity effects and interpolation) in mind. Defining weights a_{ij} in terms of the parametric forms in Equation (7.19), and then defining $w_{ij} = a_{ij}/\sum_{j \neq i} a_{ij}$, one might specify

$$Y_i \sim N(\mu_i, \tau^2)$$
$$\mu_i = \mu + e_i$$

where e_i follows a conditional prior

$$e_i \sim N\left(\sum_{j \neq i} w_{ij} e_j, \ \phi^2 / \sum_{j \neq i} a_{ij}\right)$$

This is the ICAR(1) model with non-binary weights, as in Equation (7.11b). For instance under the exponential model, $a_{ij} = \exp(-\kappa d_{ij})$, where $\kappa > 0$. As above, identifiability may be established by introducing a correlation parameter ρ, which might, for instance, be assigned a prior limited to the range [0, 1], so that

$$e_i \sim \mathrm{N}\left(\rho \sum_{j \neq i} w_{ij} e_j,\ \phi^2 / \sum_{j \neq i} a_{ij}\right) \tag{7.22}$$

Another option under an ICAR(1) prior is to centre the sampled e_i at each iteration in an MCMC run. For interpolation of Y at a new location s_{new} under this prior, one would evaluate the function a_{ij} in terms of distances between the observed locations $s_1, s_2, \ldots s_n$ and the new location s_{new}. This gives a vector of weights $a_{\text{new}} = (a_{\text{new},\, i})$ and after scaling to give w_{new}, the estimate of $e_{\text{new}} = Y_{\text{new}} - \mu$ at the new location is formed as $\Sigma_i e_i w_{\text{new},\, i}$.

Example 7.5 Nickel concentrations The principles involved in deriving a variogram are illustrated using a set of geochemical readings (nickel in ppm) at 916 stream locations in the northern part of Vancouver Island in Western Canada. The logged nickel readings are taken to lessen skewness (cf. Ecker and Gelfand, 1997). The $(916)\times(915)/2$ pairs of readings are accumulated into distance bands at a 'lag' 0.1 km apart. Many pairs exceed 5 km in separation, and the estimation here only includes 55 bands at average distances 0.05, 0.15, 0.25,... up to 5.45 km.

The observations at separation d consist of averages G_d of the differences squared

$$Z_{dij} = \sum_{dij \varepsilon (d-0.05,\ d+0.05)} (Y_i - Y_j)^2$$

for the n_d pairs of points separated by distances between $d - 0.05$ and $d + 0.05$ km. As well as the average of Z in each band, the variance V_d of the differences $(Y_i - Y_j)^2$ in each band is recorded. The means are then taken as normal with known precision n_d/V_d, with their regression means following a spherical model

$$G_d \sim \mathrm{N}(\gamma(d),\ V_d/n_d)$$
$$\gamma(d) = \nu^2 + (\sigma^2 - \nu^2)[3d/(2h) - d^3/(2h^3)] \quad \text{if } d \leq h$$
$$\gamma(d) = \sigma^2 \quad \text{if } d > h$$

A uniform prior on the ratio $\nu^2/(\sigma^2 + \nu^2)$ is adopted, with the range assigned a uniform prior with minimum 0.05 and maximum 5.

A two chain run of 5000 iterations leads to a nugget variance ν^2 averaging about 0.31, a sill variance σ^2 of 0.87 and a range averaging 1.61. The range is lower than the value of 2.13 reported by Bailey and Gattrell (1995), though these authors took wider bands (a 0.2 km lag separation) to derive the variogram.

A second model is provided by the Gaussian form with

$$\gamma(d) = \nu^2 + (\sigma^2 - \nu^2)[1 - \exp(-3d^2/r^2)]$$

This provides a lower pseudo-marginal likelihood than the spherical model, where the relevant CPO statistics are estimated using the average inverse likelihoods. The nugget variance increases to around 0.4, while the range is estimated as 1.36 with 95% credible interval (1.23, 1.50).

Example 7.6 Likelihood model for spatial covariation To illustrate the direct likelihood approach to estimating spatial covariation as a function of distance, 20 points s_i are randomly generated within the square region defined by extreme SW point (0, 0) and extreme NE point (100, 100). Distances between points are obtained as $d_{ij} = |s_i - s_j|$. The Y_i are then sampled from a multinormal density with mean 160 and covariance

$$\Sigma(d) = \tau^2 + \sigma^2 R(d)$$

where $R(d) = \exp(-3d/h)$, the range $h = 100$, the measurement error variance 20, and the scale variance $\sigma^2 = 100$. The points and their associated Y values are shown in Figure 7.2.

To reflect the WINBUGS centred Normal structure, the likelihood for the data so generated is then univariate Normal as in Equation (7.21), namely

$$Y_i \sim \mathrm{N}(\mu + e_i, \tau^2)$$

with $\mathrm{var}(u_i) = \tau^2$. In the first joint prior approach for the spatially structured errors e_i, it is assumed that

$$e \sim N_n(0, \sigma^2 R)$$

where $r_{ij} = \exp(-3d_{ij}/h)$. The maximum range for the region concerned is 141.4, and a discrete prior with 10 values from 14 to 140 is adopted. For the two variances a prior ideally expresses their interdependence, and here a prior on the sum $\varphi = 1/\tau^2 + 1/\sigma^2$ is adopted with $1/\tau^2 = U\varphi$, where $U \sim B\,(1, 1)$. Since the model includes only a mean, a residual precision exceeding 1 (variance under 1) is unlikely for the sampled data, and two priors may be considered: a G(0.1, 1) prior with mean 0.1 and a G(0.1, 0.1) prior with mean 1. With the first prior, a two chain run of 2500 iterations (with convergence after 500) is taken. This yields a reasonable estimate of the range and the variance parameters and a prediction at (10, 10) of 172.4. It is left as an exercise to experiment with different priors for φ such as the G(0.1, 0.1), and assess how predictions to (10, 10) or other points are affected.

For the conditional prior approach, the same method as for the joint prior model is used to define the inverse variances $1/\phi^2$ and $1/\tau^2$ (see Section 7.4.4). A N(0, 1) prior,

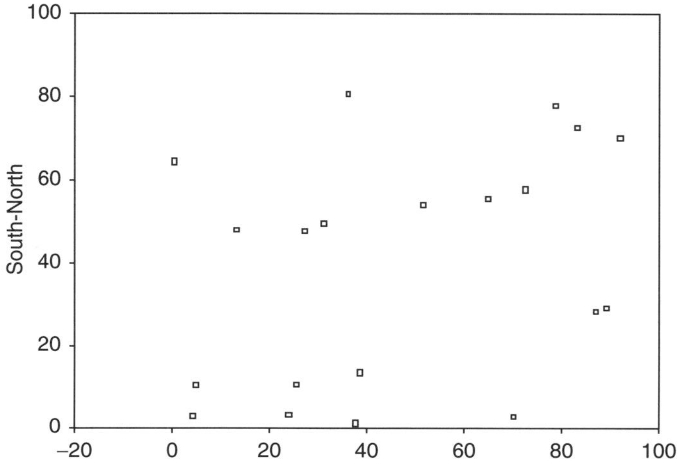

Figure 7.2 Sampled points

constrained to positive values, is adopted for κ. With $\phi \sim G(0.1, 1)$ a two chain run of 2500 iterations (with convergence at under 500 iterations) yields an estimate for τ^2 close to the generating value (posterior median 20.3), with the median of ϕ^2 estimated at 11.4 (Table 7.5). The parameter κ has a skew posterior with mean 0.12. The prediction at (10, 10) is higher than under the joint prior, namely 175.

The joint prior has a better fit than the conditional prior model according to the predictive loss method of Gelfand and Ghosh (1998), but its prediction seems to tend to 'oversmooth', since the two closest points to (10, 10) have values 172 and 182. The prediction of the conditional prior is more in line with the neighbouring observed values.

Example 7.7 Attempted suicide in East London Spatial covariance modelling for discrete outcomes is illustrated by data for attempted suicide ('parasuicide') in 66 electoral wards in East London (the East London and City Health Authority). The records of this health event are based on hsopital admissions following deliberate self-harm. Following work by Congdon (1996) and Hawton *et al.* (2001), one may relate small area variations in parasuicide to both area deprivation and the strength of community ties or their absence ('social fragmentation'). Both factors are represented by composite scores. The analysis involves actual and expected attempted suicides (Y_i and E_i) over the period 1990–92, for all ages and males and females combined, with ward locations defined by their centroids as supplied by the Office of Population Censuses and Surveys in the 1991 UK Census ward profile (and derived by population weighted averaging over the centroids of micro areas known as enumeration districts, nested within wards).

A simple Poisson regression (without spatial or unstructured errors) shows both factors apparently influencing variations in the outcome, with the coefficients on deprivation and fragmentation having posterior averages (with 95% credible intervals) of 0.047 (0.031, 0.062) and 0.049 (0.028, 0.069).

Table 7.5 Models for simulated data and prediction to new point

	Mean	St. devn.	2.5%	Median	97.5%
Joint Prior					
μ	171.3	5.8	159.0	171.3	183.2
σ^2	80.6	42.6	25.3	71.9	183.4
τ^2	19.0	15.8	2.1	15.1	58.8
h	102.1	29.7	42.0	112.0	140.0
Y.new, at (10, 10)	172.4	6.2	158.9	172.6	184.6
Conditional Prior					
μ	169.7	3.1	163.2	170.0	175.1
κ	0.12	0.05	0.02	0.12	0.23
ϕ^2	20.3	32.9	1.1	11.4	101.6
τ^2	32.1	33.6	1.0	20.3	119.9
ρ	0.81	0.20	0.18	0.88	0.99
Y.new, at (10, 10)	175.0	3.2	168.0	175.6	179.9

However, there may be unobserved and spatially varying risk factors which produce spatial dependence in the regression errors. The second coding in Program 7.7 applies the direct estimation approach using the intrinsic conditional[14] prior in Equation (7.22), and weights a_{ij} as in Equation (7.19a), to investigate the spatial structure of the errors. For identifiability, a spatial correlation parameter ρ is included so that

$$Y_i \sim \text{Poi}(E_i\theta_i)$$
$$\theta_i = \beta x_i + e_i$$
$$e_i \sim \text{N}\left(\rho \sum_{j \neq i} w_{ij} e_j,\ \phi^2 / \sum_{j \neq i} a_{ij}\right)$$
$$a_{ij} = \exp(-\kappa d_{ij})$$
$$w_{ij} = a_{ij} / \sum_{j \neq i} a_{ij}$$

A run of 2500 iterations (with convergence from 500) shows the distance decay parameter κ to average just under 2, and the error autocorrelation to average 0.74. The mean effect of fragmentation (though not deprivation) is reduced, and the credible intervals for both factors are widened. In particular, the fragmentation effect is no longer clearly positive (Table 7.6).

The third model assumes a fixed interaction matrix, defined as $c_{ij} = 1$ if ward j is among the five nearest neighbours of ward i (in terms of Cartesian distance), and $c_{ij} = 0$ otherwise. Then the ICAR(ρ) prior for spatial errors e_i is used. This model shows an enhanced mean deprivation effect, while the fragmentation effect is essentially eliminated (Table 7.6).

Table 7.6 Models for parasuicide

Distance decay model	Mean	St. devn.	2.5%	97.5%
β_1 (Intercept)	−0.13	0.11	−0.37	0.07
β_2 (Deprivation)	0.048	0.021	0.009	0.089
β_3 (Fragmentation)	0.033	0.029	−0.026	0.089
κ	1.84	0.54	0.87	3.00
ρ	0.74	0.22	0.18	0.99
ICAR (ρ)				
β_1 (Intercept)	−0.16	0.11	−0.37	0.07
β_2 (Deprivation)	0.056	0.020	0.016	0.096
β_3 (Fragmentation)	0.017	0.028	−0.036	0.074
ρ	0.75	0.18	0.32	0.99

[14] One might also consider a joint prior as in Equation (7.18b). However, computation in joint prior models involving a structured covariance matrix Σ is slowed considerably by the necessity to invert the covariance matrix during MCMC iterations. This can be avoided by discretising the prior on the parameters (apart from σ^2) in $\Sigma(d)$. Thus, for a K point grid on h in Equation (7.19a), one might pre-calculate K matrix inverses R^{-1}, where $r_{ij} = \exp(-3d_{ij}/h)$. Obtaining a suitable grid (one that encompasses the density of the parameter) may require some preliminary exploration of the problem. The set of matrix inverses is then input as extra data to a model where h is assigned a discrete prior with K bins. One may also sample directly from the relevant full conditional densities, as in Diggle *et al.* (1998), and certain options for spatial covariance matrices are now implemented in the GEOBUGS module of WINBUGS14.

In terms of the Gelfand and Ghosh (1998) criterion of footnote 5, the ICAR(ρ) model has a slight edge in fit, namely 5180 as against 5250 for the direct estimation approach (the same criterion for the simple Poisson model is 11 200). However, much can be done to improve the performance of the direct estimation method by varying the modelling of distance decay, for example using weights a_{ij} such as those in Equations (7.19b) or (7.19c) or taking a power rather than exponential decay system, with $a_{ij} = d_{ij}^{-\kappa}(\kappa > 0)$.

7.5 SPATIAL HETEROGENEITY: SPATIAL EXPANSION, GEOGRAPHICALLY WEIGHTED REGRESSION, AND MULTIVARIATE ERRORS

With geographical data, the preceding sections have shown how regression errors may well be spatially correlated, and have sought to model spatial dependencies in errors. However, an alternative or at least complementary perspective on spatial outcomes is in terms of spatial heterogeneity, meaning either heteroscedastic variances, spatially varying regression effects, or both. In the preceding worked examples, one might well allow the unstructured error such as u_i in Equation (7.9) to have a variance that differs between individual areas, or perhaps be governed by a spatial regime, involving $k < n$ subsets of areas, within which the areas have the same variance (Anselin, 2001). For instance, Example 7.2 exemplifies how the assumption of constant variance of the error terms may well be questionable, but the same approach could be applied in the other examples, and below the data in Example 7.3 is reanalysed from the heterogeneity viewpoint.

While heteroscedasticity may well be relevant, an alternative perspective on spatial heterogeneity focuses on the regression parameters themselves: linear and general linear spatial models typically assume that the structure of the model remains homogenous over the study region without any local variations in the parameter estimates. It is likely, in fact, in many applications that regression effects are not constant over the region of application (Casetti, 1992). In fact, varying regression effects over space are analogous to the more familiar state space models in time series applications which involve time varying regression coefficients.

Spatial heterogeneity in the impacts of regression predictors has been the subject of two major recent advances, namely the spatial expansion model Casetti (1972) and geographically weighted regression (LeSage, 2000). These have focused on the normal linear model (with modifications for heavy tailed errors via Student t extensions), though extension to general linear models is quite feasible. Congdon (1997) describes an ICAR(1) model for spatially varying regression coefficients with a univariate Poisson outcome, and below a variation on the prior of Leyland *et al.* (2000) is considered to the same end.

7.5.1 Spatial expansion model

The spatial expansion model of Casetti (1972, 1997) assumes that the impacts of one or more of p regressors, x_{ji} on a continuous outcome Y_i vary according to the locations (s_{1i}, s_{2i}) of the areas or points. These locations may be taken with reference to a central point (0, 0), so that one or both elements in the grid reference may be negative. One may allow for fixed (i.e. non-spatially varying) impacts $c_1, c_2, .. c_p$ of p regressors, as well as spatially varying ones $b_{ji}(j = 1, .., p)$, though the constant c_0 does not have a parallel spatially varying effect. (The latter choice is made by Lesage, 2001, on identifiability grounds.) Then for a metric univariate outcome Y over areas i one might have

$$Y_i \sim N(\mu_i, \phi) \tag{7.23a}$$

$$\mu_i = c_0 + c_1x_{1i} + c_2x_{2i} + \ldots + c_px_{pi} + b_{1i}x_{1i} + b_{2i}x_{2i} + \ldots + b_{pi}x_{pi} \tag{7.23b}$$

with the spatially varying effect (or 'base effect') modelled as

$$b_{ji} = s_{1i}\gamma_{1j} + s_{2i}\gamma_{2j} \quad j = 1, .. p \tag{7.23c}$$

There are $3p + 1$ parameters to estimate: the constant c_0, p fixed regression effects c_j, and the $2p$ base effect parameters, γ_{1j} and γ_{2j}. The combined non-spatial and spatial impact of x_j then varies by area i, and is

$$\beta_{ji} = c_j + b_{ji} \tag{7.23d}$$

with average over all areas

$$\beta_j = c_j + \bar{b}_j \tag{7.23e}$$

This model may therefore capture variations in the regression relationships, especially if clusters of adjacent observations have similar regression behaviours, or if economic relationships vary according to distance from city centres.

One way of extending these models to allow for non-constant variance is to take area specific variance parameters ϕ_i as functions of the locations (and maybe other factors), with

$$\phi_i = \exp(\gamma_0 + \gamma_1 s_{1i} + \gamma_2 s_{2i})$$

7.5.2 Geographically weighted regression

The method of Geographically Weighted Regression (GWR) also makes regressions specific to the location of point or area i (Brunsdon *et al.*, 1996, 1998). Somewhat like cross-validation with single case omission, GWR essentially consists in re-using the data n times, such that the ith regression regards the ith point as the origin. The coefficients $b_{1i}, .. b_{pi}$ for the ith regression entail distance based weightings of the ith and remaining cases (with i as the centre). These weightings define the precision parameter for each area in a Normal or Student t likelihood.

As in Equations (7.19a)–(7.19c), the weighting might be exponential, Gaussian, or spherical, with the Gaussian form being

$$a_{ik} = \exp(-d_{ik}^2/2h^2) \quad h > 0 \tag{7.24}$$

This function means that for small distances a_{ik} is close to 1, with the decay effect increasing as the bandwidth parameter h tends to zero. In all these functions, a prior may be set on h, taking account of the maximum observed inter-point or inter-area distance. Other options are profile type analyses to maximise a fit criterion. Alternatively, one may combine GWR with formal cross-validation. This is based on omitting the ith observation from the ith geographically weighted regression, and then predicting it using the remaining $n - 1$ cases.

Suppose the outcome Y_k is univariate and metric, with sample indices $k = 1, .., n$ and regressors $x_{1k}, x_{2k}, \ldots x_{pk}$ for area k. Conditioning on a choice of spatial weight function, for the ith regression, one might use the notation $y_{ik} = Y_k$ (providing n copies of the data), and then take

$$y_{ik} \sim N(\mu_{ik}, \tau_{ik}) \quad k = 1, .. n \tag{7.25a}$$

$$\tau_{ik} = \phi_i / a_{ik} \tag{7.25b}$$

$$\mu_{ik} = b_{0i} + b_{1i}x_{1k} + b_{2i}x_{2k} + \ldots + b_{pi}x_{pk} \tag{7.25c}$$

where ϕ_i is an overall variance, homogenous across areas in the ith regression. Since the a_{ik} decline with distance, nearer observations have lower variances (higher precisions). This implies a greater weighting in the ith regression such that observations near the area or point i have more influence on the parameter estimate $b_i = (b_{0i}, b_{1i}, \ldots . b_{pi})$ than observations further away.

As in the spatial expansion method, one might robustify against outlying areas (in terms of the estimated relationships between y_{ik} and x_{jk}) by taking a scale mixture (heavy tailed) approach. This allows non-constant variances with scaling factors κ_{ik} drawn from a gamma mixture G(0.5ν, 0.5ν), where ν is the degrees of freedom of the Student density. Thus

$$y_{ik} \sim N(\mu_{ik}, \tau_{ik}) \quad k = 1, .. n \tag{7.26a}$$

$$\tau_{ik} = \phi_{ik} / a_{ik} \tag{7.26b}$$

$$\phi_{ik} = \phi_i / \kappa_{ik} \tag{7.26c}$$

Lesage (2001) considers several spatial applications in terms of a choice between $\nu = 30$ or 40 (essentially a fixed variance over all areas in the ith regression so that $\phi_{ik} = \phi_i$) and ν varying between 2 and 9, leading to non-constant variances ϕ_{ik}. Under the latter, one may set a higher stage prior on ν itself. For example, taking

$$\nu \sim G(8, 2)$$

would be an informative prior consistent with expected spatial heteroscedasticity, and allow one to discriminate between non-constant variances over space as against non-constant regression relationships. On the other hand, a model with the full GWR parameterisation {namely the set of parameters h; ν; b_{ji}, $j = 0, .. p$; $i = 1, .. n$; and ϕ_{ik}, $i = 1, .. n$; $k = 1, .. n$) may become subject to relatively weak identifiability.

Although not considered in the 'classical' GWR literature there seems nothing against random effects models for the b_{ji} (the coefficient on variable j in the ith regression) as a way to pool strength and improve identifiability. Either unstructured effects could be used, with the b_{ji} referred to an overall average $\mu.b_j$ or spatially structured as considered by Lesage (2001).

7.5.3 Varying regressions effects via multivariate priors

A third approach to spatially varying regression coefficients, with particular relevance to discrete outcomes, involves extending the prior of Leyland *et al.* (2000) and Langford *et al.* (1999). Note that this is a single regression model, not involving n fold repetition like the GWR method. Thus, consider a model for a count outcome $Y_i \sim \text{Poi}(E_i\theta_i)$ such that the coefficients on two predictors x_i and z_i were spatially varying. The convolution model is then

$$\log(\theta_i) = \alpha + u_i + e_i + \beta_i x_i + \gamma_i z_i \tag{7.27}$$

where underlying the spatially structured effects e_i, β_i, and γ_i are three unstructured errors ν_{i1}, ν_{i2}, ν_{i3} which are linked to the effects in Equation (7.27) via scaled weighting systems which are here taken to be the same across the three effects:

$$e_i = \sum_{j=1}^{n} w_{ij}\nu_{j1}$$

$$\beta_i = \sum_{j=1}^{n} w_{ij}\nu_{j2}$$

$$\gamma_i = \sum_{j=1}^{n} w_{ij}\nu_{j3}$$

If also $u_i = \nu_{i4}$ then one may set a multivariate prior of dimension 4 (e.g. MVN or MVt) on the ν_{ij}, with mean zero for ν_{i1} and ν_{i4} but means β_μ and γ_μ on ν_{i2} and ν_{i3}, where β_μ and γ_μ are themselves assigned priors. In the case of Equation (7.27), let $\mu_\nu = (0, \beta_\mu, \gamma_\mu, 0)$ be the mean vector and Σ_ν be the dispersion matrix of dimension 4×4. In general, for the convolution model as above but with p covariates with spatially varying effects, Σ_ν will be of dimension $(2 + p)$.

This model is relatively simple to extend to multivariate dependence – as compared to adapting, say, the ICAR model – if allowing for interdependence of the ν_{ij} improves model fit. One might also envisage modifying the joint prior (7.18b) to allow several spatial effects (such as e_i, β_i and γ_i in Equation (27)) to both interact and have a different distance decay.

Coefficients are likely to vary when the regression association shows inconsistent strength over the region of interest. Consider the coefficients on x_i, and suppose μ_β is clearly positive. If a set of neighbouring areas have high regression coefficients (e.g. β_i above the mean μ_β), this means that in this subregion high values of x_i are consistently associated with high relative risks, and low values of x_i with low relative risks. By contrast, a set of β_i around zero in a particular subregion will reflect an inconsistent pattern: high values of x_i are sometimes associated with low relative risk, sometimes with high relative risk. This perspective may provide additional insight into the sometimes cited rationale for the spatial term e_i in the fixed coefficient model

$$\log(\nu_i) = \alpha + \beta x + u_i + e_i$$

namely that e_i proxies risk factors that are unobserved but smoothly varying in space. In fact, the varying coefficient model may clarify where in the region such unobserved risk factors are more relevant.

Example 7.8 Columbus crime: spatial expansion and GWR Here the spatial expansion and GWR methods are considered in the context of the Columbus crime data set, as analysed in Example 7.2. For the expansion method, one might take the geographically varying coefficients b_{ji} in Equation (7.23c) to be functions of the distance d_i of $\{s_{1i}, s_{2i}\}$ from the Columbus centre. The centre may be taken as area 32 in terms of the order of areas set out in Anselin (1988). This gives a base effect with just p parameters (rather than the $2p$ parameters in Equation (23c)), namely

$$b_{ji} = d_i\gamma_j \quad j = 1, .. p$$

However, suppose the original grid references, namely eastings-northings $\{s_1, s_2\}$, are retained. This option involves $2p$ parameters in the base effect model, and coefficients for area i as

$$b_{ji} = s_{1i}\gamma_{1j} + s_{2i}\gamma_{2j} \quad j = 1, .. p$$

Note that s_1 is lowest in the West and s_2 is lowest in the South.

For illustration, the total effects on crime rates of income (x_1) and house values (x_2) for the first five areas are monitored over the MCMC run. These total effects are related both to the base effects operating via the locations $\{s_{1i}, s_{2i}\}$, as well as the overall non-spatial effects, c_0 (constant), c_1 (income), c_2 (house values) after allowing for the spatial expansion. Note that a linear model with homogenous coefficients (and spatially uncorrelated errors) gives parameters and t-ratios as follows:

$$c_0 = 68.6(14.5),\ c_1 = -1.6(4.8),\ \text{and}\ c_2 = -0.27(2.7)$$

Relatively informative N(0, 1) priors are assumed for γ_{1j} and γ_{2j}. A full spatial expansion model as in Equation (7.23b) shows no non-spatial effect c_2 of house values, though a negative non-spatial effect of income (c_1) remains, and is more negative than under the homogenous coefficients model. Table 7.6 shows the combined effects, as in Equation (7.23d), for areas 1 to 5. The base effects may be interpreted as deviations from the overall effect as one moves from west to east or from south to north. Thus, the effect of income becomes less negative as one moves east, but more negative as one moves north. The effects of income in areas 1–5 are less negative than the average of -1.95 given by β_1 (see Equation (7.23e)), while those of house values are more negative than the average.

For the GWR analysis, a Gaussian distance decay function is taken, with

$$a_{ik} = \phi(d_{ik}/[s_d h])$$

where ϕ is the standard normal density, s_d is the standard deviation of all the inter-neighbourhood distances and $h = 1$ is taken as known (as indicated by a preliminary search over values between 1 and 10). As is usual in a GWR analysis, varying regression intercepts between the n regressions are allowed, as well as varying predictor effects over space (see Equation (7.25c)). The Bayesian sampling approach to estimation means one can monitor statistics such as the minimum and maximum b_{ji} for predictor j over all $i = 1, .., n$ regressions, and also the probability that the ith regression provides the maximum or minimum coefficient.

From a single chain of 20 000 iterations (500 burn in), Table 7.7 shows that the average income and house value effects, denoted bavg[ ] in Program 7.8, are similar to those obtained from the homogenous coefficient model and cited above.

Example 7.9 Lip cancers; varying regression effects and heteroscedasticity This example considers the Scottish lip cancer data of Example 7.3, but using the varying coefficient model of Equation (7.27) as one alternative explanation of the observations. The model adopted for varying coefficients is

$$\log(\theta_i) = \alpha + e_i + \beta_i x_i$$

where x_i is 0.1 times the percent of AFF workers. The other 'spatial heterogeneity' approach involves non-constant variances in the model

Table 7.6 Spatial expansion model (three chain run of 10 000 iterations, 500 burn in)

Total effects for areas 1–5	Mean	St. devn.	2.5%	Median	97.5%
Income					
1	−0.70	0.56	−1.74	−0.71	0.44
2	−0.93	0.50	−1.88	−0.95	0.09
3	−1.13	0.46	−2.00	−1.14	−0.21
4	−0.98	0.47	−1.88	−0.99	−0.03
5	−0.73	0.60	−1.83	−0.75	0.52
House values					
1	−0.66	0.15	−0.97	−0.66	−0.37
2	−0.56	0.13	−0.82	−0.56	−0.30
3	−0.48	0.12	−0.71	−0.48	−0.25
4	−0.56	0.13	−0.81	−0.56	−0.31
5	−0.63	0.16	−0.95	−0.62	−0.32
Base effects income (West to East then South to North effects)					
γ_{11}	0.105	0.054	0.004	0.106	0.211
γ_{21}	−0.047	0.034	−0.112	−0.048	0.018
Base effects house values (West to East then South to North effects)					
γ_{12}	−0.042	0.021	−0.083	−0.043	−0.001
γ_{22}	0.027	0.013	0.001	0.028	0.052
Non-spatial effects					
c_0	69.6	4.8	60.1	69.6	79.1
c_1	−3.45	1.77	−6.87	−3.29	−0.27
c_2	0.16	0.66	−1.05	0.11	1.39
Total effects					
β_1	−1.95	0.38	−2.68	−1.94	−1.18
β_2	−0.12	0.12	−0.34	−0.12	0.12

$$\log(\theta_i) = \alpha + u_i + \beta x_i$$

such that $u_i \sim N(0, 1/\phi_i)$ and $\phi_i = \phi\kappa_i$, where $\kappa_i \sim G(0.5\nu, 0.5\nu)$.

A two chain run of 50 000 iterations for the varying coefficient models (with convergence by 5000 iterations) shows a DIC of around 85.7 which is very close to that (namely 85.6) of the ICAR(ρ) model of Example 7.3. The coefficients on the AFF variable are highest in a belt of counties in the Highlands and on the Moray Firth, where variations in lip cancer mortality seem to follow the percents AFF quite closely: these are Nairn (county 13), Skye-Lochalsh (1), Inverness (19), Banff (2) and Ross and Cromarty (5). The lowest coefficients are in a set of counties in east central Scotland (just north or east of Edinburgh), namely Clackmannon (43), Falkirk (42), Kirkcaldy (25), Dunfermline (26) and two counties in the far south west of Scotland, namely

Table 7.7 GWR analysis: varying effects of income and house value

		Mean	St. devn.	2.5%	Median	97.5%
Average	Intercept	68.9	0.7	67.5	68.9	70.2
	Income	−1.60	0.05	−1.71	−1.60	−1.50
	House Value	−0.24	0.02	−0.28	−0.24	−0.21
Maximum	Intercept	83.2	3.2	78.2	82.8	90.2
	Income	−0.18	0.26	−0.60	−0.20	0.40
	House Value	0.19	0.09	0.04	0.17	0.40
Minimum	Intercept	49.1	4.6	39.3	49.4	56.8
	Income	−2.78	0.24	−3.34	−2.75	−2.39
	House Value	−0.72	0.07	−0.86	−0.71	−0.60

Annandale (55) and Stewartry (32). In these two subregions lip cancer mortality is low or average, but the percent AFF varies widely, and Clackmannon has below average mortality but a percent AFF of 16. The average coefficient on the AFF variable is 0.51 with 95% interval (0.20, 0.87), higher than under the conditional autoregressive priors.

The heteroscedastic approach employs a discrete prior with 28 bins on the value of ν. There are greater prior concentrations of the grid at lower values of ν. The posterior median for ν is 18, suggesting heavier tails than in the Normal. Other aspects of this model are that it also enhances the AFF coefficient, with a two chain run of 10 000 iterations (1000 burn in) showing a posterior mean and 95% interval of 0.68 (0.40, 0.96). There is a loss of fit (compared to the models in Example 7.3) with the DIC standing at 93, but various model developments such as adding a spatially correlated error may improve on this. The lowest κ_i are on Skye-Lochalsh, Annandale and Falkirk.

7.6 CLUSTERING IN RELATION TO KNOWN CENTRES

The above modelling strategies have considered whether measures for aggregate areas or individual outcomes exhibit a spatial structure without focusing on known cluster centres or seeking to identify unknown cluster centres. In environmental epidemiology, it is frequently the goal to assess the degree of *focused* clustering in relation to one or more hazard sites. For instance, the existence or otherwise of clustering of child leukaemia cases near nuclear processing plants or hazardous waste sites has generated a range of possible models (Cocco *et al.*, 1995; Waller *et al.*, 1994). The analysis may focus on individual disease events and the pattern of their spatial locations in relation to one or more hazard sites, or on aggregate disease counts for small areas, again in relation to point sources. In the absence of individual or aggregate measures of exposure, the distance from the source is often used as a proxy for exposure.

For testing clustering of events around a specified point source (i.e. a putative hazard site), a number of testing procedures have been suggested, but may have limited potential to model simultaneously features such as the impact of covariates, the presence of overdispersion, or generalised spatial dependence in the outcome beyond that associated with exposure to the hazard source. The basis of a more general parametric modelling strategy in relation to environmental hazard sites builds on the approaches set out by authors such as Diggle (1990), Diggle and Rowlingson (1994) and Morris and Wakefield (2000). More general modelling approaches that offer a synthesis with the work of Besag *et al.* (1991) and others in relation to unfocused clustering appear in Lawson (1994), Lawson *et al.* (1999) and Wakefield and Morris (2001).

Diggle (1990) suggested that the relative risk for disease events at location s in relation to a point source at s_0 could be represented as

$$\lambda(s) = \rho g(s) a(d)$$

where $d = |s - s_0|$ is the distance between the location and the point source, ρ is the overall region-wide rate, g(s) is the population at risk at location s, and $a(d)$ is a distance decay function expressing the postulated decline in exposure at greater distances from the source, and preferably also the (possibly elevated) relative risk of disease at the source as compared to the 'background' level.

For instance, possible distance decay functions, expressing monotonic lessening in risk as distance from the point source increases, involve the additive form

$$a(d) = 1 + \eta f(d)$$

where $f()$ might be a simple exponential function

$$f(d) = \exp(-\alpha d)$$

or a squared exponential

$$f(d) = \exp(-\alpha d^2)$$

Taking $\alpha > 0$, these functions have the property that $a()$ tends to 1 (the background risk) as d tends to infinity. Also, $1 + \eta$ can be interpreted as the relative risk at or close[15] to the source itself (where d is near 0). If in fact $\eta = 0$, or effectively so, there is no association between risk and distance from the source. Another type of model is 'hot spot' clustering, when there is uniformly elevated risk in a neighbourhood around the focus, but background risk elsewhere. Thus, for tracts at distances $d < \delta$ from the focus, the risk is $1 + \eta$, but for $d > \delta$, the risk is either just set at 1, or maybe follows a distance decay from the threshold distance:

$$a(d) = 1 + \eta \exp(-\alpha[d - \delta]^2)$$

Models where direction θ, not just distance, from the point source to the case event or tract is relevant (e.g. when prevailing winds influence the potential spread of pollution) might include terms in $\sin\theta$ and $\cos\theta$. If there is a peak risk away from the source, then a term in $\log(d)$ as well as d can be included. Lawson and Williams (1994) suggest the additive form

$$a(d, \theta) = 1 + \eta f(d, \theta)/d$$

so that risk still tends to 1 as $d \to \infty$, but $f()$ may include aspects of direction.

[15] With certain decay functions $f(d)$ may not be defined at $d = 0$, and a minimum possible distance of residence from source may be defined and used instead.

7.6.1 Areas vs. case events as data

In area studies with observations consisting of disease counts Y_i, the population risk in area i might be approximated by the expected disease total E_i, based on the population age structure in that area, and d might be the distance between the area centroid and the point source. Additionally, to account for the more usual sources of overdispersion or spatial correlation, the model for the Poisson mean μ_i might include unstructured effects u_i and spatial effects e_i as in the preceding sections. If area characteristics x_i (such as deprivation) also influence the disease risk, one has a model such as

$$\mu_i = \rho E_i \exp(\beta x_i)[1 + \mathrm{f}(d_i)] \exp(u_i + e_i)$$

or

$$\log(\mu_i) = \log \rho + \log(E_i) + \beta x_i + \log[1 + f(d_i)] + u_i + e_i$$

Often, observations consist of a set of cases and their locations. On the assumption that individuals are independent (in spatial terms) with regard to risk of disease, possibly after allowing for relevant risk factors or confounders, one may, however, model the likelihood of case events $i = 1, .. n$ at locations (s_i) in relation to a point source at s_0. The population density may be modelled via kernel methods, typically using aggregated administrative data for districts within the region. An alternative is to proxy the population distribution using a control disease unrelated to the risk from the point source (e.g. cardiovascular conditions may be assumed independent of residence near high voltage electricity lines, but certain cancers may not be).

Let the n_1 disease cases be 'successes' in terms of a Bernoulli process, with $y_i = 1$. So if the n_2 control disease cases (with $y_i = 0$) have locations s_i, $i = n_1 + 1, .. n_1 + n_2$, the probability of caseness π_i can be modelled in terms of the decay function, and possibly individual risk factors x_i (which are confounders in terms of the main interest in the relation between risk and the pollution sources). Thus,

$$y_i \sim \mathrm{Bernoulli}(\pi_i)$$

with

$$\pi_i/(1 - \pi_i) = \rho^*[1 + f(d)] \exp(\beta x_i)$$

This conditional approach (Diggle and Rowlingson, 1994; Lawson, 2001, p. 44) has the advantage of not requiring an estimate of $g(s)$. Note, however, that the interpretation of ρ^* depends upon the selection of cases and controls; specifically, $\rho^* = (a/b)\rho$, where a is the sampling proportion of cases (often 100%), b is the sampling proportion of controls, and ρ the population odds of disease (Diggle *et al.*, 2000).

7.6.2 Multiple sources

Often there are multiple sources of environmental pollution, and assuming each site contributes to the risk independently of others one may take (Biggeri and Lagazio, 1999)

$$\mu(s) = \rho g(s) a(d, \varphi) = \rho g(s) \left[1 + \sum_k \eta_k f(d, \alpha_k) \right]$$

For instance, one option is

$$\mu(s) = \rho g(s) a(d, \varphi) = \rho g(s) \left[1 + \sum_k \eta_k \exp(-\alpha_k d) \right]$$

though often either or both of η_k and α_k are taken equal. Morris and Wakefield (2000) propose a slightly different formulation, with

$$\mu(s) = \rho g(s) a(d, \varphi) = \rho g(s) \prod_k [1 + \eta_k \exp(-\alpha_k d)]$$

Lawson (2001) discusses adaptation of the multiple source model to the case where both the geographic location of the cluster centres and their number are unknown. Other issues are relevant in multi-site analysis, for example, proximities between sites themselves: risk from two sources relatively close to one another may overlap.

Example 7.10 Leukaemia cases and hazardous waste sites A focused modelling analysis for area disease counts is illustrated by cases of leukaemia during 1978–82 in 790 geographic tracts in upstate New York, as considered by Waller *et al.* (1994) in relation to 11 hazardous waste sites. The observed data in fact contain decimal parts because some cases could not be allocated with certainty to particular cells; the observed O_i vary from 0 to 9.29. For illustration, the analysis is based on counts obtained as the integer parts of $O_i + 0.5$, so $Y_i = \text{Int}[O_i + 0.5]$.

A model including a distance decay function is essentially a departure from randomness, where randomness implies that each person in the population has the same chance of a disease regardless of their place of residence (Waller *et al.*, 1994). Under randomness and without allowing for possible overdispersion, $Y_i \sim \text{Poi}(\rho P_i)$, where P_i is the population of the ith tract. Suppose, instead, the expected counts under randomness are multiplied by an additive or multiplicative function for relative risk as a function of distance from, or contiguity to, the focus.

For instance the so-called clinal clustering model assumes that

$$\mu_i = \rho P_i[1 + \eta f(d_i)] \tag{7.28}$$

where $a(d_i)$ is a distance decay function in terms of the distance of tract i from the source. To allow a 'test' of the null model

$$\mu_i = \rho P_i$$

against the full model in Equation (7.28), one may introduce a regression selection indicator, $H \sim \text{Bern}(0.5)$. Then, the likelihood involves the mean function

$$\mu_i = \rho P_i[1 + H\eta f(d_i)]$$

and the probability for the full model is the posterior probability that $H = 1$.

Here this approach is first carried out separately for each of the 11 sites, using the function $f(d) = \exp(-\alpha d)$. Note that this involves possible issues of adjusting for 'multiple tests'. This analysis employs the device of replicating the data 11 times and fitting a different Poisson model to each replicated data set, such that the mean for the kth analysis is based on distances from the kth site (Model A in Program 7.10). Since the sites are all waste disposal sites, and a single disease risk is under consideration, a single parameter α seems plausible to model the distance decay effect over the 11 sites (taking $\alpha_k = \alpha$ means that strictly the 'tests' are not separate). The excess risk η_k is allowed to vary by site, and eleven Bernoulli variables H_k model the chance that η_k is positive. So the mean for the kth analysis is

$$\mu_{ik} = \rho P_i[1 + H_k \eta_k \exp(-\alpha d_{ik})]$$

Informative G(1, 1) priors placing low weight on large values are adopted for α and η_k. A relatively diffuse G(1, 0.001) prior is taken for ρ. A two chain run of 2000 iterations (with convergence at 250) yields probabilities on the full model in Equation (7.28) similar to those implied by the classical hypothesis test procedure of Waller *et al.* (1994, p.20). Specifically, the averages of the statistics $1 - \Pr(H = 1)$ are intermediate between the significance test probabilities cited in Table 4 of Waller *et al.* (1994) for Stone's test and Rao's test (see also Lawson, 1993). The latter test is based on the sum of the deviations of the observed disease incidence in each tract from its expectation ρP_i under the null model, weighted by the exposure $f(d)$, where in Waller *et al.* (1994) and Waller and Poquette (1999) the latter is set as $f(d) = 1/d$.

Thus, the probability that $H_k = 1$ exceeds 0.995 for the first two sites (Monarch Chemicals, IBM Endicott) and is around 0.98 for the third, fourth and seventh sites (Singer, Nesco, Smith Corona). The parameter α is around 0.09 with 95% interval {0.04, 0.14}. The results here are based on allowing a free parameter in the function $f(d)$, and one could follow the Waller *et al.* approach more literally by setting $f(d) = 1/d$, though this would affect the estimates of the excess risk at the source (i.e. of $\eta = RR - 1$).

A global test of risk from one or more waste sites is implemented via the multi-site model

$$\mu_i = \rho P_i\left[1 + H\sum_k \eta_k \exp(-\alpha d)\right]$$

with $H \sim \text{Beta}(0.5)$. The probability that the Bernoulli indicator H is 1 is equivalent to the probability that there is risk from one or more sites. In fact[16], there is a negligible probability that $\Pr(H = 0)$, and so clear evidence of risk from one or more sites. The estimate of the distance decay parameter is higher under this model, at around 0.30, and the η_k vary from 1.88 (site 7) to 0.27 (site 10). Some relative risks are clearly more elevated than others, with 2.5% points for η_k away from zero in the first, fifth and seventh sites (Table 7.7).

The Poisson mean might be altered to a form with site specific selection indicators H_k

$$\mu_i = \rho P_i\left[1 + \sum_k H_k \eta_k \exp(-\alpha d)\right]$$

to account for differential risks by site. Applying this approach[17] in Model C in Program 7.10, it appears that the probabilities $\Pr(H_k = 1)$ are highest for sites 1, 5 and 7 (the probabilities are between 0.95 and 0.97), though not for site 2 (IBM Endicott) as in the 'independent tests' model A. The close proximity of sites 1 and 2 may be noted, and this may affect its estimated risk in a multi-site model.

Possible extensions of these forms of model are to add unstructured or spatially structured errors for the 790 tracts, or if they were available, to add tract covariates (e.g. income).

Example 7.11 Larynx cancer in Lancashire An example of an individual level outcome, we consider event case data consisting of 57 larynx cancer cases that occurred in a part of Lancashire over 1974–83. The controls are provided by 917 lung cancer cases in the same study region. The point source implicated in the disease is a waste incinerator

[16] As judged from a two chain run of 2000 iterations, and convergence by 250 iterations.

[17] Estimates of η_k in this model should only be based on the iterations where $G_k = 1$, since when $G_k = 0$ η_k is just a sample from the prior.

Table 7.7 Estimates of distance and site risk parameters; Global risk model

	Mean	St. devn.	2.5%	Median	97.5%
α	0.302	0.102	0.146	0.289	0.546
η_1	1.478	0.814	0.209	1.361	3.404
η_2	0.543	0.415	0.023	0.455	1.541
η_3	0.714	0.598	0.024	0.576	2.198
η_4	0.754	0.688	0.020	0.565	2.547
η_5	1.168	0.507	0.321	1.123	2.275
η_6	0.907	0.888	0.021	0.636	3.380
η_7	1.884	0.854	0.396	1.822	3.786
η_8	0.977	0.846	0.035	0.758	3.211
η_9	0.796	0.771	0.027	0.571	2.797
η_{10}	0.273	0.234	0.010	0.213	0.855
η_{11}	0.866	0.772	0.026	0.655	2.963

at location (355000, 414000). Although frequently analysed, this data set has a relatively small case total, and firm identification of exposure effects in terms of (say) distance decay may be difficult to obtain.

As above, the intensity model is

$$\lambda(s, d) = \rho^* g(s)\{1 + f(d)\}$$

or with individual confounders x_i

$$\lambda(s, d) = \rho^* \exp(\beta x_i) g(s)[1 + f(d)]$$

but in the conditional logistic model (Diggle and Rowlingson, 1994), the population intensity is 'conditioned out'. In the present example, there are no confounders. So if $Y_i = 1$ for larynx cancers and $Y_i = 0$ for lung cancers, the simplest sampling model is Bernoulli, with

$$Y_i \sim \text{Bernoulli}(\pi_i)$$
$$\pi_i/(1 - \pi_i) = \rho^*[1 + f(d)]$$

Here terms in d and $\log(d)$ are introduced in $f()$ to allow for peaked exposure and incidence away from the source. Additionally, directional measures are introduced, namely terms in $\cos(\theta)$ and $\sin(\theta)$, where θ is the angle (in radians) between source and case/control event. For instance an event due North of the source would have angle 90° and radian value $\theta = 1.57$, while an event due South of the source would have angle and radian of 270° and 4.71, respectively.

An initial 'variable selection' stage involves setting binary selection indices G_1 to G_4 on the coefficients α_1 to α_4 in the function

$$a(d, \theta) = [1 + \exp\{-\alpha_1 d + \alpha_2 \log(d) + \alpha_3 \cos(\theta) + \alpha_4 \sin(\theta)\}/\text{d}]$$

where α_1, α_2 are both taken as positive (Lawson and Williams, 1994) and N(0, 1) priors constrained to positivity are adopted. For α_3 and α_4 N(0, 10) priors are assumed. A single chain of 10 000 iterations show selection rates slightly above 0.5 for α_1 and α_3 by comparison to prior rates of 0.5 under the priors $G_j \sim$ Bern(0.5), whereas α_2 and α_4 have selection rates below 0.5. So there is (rather weak) evidence that these are potentially more important predictors of spatial variations in risk and incidence.

Investigations show that fitting a model with both $\cos(\theta)$ and linear distance is not well identified. Predictive methods (e.g. Gelfand and Ghosh, 1998) are then used to assess the alternative single predictor models

$$\begin{aligned} a(d, \theta) &= [1 + \eta_\beta \exp\{\beta \cos(\theta)\}/d] \\ a(d, \theta) &= [1 + \eta_\alpha \exp\{-\alpha d)\}/d] \end{aligned} \quad (7.29)$$

Note that the first model has a (preset) distance decay effect by default. In fact, this model identifies a reasonably well defined negative effect of β, or raised risk in the NW and SW quadrants, after taking account of the preset inverse distance effect. The parameter η_β is higher than η_α, with posterior median of 0.56 as against 0.22 for η_α. However, the models are comparable in terms of predictive fit, both around 82.5.

Figure 7.3 is a kernel plot of the relative risk effect under Equation (7.29), namely $\eta_\beta = RR - 1$, and shows it to be distinct from the zero null value, so apparently identifying a real effect. Figure 7.4 shows the posterior density of the directional parameter.

7.7 SPATIO-TEMPORAL MODELS

The above applications have considered spatial outcomes without reference to other dimensions, but the evolution over time of dependent variables for spatial outcomes has relevance in several contexts. For instance, tests for clustering may be extended to include both time and space (e.g. Rogerson, 2001), and prediction from panel data of area data may be extended to include spatial correlation (Baltagi and Dong, 2000). If age and cohort are included with time, then a broad class of spatial Age-Period-Cohort (APC) models may be used in modelling and forecasting mortality or disease incidence by area (Schmid and Knorr-Held, 2001). The focus here only illustrates the potential for Bayesian modelling, and considers especially ecological regression through time, and spatio-temporal models for innovation.

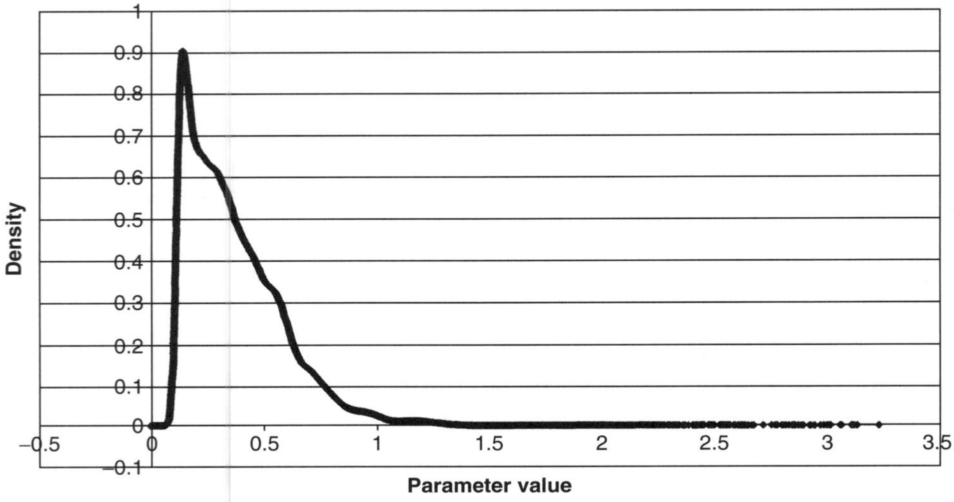

Figure 7.3 Plot of distance decay parameter

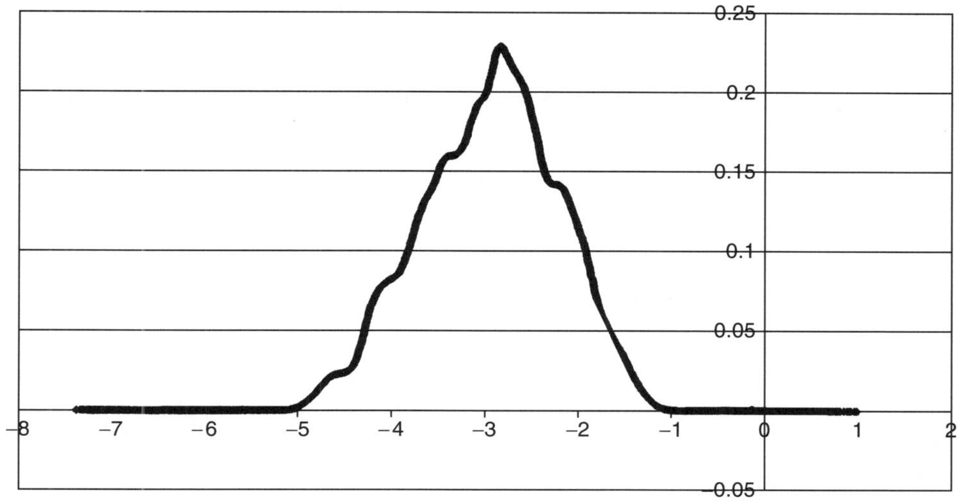

Figure 7.4 Kernel plot of directional parameter

In terms of ecological regression, Carlin and Louis (1996) discuss analysis of counts of deaths or new disease cases by area i and demographic subgroups j (age, sex, ethnicity) which are recorded at regular intervals t. This would typically generate binomial or Poisson outcomes in relation to populations at risk P_{ijt}. Thus, possible specifications are

$$Y_{ijt} \sim \text{Bin}(r_{ijt}, P_{ijt})$$

and

$$Y_{ijt} \sim \text{Poi}(E_{ijt}\,\phi_{ijt})$$

with the expected cases E_{ijt} based on applying standard rates (e.g. national rates) to the area populations P_{ijt}. The model for the Binomial rates r_{ijt} or relative risks ϕ_{ijt}, and hence the priors on the parameters of that model, would be specified in terms of two major criteria, namely the desire to stabilise rate estimates by smoothing the crude estimates, and the need for the model to reflect both spatial and temporal dependence in the observations or regression errors.

Consider a model with just areas and time, and with the impacts of different demographic structures controlled for by standardisation. Thus,

$$Y_{it} \sim \text{Poisson}(E_{it}\phi_{it})$$

and a simple model for the log relative risk $Z_{it} = \log(\phi_{it})$ would be as a sum of constant 'unstructured' effects u_i, spatially dependent effects e_i for instance with an ICAR(ρ) form as in Equation (7.12), and time parameters δ_t.

For a short panel of observations with an ICAR(ρ) model for e_i and with

$$Z_{it} = \mu + u_i + e_i + \delta_t$$

the δ_t may be best modelled as fixed effects, with $\delta_1 = 0$ for identifiability. Alternatively, the intercept μ can be omitted, and the δ_t as free parameters will then detect any trend.

If an ICAR(1) form is used for the e_i, then it will be necessary to assume both $\mu = 0$ and $\delta_1 = 0$ unless the sampled values of e_i are centred at each iteration; further questions of identifiability are considered by Waller *et al.* (1997). To allow for unstructured space-time interactions ψ_{it}, one might extend the realistic model (assuming an ICAR(ρ) model for e_i,) to

$$Z_{it} = \mu + u_i + e_i + \psi_{it}$$

where the ψ_{it} are distributed about time specific means δ_t.

7.7.1 Space-time interaction effects

In practice, more complex issues occur as discussed, for example, by Knorr-Held and Besag (1998), Gelfand *et al.* (1998) and Sun *et al.* (2000). For example, it may be that the relative balance of unstructured random variation as against spatial clustering is changing over time. This might suggests a model with clustering differentiated by time, such as

$$Z_{it} = \mu + u_i + \eta_{it}$$

where the spatially correlated risk η_{it} has a time specific variance κ_t^2 (Carlin and Louis, 1996). Modelling options for spatially structured area-time interactions η_{it} are very wide, since autocorrelations over areas may be combined with those over time. For example, defining $\eta_t = (\eta_{i1}, \eta_{i2}, \ldots . \eta_{nt})$ and a white noise errors $\nu_t = (\nu_{i1}, \nu_{i2}, \ldots . \nu_{nt})$, then instead of a simple cross-sectional spatial correlation

$$\eta_t = \lambda W \eta_t + \nu_t$$

one may have one or more time lagged spatial correlation effects, such as

$$\eta_t = \lambda_1 W \eta_t + \lambda_2 W \eta_{t-1} + \nu_t$$

Lagazio *et al.* (2001) assume spatial effects are described by contiguity and describe a 'random walk' scheme in space and time, involving smoothing forward as well as back in time, whereby

$$\eta_{it} \sim \mathrm{N}(0.5[\eta_{i,\,t-1} + \eta_{i,\,t+1}] + N_i^{-1}\left[\sum_{j\sim i} \eta_{jt} - \sum_{j\sim i} \eta_{j,\,t-1} - \sum_{j\sim i} \eta_{j,\,t+1}\right],\ \kappa^2/[2N_i])$$

where $j \sim i$ denotes those N_i areas of i which are its first order neighbours.

7.7.2 Area level trends

Another issue centres on modeling trends in the outcome (e.g. mortality) relative, say, to national levels. Such trends may be different between areas. A linear trend, uniform across all areas, can be included in a model such as

$$Z_{it} = \mu + \gamma t + u_i + e_i$$

To allow for differentiated trends between areas with some falling more than the national or regional trend, some much less, one might specify a random growth rate. Thus following Bernardinelli *et al.* (1995),

$$Z_{it} = \mu + \gamma_i t + u_i + e_i \tag{7.30}$$

where γ_i may be taken as unstructured growth rates with overall average growth rate γ. However, if trends are expected to be differentiated in a spatially distinct pattern (e.g. if the largest relative falls in the outcome are spatially clustered), then the γ_i might be assumed to be spatially dependent, for example with ICAR(ρ) or ICAR(1) form and with variance σ^2_γ (see also Sun *et al.*, 2000).

Temporally persistent differences in the outcome may be important. For example, persistent mortality differences across demographic, cultural, regional and groups have been noted by authors such as Kerkhof and Kunst (1994). Even when there is some shuffling of spatial relativities in the outcome over time, one may nevertheless ask how persistent differentials across areas are, for example, via introducing a temporal autocorrelation in the risks. This may be expressed in a panel model such as

$$\begin{aligned} Z_{it} &= \mu + \psi_{it} + u_i + e_i \\ \psi_{it} &= \rho\psi_{it-1} + \nu_{it} \end{aligned}$$

for time periods $t > 1$, where the ν_{it} are white noise errors, and

$$Z_{i1} = \mu + \psi_{i1} + u_i + e_i$$

for the first period. For stationarity, the temporal correlation parameter ρ is constrained to be between -1 and $+1$. The ψ_{it} for $t > 1$ have mean $\rho\psi_{it-1}$ and precision τ_ν. The ψ_{i1} are assumed to follow a distinct prior with mean 0 and precision τ_1.

Differences in persistence between areas under either model could be handled by making persistence an area specific random effect ρ_i, either spatially structured or unstructured. Thus,

$$\begin{aligned} Z_{it} &= \mu + \psi_{it} + u_i + e_i \\ \psi_{it} &= \rho_i\psi_{it-1} + \nu_{it} \end{aligned} \tag{7.31}$$

for time periods $t > 1$, and

$$Z_{i1} = \mu + \psi_{i1} + u_i + e_i$$

for $t = 1$. If the ρ_i are assumed spatially uncorrelated, then their prior density could be of the form $\rho_i \sim \mathrm{N}(0, \sigma^2_\rho)$ where σ^2_ρ represents variations in persistence.

7.7.3 Predictor effects in spatio-temporal models

Finally, variability in relative risks over both space and time may be caused by changing impacts of social and other risk variables. Trends in the impact of a time-specific predictor X_{it} may be modelled via

$$Z_{it} = \beta_t X_{it} + \mu + u_i + e_i$$

with β_t either fixed or random effect (e.g. modelled by a first order random walk). A model with both area and time dimensions also allows one to model differences in the importance of explanatory variates between areas, for instance via a model such as (Hsiao and Tahmiscioglu, 1997)

$$Z_{it} = \beta_i X_{it} + \delta_t + u_i + e_i$$

and this may be achieved without recourse to the special methods of Section 7.5. However, models with regression coefficients which are both time and spatially varying as in

$$Z_{it} = \beta_{it}X_{it} + \delta_t + u_i + e_i$$

would suggest using GWR or multivariate prior methods, as discussed earlier in Section 7.5.

7.7.4 Diffusion processes

Behavioural considerations may also influence the model form. In both medical geography and regional economics, diffusion models have been developed to describe the nature of spread of new cases of disease, or of new cultures or behaviours in terms of proximity to existing disease cases or cultural patterns. Here the model seeks to describe a process of contagion or imitation, namely of dependence among neighbouring values of the outcome, and error terms may possibly be taken as independent, once the autodependence in the outcomes is modelled satisfactorily. To cite Dubin (1995) in the case of adoption of innovations by firms, 'just as a disease spreads by contact with infected individuals, an innovation becomes adopted as more firms become familiar with it, by observing prior adopters'. If the outcome is absorbing, or at least considered irreversible for modelling purposes, then the process of diffusion continues until all potential subjects exhibit the behaviour concerned, and the process approximates S-shaped or logistic diffusion.

Dubin considers a dynamic logit model of diffusion to describe the adoption of innovations by firms or entrepreneurs. The observed outcome is binary Y_{it}, equalling 1 if the innovation is adopted by firm i at time t, and $Y_{it} = 0$ otherwise. Note that it is not possible for $Y_{i,\, t+k}$ $(k > 0)$ to be zero of Y_{it} is 1. This outcome is generated by an underlying utility or profit level $Y_{it}{}^*$, which is a function of the firm's characteristics and its distance from earlier adopters:

$$Y_{it}{}^* = X_{it}\beta + \sum_j \rho_{ij} Y_{j,\, t-1} + u_{it} \tag{7.32a}$$

where $Y_{it}{}^*$ is the expected profit from the innovation and the u_{it} are without spatial or time dependence. The influence ρ_{ij} of prior adopters is a function of inter-firm distance d_{ij}, such that

$$\rho_{ij} = \alpha_1 \exp(-\alpha_2 d_{ij}) \tag{7.32b}$$

with $\rho_{ii} = 0$. Here α_1 expresses the impact on the chances of adopting an innovation of the presence of adjacent or nearby adopters (with d_{ij} small), and $\alpha_2 > 0$ expresses the attenuation of this impact with distance.

Example 7.12 Adoption of innovations In a similar way to Dubin (1995), we simulate data over $T = 5$ periods for 50 firms using a grid of 100×100 km, so that the maximum possible distance between firms is 140 km. The first period data are generated with

$$Y_{i1}{}^* = -3 + 2x_{1i} - 3x_{2i} \tag{7.33}$$

where x_1 and x_2 are standard normal. Grid references (s_{1i}, s_{2i}) are randomly selected from the interval (0, 100). 15 out of the 50 firms are adopters at time 1 (i.e. have positive $Y_{i1}{}^*$) using this method. At the second stage the observations Y_{it}, at times $t > 1$, are generated according to the 'profits model' in Equation (7.32), taking $\alpha_1 = 0.5$ and $\alpha_2 = 0.02$ and distances defined between pairs (s_{1i}, s_{2i}). In this example, if it is known that $Y_1 = 0$ then $Y_0 = 0$ necessarily, so there are partly observed 'initial conditions'. For firms with $Y_{i1} = 1$, one may simulate Y_{i0} using an additional parameter π_0.

Then with the 50×5 simulated data points, the goal is to re-estimate the generating parameters. A beta prior B(1, 1) is adopted for π_0, and N(0, 1) priors on the parameters α_j of the influence function ρ_{ij} (with only positive values allowed). Non-informative priors are taken on the parameters of β. On this basis, a two chain run of 2500 iterations (and convergence from 250) shows the β parameters are re-estimated in such a way as to correspond to those in (7.33), and the distance decay parameter α_2 is also closely reproduced (Table 7.8).

Especially in processes where transitions from 1 to 0 as well as from 0 to 1 are possible, one might also consider space and time decay dependence via functions such as

$$\rho_{ijt} = \alpha_1 \exp(-\alpha_2 d_{ij} - \alpha_3 t)$$

with the sign and size of α_3 reflecting the path of the process over time.

In the present application (despite knowing that the mode of data generation assumes α_2 fixed over time), one may more generally envisage distance decay varying over time, since the balance between adopters and non-adopters, and hence the spatial distribution of the two categories, changes through time. Therefore, an alternative model here takes α_2 to be time varying with

$$Y_{it}^* = X_{it}\beta + \sum_j \rho_{ijt} Y_{j,\, t-1} + u_{it}$$

$$\rho_{ijt} = \alpha_1 \exp(-\alpha_{2t} d_{ij})$$

In fact the predictive loss criterion (footnote 5) with $w = 1$ shows the constant decay effect model to be slightly preferred with $D.z$ at 22.1 compared to 23.6 under time varying decay. The coefficients α_{2t} are much higher in the first two periods[18], with means 0.8 and 0.33 (though both parameters have skewed densities), as compared to means for the last three periods of 0.041, 0.035 and 0.031, respectively. The α_1 coefficient is elevated to 0.9. A suggested exercise is to fit the model with both distance function parameters time varying:

$$\rho_{ijt} = \alpha_{1t} \exp(-\alpha_{2t} d_{ij})$$

Example 7.13 Changing suicide patterns in the London boroughs This analysis considers spatio-temporal models in a disease mapping application with event count data. Specifically, the focus is on trends in period specific total suicide mortality (i.e. for all

Table 7.8 Profits model for spatially driven innovation

	Mean	St. devn.	2.50%	Median	97.50%
α_1	0.58	0.20	0.25	0.55	1.05
α_2	0.019	0.009	0.003	0.019	0.037
β_1	−3.10	0.61	−4.37	−3.07	−1.99
β_2	1.94	0.35	1.30	1.93	2.65
β_3	−3.28	0.53	−4.37	−3.26	−2.32
π_0	0.032	0.030	0.002	0.026	0.103

[18] Convergence occurs later in this model, at around 500 iterations in a two chain run of 2500 iterations.

ages and for males and females combined) for the 33 boroughs in London and eight periods of two years each between 1979 and 1994. Expected deaths are based on national (England and Wales) age-sex specific death rates for each period. There is evidence for changing suicide mortality in London over these periods relative to national levels, and for shuffling of relativities within London. Thus, London's suicide rate seemed to be falling against national levels with an especially sharp fall in central London areas.

Here two models among those discussed by Congdon (2001) are considered, namely model (7.30), allowing for differential trends to be spatially structured, and a differential persistence model as in Equation (7.31) with unstructured borough persistence effects ρ_i. The latter model was estimated subject to the assumption of stationarity in the individual borough autocorrelations. Fit is assessed by the pseudo marginal likelihood and via the Expected Predictive Deviance (EPD) criterion. This is based on 'predicting' new or replicate data from the posterior parameters. The better the new data match the existing data, the better the fit is judged to be (Carlin and Louis, 1996, Section 6.4). Specifically, Poisson counts $Y_{it.\text{new}}$ are sampled from the posterior distribution defined by the predicted Poisson means and compared with the observed Y_{it} via the usual Poisson deviance measure. As for the usual deviance, the EPD is lower for better fitting models.

Estimation of the differential growth model in (7.30) using a two chain run of 5000 iterations shows an average decline of 0.02 in the relative suicide risk in London boroughs in each period (measured by G <- mean(s[ ]) in Model A in Program 7.13). The steepest decline, at 0.054 per period, is in the central London borough of Westminster (area 33).

For the differential persistence model, the posterior estimate of σ_ρ at around 0.65 (with 95% interval from 0.5 to 0.85) suggests some variation in the extent to which suicide risk within boroughs is temporally autocorrelated. The average correlation is around 0.44, but it varies in individual boroughs from −0.15 to 0.92. The highest continuity was in low suicide suburban boroughs which tended to remain at low levels throughout the span of the study, and in central London boroughs where suicide remained consistently high, though the excess above the overall average was falling through time.

In selecting between the linear growth model and the persistence models, there is some conflict between the model choice measures. Simple fit (deviance and EPD) unequivocally favours the latter, with the mean EPD at 525 compared to 584 for the differential growth model. However, the marginal likelihood based on totaling the CPO criterion over all periods and areas shows the differential growth model with a marginal likelihood around −946 as against −963 for the spatial persistence model. A suggested exercise is to assess the effective parameters in both models and use the DIC criterion as a further method of selection. Further model options are to assume spatially correlated persistence, and to relax the stationarity assumption on borough autocorrelations.

7.8 REVIEW

The chapter has sketched some of the major spatial applications where Bayesian ideas and modelling principles have been important and beneficial, and the growing role of Bayesian spatial modelling is closely linked to improved and simplified estimation via

MCMC methods. It seems inevitable that Bayesian approaches will play a central role in future developments, in areas such as

(a) direct modelling of spatial covariance matrices, drawing on geostatistical techniques but applying them in new areas (e.g. disease mapping);
(b) spatial models that allow for discontinuities or outliers, developing on the work of authors such as Lawson and Clark (2002) and Lesage (2000);
(c) models that allow for spatially varying impacts of covariates, including single equation regression methods or extensions of multi-regression GWR approaches;
(d) spatial model assessment based on cross validatory principles (Congdon, 2002b);
(e) focused cluster models where the outcome is ordinal (e.g. severity of disease) or multivariate;
(f) spatial and space-time models in multivariate outcomes.

REFERENCES

Anselin, L. (1988) *Spatial Econometrics: Methods and Models*. Dordrecht: Kluwer Academic.

Anselin, L. (2001) Spatial econometrics. In: Baltagi, B. (ed.), *A Companion to Theoretical Econometrics*. Oxford: Basil Blackwell, pp. 310–330.

Anselin, L. and Florax, R. (eds.) (2002) *Advances in Spatial Econometrics*. Berlin: Springer-Verlag.

Bailey, T. and Gatrell, A. (1995). *Interactive Spatial Data Analysis*. Harlow: Longman Scientific and Technical.

Baltagi, B. and Dong, L. (2002) Prediction in the panel data model with spatial correlation. In: Anselin, L. and Florax, R. (eds.), *Advances in Spatial Econometrics*. Berlin: Springer-Verlag.

Bell, B. and Broemeling, L. (2000). A Bayesian analysis for spatial processes with application to disease mapping. *Stat. Med.* **19**, 957–974.

Bernardinelli, L. and Montomoli, C. (1992) Empirical Bayes versus fully Bayesian analysis of geographical variation in disease risk. *Stat. Med.* **11**, 983–1007.

Bernardinelli, L., Clayton, D., Pascutto, C., Montomoli, C. and Ghislandi, M. (1995a) Bayesian analysis of space-time variations in disease risk. *Stat. Med.* **11**, 983–1007.

Bernardinelli, L., Clayton, D. and Montomoli, C. (1995). Bayesian estimates of disease maps: how important are priors? *Stat. Med.* **14**, 2411–2432.

Besag, J. and Kooperberg, C. (1995) On conditional and intrinsic autoregressions. *Biometrika* **82**(4), 733–746.

Besag, J., York, J. and Mollié, A. (1991) Bayesian image restoration, with two applications in spatial statistics. *Ann. Inst. Stat. Math.* **43**, 1–20.

Best, N. (1999) Bayesian ecological modelling. In: Lawson, A. *et al.* (eds.), *Disease Mapping and Risk Assessment for Public Health*. Chichester: Wiley.

Best, N., Arnold, R., Thomas, A., Waller, L. and Conlon, E. (1999) Bayesian models for spatially correlated disease and exposure data. In: Bernardo, J. *et al.* (eds.), *Bayesian Statistics 6*. Oxford: Oxford University Press, pp. 131–156.

Biggeri, A. and Lagazio, C. (1999) Case-control analysis of risk around putative sources. In: Lawson, A. *et al.* (eds.), *Disease Mapping and Risk Assessment for Public Health*. Chichester: Wiley.

Brunsdon, C., Fotheringham, A. and Charlton, M. (1996) Geographically weighted regression: a method for exploring spatial nonstationarity. *Geograph. Anal.* **28**(4), 281–298.

Brunsdon, C., Fotheringham, A. and Charlton, M. (1998) Geographically weighted regression – modelling spatial non-stationarity. *J. Roy. Stat. Soc., Ser. D–The Statistician* **47**(3), 431–443.

Carlin, B. and Louis, T. (1996) *Bayes and Empirical Bayes Methods for Data Analysis*. Boca Raton, FL: Chapman & Hall/CRC.

Casetti, E. (1972) Generating models by the expansion method: applications to geographic research. *Geograph. Anal.* **4**, 81–91.

Casetti, E. (1992) Bayesian regression and the expansion method. *Geograph. Anal.* **24**, 58–74.

Casetti, E. (1997) The expansion method, mathematical modeling, and spatial econometrics. *Int. Regional Sci. Rev.* **20**(1–2), 9–33.

Cliff, A. and Ord, J. (1981) *Spatial Processes: Models and Applications*. Pion.

Cocco, P., Bernardinelli, L., Biddau, P., Montomoli, C., Murgia, G., Rapallo, M., Targhetta, R., Capocaccia, R., Fadda, D. and Frova, L. (1995) Childhood Acute Lymphoblastic Leukemia: a cluster in Southwestern Sardinia. *Int. J. Occup. Environ. Health*, **1**(3), 232–238.

Congdon, P. (1996) The incidence of suicide and parasuicide: a small area study. *Urban Stud.* **33**, 137–158.

Congdon, P. (1997) Bayesian models for the spatial structure of rare health outcomes: a study of suicide using the BUGS program. *J. Health and Place* **3**(4), 229–247.

Congdon, P. (2001), Bayesian models for suicide monitoring. *Euro. J. Population* **15**, 1–34.

Congdon, P. (2002a) A model for mental health needs and resourcing in small geographic areas: a multivariate spatial perspective. *Geograph. Anal.* **34**(2), 168–186.

Congdon, P. (2002b) A life table approach to small area health indicators for health resourcing. *Stat. Modelling* **2**, 1–25.

Cook, D. and Pocock, S. (1983) Multiple regression in geographical mortality studies, with allowance for spatially correlated errors. *Biometrics* **39**, 361–371.

Cressie, N. (1993) *Statistics for Spatial Data*. New York: Wiley.

Cuzick, J. and Edwards, R. (1990) Spatial clustering for inhomogeneous populations. *J. Roy. Stat. Soc., Ser. B* **52**, 73–104.

De Graaff, T., Florax, R., Nijkamp, P. and Reggiani, A. (2001) A general misspecification test for spatial regression models: dependence, heterogeneity, and nonlinearity. *J. Regional Sci.* **41**, 255–276.

Diggle, P. (1990) *Time Series: a Biostatistical Introduction*, London: Oxford University Press.

Diggle, P. J. and Rowlingson, B. (1994) A conditional approach to point process modelling of elevated risk. *J. Roy. Stat. Soc. A* **157**, 433–440.

Diggle, P., Elliott, P., Morris, S. and Shaddick, G. (1997). Regression modelling of disease risk in relation to point sources. *J. Roy. Stat. Soc. A* **160**, 491–505.

Diggle, P., Tawn, J. and Moyeed, R. (1998) Model-based geostatistics. *J. Roy. Stat. Soc., Ser. C, Appl. Stat.* **47**(3), 299–350.

Diggle, P., Morris, S. and Wakefield, J. (2000) Point-source modelling using matched case-control data. *Biostatistics* **1**, 89–115.

Doreian, P. (1982) Maximum-likelihood methods for linear-models – spatial effect and spatial disturbance terms. *Sociological Meth. Res.* **10**(3), 243–269.

Dubin, R. (1995) Estimating logit models with spatial dependence. In: Anselin, L. and Florax, R. (eds.), *New Directions in Spatial Econometrics*. Berlin: Springer-Verlag.

Ecker, M. and Gelfand, A. (1997) Bayesian variogram modeling for an isotropic spatial process. *J. Agric., Biol. and Environ. Stat.* **2**(4), 347–369.

Elliott, P., Wakefield, J., Best, N. and Briggs, D. (eds.) (2000) *Spatial Epidemiology; Methods and Applications*. Oxford: Oxford University Press.

Fotheringham, A., Brunsdon, C. and Charlton, M. (2000) *Quantitative Geography*. London: Sage.

Gelfand, A. and Dey, D. (1994) Bayesian model choice: asymptotic and exact calculations. *J. Roy. Stat. Soc. B* **56**, 501–514.

Gelfand, A. (1996) Model determination using sampling-based methods. In: Gilks, W., Richardson, S. and Spiegelhalter, D. (eds.), *Markov Chain Monte Carlo in Practice*. London: Chapman & Hall, pp. 145–161.

Gelfand, A., Ghosh, S., Knight, J. and Sirmans, C. (1998) Spatio-temporal modeling of residential sales markets. *J. Bus. & Economic Stat.* **16**, 312–321.

Ghosh, M., Natarajan, K., Stroud, T. and Carlin, B. (1998) Generalized linear models for small-area estimation. *J. Am. Stat. Assoc.* **93**, 273–282.

Haggett, P., Cliff, A. and Frey, A. (1977) *Locational Analysis in Human Geography, 2nd ed.* London: Edward Arnold.

Handcock, M. and Stein, M. (1993) A Bayesian analysis of Kriging. *Technometrics* **35**, 403–410.

Hawton, K., Harriss, L. and Hodder, K. (2001) The influence of the economic and social environment on deliberate self-harm and suicide: an ecological and person-based study. *Psychol. Med.* **31**, 827–836.

Hepple, L. (1995) Bayesian techniques in spatial and network econometrics: model comparison and posterior odds. *Environment and Planning A* **27**, 447–469.

Hill, E., Ding, L. and Waller, L. (2000) A comparison of three tests to detect general clustering of a rare disease in Santa Clara County. *California, Stat. in Med.* **19**, 1363–1378.

Hsiao, C. and Tahmiscioglu, A. (1997) A panel analysis of liquidity constraints and firm investments. *J. Am. Stat. Assoc.* 455–465.

Kelejian, H. H. and Robinson, D. P. (1993) A suggested method of estimation for spatial interdependent models with autocorrelated errors, and an application to a county expenditure model. *Papers in Regional Sci.* **72**, 297–312.

Kerkhof, A. and Kunst, A. (1994) A European perspective on suicidal behaviour. In: *The Prevention of Suicide*. London: Department of Health, HMSO.

Knorr-Held, L. and Besag, J. (1998) Modelling risk from a disease in time and space. *Stat. in Med.* **17**, 2045–2060.

Knorr-Held, L. and Rasser, G. (2000) Bayesian detection of clusters and discontinuities in disease maps. *Biometrics* **56**, 13–21.

Lagazio, C., Dreassi, E. and Bernardinelli, A. (2001) A hierarchical Bayesian model for space-time variation of disease risk. *Stat. Modelling*, **1**.

Langford, I., Leyland, A., Rasbash, J. and Goldstein, H. (1999) Multilevel modelling of the geographical distributions of rare diseases. *J. Roy. Stat. Soc., Ser. C* **48**, 253–268.

Lawson, A., Biggeri, A., Böhning, D., Lesaffre, E., Viel, J. and Bertollini, B. (1999) *Disease Mapping and Risk Assessment for Public Health*. New York: Wiley.

Lawson, A. (2001) *Statistical Methods in Spatial Epidemiology*. New York: Wiley.

Lawson, A. and Clark, A. (2002) Spatial mixture relative risk models applied to disease mapping. *Stat. in Med.* **21**, 359–370.

Lawson, A., Biggeri, A. and Williams, F. (1999) A review of modelling approaches in health risk assessment around putative sources. In: Lawson, A., Biggeri, A., Bohning, D., Lesaffre, E. *et al.* (eds.), *Disease Mapping and Risk Assessment for Public Health*. New York: Wiley.

Lawson, A. and Williams, F. (1994) Armadale: a case study in environmental epidemiology. *J. Roy. Stat. Soc. A* **157**, 285–298.

Lesage, J. (1997) Bayesian estimation of spatial autoregressive models. *Int. Regional Sci. Rev.* **20**, 113–130.

Lesage, J. (2000) Bayesian estimation of limited dependent variable spatial autoregressive models. *Geograph. Anal.* **32**(1).

Lesage, J. (2001) A family of geographically weighted regression models. Manuscript, Department of Economics, University of Toledo.

Lesage, J. (1999) The Theory and Practice of Spatial Econometrics. Web Manuscript (http://www.spatial-econometrics.com/). Department of Economics, University of Toledo.

Leyland, A., Langford, I., Rasbash, J. and Goldstein, H. (2000) Multivariate spatial models for event data. *Stat. in Med.* **19**, 2469–2478.

MacMillen, D. (1995) Spatial effects in probit models, a Monte Carlo investigation. In: Anselin, L. and Florax, R. (eds.), *New Directions in Spatial Econometrics*. Heidelberg: Springer-Verlag, pp. 189–228.

Mardia, K. and Marshall, R. (1984) Maximum likelihood estimation of models for residual covariance in spatial regression. *Biometrika* **71**, 135–146.

Militino, A., Ugarte, M. and Dean, C. (2001), The use of mixture models for identifying high risks in disease mapping. *Stat. in Med.* **20**, 2035–2049.

Mollié, A. (2000) Bayesian mapping of Hodgkins disease in France. In: Elliott, P. *et al.* (eds.), *Spatial Epidemiology; Methods and Applications*. Oxford: Oxford University Press, pp. 267–285.

Morris, S. and Wakefield, J. (2000) Assessing of disease risk in relation to a pre-specified source. In: Elliott, P., Wakefield, J. C., Best, N. G. and Briggs, D. (eds.), *Spatial Epidemiology: Methods and Applications*. Oxford: Oxford University Press, pp. 152–184.

Mugglin, A., Carlin, B., Zhu, L. and Conlon, E. (1999) Bayesian areal interpolation. *Environment and Planning* **31A**, 1337–1352.

Mugglin, A. and Carlin, B. (1998). Hierarchical modeling in geographic information systems: population interpolation over incompatible zones. *J. Agricultural, Biol. and Environmental Stat.* **3**, 111–130.

Oliver, M., Muir, K., Webster, R., Parkes, S., Cameron, A., Stevens, M. and Mann, J. (1992) A geostatistical approach to the analysis of pattern in rare disease. *J. Public Health Med.* **14**, 280–289.

Ord, K. (1975) Estimation methods for models of spatial interaction. *J. Am. Stat. Assoc.* **70**, 120–126.

Richardson, S. (1992) Statistical methods for geographical correlation studies. In: Elliott, P., Cuzick, J., English, D. and Stern, R. (eds.), *Geographical and Environmental Epidemiology: Methods for Small Area Studies*. Oxford: Oxford University Press.

Richardson, S., Guihenneuc, C. and Lasserre, V. (1992) Spatial linear models with autocorrelated error structure. *The Statistician* **41**, 539–557.

Richardson, S. and Monfort, C. (2000) Ecological correlation studies. In: Elliott, P., Wakefield, J., Best, N. and Briggs, D. (eds.) *Spatial Epidemiology; Methods and Applications*. Oxford: Oxford University Press.

Ripley, B. (1981) *Spatial Statistics*. New York: Wiley.

Rogerson, P. (2001) Monitoring point patterns for the development of space–time clusters. *J. Roy. Stat. Soc., Ser. A (Statistics in Society)* **164**(1), 87–96.

Schmid, V. and Knorr-Held, L. (2001) Spatial extensions of age-period-cohort models. Manuscript, Department of Statistics, Ludwig-Maximilans University.

Stern, H. and Cressie, N. (1999) Inferences for extremes in disease mapping. In: Lawson, A., Biggeri, A., Bohning, D., Lesaffre, E. *et al.* (eds.), *Disease Mapping and Risk Assessment for Public Health*. New York: Wiley, pp. 61–82.

Sun, D., Tsutakawa, R., and Speckman, P. (1999) Posterior distribution of hierarchical models using CAR(1) distributions. *Biometrika* **86**(2), 341–350.

Sun, D., Tsutakawa, R., Kim, H. and He, Z. (2000) Spatio-temporal interaction with disease mapping. *Stat. in Med.* **19**(15), 2015–2035.

Turnbull, B., Iwano, E., Burnett, W., Howe, H. and Clark, L. (1990) Monitoring for clusters of disease: application to leukemia incidence in upstate New York. *Am. J. Epidemiology* **132**, 136–143.

Wakefield, J. and Morris, S. (2001) The Bayesian modelling of disease risk in relation to a point source. *J. Am. Stat. Assoc.* **96**, 77–91.

Wakefield, J., Best, N. and Waller, L. (2000) Bayesian approaches to disease mapping. In: Elliott, P. *et al.* (eds.), *Spatial Epidemiology; Methods and Applications*. Oxford: Oxford University Press.

Waller, L., Turnbull, B., Clark, L. and Nasca, P. (1994) *Spatial Pattern Analyses to Detect Rare Disease Clusters*. In: *Case Studies in Biometry*. Somerset, NJ: Wiley-Interscience, pp. 3–23.

Waller, L., Carlin, B. and Xia, H. (1997) Structuring correlation within hierarchical spatio-temporal models for disease rates. In: Gregoire, T., Brillinger, D., Diggle, P., Russek-Cohen, E., Warren, W. and Wolfinger, R. (eds.), *Modelling Longitudinal and Spatially Correlated Data: Methods, Applications, and Future Directions*. New York: Springer-Verlag, pp. 309–319.

Waller, L. and Poquette, C. (1999) The power of focussed score tests under misspecified cluster models. In: Lawson, A., Biggeri, A., Bohning, D., Lesaffre, E. *et al.* (eds.), *Disease Mapping and Risk Assessment for Public Health*. New York: Wiley, pp. 257–269.

7.9 EXERCISES

1. In Example 7.1, fit the spatial errors model using the weights based on distance/shared boundary rather than contiguity. How does fit (e.g. via the pseudo marginal likelihood based on CPOs) compare to that cited previously?

2. In Example 7.2, fit the spatial autoregression model (7.5) under both Student t and homoscedastic Normal errors u, and assess inferences in terms of the impact of outlier areas.

3. In Example 7.3, compare the fit of the four models considered using the deviance information criterion of Spieglehalter *et al.* (1998).

4. In Example 7.3, fit the ICAR(1) model including unstructured as well as spatial errors and under the options with and without the AFF variable. How does the ratio of marginal variances (spatial to unstructured variances) change?

5. Fit the Leyland *et al.* spatial model (see Example 7.4) to the lip cancer data of Example 7.3. How does this affect the coefficient for the occupation variable?

6. The following are 400 microwave backscatter measures in a large field in East Anglia, UK, on a 20×20 grid with 12 m intervals between points:

   22 39 38 67 46 1 46 73 95 51 92 68 33 64 77 73 82 54 61 97
   43 36 58 32 35 35 88 56 73 64 74 62 55 73 105 80 90 21 57 100
   45 15 34 28 35 89 113 45 54 44 49 59 94 58 85 112 77 83 48 75
   48 14 14 83 55 81 73 33 58 74 81 78 66 96 69 102 94 86 96 80
   18 24 51 88 65 41 47 20 46 94 76 97 73 94 41 95 100 62 101 75
   66 53 40 50 29 19 67 50 57 85 57 88 69 106 88 79 98 69 82 95
   113 50 24 11 50 64 90 80 63 68 52 68 79 90 74 40 73 87 100 86
   103 31 45 29 56 115 132 78 82 54 70 51 40 69 80 71 88 95 94 68
   83 70 104 81 85 120 186 87 85 86 34 30 49 61 103 99 94 87 90 104
   72 78 186 132 111 153 157 119 90 82 80 95 79 57 79 57 68 105 63 99
   82 164 157 138 136 155 157 109 90 104 115 101 86 36 98 66 57 73 94 97
   107 166 157 101 101 93 157 85 82 80 111 82 77 104 67 70 79 57 76 81
   105 87 103 64 58 77 97 87 81 65 85 83 44 97 67 98 99 50 83 81
   145 116 91 94 34 49 73 64 46 83 60 69 90 89 63 89 84 67 47 86
   135 100 115 100 46 44 81 51 27 81 81 62 95 96 67 82 77 68 79 80
   78 127 115 90 47 92 103 79 58 95 54 70 93 105 68 100 39 65 98 96
   76 140 100 95 63 78 98 83 66 56 79 71 92 85 87 92 97 74 102 78
   82 119 69 83 78 40 73 81 77 93 61 63 78 62 87 61 76 75 98 23
   77 53 58 79 64 83 66 64 94 89 24 72 90 68 55 82 92 81 63 64
   86 63 78 84 67 78 63 71 91 60 80 54 80 98 65 80 99 61 73 37

   Group the 400.399/2 pairs of points into bands according to separations 0–12,12–24, ... etc. up to a maximum distance band, and derive the necessary statistics for a variogram (this may be easier done outside BUGS). Then fit exponential and spherical variograms to the series, as in Example 7.5.

6. In Example 7.7, compare the fit of the two models using the DIC criterion. How does this compare to the Gelfand–Ghosh criterion? Employing both criteria, fit a model where a_{ij} is a power function, namely $a_{ij} = d_{ij}^{-\kappa}$.

7. In Example 7.8, add code to find which areas have the maximum coefficients under the GWR model. How does this add to knowledge about possible outlier areas? Also, consider the effect of taking Student t errors with known degrees of freedom (e.g. $\nu = 4$).

8. In Example 7.9, combine the heteroscedastic error model with an ICAR(ρ) spatial error. How does this affect the AFF coefficient and the fit via DIC?

9. In Example 7.11, fit the three parameter decay and direction function

$$f(d, \theta) = [1 + \eta \exp\{\beta \cos(\theta)\}/d^{\alpha}]$$

   with α positive.

10. In Example 7.12, fit the time varying spatial effect

$$\rho_{ijt} = \alpha_{1t} \exp(-\alpha_{2t} d_{ij})$$

and assess fit against the constant parameter model

$$\rho_{ij} = \alpha_1 \exp(-\alpha_2 d_{ij})$$

11. Fit the spatio-temporal model for London borough suicides of Example 7.13 assuming spatially correlated borough specific persistence (temporal correlation) parameters.

CHAPTER 8

Structural Equation and Latent Variable Models

8.1 INTRODUCTION

Structural equation modelling describes multiple equation models that include latent or unmeasured variables ('factors' or 'constructs') for which multiple observed indicators, continuous or discrete are available (Bollen, 1998), and that allow for measurement error in the latent constructs. They have found a major application in areas such as psychology, education, marketing and sociology where underlying constructs (depression, product appeal, teacher style, anomie, authoritarianism, etc.) are not possible to measure directly.

Instead, the observed data on a large number of indicators are used to define (i.e. serve as proxies for) the underlying constructs, which are required to be fewer in number than the observed variables to ensure identifiability (Bartholomew, 1984). Depending on the application the observed variables may be known as items (e.g. in psychometric tests), as indicators or as manifest variables. One might assume that conditional on the constructs, these observed indicators are independent – so that the constructs account for the observed correlations between the indicators (this is often called the 'local independence' property). However, as Bollen (2002) points out, this is not an intrinsic feature of structural equation models.

Structural equation models include both measurement models, or confirmatory factor analysis models, confined to representing the constructs as functions of the indicators, and fully simultaneous models allowing interdependence between the constructs. The canonical structural equation model takes the LISREL form (Joreskog, 1973), with a model relating endogenous constructs ψ to each other and to exogenous constructs ξ and a measurement error model linking observed indicators Y and X to the latent variables. Thus for the ith subject, the structural model is

$$\psi_i = A + B\psi_i + G\xi_i + w_i \tag{8.1a}$$

where ψ_i is a $p \times 1$ vector of endogenous constructs, ξ_i is a $q \times 1$ vector of exogenous constructs, w_i is a $p \times 1$ vector of errors on the endogenous constructs, B is a $p \times p$ parameter matrix describing interrelations between the endogenous constructs ψ, G is a $p \times q$ parameter matrix describing the effect of exogenous on endogenous constructs,

Applied Bayesian Modelling P. Congdon
 ISBN: 0-471-48695-7

and A is a $p \times 1$ intercept. In any particular model, there may be no ξ variables, and some of the B and G coefficients are typically assigned zero or other default values. The links between observed indicators and the constructs are defined by the measurement model or models:

$$Y_i = \kappa_y + \Lambda_y \psi_i + u_i^{Y} \qquad (8.1b)$$

$$X_i = \kappa_x + \Lambda_x \xi_i + u_i^{X} \qquad (8.1c)$$

where Y_i is an $M \times 1$ vector of indicators describing the endogenous construct vector ψ, and X_i is an $L \times 1$ vector of indicators that proxy the exogenous construct vector ξ; κ_y and κ_x are $M \times 1$ and $L \times 1$ intercepts; and Λ_y and Λ_x are $M \times p$ and $L \times q$ matrices of loading coefficients describing how the indicators determine the scores of an individual on the latent constructs. Typically, restrictions are applied on some loadings to ensure identifiability. A particular analysis may involve just a measurement model, in which case the distinction between different types of indicator is not relevant, and the measurement model reduces to

$$Y_i = \kappa + \Lambda \psi_i + u_i \qquad (8.1d)$$

The focus in structural equation models is on confirming the nature of the underlying constructs as postulated by substantive theory, or on testing causal hypotheses based on theory, rather than on exploratory data analysis. Thus, confirmatory factor analysis, a particular from of structural equation model, differs from exploratory factor analysis in postulating a restricted loading structure in which only certain loadings are free to be estimated (see Section 8.2). The nature of the linkages from the latent variables to the observed indicators is often defined on a priori theoretical grounds, so the 'prior' of a Bayes model for a confirmatory factor analysis may have a large subject matter element and is to that extent clearly 'informative'.

As an example of this type of model and of the identifiability constraints on the parameters, suppose five indicators $Y_1, \ldots, Y_5$ are taken to be measures of two constructs ψ_1 and ψ_2. Further, suppose indicators Y_1, Y_2, Y_3 (e.g. measures of spatial ability) have loadings λ_1, λ_2 and λ_3 on construct ψ_1 while indicators Y_4 and Y_5 (e.g. measures of linguistic skills) have loadings λ_4, λ_5 on ψ_2. However, there are no loadings of Y_1, Y_2 and Y_3 on ψ_2 or of Y_4 and Y_5 on ψ_1. This is a hypothesized structure and one open to assessment: it may be that some indicators in fact show significant 'cross-loading', with Y_1 showing a non-zero link to ψ_2 for example. It is also open to question whether ψ_1 and ψ_2 (spatial and linguistic ability) are correlated or orthogonal.

Since ψ_1 and ψ_2 have arbitrary location and scale, one option to gain identifiability is to define them to be in standard form, with zero means and variances of unity (Bentler and Weeks, 1980). If correlation between the two constructs is allowed, then they might be taken to be bivariate normal with variances 1 in the diagonal of the covariance matrix, but with an off-diagonal correlation parameter. Note, though, that there is no necessary restriction to assuming the constructs in a structural equation model to be Normally distributed.

Under this predefined scale option the loadings λ_1, λ_2 and λ_3 relating Y_1, Y_2 and Y_3 to ψ_1, and λ_4 and λ_5 relating the verbal test scores to ψ_2, are all then free parameters. So letting $\psi_i = (\psi_{i1}, \psi_{i2})$, one might have

$$\psi_i \sim N_2\left(\begin{bmatrix} 0 \\ 0 \end{bmatrix}, \begin{bmatrix} 1 & \rho \\ \rho & 1 \end{bmatrix}\right)$$

$$Y_{1i} = \kappa_1 + \lambda_1\psi_{1i} + u_{1i}$$
$$Y_{2i} = \kappa_2 + \lambda_2\psi_{1i} + u_{2i}$$
$$Y_{3i} = \kappa_3 + \lambda_3\psi_{1i} + u_{3i}$$
$$Y_{4i} = \kappa_4 + \lambda_4\psi_{2i} + u_{4i}$$
$$Y_{5i} = \kappa_5 + \lambda_5\psi_{2i} + u_{5i}$$

Since the constructs are intended to summarise information in the indicators relatively informative priors, e.g. N(1, 1) or N(0, 1), may be used for the loadings; for example, see Johnson and Albert (1999) in the context of item analysis. Often a scaling of the observed indicators (e.g. centred or standardised X and Y) is useful also in identification. Note that ρ is identifiable in the above example, because all cross loadings have been set at zero; for example, there are no loadings of Y_1, Y_2 and Y_3 on ψ_2.

In general, if the ψ_{mi} are taken to be uncorrelated and to have variance 1, and all $M.p = 10$ possible loadings are of interest, then $p(p-1)/2 = 1$ restrictions are required for identifiability (e.g. Everitt, 1984). If λ_{mj} is the loading on the mth indicator on the jth factor, then one might for example impose zero, unity or equality constraints (for the above these might be $\lambda_{51} = 0$ or $\lambda_{11} = \lambda_{42}$). Note that a Bayesian approach may circumvent the need for exact constraints (see below).

An alternative parameterisation fixes the scale of the constructs by selecting one loading corresponding to each factor – here one among the loadings $\{\lambda_1, \lambda_2, \lambda_3\}$ and one among $\{\lambda_4, \lambda_5\}$ – and setting them to a predetermined non-zero value, usually 1. This is similar to the corner constraint used for categorical predictors in log-linear regression, except that it takes a multiplicative rather than additive form. In the example considered here, the variances of ψ_1 and ψ_2 are then free parameters related to those of certain observed indicators. Suppose $\lambda_1 = \lambda_4 = 1$, so that

$$Y_{1i} = \kappa_1 + \quad \psi_{1i} + u_{1i}$$
$$Y_{2i} = \kappa_2 + \lambda_2\psi_{1i} + u_{2i}$$
$$Y_{3i} = \kappa_3 + \lambda_3\psi_{1i} + u_{3i}$$
$$Y_{4i} = \kappa_4 + \quad \psi_{2i} + u_{4i}$$
$$Y_{5i} = \kappa_5 + \lambda_5\psi_{2i} + u_{5i}$$

with the ψ again being bivariate normal with zero means but all three parameters in the dispersion matrix free.

It may be noted that fixing a loading λ_1 to, say, 1 has utility in preventing 'relabelling' of the construct scores ψ_1 during MCMC sampling. Since $Y_1 - Y_3$ in this example are positive measures of spatial ability setting $\lambda_1 = 1$ means the construct ψ_1 will be a positive measure of this ability. If, however, one adopted the predetermined variance identifiability constraint such as $\psi_1 \sim N(0, 1)$ where all the λ_j are free, it may be necessary, to prevent label switching, to set a prior on one or possibly more loadings constrained to positivity, e.g.

$$\lambda_1 \sim N(1, 1)\ I(0,)$$

8.1.1 Extensions to other applications

This framework may be extended to panel data where multiple items Y_{imt} are observed for subjects, variables $m = 1, .., M$ and over times $t = 1, .., T$; here the constructs may also be modelled as time varying (Section 8.4). Repetitions may also occur over groups g such that the data have the form Y_{img}, where g is the group, m the indicator, and i the subject (Section 8.2). Latent construct models with ψ_j continuous may equally be applied to explain multiple discrete indicators, for example collections of binary tests or items (Bartholomew and Knott, 1999). One may also consider latent constructs which are discrete, and so connections to discrete mixtures (Chapter 2) and discrete mixture regressions (Chapter 3) become apparent. Latent class analysis (Section 8.3) is often applied in psychology and attitude research, and it may be that the latent construct in a particular application can either be modelled as discrete or as a continuous 'trait', and the choice between these options is made on grounds of statistical fit (Langeheine and Rost, 1988).

8.1.2 Benefits of Bayesian approach

Classical structural equation methods for continuous indicators Y that are based on multivariate Normality of the indicators usually involve minimising a discrepancy between the observed covariance and the predicted covariance under the model – since under multivariate Normality the covariance matrix is sufficient for the data, and in particular the associations between indicators. This is the default approach in computer packages such as LISREL or EQS, with the input consisting of $M \times M$ covariance matrices rather than the full $n \times M$ array of observations Y_{im}. Browne (1984) generalises the multivariate Normal theory to any multivariate distribution for continuous variables, so permitting analysis of covariance structures only, providing non-Normality (e.g. kurtosis) is not pronounced. It is, however, becoming more common to analyse raw data so that appropriate techniques allowing for non-Normality (e.g. excessive kurtosis) or potential outliers can be applied. This may also be a more flexible option if the indicators contain a mix of discrete and continuous measures.

Considerations of robustness, and other questions such as the ease with which parameter restrictions may be imposed and predictions made for new cases, may point to a Bayes approach which retains the full observation set as input – this indeed is easier to implement in WINBUGS. Comparisons between different models might involve variations on the measurement model, the structural model, or both. A Bayesian perspective may have advantages here if hypotheses are not be restricted to nested alternatives (Fornell and Rust, 1989).

Scheines *et al.* (1999) discuss how a Bayes approach may assist in estimating what are conventionally described as unidentifiable models, for instance regressions in which all predictors are measured with error and the strategy of instrumental variable estimation is not feasible. Lee (1992) describes methods to restate the usual classical identifiability restrictions as stochastic. For instance, one might assume both a free dispersion matrix for factors ψ_{i1} and ψ_{i2} and free loadings $\lambda_1, \ldots, \lambda_5$, but informative priors on λ_1 and λ_4 would be taken based on the classical constraint; for instance, taking

$$\lambda_1 \sim N(1, 0.1)$$
$$\lambda_4 \sim N(1, 0.1)$$

allows limited movement around the usual constraints. If all ten loadings were of interest and an identity dispersion matrix assumed for ψ_{ij}, j = 1, 2, then an identifiability constraint such as $\lambda_{11} = \lambda_{42}$ might be expressed as

$$\lambda_{11} = \alpha\lambda_{42} + e$$

where α is a parameter centred at 1 and e is a low variance error.

More generally, the Bayes method has potential advantages over maximum likelihood estimates for describing the densities of the parameters of structural equation models. The sampling distribution of parameters in such models is only asymptotically Normal (Bollen, 1989), so that maximum likelihood standard errors calculated assuming the distribution of the estimator is Normal may be distorted for small sample sizes (Boomsma, 1983). Bootstrap techniques are one option to overcome this problem (Bollen, 1996). However, in the Bayesian approach MCMC samples are taken from the true posterior regardless of sample size, and so standard errors calculated from MCMC output are more reliable for small samples, or when there are other sources of non-normality.

On the other hand, a maximum likelihood solution generally converges rapidly to a solution with a clear labelling of constructs. As mentioned above, precautions (in the form of constrained priors on loadings, say) may need to be taken in a Bayesian MCMC sampling context to ensure unique labelling of constructs, and prevent label-switching during sampling.

8.2 CONFIRMATORY FACTOR ANALYSIS WITH A SINGLE GROUP

The advantages of confirmatory models with prior information from a subject matter base as compared to exploratory factor analysis (where all loadings are free to be estimated) have been argued by Hertzog (1989), in terms of the interpretative ambiguity of exploratory analysis. Thus, using an orthogonal rotation or oblique rotation in an exploratory factor analysis will give different answers as to factor structures, and various often arbitrary methodological choices in oblique rotation will lead to different answers about the correlations between the latent constructs. As well as having a clear and unambiguous factor structure, which permits evaluation of alternative hypotheses (e.g. correlated or orthogonal factors), confirmatory factor analysis permits assessment of the construct validity of measurement scales.

An important question in confirmatory factor analysis is the stability or 'invariance' of structural relationships and measurement models across groups or over time. As Mutran (1989) points out, the availability of panel or group data allows a new set of testable hypotheses; for example, should the effect of each indicator on the underlying construct or constructs be allowed to change and, if so, is the same construct being studied? If the measurement model is allowed to vary or change, one may assess whether certain indicators of the latent construct ψ lose or gain reliability. Note that for many groups, a multi-level factor analysis with appropriate random effects is more relevant (Section 8.4.4).

Among the possible parameterisations for testing multi-group invariance (assuming a few groups only) in latent structure, means and covariances is one adopted by Byrne *et al.* (1989). They consider $M = 11$ indicators of adolescent self-concept in terms of $p = 4$ factors: General Self-Concept (GSC), Academic Self-Concept (ASC), Language Skills (LSC), and Mathematics (MSC). Further, the relation between indicators and concepts may differ by pupil group: there are $G = 2$ groups, namely $n_1 = 582$ 'high track' students and $n_2 = 248$ 'low track' students. Byrne *et al.* assume that the measurement model for the mth indicator in the gth group is parameterised as

$$Y_{img} = \kappa_m + \lambda_m \psi_{ijg} + u_{img}$$

where $i = 1, .. n_g$, $j = C[j]$ is that single factor among p factors relevant to predicting the mth indicator, and the u_{img} are independent distributed for variable m and group g. Thus, only one factor is taken to explain any particular item, and the intercepts and item-factor loadings in the measurement model do not differ by group. The factor construct model for the reference group (e.g. $g = 1$) is

$$\psi_{ij1} = \delta_{ij}$$

for the $j = 1, .. p$ factors and the students $i = 1, .. n_g$. The δ_{ij} may be taken as independent with means zero and variance σ_j^2, or to follow a multivariate prior with covariance matrix Σ. For the remaining groups, the factor model is

$$\psi_{ijg} = A_{gj} + \delta_{ij}$$

where A_{gj} is the intercept on the jth factor for the gth group.

Example 8.1 Confirmatory factor analysis of adolescent self-concept and extension to a group factor model As in the above discussion, we follow Byrne *et al.* (1989) and Byrne and Shavelson (1986) in taking 11 indicators from the Affective Perception Inventory (Soares and Soares, 1979) as representing four correlated ability concepts. The first, fifth and sixth indicators load on the first general self-concept factor, denoted ψ_1. Indicators Y_2 and Y_7 load on the ASC factor ψ_2, indicators Y_3, Y_8 and Y_9 on the LSC factor ψ_3, and Y_4, Y_{10} and Y_{11} on the remaining Mathematics Self-Concept (MSC) ψ_4. There are thus no multiple loadings, e.g. of Y_1 on both ψ_1 and ψ_2.

As one device to gain identifiability, the loading λ_1 of Y_1 on ψ_1 is set to 1, the loading (λ_2) of Y_2 on ψ_2 to 1, and those of Y_3 on ψ_3 and of Y_4 on ψ_4 (namely, λ_3 and λ_4) to 1 also. Further, the observations (for $n = 996$ subjects) are centred, so eliminating intercepts. Hence, the model is

$$\begin{aligned}
Y_{i1} &= \psi_{i1} + u_{i1}\\
Y_{i2} &= \psi_{i2} + u_{i2}\\
Y_{i3} &= \psi_{i3} + u_{i3}\\
Y_{i4} &= \psi_{i4} + u_{i4}\\
Y_{i5} &= \lambda_5\psi_{i1} + u_{i5}\\
Y_{i6} &= \lambda_6\psi_{i1} + u_{i6}\\
Y_{i7} &= \lambda_7\psi_{i2} + u_{i7}\\
Y_{i8} &= \lambda_8\psi_{i3} + u_{i8}\\
Y_{i9} &= \lambda_9\psi_{i3} + u_{i9}\\
Y_{i,10} &= \lambda_{10}\psi_{i4} + u_{i,10}\\
Y_{i,11} &= \lambda_{11}\psi_{i4} + u_{i,11}
\end{aligned}$$

The factor scores ψ_{im}, $m = 1, .., 4$, i $= 1, .., n$, are taken to be multivariate Normal with zero means and covariance matrix Σ, while the errors u_{ij}, $j = 1, .., M$, $i = 1, .., n$, are taken to be independent Normal with zero means. The estimated loadings (Table 8.1) for this model, obtained from a two chain 5000 iteration run (and convergence from 250 iterations), are close to those reported by Byrne and Shavelson.

The covariances between the factors are all positive, and may be taken to broadly support the idea of a unidimensional self-concept factor; the correlation between ASC and LSC exceeds 0.7, and that between ASC and MSC exceeds 0.6. One might therefore entertain a higher order factor model with a single factor underlying the four specialised factors. So a particular form of Equation (8.1a) obtains

$$\psi_{ij} = \alpha_j + \beta_j \eta_i + w_{ij}$$

where η_i denotes the single second order factor scores, assumed to be N(0, 1). One rationale for higher order factors is that they explain the associations between lower order factors, and so eliminate the need for correlated measurement errors. Here identifiability was improved by taking the β_j to be confined to positive values, and taking $\alpha_j = 0$, since initial analysis suggested none of the intercepts to be non-zero. With these constraints, convergence was obtained at around 2000 iterations in a two chain run of 5000 iterations.

The predictive loss criterion of Gelfand and Ghosh (1998) gives virtually identical results for the double order model (around 5335) as for the single order model, though the former has the advantage of providing extra substantive insight. The posterior means of the β_j are, respectively, 0.43, 0.58, 0.54 and 0.42, with β_4 having the highest

Table 8.1 Self-concept: covariation between correlated factors and loadings relating indicators ($M = 11$) to factors (seven free loadings)

Variances of Factors (Σ_{jj}) and Covariances between Factors ($\Sigma_{jk}, j \neq k$)				
	Mean	2.5%	Median	97.5%
Σ_{11}	0.74	0.66	0.74	0.83
Σ_{12}	0.24	0.19	0.24	0.29
Σ_{13}	0.21	0.16	0.21	0.25
Σ_{14}	0.21	0.16	0.21	0.27
Σ_{22}	0.67	0.6	0.67	0.77
Σ_{23}	0.42	0.37	0.41	0.48
Σ_{24}	0.5	0.44	0.49	0.57
Σ_{33}	0.5	0.51	0.58	0.66
Σ_{34}	0.09	0.04	0.09	0.14
Σ_{44}	0.87	0.78	0.87	0.96
Indicator-factor loadings				
λ_5	0.82	0.76	0.82	0.88
λ_6	1.06	0.98	1.05	1.12
λ_7	1.03	0.96	1.03	1.1
λ_8	1.11	1.04	1.11	1.18
λ_9	1	0.92	1	1.1
λ_{10}	0.98	0.94	0.98	1.03
λ_{11}	0.95	0.91	0.95	0.99

posterior variance – suggesting that the mathematics self-concept is least aligned with a putative single self-concept. The λ coefficients in the higher order model are very similar to those in Table 8.1.

Continuing the data and indicator format as above, Byrne *et al.* (1989) compare two groups (denoted high and low track) with a low track group as reference. Then the mth indicator for the ith subject in the gth group is modelled as

$$Y_{img} = \kappa_m + \lambda_m \psi_{ijg} + u_{img}$$

where the errors u are independently Normal, the reference group scores are

$$\psi_{ij2} = \delta_{ij} \qquad j = 1, 4; \; i = 1, .. 248$$

and the other group's scores are given by

$$\psi_{ij1} = \delta_{ij} + A_j \qquad j = 1, 4; \; i = 1, .. 582$$

with the δ_{ij} taken to be multivariate Normal. In line with factorial invariance, the loadings are not group specific and as above loadings $\lambda_1, \ldots \lambda_4$ are preset and so the covariance matrix Σ for the δ_{ij} may be estimated. Hence, there are 15 intercepts (11 for the measurement model, and 4 for the factor model) together with seven loadings to estimate. The differential intercepts A_j in the high track group are the major focus of interest.

A two chain run of 10 000 iterations (convergent from around 1000) accordingly shows higher intercepts on the Academic, Linguistic and Mathematics Self Concepts in the 'High Track' students (see Table 8.2). This model adequately represents the data since a posterior predictive check is 0.54, based on comparing error sum of squares of

Table 8.2 Self-concept: group comparisons under invariance, high track intercepts

	Mean	2.5%	97.5%
A_1	−0.87	−2.83	1.16
A_2	10.18	8.77	11.58
A_3	4.47	3.33	5.61
A_4	7.48	5.24	9.73
Correlations between factors			
Corr[ψ1, ψ2]	0.38	0.31	0.46
Corr[ψ1, ψ4]	0.30	0.22	0.37
Corr[ψ1, ψ3]	0.24	0.17	0.31
Corr[ψ2, ψ3]	0.61	0.54	0.68
Corr[ψ2, ψ4]	0.62	0.56	0.67
Corr[ψ3, ψ4]	−0.02	−0.10	0.05
Factor loadings			
λ_5	0.52	0.48	0.56
λ_6	0.36	0.34	0.39
λ_7	0.52	0.47	0.56
λ_8	1.36	1.25	1.48
λ_9	0.67	0.60	0.74
λ_{10}	0.68	0.65	0.70
λ_{11}	0.44	0.42	0.46

actual and samples of replicate data. As in the single group analysis, the ASC factor is highly correlated over 0.5 with both LSC and MSC, but the latter two are effectively independent.

Another model allows ψ_{ij1} and ψ_{ij2} to be drawn from different multivariate Normal densities, and for the indicators to have different loadings λ_{mg} and intercepts κ_{mg} within groups (Model F in Program 8.1). There is an improved predictive criterion compared to the previous 'invariance' model (about 429 000 vs. 438 000).

Example 8.2 Mental ability indicators: robust CFA Yuan and Bentler (1998) illustrate the implications of non-Normality, including possible outliers that distort inferences, using mental ability test data analysed by Joreskog (1970) and first presented by Holzinger and Swineford (1939). There are in full 26 test items defined for 145 children in the seventh and eighth grades of two schools. Yuan and Bentler focus on 9 of the 26 tests, namely

1. Visual perception
2. Cubes
3. Lozenges
4. Paragraph comprehension
5. Sentence comprehension
6. Word meanings
7. Addition
8. Counting dots
9. Straight curved capitals

They postulate three factors, the first spatial ability taken to explain X_1-X_3; the second, verbal ability, explaining X_4-X_6; and the third, speed in tasks, designed to explain X_7-X_9.

For these data, Yuan and Bentler use a number of robust frequentist techniques and densities such as M–estimation and Multivariate t density models, which place low weights on exceptional data points. They find an improved fit for these methods over Normal theory based maximum likelihood. Here a scale mixture form of the Student t with degrees of freedom ν, with precisions for case i and variable m defined by

$$\phi_{im} = \Phi_m w_{im}$$

where the $\Phi_m = 1/s_m^2$ is an overall precision for variable m, and the weights w_{im} are drawn from Gamma ($\nu_m/2$, $\nu_m/2$) densities. The degrees of freedom parameters are drawn from an exponential prior with parameters η_m, and are constrained to be above 1. The η_m parameters are drawn from a uniform (0.01, 1) prior corresponding approximately to means 1 and 100 for ν_m.

The means μ_{im} for indicators X_1 to X_9 are given by

$$\mu_{im} = \kappa_m + \lambda_{m,\, C[m]} \Psi_{i,\, C[m]}$$

where $\{C = 1, 1, 1, 2, 2, 2, 3, 3, 3\}$ is the factor index for variable m. The scale of the constructs is defined by fixing $\lambda_{11} = \lambda_{42} = \lambda_{73} = 1$, so that the full covariance matrix between factors is estimated. The loadings under the alternative method when their variance is set at 1 (and the covariance matrix becomes a correlation matrix) are derived in parallel, namely

Table 8.3 Mental ability: loadings and degrees of freedom estimates

	Loadings				Degrees of Freedom		
	Mean	2.5%	97.5%		Mean	2.5%	97.5%
ζ_{11}	4.6	2.8	5.9	ν_1	20.4	2.4	87.9
ζ_{21}	2.5	1.1	4.3	ν_2	30.6	3.0	225.4
ζ_{31}	6.3	2.7	11.1	ν_3	65.4	4.1	307.7
ζ_{42}	1.6	1.2	1.9	ν_4	33.4	4.4	167.5
ζ_{52}	2.1	1.4	2.8	ν_5	27.9	3.7	127.6
ζ_{62}	3.5	2.4	4.8	ν_6	18.7	2.4	111.2
ζ_{73}	17.1	14.0	20.1	ν_7	53.7	5.3	214.4
ζ_{83}	19.0	13.4	27.0	ν_8	2.5	1.0	6.9
ζ_{93}	24.4	16.3	36.8	ν_9	26.9	4.3	120.2

Correlations between factors

r_{12}	0.56	0.38	0.72
r_{13}	0.39	0.20	0.56
r_{23}	0.23	0.05	0.39

$$\zeta_{mj} = \lambda_{mj} S_m$$

A simple measure of fit is provided by the square root of the Mean Error sum of squares (MSE). A posterior predictive check, based on comparing observed mean square error and that obtained with replicated item scores (sampling from the posterior means under the model) shows acceptable fit. The check criterion comparing the new MSE with the observed one averages about 0.54. The predictions may also be used in criteria such as those of Gelfand and Ghosh (1998) and Ibrahim *et al.* (2001).

The loadings and the relativities between them (in terms of ratios $\lambda_{mj}/\lambda_{mk}$, where $j, k = 1, 3$) are similar to those of Yuan and Bentler. However, the credible intervals are wider than those implied by the standard errors of the robust procedures in Yuan and Bentler (1998, Table 1). For example, the standard errors of ζ_{83} over the five robust procedures vary from 1.67–1.82, as against a posterior standard deviation of around 3.2. Asymmetry in the densities of these loadings is also apparent, and would not be allowed for in many frequentist procedures. The degrees of freedom specific to each indicator show the 'Counting Dots' item as the most dubious in terms of Normality. Correlations between the factors suggest verbal and spatial ability to be positively associated.

The weights w_{im} show clear outliers for some variables. On 'Counting Dots', the 24th and 106th subjects seem to be extreme outliers, with weights w_{i8} below 0.005; these are also identified by methods applied by Yuan and Bentler. The profile of remaining cases on this variable is much less subject to skew; the next weight (for subject 35) is nearly 0.8.

Example 8.3 Nonlinear and interactive latent variable effects The question of non-linear effects of latent variables or of interactions between them has been raised in applications of structural equation models (e.g. Bollen and Paxton, 1998). Arminger and Muthen (1998) and Zhu and Lee (1999) consider Bayesian approaches for possible schemes involving one or more latent variables $\xi_{i1}, .. \xi_{iq}$, for which indicators $X_{i1}, .. X_{iL}$ are available. If, also, $\psi_{i1}, .. \psi_{ip}$ have indicators $Y_{i1}, .. Y_{iM}$, then one may have a structural model whereby the endogenous constructs are predicted by powers of, or interactions between, the ξ_{ik}. For instance, if $p = 1$ and $q = 2$, one might have

$$\psi_i = \alpha + \gamma_1 \xi_{i1} + \gamma_2 \xi_{i1}^2 + \gamma_3 \xi_{i2} + \gamma_4 \xi_{i2}^2 + \gamma_5 \xi_{i1} \xi_{i2}$$

Arminger and Muthen (1998, p. 285) consider simulated data for $n = 100$ subjects based on a measurement model relating five indicators to two constructs, ξ_1 and ξ_2, and a structural model for a single observed and centred response Y_i modelled as a function of

(a) main terms in ξ_1 and ξ_2,
(b) an interaction between them, and
(c) a random error, w_i, independent of the constructs. So

$$Y_i = \gamma_1 \xi_{1i} + \gamma_2 \xi_{2i} + \gamma_3 \xi_{1i} \xi_{2i} + w_i$$

This type of model has applications in performance testing (see Early *et al.*, 1990), where ξ_1 might be task complexity and ξ_2 goal specificity, with γ_1 and γ_2 expected to be positive, but γ_3 negative on theoretical grounds.

The measurement model for the simulation involves loadings λ_{mk} from items $X_m (m = 1, .. 5)$ to factors k as follows: $\lambda_{11} = 1$ (preset), $\lambda_{21} = 0.7$, $\lambda_{31} = -0.5$, $\lambda_{42} = 1$ (preset), $\lambda_{52} = 1.6$. The coefficients in the structural model are

$$\{\gamma_1, \gamma_2, \gamma_3\} = (0.8, 1.7, 0.5)$$

The remaining parameters used to simulate the data are contained in Program 8.3, with the dispersion matrix Σ of the constructs ξ_{ik} including a positive correlation between the two factors.

Satisfactory convergence in this and similar small sample examples may be assisted by good starting values (e.g. from maximum likelihood analysis) and by further prior information on the structural or measurement model parameters (e.g. that the interaction parameter is expected to be positive or negative).

Here N(0, 1) priors on the free loadings λ and coefficients γ are adopted without constraints to positive or negative values (with the loadings $\lambda_{11} = \lambda_{42} = 1$ being predetermined). Priors on precisions are as in Arminger and Muthen. Some guidance as to the labelling of the constructs (in terms of ensuring identifiability without label switching in MCMC sampling) is provided by the preset loadings λ_{11} and λ_{42}. Clear convergence, and incidentally convergence towards the parameters which reasonably approximate the generating parameters (Table 8.4), is apparent by around 5000 iterations in a two chain run of 25 000 iterations.

An inconsistency with the mode of generating the data is apparent if the assumed model is taken as

$$Y_i = \gamma_1 \xi_{i1} + \gamma_2 \xi_{i2} + \gamma_3 \xi_{i1} \xi_{i2} + \gamma_4 \xi_{i1}^2 + \gamma_5 \xi_{i2}^2$$

This model (Model B in Program 8.3), estimated with the same priors as above (and with a 25 000 iteration run), shows 'significant' γ_4 and γ_5, with 95% credible intervals (0, 0.2) and (−0.56, −0.07). Estimates of the other coefficients are similar to those of Model A. The predictive criterion of Gelfand and Ghosh, here relevant for predicting both indicator sets X and Y (via the variables *Z.X* and *Z.Y* in Program 8.3), shows an improved prediction for the extended model.

Table 8.4 Nonlinear factor effects, parameter summary

	Mean	St. devn.	2.5%	Median	97.5%
Σ_{11}	1.66	0.25	1.24	1.64	2.22
Σ_{12}	0.27	0.13	0.03	0.26	0.54
Σ_{22}	0.27	0.13	0.03	0.26	0.54
γ_1	0.66	0.09	0.49	0.66	0.84
γ_2	1.73	0.21	1.37	1.72	2.18
γ_3	0.51	0.11	0.31	0.50	0.74
λ_{21}	0.63	0.05	0.53	0.63	0.74
λ_{31}	−0.42	0.06	−0.54	−0.42	−0.32
λ_{52}	1.36	0.16	1.08	1.36	1.68

8.3 LATENT TRAIT AND LATENT CLASS ANALYSIS FOR DISCRETE OUTCOMES

As discussed above, the Normal linear model relating M continuous outcomes Y to p underlying continuous factors ψ has the form

$$Y_{im} = \kappa_m + \lambda_{m1}\psi_{i1} + \lambda_{m2}\psi_{i2} + \ldots + \lambda_{mP}\psi_{ip} \ldots + u_{im} \qquad m = 1, \ldots, M$$

for $i = 1, \ldots n$ subjects. For identifiability, one option is to assume the factors $\psi_i = (\psi_{i1}, \ldots \psi_{ip})$ have zero mean and variance unity. The residual error terms u are usually taken to be conditionally independent, since the ψ_{ij} are intended to explain the correlations among the observed items.

A similar framework may be postulated for observations on M discrete items (e.g. binary or ordinal data), which are to be explained by p metric factors. For example, again assuming conditional independence, suppose we have M binary items and that there is $p = 1$ latent trait. Then a typical model analogous to that above has the form

$$Y_{im} \sim \text{Bern}(\pi_{im}) \qquad i = 1, \ldots n;\ m = 1, \ldots M$$

$$\text{logit}(\pi_{im}) = \kappa_m + \lambda_m\psi_i$$

where the ψ_i are again standard Normal factor scores (sometimes called 'latent traits' in psychological or educational applications), where the κ_m represent the success rate on item m, and where the λ_m are loadings relating the mth item to scores on the latent trait. A special variant of this is the Rasch model in educational testing, where $\lambda_m = 1$ for all items,

$$\text{logit}(\pi_{im}) = \kappa_m + \psi_i$$

where κ_m represents the 'difficulty' or discriminatory power of the mth item, and the ψ_i are interpreted as abilities or frailties of the subjects.

A number of modelling schemes have been proposed for other types of multivariate discrete outcomes, including ordered multinomial outcomes and models for correlated count outcomes (Chib and Winkelmann, 2000). Options include non-conjugate models with multivariate Normal or Student t error densities within a log link (Poisson) or logit link (binomial) framework, or models involving latent continuous observations Y^* which parallel each observed discrete variable. However, alternatively, a multivariate error structure of dimension M might be replaced by a factor structure with $p < M$ constructs.

For example, suppose the observations consist of a mixture of M-H continuous variables and H ordinal outcomes containing $r_1, r_2, .. r_H$ categories, respectively. For simplicity, assume $r = r_1 = r_2 = .. = r_H$, and that the observed ranks are denoted Y_{im}, $i = 1, .. n$, $m = 1, H$. To model correlation among these variables or introduce regression effects, one may define (Chapter 3) latent variables Y^*_{im} and cut points δ_{mk} on their range, such that

$$Y_{im} = k \quad \text{if} \quad \delta_{m,\,k-1} \leq Y^*_{im} < \delta_{mk}$$

with $r - 1$ cut-points, defined as

$$-\infty \leq \delta_{m1} \leq \ldots \delta_{m,\,r-1} \leq \infty$$

As noted above, the Y^*_{im} might be taken to be multivariate Normal or Student t of dimension H. For binary outcomes, Y^*_i is constrained to be positive when $Y_i = 1$, and sampling Y^* from a Normal with variance 1 is equivalent to assuming a probit link for the probability that $Y = 1$. With a large number of ordinal or binary outcomes, observed together with metric outcomes, one may consider modelling the data in terms of a smaller number of constructs: then the combined set of variables $(Y^*_1, .. Y^*_H, Y_{H+1}, \ldots Y_M)$ is expressed in terms of the usual LISREL model in Equation (8.1), but with variance structures for the Y^* variables defined by identifiability.

8.3.1 Latent class models

In some circumstances, it may be more plausible to treat the p latent variables as categoric rather than metric. Thus, subjects are classified into one of K classes if $p = 1$, or cross-classified into one of $K_1 \times K_2$ classes if $p = 2$, rather than being located on a continuous scale or scales. For example, Langeheine (1994) considers a longitudinal setting where observed items relate to children at different ages, with the changing observations on the items taken to represent stage theories of developmental psychology which postulate distinct stages (i.e. discrete categories) of intellectual development. In other circumstances, there may be no substantive rationale for preferring a latent trait or latent class model, but both provide adequate fit to the observed data – so leading to model indeterminacy (Bartholomew and Knott, 1999, Chapter 6).

Let the prior probabilities on the K classes of a single latent category ψ be denoted η_k, with $\Sigma\eta_k = 1$. The 'independent variable' ψ is now comparable to a categorical factor in log-linear models, with the first category providing a reference category under a 'corner constraint'. The above example for M binary items may now be expressed

$$\begin{aligned} &\mathrm{Y}_{im} \sim \mathrm{Bern}(\pi_{im}) \\ &\mathrm{logit}(\pi_{im}) = \kappa_m + \lambda_{m2}\delta[\psi_i = 2] + \ldots + \lambda_{mK}\delta[\psi_i = K] \end{aligned} \quad (8.2a)$$

where $\delta[\psi_i = k]$ is 1 if the ith subject is allocated to the kth category of ψ and $\delta[\psi_i = k] = 0$ if ψ_i is not allocated to the kth category. Depending on whether $K = 2$ or $K > 2$, the subjects are allocated to classes according to

$$\psi_i \sim \mathrm{Bern}(\eta)$$

with η assigned a beta prior (say), or according to

$$\psi_i \sim \mathrm{Categorical}(\eta)$$

where η is of dimension K and may be assigned a Dirichlet prior.

An alternative, but ultimately equivalent, modelling framework takes

$$Y_{im} \sim \text{Bern}(\pi_{\psi_i,\, m}) \qquad (8.2b)$$

where ψ_i has K categories of 'caseness' and π_{km} are the probabilities of items m for subject i according to that subject's caseness. With priors on ψ_i as above, beta priors may then be assigned to the π_{km} or a link function used such as

$$\text{logit}(\pi_{km}) = \theta_{km} \qquad k = 1, .., K \qquad (8.2c)$$

where Normal priors, perhaps constrained to produce consistent labels during MCMC sampling, are adopted for the θ_{km}.

Latent categories may be useful in medical diagnosis where the observed indicators Y are various tests or criteria of illness, with none being certain or 'gold standard' indicators of the presence of a disease. Thus, following Rindskopf and Rindskopf (1986), one may assess whether a single latent categorisation (e.g. if $K = 2$, the latent categories might be ill vs. not ill) underlies several observed binary diagnostic items, none of which provide a 'gold standard' test. This type of model is illustrated with psychiatric caseness data from Dunn (1999).

Example 8.4 Psychiatric caseness Dunn (1999) presents data on three binary diagnostic items applied to $n = 103$ patients. The binary responses are taken from dichotomising more extensive scales, and are denoted CIS (Clinical Interview Schedule), GHQ (General Health Questionnaire) and HADS (Hospital Anxiety and Depression Scale) (see Dunn, 1999, p. 7). Let G_{hij} denote the totals of patients in category h of CIS, i of GHQ, and j of HADS, where h, i and $j = 1$ for less ill, and h, i and $j = 2$ for more ill.

Then as in Dunn (1999), one may apply a log-linear model incorporating the classification LC representing the patients latent diagnostic status (assumed to have $K = 2$ states, with index $k = 1, 2$). In terms of the notation of Section 8.3.1, LC corresponds to the latent caseness category ψ_{hij}. As in a more general latent mixture regression (Chapter 3), there are questions around label switching in MCMC sampling, and so of adopting priors which ensure consistent labelling. The likelihood is multinomial

$$G_{hij} \sim \text{Mult}(\pi_{hij},\, n)$$

with

$$\pi_{hij} = \Sigma_k \mu_{hijk} / \Sigma_{hijk} \mu_{hijk}$$

and $\log(\mu_{hijk})$ modelled in terms of (a) the main effects in the items and the latent variable, and (b) the interaction effects between the items and the latent variable. Specifically,

$$\log \mu_{hijk} = C + \beta_{1,\, h} + \beta_{2,\, i} + \beta_{3,\, j} + \beta_{4,\, k} + \alpha_{1,\, h,\, k} + \alpha_{2,\, i,\, k} + \alpha_{3,\, j,\, k}$$

As usual in a log-linear model, there are corner constraints on the parameters (e.g. $\beta_{11} = \beta_{21} = \beta_{31} = \beta_{42} = 0$).

An alternative analysis involves disaggregating the observed data to individual level, and adopting Bernoulli sampling, especially if there are continuous covariates which assist in predicting the latent diagnosis.

In adopting the aggregate approach, a prior N(0, 10) confined to positive values is adopted for the free parameter $\beta_{4,\, 2}$ so that category two of the latent class variable identifies more ill patients (the 'cases'). A two chain run of 10 000 iterations (with convergence from around 4000 iterations) then produce similar results to those cited by Dunn (1999, p. 42) for the main effects of CIS, GHQ and HADS items, and of the latent diagnosis LC, and for the interactions CIS × LC, GHQ × LC and HADS × LC. The posterior medians of logged odds ratios describing the interaction between LC and the three items actually used are given in Table 8.5 (i.e. the interaction parameters CIS.LC, GHQ.LC and HAD.LC). These are 3.67, 4.08 and 3.86, respectively, and indicate a consistency between the three items.

Table 8.5 Psychiatric caseness: items (CIS, HAD, GHQ) and Latent Class (LC)

	Mean	St. devn.	2.5%	Median	97.5%
Aggregated (log-linear) model					
CIS.LC	3.73	0.84	2.32	3.67	5.70
GHQ.LC	4.23	1.05	2.64	4.08	6.61
HAD.LC	3.99	0.97	2.45	3.86	6.34
CIS	−2.08	0.73	−3.88	−1.98	−0.97
HAD	−2.32	0.83	−4.31	−2.19	−1.10
GHQ	−2.86	0.98	−5.04	−2.69	−1.45
LC	−4.71	0.85	−6.62	−4.63	−3.26
Disaggregated model Class Probabilities (Prob LC = k)					
η_1	0.42	0.05	0.31	0.42	0.49
η_2	0.58	0.05	0.51	0.58	0.69
Marginal Tables					
(1) CIS by LC					
Both No	39	3	34	39	44
CIS Yes, LC No	5	2	0	5	10
LC Yes, CIS No	10	3	5	10	15
LC Yes, CIS Yes	49	2	44	49	54
(2) GHQ by LC					
Both No	42	3	35	42	48
GHQ Yes, LC No	2	2	0	2	6
LC Yes, GHQ No	12	3	6	12	19
LC Yes, GHQ Yes	47	2	43	47	49
(3) HAD by LC					
Both No	40	3	34	40	45
HAD Yes, LC No	4	2	0	4	8
LC Yes, HAD No	10	3	5	10	16
LC Yes, HAD Yes	49	2	45	49	53
Probabilities π_{km} by LC ($k = 1, 2$) and Item ($m = 1, 3$)					
π_{11}	0.11	0.06	0.00	0.10	0.25
π_{12}	0.05	0.05	0.00	0.04	0.17
π_{13}	0.08	0.06	0.00	0.08	0.22
π_{21}	0.83	0.06	0.71	0.84	0.94
π_{22}	0.79	0.07	0.65	0.79	0.92
π_{23}	0.83	0.06	0.71	0.84	0.94

Marginal tables cross-tabulating patients by their caseness on each item and caseness on LC can also be obtained. If the categorisation by LC is regarded as the 'true' diagnosis, then the sensitivity of each item can be obtained with regard to detecting the true diagnosis. The overall probability of being a case (η_2) is estimated as 0.56.

A suggested exercise is to include second order interactions with LC (via parameters $\gamma_{1hik}, \gamma_{2hjk}, \gamma_{3ijk}$) and assess fit via a predictive criterion – for instance, using samples of new cross-classified data $G_{new,\,hij}$. Note that this model yields a higher probability of being a case.

The second analysis uses disaggregated item data with Bernoulli sampling, and the model in Equations (8.2b)–(8.2c). The prior constraint $\eta_2 > \eta_1$ is set for identifiability. One possible set of outputs from this analysis is the marginal cross-tabulations of item by LC classification. For instance, in the LC by GHQ table, the LC variable classifies 59 patients as cases, whereas there are 12 patients classified as well by the GHQ but who are cases according to LC. On this basis, the sensitivity of the GHQ (with LC regarded as the 'true' diagnosis) is obtained as 47/59 = 80%.

The sensitivities are equivalently the probabilities π_{2m} of being classed as more ill on the three indicators given $\psi_i = 2$. They show that the CIS scale has a high sensitivity (83%), but also a relatively high chance of classifying someone as ill when they are well according to LC (i.e. a false positive).

Finally, a latent trait analysis with a single continuous factor is carried out, namely

$$Y_{im} \sim \text{Bern}(\pi_{im}) \qquad i = 1,..n;\ m = 1,..M$$
$$\text{logit}(\pi_{im}) = \kappa_m + \lambda_m\psi_i$$

with all λ_m as free parameters and $\psi_i \sim$ N(0, 1). N(0, 10) and N(1, 10) priors are assigned to κ_m and λ_m, respectively. Sampling new data $Z_{im} \sim \text{Bern}(\pi_{im})$ and accumulating the subjects according to whether $Z_{im} = 0$ or 1 (m = 1, 2, 3) forms a predicted cross-classification $G_{hij,\,new}$. These can be compared with the actual counts by a predictive fit criterion (Gelfand and Ghosh, 1998). Another possibility with the latent trait model is the opportunity to choose $K \geq 2$ cut points on the latent trait ψ_i; for instance, taking $K = 3$ might correspond to the divisions: well, some symptoms, and definitely ill. Here $K = 2$ is chosen with the cut point at zero, and H[1:103] is the resulting classifier in Program 8.4.

This model appears to have an edge on the latent class model, with the predictive criterion (when $w = 1$) being around 91 compared to 111 under the LCA model; its advantage is in providing more precise predictions. From a 10 000 iteration run in two chains, the posterior means and 95% credible intervals of $\Lambda = (\lambda_1, \lambda_2, \lambda_3)$ are 3.3 (1.7, 6.5), 3.9 (1.9, 7.5), and 3.7 (1.8, 7.2). The selected cut point on the latent trait leads to similar marginal tables as those in Table 8.5, though slightly fewer patients are classed as cases (namely those with positive ψ_i).

Example 8.5 AIDS risks To illustrate the application of a latent trait structure for modelling associations between a mix of continuous and discrete outcomes, data from Shi and Lee (2000) on sexual practices and attitudes of female sex workers in the Phillipines is used (see also Song and Lee, 2001). Three indicators (Y_1, Y_2, Y_3) are continuous and three (Y_4, Y_5, Y_6) ordinal. The last three relate to AIDS risk from intercourse with an AIDS infectee, a stranger or a drug user, and are positive measures of 'caution' or 'worry' about contracting AIDS. They all have $r = 5$ ranked categories, from 1 = no risk to 5 = high risk. By contrast, the first three measures are positive

Table 8.6 AIDS risk: bivariate factor model

	Mean	St. devn.	2.5%	97.5%
Cut points				
δ_{11}	−1.57	0.09	−1.76	−1.40
δ_{12}	−0.71	0.08	−0.86	−0.56
δ_{13}	0.42	0.07	0.27	0.57
δ_{14}	0.95	0.08	0.79	1.11
δ_{21}	−4.33	0.32	−5.02	−3.77
δ_{22}	−3.24	0.24	−3.78	−2.84
δ_{23}	−1.39	0.13	−1.68	−1.17
δ_{24}	−1.04	0.11	−1.28	−0.85
δ_{31}	−3.30	0.29	−3.99	−2.85
δ_{32}	−2.22	0.21	−2.73	−1.89
δ_{33}	−0.20	0.09	−0.37	−0.02
δ_{34}	0.25	0.09	0.08	0.44
Loadings				
λ_{11}	0.31	0.06	0.19	0.43
λ_{21}	0.66	0.13	0.46	0.96
λ_{31}	0.39	0.08	0.25	0.55
λ_{42}	0.56	0.25	0.22	1.12
λ_{52}	1.16	0.43	0.49	2.17
λ_{62}	1.36	0.49	0.62	2.42
Factor correlation				
ω	−0.15	0.09	−0.38	−0.01

measures of 'recklessness': relating to frequencies of actual intercourse, of 'hand jobs' and of 'blow jobs'.

Following Shi and Lee, two latent variables ψ_{ij}, $i = 1, .. n$, $j = 1, 2$ may be proposed with the 6×2 matrix Λ of loadings of $(Y_1, \ldots, Y_6)$ on the factors taking the following structure:

$$\Lambda' = \begin{matrix} \lambda_{11} & \lambda_{21} & \lambda_{31} & 0 & 0 & 0 \\ 0 & 0 & 0 & \lambda_{42} & \lambda_{52} & \lambda_{62} \end{matrix}$$

where λ_{m1}, $m = 1, .. 3$ are free loadings relating $Y_1 - Y_3$ to ψ_{i1} and λ_{m2}, $m = 4, .. 6$ relate $Y_4 - Y_6$ to ψ_{i2}. Thus, if the loadings are consistently positive, the factors are interpretable as overall levels of worry about AIDS and recklessness, respectively. The ψ_{ij} are bivariate Normal with zero means, and their covariation described by (and only identifiable as far as) a correlation matrix,

$$\Omega = \begin{pmatrix} 1 & \omega \\ \omega & 1 \end{pmatrix}$$

Note that one might consider a single factor model as an alternative, since conceptually the factors seem to overlap.

The three continuous indicators Y_{im} are taken to be Normal and, despite being standardised (as they are by Shi and Lee), taken to have means

$$\kappa_m + \lambda_{m1}\psi_{i1}$$

with variances ϕ_m. Observed values on the three ordinal indicators Y_{im} are determined by underlying latent variables Y^*_{im}, with $r-1$ cut points $\delta_{mk}(m = 4, 5, 6; k = 1,..,r-1)$. Thus,

$$Y_{im} \sim \text{Categorical}\ (P_{i,\ m,\ 1:r}) \quad m = 4, 5, 6$$

where

$$\begin{aligned} P_{i,\ m,\ 1} &= Q_{i,\ m,\ 1} \\ P_{i,\ m,\ k} &= Q_{i,\ m,\ k} - Q_{i,\ m,\ k-1} \qquad k = 2,..,r-1 \\ P_{i,\ m,\ r} &= 1 - Q_{i,\ m,\ r-1} \end{aligned}$$

and

$$\text{logit}\ [Q_{i,\ m,\ k}] = \delta_{mk} - \mu_{im}$$

The subject means μ_{im} for the ordinal variables are given by

$$\begin{aligned} \mu_{i4} &= \lambda_{42}\psi_{i2} \\ \mu_{i5} &= \lambda_{52}\psi_{i2} \\ \mu_{i6} &= \lambda_{62}\psi_{i2} \end{aligned}$$

Shi and Lee obtain a negative correlation ω of -0.17 between ψ_1 and ψ_2 and positive loadings on all of λ_{m1} and λ_{m2}. Here a constraint of positivity on the loadings is applied for identifiability. Another option might be an exchangeable prior $\lambda_{mk} \sim N(\mu_\lambda, \sigma^2_\lambda)$, where μ_λ is constrained to be positive.

A two-chain run of 2500 iterations (with convergence after 1250) reproduces the structure obtained by Shi and Lee (see Table 8.6), namely positive loadings on the first factor (worry about aids) and on the second indicative of recklessness, and accordingly a negative correlation of around -0.15 between the factors. The density of ω shows some negative skewness, so the ratio of the posterior mean to posterior standard deviation may mislead in terms of 'significance'.

8.4 LATENT VARIABLES IN PANEL AND CLUSTERED DATA ANALYSIS

The structural equations approach, whether with latent traits or classes, may be applied to repeated data over time. For instance, Muthen (1997) stresses the role of latent variable methods in panel and multi-level models with random coefficients and variance components, as against the conventional psychometric application in terms of constructs measured with error. Molenaar (1999) cites some of the benefits of longitudinal data as a way to understanding causal mechanisms, and also offering ways of controlling the processes under scrutiny, depending on interventions made or not made. The elements of procedures to establish causality are illustrated by latent or observed variables ψ_{1t} and ψ_{2t} at times t with coefficients β_{ij} relating ψ_{it} to and $\psi_{j,\ t-1}$. Then, if ψ_1 is a cause of ψ_2, but not vice versa, one expects $\beta_{12} = 0$ and $\beta_{21} \neq 0$; such procedures are formalised as Granger causality (Granger, 1969). A variety of this type of inference, involving causal inter-relations of latent constructs over time, is illustrated by the well known alienation study of Wheaton *et al.* (1977), considered in Example 8.11.

8.4.1 Latent trait models for continuous data

Corresponding to the canonical cross-sectional LISREL type specification in Equation (8.1) suppose the data consist of time varying outputs possibly multivariate $\{Y_{imt}\}$, and time varying inputs $\{X_{ilt}\}$, which are respectively indicators for time varying endogenous construct ψ_{it} and a time varying exogenous construct ξ_{it}. Then the measurement model (for $t = 1, .., T$) has the form

$$Y_{imt} = \kappa^y_{jt} + \lambda_{1m}\psi_{it} + u^y_{ijt} \qquad m = 1, ..M \tag{8.3a}$$

$$X_{ijt} = \kappa^x_{jt} + \lambda_{2j}\xi_{it} + u^x_{ijt} \qquad j = 1, ..L \tag{8.3b}$$

and the associated structural model could be simply

$$\psi_{it} = \alpha + \gamma\xi_{it} + w_{it} \tag{8.3c}$$

As in Chapter 6, one may also propose permanent subject effects. Accordingly, Longford and Muthen (1992) suggest a measurement model with both constant latent effects ψ_{1i} and time varying latent effects ψ_{2it} for multiple observed items Y_{imt}:

$$Y_{imt} = \kappa_m + \lambda_{m1}\psi_{1i} + \lambda_{m2}\psi_{2it} + u_{imt} \tag{8.3d}$$

though this model raises identifiability issues. For example, if no constraints are placed on the M pairs of loadings $\{\lambda_{m1}, \lambda_{m2}\}$, then both construct variances must be fixed. If the measurement model included only stable traits, then one might consider changing loadings λ_{m1t}.

As a generalisation of Equation (8.3c), various autoregressive structural models are possible (Molenaar, 1999; Hershberger *et al.*, 1996), for example, an AR(1) lag in both ψ and ξ

$$\psi_{it} = \alpha + \beta\psi_{it-1} + \gamma\xi_{it-1} + w_{it}$$

Molenaar (1999) outlines state-space models for a multivariate latent vector $\psi_{it} = (\psi_{i1t}, \psi_{i2t}, \dots \psi_{ipt})$ of dimension $p < M$, and the joint evolution of $\{Y_{imt}, m = 1, M\}$ and $\{\psi_{ijt}, j = 1, ..p\}$ is described by the matrix model

$$\psi_{it} = B_t\psi_{it-1} + w_{it} \tag{8.4a}$$

$$\mathbf{Y}_{it} = \Lambda_t\psi_{it} + u_{it} \tag{8.4b}$$

where B_t is $p \times p$ and Λ_t is $M \times p$.

8.4.2 Latent class models through time

The latent class models of Section 8.3.1 may similarly be extended to models where repetition is over times instead of items, or possibly over both times and items (see Hagenaars (1994) and Langeheine (1994)). For the latent class model, the addition of a time dimension means changes in underlying state may occur, and in this context a latent class model may have the edge over a latent trait model in terms of substantive interpretability. Consider repeated observations on a binary item for subjects i:

$$Y_{it} \sim \text{Bern}(\pi_{it})$$

or equivalently, defining the observations as (1, 2) instead of (0, 1):

$$Y_{it} \sim \text{Categorical}(\rho_{i,\, t,\, 1:2})$$

where ρ_{it1} and ρ_{it2} are the probabilities of $Y_{it} = 1$, and $Y_{it} = 2$, respectively, with $\rho_{it1} = \pi_{it}$. More generally, an observed variable with R categories will have likelihood

$$Y_{it} \sim \text{Categorical}(\rho_{i,\, t,\, 1:R})$$

A latent class model for this type of observational series is often specified by assuming Markov dependence between the observed and/or latent states at times t and $t-1$. Under a first order Markov model, the state occupied at time t depends only upon the previous state, and not on any earlier ones. Then one possible model involves a Markov chain defined by the observed categories, but mixes transition behaviour over a latent state ψ_i with K categories (van de Pol and Langeheine, 1990). It is necessary to consider the first period observation Y_{i1} separately from others, as it has no (observed) antecedent, and so initial allocation probabilities $\delta_k (k = 1, .. K)$ may be assumed, which are also differentiated by the latent state ψ. So for subjects $i = 1, .. n$, one might specify

$$\psi_i \sim \text{Categorical}(\eta_{1:K}) \tag{8.5a}$$

$$Y_{i1} \sim \text{Categorical}(\delta_{\psi_i,\, 1:R}) \tag{8.5b}$$

$$Y_{it} \sim \text{Categorical}(\rho_{\psi_i,\, Y_{i,t-1},\, 1:R}) \quad t > 1 \tag{8.5c}$$

Another option models the Y_{it} as independent of previous observed category $Y_{i,\, t-1}$, but involves transitions on a latent Markov chain, defined by a variable ψ_{it} with K states. So the multinomial probabilities η are now specific for $\psi_{i,\, t-1}$, with

$$\psi_{it} \sim \text{Categorical}(\eta_{\psi_{i,\, t-1},\, 1:K}) \quad t > 1 \tag{8.6a}$$

and the observations have multinomial probabilities defined by the selected category of ψ_{it}

$$Y_{it} \sim \text{Categorical}(\rho_{\psi_{i,\, t},\, 1:R}) \tag{8.6b}$$

The first period latent state is modelled as

$$\psi_{i1} \sim \text{Categorical}(\delta_{1:K}) \tag{8.6c}$$

A higher level of generality, analogous to Equation (8.3d) for metric data, would be provided by a model with mixing over a constant latent variable ψ with K categories, and a latent transition variable ζ_{it} with L categories defined by the mixing variable. So

$$\begin{aligned}
\psi_i &\sim \text{Categorical}(\eta_{1:K}) \\
\zeta_{i1} &\sim \text{categorical}(\kappa_{\psi_i,\, 1:L}) \\
\zeta_{it} &\sim \text{categorical}(\zeta_{\psi_i},\, \zeta_{it-1,\, 1:L}) \quad t > 1 \\
Y_{it} &\sim \text{categorical}(\rho_{\psi_i},\, \zeta_{it,\, 1:R})
\end{aligned}$$

8.4.3 Latent trait models for time varying discrete outcomes

The time varying factor generalisation to multiple discrete responses Y_{ijt} and predictors X_{ijt} may, especially for binomial or ordinal data, involve latent continuous underlying variables Y^*_{ijt} and X^*_{ijt}. The model in these latent variables then resembles measurement and structural models adopted for continuous data, as in Equations (8.3a)–(8.3c).

Thus, Palta and Lin (1999) propose a model for observations on M longitudinal binary and ordinal items which allows for measurement error by introducing a single latent construct (or possibly p latent constructs, where p is less than M). Their empirical example consider the case where all outcomes y are binary, specifically $M = 2$ binary items relating to tiredness. Then, as usual for binary outcomes, suppose there is an underlying latent scale Y^* for each item:

$$\begin{aligned} Y_{i1t} &= 1 \text{ if } \quad Y^*_{i1t} > \tau_1 \\ &= 0 \text{ if } \quad Y^*_{i1t} \leq \tau_1 \end{aligned} \tag{8.7a}$$

$$\begin{aligned} Y_{i2t} &= 1 \text{ if } \quad Y^*_{i2t} > \tau_2 \\ &= 0 \text{ if } \quad Y^*_{i2t} \leq \tau_2 \end{aligned} \tag{8.7b}$$

The latent scale is in turn related to an underlying time-varying continuous construct ξ_{it} in a structural model:

$$Y^*_{i1t} = \beta_1 \xi_{it} + w_{i1t} \tag{8.7c}$$

$$Y^*_{i2t} = \beta_2 \xi_{it} + w_{i2t} \tag{8.7d}$$

where $\beta_1 = 1$ for identifiability. The structural model then relates the ξ_{it} to observed covariates X_{ijt} (taken as free of measurement error):

$$\xi_{it} = \kappa + \lambda_1 X_{i1t} + \lambda_2 X_{i2t} + \ldots e_{it} \tag{8.7e}$$

In the Palta–Lin model, the e_{it} are taken to be autocorrelated, with

$$\begin{aligned} e_{it} &= \rho_1 e_{i,\ t-1} + u^{(1)}_{it} \\ u^{(1)}_{it} &\sim \text{N}(0,\ \sigma^2_1) \end{aligned} \tag{8.7f}$$

while the errors $w_{ikt}(k = 1, .. K)$ are also autocorrelated, but with the same variance for all k:

$$\begin{aligned} w_{ikt} &= \rho_2 w_{ik,\ t-1} + u^{(2)}_{ikt} \\ u^{(2)}_{ikt} &\sim \text{N}(0,\ \sigma^2_2) \end{aligned}$$

Because, for binary data, the scale of the latent Y^*_{ikt} is arbitrary, a fixed scale assumption such as var(w)=1 is typically assumed. Here the constraint $\sigma^2_1 + \sigma^2_2 = 1$ is adopted to fix the scale. So that κ can be identified, the thresholds τ_1 and τ_2 must also be set to zero.

8.4.4 Latent trait models for clustered metric data

Panel data is a particular type of clustered design, and the principle of multivariate data reduction applies to other types of data which are hierarchically structured. Thus, consider cross-sectional data with level 1 units (e.g. pupils) clustered by higher level units (e.g. schools at level 2), with latent variables operating at each level. Assume a two-level model with p_1 factors ψ at level 1 and p_2 factors φ at level 2 (clusters), and a continuous outcomes Y_{ijm} for clusters $j = 1, .. J$, individuals $i = 1, \ldots n_j$ within clusters, and variables $m = 1, .. M$.

Assume further for illustration that $p = p_1 = p_2 = 2$. Then one might take

$$Y_{ijm} = \kappa_m + \lambda_{m1}\varphi_{1j} + \lambda_{m2}\varphi_{2j} + \gamma_{m1}\psi_{1ij} + \gamma_{m2}\psi_{2ij} + u_{2jm} + u_{1ijm} \quad (8.8)$$

where the level 1 errors u_{1ijm} are Normal with variances σ_m^2 and the level 2 error is MVN of order 2 with dispersion matrix Σ. The priors adopted for the variances/dispersions of the factor scores $\{\varphi_{1j}, \varphi_{2j}, \psi_{1ij}, \psi_{2ij}\}$ depend in part upon the assumptions made on relationships between the loadings at different levels.

Thus, the dispersion matrices Φ_1 and Φ_2 of the constructs $\psi = (\psi_{1ij}, \psi_{2ij})$ and $\varphi = (\varphi_{1j}, \varphi_{2j})$, respectively, are assumed to be identity matrices if the $M \times 2$ loadings $\Lambda = \{\lambda_{m2}, \lambda_{m2}\}$ at level 2 are estimated independently of the $M \times 2$ loadings $\Gamma = \{\gamma_{m1}, \gamma_{m2}\}$ at level 1. For $p_1 = p_2 = 2$, factors[1] at each level, there also needs to be one constraint on the level 2 loadings (e.g. setting $\lambda_{11} = 1$) and one on the level 1 loadings (e.g. setting $\gamma_{11} = 1$) for identifiability. Setting structural relationships between the loadings at different levels (or setting extra loadings to fixed values) makes certain dispersion parameters estimable. For example, one might take $\Lambda = \Gamma$. Depending on the problem, further constraints may be needed to ensure identification under repeated sampling in a fully Bayesian model.

8.4.5 Latent trait models for mixed outcomes

Analogous models including mixtures of discrete and continuous outcome variables have been proposed (e.g. Dunson, 2000). Thus, consider a set of observations Y_{ijm} on variables $m = 1, .. M$, for clusters $j = 1, .. J$ and sub-units $i = 1, .. n_j$ within clusters. Linked to the observations are latent variables Y^*_{ijm} drawn from densities in the exponential family (e.g. normal, Poisson, gamma), with means $\theta_{ijm} = E(Y^*_{ijm})$ predicted by

$$h(\theta_{ijm}) = \beta X_{ijm} + \varphi_j V_{ijm} + \psi_{ij} W_{ijm}$$

where h is a link function. X_{ijm} is an $M \times 1$ covariate vector with impact summarised by a population level regression parameter β, and the φ_j and ψ_{ij} are, respectively, vectors of cluster latent variables and latent effects specific to cluster and sub-unit. V_{ijm} and W_{ijm} are vectors of covariates, and may be subsets of the X_{ijm}, but often are just constants, with $V_{ijm} = W_{ijm} = 1$. The latent variables Y^*_{ijm} and Y^*_{ijl} $(m \neq l)$ are usually assumed independent conditionally on φ_j and ψ_{ij}. Frequently, the Y^*_{ijm} are Normal with identity link and hence expectation

$$\theta_{ijm} = \beta X_{ijm} + \varphi_j V_{ijm} + \psi_{ij} W_{ijm}$$

and diagonal covariance matrix of dimension $M \times M$.

In the case $V_{ijm} = W_{ijm} = 1$, an alternative formulation for θ_{ijm} takes the φ_j and ψ_{ij} as having known variances (e.g. unity), and introduces factor loadings λ_m and γ_m specific to variable m. For example, with a single factor at cluster and cluster-subject level

[1] For p_1 factors at level 1 and p_2 at level 2 and with Λ estimated independently of Γ, there are $p_1(p_1 - 1)/2$ constraints needed on Λ and $p_2(p_2 - 1)/2$ on Γ. Informative priors may be an alternative to deterministic constraints.

$$\theta_{ijm} = \beta X_{ijm} + \lambda_m \varphi_j + \gamma_m \psi_{ij}$$

As an example where all the observations Y_{ijm} are all binary, the Y^*_{ijm} could be taken as latent Normal variables, such that

$$Y_{ijm} = 1 \quad \text{if} \quad Y^*_{ijm} > 0$$

For identifiability the variances of Y^*_{ijm} are taken as unity, so that the probability of an event, i.e. $\pi_{ijm} = \Pr(Y_{ijm} = 1)$, is

$$\Phi(\beta X_{ijm} + \varphi_j V_{ijm} + \psi_{ij} W_{ijm})$$

where Φ is the distribution function of a standard Normal variable. The Poisson is an alternative latent density in this example, with

$$\begin{aligned} Y_{ijm} &= 1 \quad \text{if } Y^*_{ijm} \geq h_m \\ Y_{ijm} &= 0 \quad \text{if } Y^*_{ijm} < h_m \end{aligned} \tag{8.9a}$$

where h_m is a threshold count (e.g. unity), and where

$$Y^*_{ijm} \sim \text{Poi}(\theta_{ijm}) \tag{8.9b}$$

and

$$\log(\theta_{ijm}) = \beta X_{ijm} + \varphi_j V_{ijm} + \psi_{ij} W_{ijm} \tag{8.9c}$$

The 'hits' variable h_m may be a free parameter, and may differ between variables m. If $h_m = 1$ then Model 9 is equivalent to complementary log-log link for $\Pr(Y_{ijm} = 1) = \pi_{ijm}$, namely

$$\log(-\log(1 - \pi_{ijm})) = \beta X_{ijm} + \varphi_j V_{ijm} + \psi_{ij} W_{ijm}$$

Another possibility is a 'no hits' mechanism in (8.9a) defined by

$$Y_{ijm} = 1 \text{ if } Y^*_{ijm} = 0$$

If the Y_{ijm} consisted of M_1 binary variables and M-M_1 continuous variables, then (cf. Muthen, 1984) one sets observed and latent variables identically equal

$$Y^*_{ijm} = Y_{ijm} \quad m = M_1 + 1, .. M$$

while one of the latent variable options above for the binary outcomes $m = 1, .. M_1$, is used, such as

$$Y_{ijm} = 1 \quad \text{if} \quad Y^*_{ijm} > 0$$

A diagonal dispersion matrix $V = \text{cov}(Y^*)$ will then have M-M_1 free variance parameters. Extensions to the case where the set of the Y_{ijm} includes polytomous outcomes, with categories ordered or otherwise, can be made.

Example 8.6 Changes in depression state The first two of the models in Section 8.4.2 for discrete longitudinal series, as in Equations (8.5a)–(8.5c) and (8.6a)–(8.6c), were applied to data on 752 subjects for $T = 4$ periods (Morgan *et al.*, 1983). The data were binary ($R = 2$), with 0 = 'not depressed' and 1 = 'depressed', coded in the categorical form (1, 2) in Program 8.6. Following Langeheine and van de Pol (1990), two latent states are assumed on the underlying mixture or latent transition variables.

The mixed Markov model with mixing only over a discrete latent variable ψ_i, but no latent transitions is applied first, as in Equation (8.5) (Model A). To gain identifiability, the following constraints are made:

$$\eta_2 > \eta_1$$
$$\delta_{12} > \delta_{11}$$
$$\delta_{21} > \delta_{22}$$

where $\delta_{\psi_i, j}, j = 1, R$ defines the multinomial likelihood for the initial observations Y_{i1}. To set these constraints gamma priors are used, and then the property that Dirichlet variables can be obtained[2] as ratios to the sum of the gamma variables. These constraints are based on the maximum likelihood solution reported by Langeheine and van de Pol (1990, Table 4, Model D).

A two chain[3] run of 5000 iterations shows early convergence under the above constraints. The posterior parameter estimates suggest a small group ($\eta_1 = 0.15$) with an initially high chance of being depressed ($\delta_{12} = 0.65$), but around 40% chances of becoming non-depressed ($\rho_{12} = 0.40$), as in Table 8.7. The larger latent group ($\eta_2 = 0.85$) has a high initial probability of being non-depressed and high rate of staying so over time ($\rho_{211} = 0.94$).

To ensure identifiability of Model B, namely the latent Markov chain model in Equations (8.6a)–(8.6c), an alternative strategy to constraining parameters is adopted; specifically, it is assumed that one individual with the pattern 'no depressed' at all four periods is in latent class 1, and one individual with the response 'depressed' at all periods is in state 2. With this form of (data based) prior there is early convergence in a two chain run of 5000 iterations.

The substantive pattern identified under this model (Table 8.8) is in a sense more clear cut than Model A, since it identifies a predominantly non-depressive latent class (defined by ψ_1) with high initial probability $\delta_1 = 0.80$, which has virtually no chance

Table 8.7 Depression state, Model A parameters

	Mean	St. devn.	2.50%	97.50%
δ_{11}	0.346	0.079	0.193	0.495
δ_{12}	0.654	0.079	0.505	0.807
δ_{21}	0.907	0.020	0.871	0.948
δ_{22}	0.093	0.020	0.052	0.129
η_1	0.150	0.041	0.086	0.247
η_2	0.850	0.041	0.754	0.914
ρ_{111}	0.419	0.135	0.139	0.661
ρ_{112}	0.581	0.135	0.339	0.861
ρ_{121}	0.398	0.053	0.292	0.502
ρ_{122}	0.603	0.053	0.498	0.708
ρ_{211}	0.938	0.009	0.920	0.956
ρ_{212}	0.062	0.009	0.044	0.080
ρ_{221}	0.883	0.059	0.769	0.990
ρ_{222}	0.117	0.059	0.010	0.231

[2] If $x_1 \sim G(w, 1)$ and $x_2 \sim G(w, 1)$, $y_1 = x_1/\Sigma x_j$, $y_2 = x_2/\Sigma x_j$, then $\{y_1, y_2\}$ are Dirichlet with weight vector (w, w).
[3] Null initial values in one chain, and the other based on Langeheine and van de Pol.

Table 8.8 Depression state, Model B

	Mean	St. devn.	2.5%	97.5%
δ_1	0.804	0.033	0.733	0.868
δ_2	0.196	0.033	0.138	0.261
η_{11}	0.990	0.009	0.967	1.006
η_{12}	0.010	0.009	0.000	0.027
η_{21}	0.165	0.048	0.075	0.259
η_{22}	0.835	0.048	0.737	0.929
ρ_{11}	0.945	0.010	0.927	0.965
ρ_{12}	0.055	0.010	0.034	0.074
ρ_{21}	0.359	0.048	0.264	0.453
ρ_{22}	0.641	0.048	0.549	0.735

($\eta_{12} = 0.01$) of moving to the other latent state defined by ψ_2. The conditional probability of being non-depressed, given $\psi_1 = 1$ is 0.945. The other latent transition variable is more mixed in substantive terms, and includes a small group of non-depressed who are not certain to stay so.

Example 8.7 Ante-natal knowledge This example applies the model of Palt and Lin (Section 8.4.3) to data from an ante-natal study reported by Hand and Crowder (1996, p. 205). The observations consist of four originally continuous knowledge scales observed for 21 women before and after a course. There are nine treatment subjects who received the course, and 12 control subjects. The four variates are scales with levels 0–5, 0–20, 0–30 and 0–5.

For illustrative purposes, the original data are dichotomised with $Y_1 = 1$ if the first scale is 5, 0 otherwise, $Y_2 = 1$ if the second scale exceeds 15, $Y_3 = 1$ if the third scale exceeds 22, and $Y_4 = 1$ if the fourth scale exceeds 4. Hence, the model (for $k = 1, 4$; $t = 1, 2$) is

$$\begin{aligned}
Y_{ikt} &= 1 \quad \text{if } Y^*_{ikt} > 0\\
Y_{ikt} &= 0 \quad \text{if } Y^*_{ikt} \leq 0\\
Y^*_{ikt} &= \beta_k \xi_{it} + w_{ikt}\\
\xi_{it} &= \lambda_1 + \lambda_2 x_i + e_{it}\\
e_{it} &= \rho_1 e_{i,\,t-1} + u^{(1)}_{it}\\
w_{ikt} &= \rho_2 w_{i,\,k,\,t-1} + u^{(2)}_{ikt}
\end{aligned}$$

with $\sigma_1^2 = \text{var}(u_1)$, $\sigma_2^2 = \text{var}(u_2)$. The only covariate is the fixed treatment variable x (i.e. the course on knowledge) with coefficient λ_2. N(0, 10) priors are adopted for $\beta = (\beta_2, \beta_3, \beta_4)$ and $\lambda = (\lambda_1, \lambda_2)$, and a G(1, 1) prior for $1/\sigma_2^2 \cdot \sigma_1^2$ is then obtained via the constraint $\sigma_1^2 + \sigma_2^2 = 1$.

A two chain run of 10 000 iterations (with convergence from 1500) shows no evidence of a treatment effect on the underlying knowledge scale ξ_{it} over patients i and periods t (Table 8.9). The scores on this scale for individual women show improvements among the control group (e.g. compare ξ_2 with ξ_1 for subjects 4 and 6), as well as the course group. There is a high intra-cluster correlation ρ_2 governing measurement errors w_{ikt} on

Table 8.9 Ante-natal knowledge posterior parameter summaries

	Mean	St. devn.	2.5%	97.5%
$\xi_{1,1}$	0.59	0.49	−0.26	1.66
$\xi_{1,2}$	0.58	0.47	−0.24	1.59
$\xi_{2,1}$	0.13	0.39	−0.66	0.94
$\xi_{2,2}$	0.77	0.47	−0.03	1.80
$\xi_{3,1}$	−0.70	0.49	−1.82	0.14
$\xi_{3,2}$	0.47	0.49	−0.34	1.58
$\xi_{4,1}$	−0.77	0.51	−1.87	0.04
$\xi_{4,2}$	0.88	0.57	−0.02	2.24
$\xi_{5,1}$	−0.24	0.40	−1.11	0.46
$\xi_{5,2}$	0.96	0.55	0.07	2.26
$\xi_{6,1}$	−0.75	0.47	−1.83	0.07
$\xi_{6,2}$	0.82	0.50	0.00	1.95
$\xi_{7,1}$	−0.39	0.42	−1.32	0.34
$\xi_{7,2}$	1.10	0.57	0.16	2.34
$\xi_{8,1}$	−0.14	0.39	−0.94	0.61
$\xi_{8,2}$	0.44	0.45	−0.35	1.42
$\xi_{9,1}$	−0.67	0.47	−1.69	0.12
$\xi_{9,2}$	1.05	0.64	0.13	2.67
$\xi_{10,1}$	−0.24	0.39	−1.07	0.49
$\xi_{10,2}$	0.42	0.45	−0.34	1.45
$\xi_{11,1}$	−0.73	0.47	−1.78	0.10
$\xi_{11,2}$	0.50	0.51	−0.35	1.63
$\xi_{12,1}$	0.06	0.37	−0.67	0.82
$\xi_{12,2}$	0.93	0.55	0.07	2.20
$\xi_{13,1}$	−0.63	0.49	−1.69	0.24
$\xi_{13,2}$	−0.65	0.53	−1.86	0.27
$\xi_{14,1}$	−0.87	0.54	−2.15	0.00
$\xi_{14,2}$	0.40	0.44	−0.39	1.34
$\xi_{15,1}$	−0.88	0.55	−2.20	−0.03
$\xi_{15,2}$	0.36	0.42	−0.40	1.29
$\xi_{16,1}$	−0.16	0.42	−1.01	0.63
$\xi_{16,2}$	−0.17	0.41	−1.02	0.61
$\xi_{17,1}$	−0.03	0.42	−0.93	0.74
$\xi_{17,2}$	0.93	0.55	−0.02	2.12
$\xi_{18,1}$	−0.80	0.52	−2.03	0.06
$\xi_{18,2}$	0.03	0.40	−0.75	0.85
$\xi_{19,1}$	0.57	0.50	−0.31	1.65
$\xi_{19,2}$	0.58	0.53	−0.34	1.76
$\xi_{20,1}$	−0.39	0.42	−1.31	0.37
$\xi_{20,2}$	0.14	0.39	−0.59	0.99
$\xi_{21,1}$	−0.06	0.40	−0.90	0.70
$\xi_{21,2}$	0.74	0.51	−0.11	1.93
λ_1	0.21	0.22	−0.21	0.67
λ_2	−0.05	0.24	−0.52	0.41
β_2	2.40	1.09	0.80	4.97
β_3	2.93	1.31	0.97	5.93
β_4	4.04	1.48	1.44	7.34
ρ_1	−0.20	0.31	−0.75	0.42
ρ_2	0.89	0.10	0.64	0.98

the same item at different time points, but the autocorrelation in the latent construct is lower (in fact, biased to negative values).

Example 8.8 Factor structures at two levels This example replicates the analysis by Longford and Muthen (1992), in which metric data Y_{ijm} are generated for $i = 1,..10$ subjects within $j = 1,..20$ clusters for $m = 1,..5$ variables, as in Equation (8.8), namely

$$Y_{ijm} = \kappa_m + \lambda_{m1}\varphi_{1j} + \lambda_{m2}\varphi_{2j} + \gamma_{m1}\psi_{1ij} + \gamma_{m2}\psi_{2ij} + u_{2jm} + u_{1ijm}$$

This example illustrates how identification of the assumed parameters from the data thus generated requires constrained priors on the loadings to ensure consistent labelling of the constructs during sampling. In the Longford and Muthen simulation, the means $\kappa = \{\kappa_1,..\kappa_5)$ are zero, and the level 1 variances $\text{Var}(u_{1ijm})$ are 1. The level 2 variances $\text{Var}(u_{2jm})$ are 0.2. Also in the simulation, the loadings at level 1 and 2 are taken to be the same, i.e.

$$\Lambda = \Gamma = \begin{pmatrix} 1 & 1 & 1 & 1 & 1 \\ 1 & -1 & 0 & -1 & 1 \end{pmatrix}^T$$

The dispersion matrices of the factor scores $\{\psi_1, \psi_2\}$ at level 1 and $\{\varphi_1, \varphi_2\}$ at level 2 are, respectively,

$$\Phi_1 = \begin{pmatrix} 1 & 0 \\ 0 & 1 \end{pmatrix}$$

and

$$\Phi_2 = \begin{pmatrix} 1.25 & 1 \\ 1 & 1.25 \end{pmatrix}$$

In estimation of an appropriate model from the data thus generated (i.e. coming to the data without knowing how it was generated), one might adopt several alternative prior model forms. The assumption $\Lambda = \Gamma$ might in fact be taken on pragmatic grounds to improve identifiability of the level 2 loading matrix. Here for illustration this is not assumed.

Then with Λ and Γ independent, minimal identifiability requires one of the loadings at each level must take a preset value. Here it is assumed that $\gamma_{11} = \lambda_{11} = 1$. The level 1 and 2 factor variances are taken as preset at one, and with no correlation between φ_{1j} and φ_{2j}. Given the small cluster sizes, the observation variances, and level 2 loadings may all not be reproduced that closely. Identifiability was further ensured by assuming the first factor at each level is 'unipolar' (has consistently positive loadings in relation to the indicators Y), and by defining the second factor at each level as bipolar, for instance constraining γ_{12} to be positive and γ_{22} to be negative. In any particular confirmatory factor analysis, such assumptions would require a substantive basis.

On this basis a two chain run of 5000 iterations shows convergence at under 1500 iterations, and shows estimated level 1 loadings reasonably close to the theoretical values (Table 8.10). The observational variances are also reasonably closely estimated. The level 2 loadings also broadly reproduce the features of the theoretical values.

In the absence of knowledge of the mode of data generation, one might alternatively (a) adopt a single level 2 factor, while still retaining a bivariate factor structure at level 1, or (b) retain a bivariate level 2 factor but take $\Lambda = \Gamma$.

Table 8.10 Two level factor structure, parameter summary

	Mean	St. devn.	2.5%	97.5%
Level 1 loadings				
γ_{12}	1.19	0.29	0.55	1.62
γ_{21}	1.09	0.18	0.70	1.40
γ_{22}	−0.92	0.21	−1.38	−0.59
γ_{31}	0.82	0.10	0.61	1.02
γ_{32}	−0.10	0.16	−0.44	0.16
γ_{41}	0.85	0.19	0.42	1.18
γ_{42}	−0.87	0.17	−1.22	−0.57
γ_{51}	0.89	0.25	0.54	1.46
γ_{52}	1.08	0.16	0.74	1.39
Level 2 loadings				
λ_{12}	1.13	0.35	0.42	1.84
λ_{21}	0.60	0.27	0.11	1.21
λ_{22}	−0.46	0.27	−1.10	−0.04
λ_{31}	0.47	0.22	0.08	0.91
λ_{32}	0.22	0.27	−0.31	0.71
λ_{41}	0.36	0.21	0.03	0.82
λ_{42}	−0.38	0.23	−0.86	0.02
λ_{51}	1.09	0.27	0.52	1.58
λ_{52}	0.89	0.39	0.08	1.61
Level 1 variances				
$\mathrm{Var}(u_{11})$	0.71	0.49	0.06	1.73
$\mathrm{Var}(u_{12})$	0.43	0.24	0.06	0.93
$\mathrm{Var}(u_{13})$	0.94	0.12	0.72	1.20
$\mathrm{Var}(u_{14})$	1.45	0.24	0.99	1.93
$\mathrm{Var}(u_{15})$	1.18	0.48	0.10	1.87
Level 2 variances				
$\mathrm{Var}(u_{21})$	0.13	0.11	0.03	0.43
$\mathrm{Var}(u_{22})$	0.15	0.12	0.03	0.47
$\mathrm{Var}(u_{23})$	0.26	0.15	0.05	0.62
$\mathrm{Var}(u_{24})$	0.10	0.07	0.02	0.27
$\mathrm{Var}(u_{25})$	0.22	0.17	0.03	0.65

Example 8.9 Toxicity in mice This example uses simulated data on reproductive toxicity in mice, drawing on the work of Dunson (2000) concerning the toxicological impacts of the solvent ethylene glycol monomethyl ether (EGMME). Dunson analyses data on litters i from parental pairs j, with $n - n_c = 132$ litters born to pairs exposed to EGMME and $n_c = 134$ litters to control pairs not exposed. There are up to five litters per pair, and two outcomes for each litter, namely Y_{ij1} binary and Y_{ij2} Poisson, with $Y_{ij1} = 1$ if the birth was delayed (i.e. prolonged birth interval) and Y_{ij2} relating to litter size. Litter size has an effective maximum of 20.

The observed variables are linked to underlying Poisson variables Y^*_{ij1} and $Y^*_{ij2} = (Y^*_{ij21}, Y^*_{ij22}, \ldots . Y^*_{ij2M})$, with $M = 20$. Let $X_i = 1$ for exposed pairs and $X_i = 0$ otherwise. Then the Poisson means are

$$E(Y^*_{ij1}) = \exp(\beta_{1i} + \beta_2 X_i + \varphi_{1j} + \psi_{ij})$$

and for $m = 1, \ldots M$, where $M = 20$ is the maximum litter size

$$E(Y^*_{ij2m}) = \exp(\beta_{3i} + \beta_4 X_i + \varphi_{2j} + \psi_{ij})$$

The β_{1i} and β_{3i} are intercepts specific to each of the five possible litters per pair; thus β_{11} is the intercept specific to the first litter (when $i = 1$), and so on. β_2 and β_4 are exposure effects on times between births and on litter size. The correlations between outcomes are modelled via the common error ψ_{ij}. The observed indicators are defined according to hits and no hits mechanisms, respectively, which is cumulated in the case of Y_{ij2}. Thus

$$Y_{ij1} = \delta(Y^*_{ij1} \geq 1)$$
$$Y_{ij2} = \sum_{m=1}^{M} \delta(Y^*_{ij2m} = 0)$$

where $\delta(u)$ equals 1 if condition u holds, and zero otherwise. Note that $Y^*_2 = 0$ represents 'no defect preventing successful birth', so that cumulating over $\delta(Y^*_2 = 0)$ gives the number of mice born. However, the actual complementary log-log model involves the chance $Y^*_2 \geq 1$ of a defect at each m.

Here data are simulated on 270 litters (five litters for $n_c = 27$ control pairs, and five litters for $n_e = 27$ exposed pairs) using the parameters supplied by Dunson (2000, p. 364). In re-estimating the model, N(0, 1) priors are assumed on the parameters β_2 and β_4, together with the informative priors of Dunson (2000) on β_{1i} and β_{3i} and the precisions of φ_{1j}, φ_{2j} and ψ_{ij}. The complementary log-log link is used to reproduce the hits mechanism. A two chain run of 2000 iterations, with convergence after 250, gives estimates for β_2 and β_4 parallel to those of Dunson (2000, p. 364); the posterior means and 95% credible intervals for these parameters are 1.92 (1.44, 2.40) and 0.32 (0.15, 0.48), respectively.

The interpretation is that exposure to the EGMEE delays births and reduces litter sizes. There are several possible sensitivity analyses, including the extent of stability in parameter estimates under less informative priors on the precisions.

Here an alternative model including explicit loadings on the constructs φ_{1j} and φ_{2j} is also investigated. In this model, the two constructs have variance 1, but are allowed to be correlated (with parameter ω). On these assumptions, the loadings $\{\lambda_1, \ldots, \lambda_4)$ in the following model may be identified:

$$Y_{ij1} = \delta(Y^*_{ij1} \geq 1)$$
$$Y_{ij2} = \sum_{m=1}^{M} \delta(Y^*_{ij2m} = 0)$$
$$E(Y^*_{ij1}) = \exp(\beta_{1i} + \lambda_1 \varphi_{1j} + \psi_{ij}) \quad (i = 1, 5; j = 1, n_c)$$
$$E(Y^*_{ij1}) = \exp(\beta_{1i} + \beta_2 + \lambda_2 \varphi_{1j} + \psi_{ij}) \quad (i = 1, 5; j = n_c + 1, n)$$
$$E(Y^*_{ij2m}) = \exp(\beta_{3i} + \lambda_3 \varphi_{2j} + \psi_{ij}) \quad (i = 1, 5; j = 1, n_c; m = 1, \ldots M)$$
$$E(Y^*_{ij2m}) = \exp(\beta_{3i} + \beta_4 + \lambda_4 \varphi_{2j} + \psi_{ij}) \quad (i = 1, 5; j = 1, n_c + 1, n; m = 1, \ldots M)$$

λ_2 and λ_4 are constrained to be positive for identifiability.

Table 8.11 Factor model for birth outcomes

	Mean	St. devn.	2.5%	97.5%
β_2	1.92	0.23	1.48	2.38
β_4	0.27	0.08	0.12	0.43
λ_1	0.00	0.31	−0.59	0.65
λ_2	0.41	0.20	0.09	0.87
λ_3	0.82	0.38	0.22	1.69
λ_4	1.19	0.52	0.36	2.50
ω	0.34	0.23	0.09	0.97

A two chain run of 2000 iterations shows there is a positive correlation coefficient between the factors (albeit for these simulated data). There are also higher loadings λ_2 and λ_4 for the exposed group (as compared to λ_1 and λ_3, respectively) on the parent level factors which represent chances of birth delay and birth defect, respectively. In real applications, this might represent excess risk beyond that represented by the simple dummy for exposure to EGMEE.

8.5 LATENT STRUCTURE ANALYSIS FOR MISSING DATA

In structural equation models, including confirmatory factor models, it may be that latent variables rather than (or as well as) observed indicators contribute to predicting or understanding missingness mechanisms. Sample selection or selective attrition that lead to missing data may be more clearly related to the constructs than to any combination of the possible fallible proxies for such constructs. As above (Chapter 6), assume the full set of observed and missing indicator data is denoted $Y = \{Y_{\text{obs}}, Y_{\text{mis}}\}$where Y_{mis} is of dimension M, and that the observed data includes an $n \times M$ matrix of binary indicators R_{im} corresponding to whether Y_{ij} is missing ($R_{im} = 1$) or observed ($R_{im} = 0$).

Maximum likelihood and EM approaches to missing data in structural equation and factor analysis models are considered by Rovine (1994), Arbuckle (1996) and Allison (1987). As noted by Arbuckle (1996), the methods developed are often based on the missing at random assumption, or assume special patterns of missingess, such as the monotone pattern. Under monotone missingness, one might have completely observed variable X for all n subjects, a variable Y observed for only n_1 subjects, and a variable Z fully observed for only n_2 subjects and observed only when Y is (the n_2 subjects are then a subsample of the n_1). Then the likelihood may be written

$$\prod_{i=1}^{n} f(X_i|\theta) \prod_{i=1}^{n_1} f(Y_i|X_i, \theta) \prod_{i=1}^{n_2} f(Z_i|Y_i, X_i, \theta)$$

Under the MAR assumption, the distribution of R depends only upon the observed data, so

$$f(R|Y, \omega) = f(R|Y_{\text{obs}}, \omega)$$

whereas in many situations (e.g. attrition in panel studies) the attrition may depend upon the values of the indicators that would have been observed in later waves. An example of this non-random missingness is based on the Wheaton *et al.* (1977) study into alienation, considered in Example 8.11. Since the interest is in accounting for missingness via latent constructs ψ based on the entire Y matrix, a non-random missingness model might take the form $f(R|\psi, \omega)$.

Example 8.10 Fathers' occupation and education Allison (1987) considers a study by Bielby *et al.* (1977) which aimed to find the correlation between father's occupational status and education for black men in the US. With a sample of 2020 black males, Bielby *et al.* found a correlation of 0.433, but realised this might be attenuated by measurement error. They therefore re-interviewed a random sub-sample of 348 subjects approximately three weeks later, and obtained replicate measures on status and education. Let y_1 and y_3 denote the first measures (on all 2020 subjects) relating to status and education, respectively. For the sub-sample, observations are also obtained on y_2 and y_4, repeat measures of status and education, respectively. For the 1672 subjects remaining of the original sample, these two variables are then missing.

On this basis, one may assume, following Allison, that the missing data are missing completely at random – though sampling mechanisms may generate chance associations which invalidate the intention of the design. Since R_{i2} and R_{i4} are either both 1 or both zero, a single response indicator $R_i = 1$ for y_2 and y_4 present and $R_i = 0$, otherwise may be adopted.

For the 348 complete data subsample, the observed means are {16.62, 17.39, 6.65, 6.75} and the variance-covariance matrix is

180.9	126.8	24.0	22.9
126.8	217.6	30.2	30.5
24.0	30.2	16.2	14.4
22.9	30.5	14.4	15.1

while for the larger group the means on y_1 and y_3 are 17 and 6.8, with variance-covariance matrix

217.3	25.6
25.6	16.2

Bielby *et al.* and Allison assume that the data were generated by two underlying factors ('true' occupational and educational status) with

$$Y_{1i} = \kappa_1 + \lambda_1\psi_{1i} \tag{8.10a}$$

$$Y_{2i} = \kappa_2 + \lambda_2\psi_{1i} \tag{8.10b}$$

$$Y_{3i} = \kappa_3 + \lambda_3\psi_{2i} \tag{8.10c}$$

$$Y_{4i} = \kappa_4 + \lambda_4\psi_{2i} \tag{8.10d}$$

Both Bielby *et al.* and Allison take ψ_1 and ψ_2 to have a free dispersion matrix which allows for covariation between the factors. This option means a constraint $\lambda_1 = \lambda_3 = 1$ is needed for identifiability. Alternatively, one might take ψ_1 and ψ_2 to be standardised variables with their dispersion matrix containing a single unknown correlation parameter ρ. In this case, all the λ parameters are identifiable. If in fact missingness is not MCAR, then it will be related either to the known observations Y_1 and Y_3, or to the partially unknown observations Y_2 and Y_4, or to the factor scores, ψ_1 and ψ_2.

Table 8.12 Fathers' occupation and education

	Mean	St. devn.	2.5%	Median	97.5%
Coefficients in missingness model					
β_0	1.362	0.223	0.933	1.359	1.805
β_1	0.005	0.009	−0.013	0.005	0.022
β_2	0.019	0.031	−0.041	0.019	0.079
Correlation between factors					
ρ	0.608	0.047	0.511	0.609	0.696
Free loadings					
λ_2	1.869	0.140	1.627	1.857	2.129
λ_4	1.024	0.047	0.938	1.022	1.123

The analysis of Allison is replicated here with an original sample size of 505, and later subsample of size 87. An open mind on the response mechanism is retained, and it is assumed that the probability of non-response $\pi_i = P(R_i = 1)$ may be related (via a logit link) to Y_1 and Y_3 in line with MAR response. The response model is then

$$\text{logit}(\pi_i) = \beta_0 + \beta_1 Y_{1i} + \beta_2 Y_{3i} \tag{8.11}$$

and priors $\beta_j \sim N(0, 1)$ are assumed. The measurement model in Equation (8.10) is applied to all 592 subjects, regardless of observation status.

Parameter summaries are based a single chain run taken to 120 000 iterations (with 5000 burn in) for estimates of $\{\beta_0, \beta_1, \beta_2\}$, $\{\lambda_2, \lambda_4\}$ and the correlation ρ between the two factors (and hence between true social and educational status). The diagnostics of Raftery and Lewis (1992) suggest this number is required because of a high autocorrelation in the samples, especially of λ_2 and λ_4. The inter-factor correlation of 0.61 (Table 8.12) compares to the estimate of 0.62 cited by Allison (1987, p. 86). There is in fact no evidence of departure from MCAR in the data as sampled, in the sense that β_1 and β_2 in Model (8.11) are not different from zero (their 95% credible intervals straddle zero). The missingness model is though subject to possible revision, for example taking only Y_3 as the predictor. Subject to identifiability, response models including $\{Y_2, Y_4\}$ or $\{\psi_1, \psi_2\}$ may also be investigated.

Example 8.11 Alienation over time This example considers adaptations of the data used in a structural equation model of alienation over time as described by Wheaton *et al.* (1977), originally with $n = 932$ subjects. In a reworked analysis of simulated data from this study reported by Muthen *et al.* (1987), there are six indicators of two constructs (social status and alienation) at time 1, and three indicators of alienation at time 2. A slightly smaller number of subjects (600) was assumed.

Denote social status and alienation at time 1 by ξ_1 and ψ_1, and alienation at time 2 by ψ_2. The original indicators for $i = 1, .. 600$ subjects are standardised, with three indicators $X_1 - X_3$ at time 1 related to the social status (exogenous) construct as follows:

$$X_{1i} = \lambda_{11}\xi_i + u_{1i}$$

$$X_{2i} = \lambda_{21}\xi_i + u_{2i}$$
$$X_{3i} = \lambda_{31}\xi_i + u_{3i}$$

where u_1, u_2 and u_3 are independently univariate Normal. The three indicators of alienation are denoted Y_{11}, Y_{21} and Y_{31} at time 1 and Y_{12}, Y_{22} and Y_{32} at time 2. They are related to the alienation construct at times 1 and 2 as follows:

$$Y_{11i} = \lambda_{12}\psi_{1i} + u_{4i}$$
$$Y_{21i} = \lambda_{22}\psi_{1i} + u_{5i}$$
$$Y_{31i} = \lambda_{32}\psi_{1i} + u_{6i}$$

and

$$Y_{12i} = \lambda_{13}\psi_{2i} + u_{7i}$$
$$Y_{22i} = \lambda_{23}\psi_{2i} + u_{8i}$$
$$Y_{32i} = \lambda_{33}\psi_{2i} + u_{9i}$$

The constructs themselves are related first by a cross-sectional model at time 1, namely

$$\psi_{1i} = \beta_{11}\xi_i + w_{1i} \tag{8.12a}$$

and by a longitudinal model relating time 2 to time 1, namely

$$\psi_{2i} = \beta_{21}\xi_i + \beta_{22}\psi_{1i} + w_{2i} \tag{8.12b}$$

Thus, alienation at time 2 depends upon alienation at time 1 and status at time 1.

Muthen *et al.* use various models to simulate missingness, which is confined to the wave 2 indicators of alienation, and applies to all items for non-responding subjects. Thus, let R_i be a binary indicator of whether a subject is missing at wave 2 (i.e. unit rather than item non-response) with $R_i = 1$ for response present and $R_i = 0$ for response missing (this coding for R is used to be consistent with Muthen *et al.*). Underlying this binary indicator is a latent continuous variable R_i^*, which is zero if $R_i^* < \tau$.

One model for the R_i^* assumes they are related only to fully observed (i.e. first wave) data X_{ki} and Y_{k1i}:

$$\begin{aligned} R_i^* = {} & 0.667\omega^*(X_{1i} + X_{2i} + X_{3i}) \\ & - 0.333\omega^*(Y_{11i} + Y_{21i} + Y_{31i}) + \delta_i \end{aligned} \tag{8.13}$$

where $\delta_i \sim N(0, 1)$ and $\omega^* = 0.329$. The cut off τ is taken as -0.675. This is missingness at random (only depending on observed data), and leads to a missingness rate of around 25%.

Another choice makes missingness depend upon both the wave 1 and 2 outcomes whether observed or not, so that the missingness mechanism is non-ignorable. Thus,

$$\begin{aligned} R_i^* = {} & 0.667\omega^*(X_{1i} + X_{2i} + X_{3i}) \\ & - 0.333\omega^*(Y_{11i} + Y_{21i} + Y_{31i} + Y_{12i} + Y_{22i} + Y_{32i}) + \delta_i \end{aligned} \tag{8.14}$$

In this case, Muthen *et al.* varied the degree of selectivity by setting ω^* at 0.33, 0.27 or 0.19.

A further option, also non-ignorable, makes missingness depend upon the latent factors so that

$$R_i^* = 0.667\omega^*\xi_i - 0.333\omega^*(\psi_{1i} + \psi_{2i}) + \delta_i \tag{8.15}$$

with $\omega^* = 0.619$. This is non-ignorable, because ψ_2 is defined both by observed and missing data at phase 2.

Accordingly, values of $(X_j, j = 1, 3)$ $\{Y_{j1}, j = 1, 3\}$ and $\{Y_{j2}, j = 1, 3\}$ are generated and sampled data at wave 2 then removed according to the missingness model. One may then compare (a) the estimates of the parameters $\{\Lambda, \beta, \text{var}(w_j), \text{var}(u_m)\}$ using the original data with no imputed non-response (b) the parameters obtained when adopting a missingness model based only on a MAR mechanism, and (c) the parameters obtained adopting a missingness model based on the latent factors, for example as in Equation (8.15). Under (b), logit models for R_i or R_i^* depend upon the fully observed observations at wave 1, and under (c) such models depending on the constructs ψ_1, ψ_2 and ξ.

Using the 9×9 correlation matrix provided by Muthen *et al.*, a full data set may be generated and missingness then imputed according to Equation (8.13), (8.14) or (8.15). We adopt the option in Equation (8.14), where missingness is related to all indicators, whether subject to non-response at wave 2 or not, and take $\omega^* = 0.27$. To generate the data, it is necessary to sample all the $\{X_{ji}, Y_{jti}\}$ and then 'remove' the sampled data for missing cases where R_{ij}^* is under the threshold. The form of the missingness models (8.13)–(8.15) means there is either complete non-response at wave 2 or complete response at unit level on all three indices. So individual item response indices R_{ij} may be replaced by a single unit response index, $R_i = 0$ for all missing observations at wave 2, and $R_i = 1$ otherwise. There are 169 of the 600 observations with missingness at wave 3, a rate of 28%.

Under the response mechanism in Equation (8.14), attrition is greater for lower status persons and more alienated persons: so missingess might be expected to be greater for subjects with higher scores on ψ_1 and ψ_2. In Model B in Program 8.11, the response model relates $\pi_i = \Pr(R_i = 1)$ to the factor scores, namely,

$$\text{logit}(\pi_i) = \omega_0 + \omega_1\xi_i + \omega_2\psi_{1i} + \omega_3\psi_{2i} \tag{8.16}$$

We then obtain the expected negative impacts on response of alienation (ψ_1 and ψ_2) and a positive impact of status, ξ (see Table 8.13 obtained from iterations 500–5000 of a two chain run). The impact of ψ_2 is as might be expected, less precisely estimated than that of ψ_1 Other models relating π_i to (say) just ξ and ψ_2 might be tried. The coefficients of the structural model (8.12) are close to the parameters obtained from the fully observed sample, though the negatively signed impact β_{21} of social status ξ on alienation ψ_2 at time 2 is enhanced.

Instead one might assume a model with no information to predict missingess, i.e.

$$\text{logit}(\pi_i) = \omega_0 \tag{8.17}$$

(This is pi.1[] in Model B in Program 8.11.) In the present case, and with the particular sample of data from the covariance matrix of Muthen *et al.*, this produces very similar estimates of structural and measurement coefficients to the non-ignorable model.

Both models in turn provide similar estimates of the parameters to those based on the fully observed data set of 600×9 variables (Model C in Program 8.11). Model (8.16) allowing for non-ignorable missingness provides an estimate for λ_{23} closer to the full data parameter, but β_{21} is better estimated under the MCAR model (8.17). So for this particular sampled data set, there is no benefit in using a missingness model linked to values on the latent constructs. However, to draw firm conclusions about the benefits of ignorable vs. non-ignorable missingess it would be necessary to repeat this analysis with a large number of replicate data sets.

Table 8.13 Alienation study missing data, parameter summary

	Mean	St. devn.	2.50%	Median	97.50%
Missingness coefficients					
ω_1	1.13	0.12	0.91	1.12	1.37
ω_2	0.70	0.26	0.19	0.70	1.21
ω_3	−0.61	0.32	−1.24	−0.61	0.03
ω_4	−0.50	0.47	−1.48	−0.48	0.42
Structural model coefficients					
β_{11}	−0.57	0.06	−0.69	−0.57	−0.46
β_{21}	−0.31	0.07	−0.46	−0.31	−0.17
β_{22}	0.54	0.07	0.40	0.54	0.69
Measurement model coefficients					
λ_{11}	1.00				
λ_{22}	1.06	0.07	0.93	1.05	1.20
λ_{31}	0.71	0.06	0.59	0.70	0.82
λ_{12}	1.00				
λ_{22}	1.00	0.06	0.88	0.99	1.11
λ_{32}	0.72	0.05	0.62	0.72	0.83
λ_{13}	1.00				
λ_{23}	0.89	0.09	0.73	0.89	1.08
λ_{33}	0.56	0.08	0.41	0.56	0.72

8.6 REVIEW

Despite some long-standing Bayesian discussion of certain aspects of factor analysis and structural equation modelling (e.g. Lee, 1981; Press and Shigemasu, 1989), recent MCMC applications have occurred at a relatively low rate compared to other areas. This may in part reflect the availability of quality software adopting a maximum likelihood solution. Some of the possible advantages of Bayesian analysis are suggested by Scheines *et al.* (1999) and Lee (1992) in terms of modifying formal deterministic constraints to allow for stochastic uncertainty. Recent developments introducing structural equation concepts into multi-level analysis are discussed by Jedidi and Ansari (2001), who include an application of the Monte Carlo estimate (Equation (2.13b) in Chapter 2) of the CPO to derive pseudo Bayes factors.

However, considerable issues in the application of repeated sampling estimation remain relatively unexplored: whereas label switching in discrete mixture regression is well documented (see Chapters 2 and 3), the same phenomenon occurs in models of continuous latent traits. Similarly in latent class models with two or more latent class variables (e.g. latent class panel models as in Section 8.4.2) there are complex questions around consistent labelling of all such variables and how far constraints might restrict the solution. Bayesian SEM applications with discrete data are also relatively few.

REFERENCES

Allison, P. (1987) Estimation of linear models with incomplete data. In Clogg, C. (ed.), *Sociological Methodology*, Oxford: Blackwells, pp. 71–103.

Andrews, F. (1984) Construct validity and error components of survey measures: a structural modeling approach. *Public Opinion Quart*. **48**, 409–442.

Arbuckle, J. (1996) Full information estimation in the presence of incomplete data. In: Marocoulides, G. and Schumacker, R. (eds.), *Advanced Structural Equation Modelling*. Lawrence Erlbaum.

Arminger, G. and Muthen, B. (1998) A Bayesian approach to nonlinear latent variable models using the Gibbs sampler and the Metropolis-Hastings algorithm. *Psychometrika* **63**, 271–300.

Bartholomew, D. (1984) The foundations of factor analysis. *Biometrika* **71**, 221–232.

Bartholomew, D. and Knott, M. (1999) *Latent Variable Models and Factor Analysis* (Kendall's Library of Statistics, 7). London: Arnold.

Bentler, P. and Weeks, D. (1980) Linear structural equations with latent variables. *Psychometrika* **45**, 289–308.

Bielby, W., Hauser, R. and Featherman, D. (1977) Response errors of black and nonblack males in models of the intergenerational transmission of socioeconomic status. *Am. J. Sociology* **82**, 1242–1288.

Bollen, K. (1989) *Structural Equations With Latent Variables*. New York: Wiley.

Bollen, K. (1996) Bootstrapping techniques in analysis of mean and covariance structures. In: Marocoulides, G. and Schumacker, R. (eds.), *Advanced Structural Equation Modelling*. Lawrence Erlbaum.

Bollen, K. (1998) Structural equation models. In: *Encyclopaedia of Biostatistics*. New York: Wiley, pp. 4363–4372.

Bollen, K. (2001) Latent variables in psychology and the social sciences. *Ann. Rev. Psychol.* **53**, 605–634.

Bollen, K. and Paxton, P. (1998) Interactions of latent variables in structural equation models. *Structural Equation Modeling* **5**, 267–293.

Boomsma, A. (1983). On the robustness of LISREL (maximum likelihood estimation) against small sample size and non-normality. Amsterdam: Sociometric Research Foundation.

Browne, M. (1984) Asymptotically distribution-free methods for the analysis of covariance structures. *Br. J. Math. Stat. Psychol.* **37**, 62–83.

Byrne, B. and Shavelson, R. (1986) On the structure of adolescent self-concept. *J. Educat. Psychol.* **78**, 474–481.

Byrne, B., Shavelson, R. and Muthen, B. (1989) Testing for the equivalence of factor covariance and mean structures: the issue of partial measurement invariance. *Psychol. Bull.* **105**, 456–466.

Chib, S. and Winkelmann, R. (2000) Markov Chain Monte Carlo Analysis of Correlated Count Data. Technical report, John M. Olin School of Business, Washington University.

Dunn, G. (1999) *Statistics in Psychiatry*. London: Arnold.

Early, P., Lee, C. and Hanson, L. (1990) Joint moderating effects of job experience and task component complexity: relations among goal setting, task strategies and performance. *J Organizat. Behaviour* **11**, 3–15.

Everitt, B. (1984) *An Introduction to Latent Variable Models*. London: Chapman & Hall.

Fornell, C. and Rust, R. (1989) Incorporating prior theory in covariance structure analysis: a Bayesian approach. *Psychometrika* **54**, 249–259.

Gelman, A., Carlin, J., Stern, H. and Rubin, D. (1995) *Bayesian Data Analysis*. London: Chapman & Hall.

Granger, C. (1969) Investigating causal relations by econometric methods and cross-spectral methods. *Econometrica* **34**, 424–438.

Hagenaars, J. (1994) Latent variables in log-linear models of repeated observations. In: Von Eye, A. and Clogg, C. (eds.), *Latent Variables Analysis:Applications for Developmental Research*. London: Sage.

Hand, D. and Crowder, M. (1996) *Practical Longitudinal Data Analysis*. London: Chapman & Hall.

Hershberger, M. P. and Corneal, S. (1996) A hierarchy of univariate and multivariate structural time series models. In: Marocoulides, G. and Schumacker, R. (eds.), *Advanced Structural Equation Modelling*. Lawrence Erlbaum.

Hertzog, C. (1989) Using confirmatory factor analysis for scale development and validation. In: Lawton, M. and Herzog, A. (eds.), *Special Research Methods for Gerontology*. Baywood Publishers, pp. 281–306.

Holzinger, K. J. and Swineford, F. (1939) A study in factor analysis: the stability of a bi-factor solution. Supplementary Educational Monographs. Chicago, IL: The University of Chicago.

Ibrahim, J., Chen, M. and Sinha, D. (2001b) Criterion-based methods for Bayesian model assessment. *Statistica Sinica* **11**, 419–443.

Jedidi, K. and Ansari, A. (2001) Bayesian structural equation models for multilevel data. In: Marcoulides, G. and Schumacker, R. (eds.), *Structural Equation Modeling and Factor Analysis: New Developments and Techniques in Structural Equation Modeling*. Laurence Erlbaum.

Johnson, V. and Albert, J. (1999) *Ordinal Data Modeling*. New York: Springer-Verlag.

Joreskog, K. (1970) A general method for analysis of covariance structures. *Biometrika* **57**, 239–251.

Joreskog, K. G. (1973) A general method for estimating as linear structural equation system. In: Goldberger, A. S. and Duncan, O. D. (eds.), *Structural Equation Models in the Social Sciences*. New York: Seminar Press, pp. 85–112.

Langeheine, R. (1994) Latent variable Markov models. In: Von Eye, A. and Clogg, C. (eds.), *Latent Variables Analysis:Applications for Developmental Research*. London: Sage.

Langeheine, R. and van de Pol, F. (1990) A unifying framework for Markov modeling in discrete space and discrete time. *Sociological Methods & Res.* **18**, 416–441.

Lee, S. (1991) A Bayesian approach to confirmatory factor analysis. *Psychometrika* **46**, 153–160.

Langeheine, R., and Pol, F. van de. 1990. A unifying framework for Markov modelling in discrete space and discretetime. Sociological Methods and Research, **18**, 416–441.

Langeheine, R. and J. Rost, J. (eds.) (1988) *Latent Trait and Latent Class Models*. New York: Plenum.

Langeheine, R. (1994) Latent variables Markov models. In: *Latent Variables Analysis. Applications for Developmental Research*. Newbury Park, CA: Sage, pp. 373–395.

Lee, S. (1981) A Bayesian approach to confirmatory factor analysis. *Psychometroka* **46**, 153–160.

Lee, S. (1992) Bayesian analysis of stochastic constraints in structural equation models. *Br. J. Math. Stat. Psychology*. **45**, 93–107.

Lee, S. and Press, S. (1998) Robustness of Bayesian factor analysis estimates. *Commun. Stat., Theory Methods* **27**(8), 1871–1893.

Longford, N. and Muthen, B. O. (1992) Factor analysis for clustered observations. *Psychometrika* **57**(4), 581–597.

Molenaar, P. (1999) Longitudinal analysis. In: Ader, H. and Mellenbergh, G. (eds.), *Research and Methodology in the Social, Behavioural and Life Sciences*. London: Sage, pp. 143–167.

Muthen, B. (1984) A general structural equation model with dichotomous, ordered categorical, and continuous latent variable indicators. *Psychometrika* **49**, 115–132.

Muthen, B. (1997) Latent variable modelling of longitudinal and multilevel data. In: Raftery, A. (ed.), *Sociological Methodology*. Boston: Blackwell, pp. 453–480.

Muthen, B., Kaplan, D. and Hollis, M. (1987) On structural equation modeling with data that are not missing completely at random. *Psychometrika* **52**, 431–462.

Mutran, E. (1989) An example of structural modeling in multiple-occasion research. In: Lawton, M. and Herzog, A. (eds.), *Special Research Methods for Gerontology*. Baywood Publishers, pp. 265–279.

Palta, M. and Lin, C. (1999) Latent variables, measurement error and methods for analyzing longitudinal binary and ordinal data. *Stat. in Med.* **18**, 385–396.

Press, S. J. and Shigemasu, K. (1989) Bayesian inference in factor analysis. In: Glesser, L. *et al.* (eds.), *Contributions to Probability and Statistics, Essays in honor of Ingram Olkin*. New York: Springer Verlog.

Raftery, A. E. and Lewis, S. M. (1992) One long run with diagnostics: Implementation strategies for Markov chain Monte Carlo. *Stat. Sci.* **7**, 493–497.

Rindskopf, D. and Rindskopf, W. (1986) The value of latent class analysis in medical diagnosis. *Stat. in Med.* **5**, 21–27.

Rovine, M. (1994) Latent variable models and missing data analysis. In: *Latent Variables Analysis. Applications for Developmental Research*. Newbury Park, CA: Sage.

Scheines, R., Hoijtink, H. and Boomsma, A. (1999) Bayesian estimation and testing of structural equation models. *Psychometrika* **64**, 37–52.

Shi, J. and Lee, S. (2000) Latent variable models with mixed continuous and polytomous data. *J. Roy. Stat. Soc., Ser. B.* **62**.

Soares, A. and Soares, L. (1979) *The Affective Perception Inventory*. CT: Trumbell.

Song, X. and Lee, S. (2001) Bayesian estimation and test for factor analysis model with continuous and polytomous data in several populations. *Br. J. Math. Stat. Psychol.* **54**, 237–263.

Van de Pol, F. and Langeheine, R. (1990) Mixed Markov latent class models. In: Clogg, C. C. (ed.), *Sociological Methodology 1990*. Oxford: Basil Blackwell.

Wheaton, B., Muthen, B., Alwin, D. and Summers, G. (1977) Assessing reliability and stability in panel models. *Sociological Methodology* 84–136.

Yuan, K.-H. and Bentler, P. M. (1998) Structural equation modeling with robust covariances. *Sociological Methodology* **28**, 363–96.

Zhu, H. and Lee, S. (1999) Statistical analysis of nonlinear factor analysis models. *Br. J. Math. Stat. Psychol.* **52**, 225–242.

EXERCISES

1. In Example 8.2, try the more usual Normal density assumption for the indicators (equivalent to $W_{im} = 1$ by default for all i) and assess fit against the Student t model using the Gelfand-Ghosh or DIC criterion. The latter involves the likelihood combined over all indicators X_m.

2. In Example 8.3, try estimating both models with a non-zero intercept γ_0 in the structural model (with an N(0, 1) prior, say). Does this affect the relative performance of the two models?

3. In Example 8.4, add second order interaction parameters between the items and LC in the log-linear model for the G_{hij}, and compare fit with the model confined to first order interactions.

4. In Example 8.5, try a single factor model and evaluate its fit against the two factor model.

5. In the Ante-natal Knowledge example, try an analysis without preliminary dichotomisation but regrouping the data into ordinal scales.

6. In Example 8.8, try alternative ways to possibly improve identifiability, namely (a) assuming that level 1 and 2 loadings are the same, and (b) that only one factor is relevant at level 2. Set up the Normal likelihood calculation and assess changes in DIC or a predictive criterion.

7. In Example 8.10, repeat the analysis with only Y_3 included in the missingness model. Assess the change in DIC as compared to the model used in the Example, where the relevant deviances are for both response indicators and observed outcomes.

CHAPTER 9

Survival and Event History Models

9.1 INTRODUCTION

Processes in the lifecycle of individuals including marriage and family formation, changes in health status, changes in job or residence may be represented as event histories. These record the timing of changes of state, and associated durations of stay, in series of events such as marriage and divorce, job quits and promotions. Many applications of event history models are to non-repeatable events such as mortality, and this type of application is often called survival analysis. Survival and event history models have grown in importance in clinical applications (e.g. in clinical trials), in terms of survival after alternative treatments, and in studies of times to disease recurrence and remission, or response times to stimuli.

For non-renewable events the stochastic variable is the time from entry into observation until the event in question. So for human survival, observation commences at birth and the survival duration is defined by age at death. For renewable events, the dependent variable is the duration between the previous event and the following event. We may be interested in differences either in the rate at which the event occurs (the hazard rate), or in average inter-event times. Such heterogeneity in outcome rate or inter-event durations may be between population sub-groups, between individuals as defined by combinations of covariates, or as in medical intervention studies, by different therapies. Thus, in a clinical trial we might be interested in differences in patient survival or relapse times according to treatment.

Whereas parametric representations of duration of stay effects predominated in early applications, the current emphasis includes semiparametric models, where the shape of the hazard function is essentially left unspecified. These include the Cox proportional hazards model (Cox, 1972) and recent extensions within a Bayesian perspective such as gamma process priors either on the integrated hazard or hazard itself (Kalbflesich, 1978; Clayton, 1991; Chen *et al.*, 2000). While the shape of the hazard function in time is often of secondary interest, characteristics of this shape may have substantive implications (Gordon and Molho, 1995).

Among the major problems that occur in survival and inter-event time modelling is a form of data missingness known as 'censoring'. A duration is censored if a respondent withdraws from a study for reasons other than the terminating event, or if a subject does

Applied Bayesian Modelling P. Congdon

not undergo the event before the end of the observation period. Thus, we know only that they have yet to undergo the event at the time observation ceases. This is known as 'right censoring', in that the observed incomplete duration is necessarily less than the unknown full duration until the event. Other types of censoring, not considered in the examples below, are left censoring and interval censoring. In the first, subjects are known to have undergone the event but the time at which it occurred is unknown, while in the second it is known only that an event occurred within an interval, not the exact time within the interval.

Another complication arises through unobserved variations in the propensity to experience the event between individual subjects, population groups, or clusters of subjects. These are known as 'frailty' in medical and mortality applications (Lewis and Raftery, 1995). If repeated durations are observed on an individual, such as durations of stay in a series of jobs, or multiple event times for patients (Sinha and Dey, 1997), then the cluster is the individual employee or patient. The unobserved heterogeneity is then analogous to the constant subject effect in a panel model. Given the nature of the dependent variable, namely the length of time until an event occurs, unmeasured differences lead to a selection effect. For non-renewable events such as human mortality, high risk individuals die early and the remainder will tend to have lower risk. This will mean the hazard rate will rise less rapidly than it should.

A third major complication occurs in the presence of time varying covariates and here some recent approaches to survival models including counting processes (Andersen *et al.*, 1993; Fleming and Harrington, 1991) are relatively flexible in incorporating such effects. In event history applications, counting processes also allow one to model the effect of previous moves or durations in a subject's history (Lindsey, 2001).

Survival model assessment from a Bayesian perspective has been considered by Ibrahim *et al.* (2001a, 2001b) and Sahu *et al.* (1997), who consider predictive loss criteria based on sampling new data; and by Volinsky and Raftery (2000), who consider the appropriate form of the Bayesian Information Criterion (BIC). Pseudo-Bayes factors may also be obtained via harmonic mean estimates of the CPO (Sahu *et al.*, 1997; Kuo and Peng, 2000, p. 261) based on the full data.

Volinsky and Raftery suggest that the multiplier for the number of parameters be not $\log(n)$ but $\log(d)$, where n and d are, respectively, the total subjects and the observed number of uncensored subjects. Then if ℓ is the log-likelihood at the maximum likelihood solution and p the number of parameters

$$\text{BIC} = -2\ell + p\log(d)$$

Another version of this criterion, namely the Schwarz Bayesian Criterion (SBC), is proposed by Klugman (1992). This includes the value of the prior $\pi(\bar{\theta})$ at the posterior mean $\bar{\theta}$, and log-likelihood $\ell(\bar{\theta})$, so that

$$\text{SBC} = \ell(\bar{\theta}) + \pi(\bar{\theta}) - p\log(n/\pi)$$

where the last term on the right-hand side involves $\pi = 3.1416$. The AIC, BIC and SBC rely on knowing the number of parameters in different models, but the often high level of missing data through censoring means, for instance, that the true number of parameters is unknown and the method of Spiegelhalter *et al.* (2001) might therefore be relevant.

Predictive loss methods may be illustrated by an adaptation of the Gelfand and Ghosh (1998) approach; thus, let t_i be the observed times, uncensored and censored, θ the parameters, and z_i the 'new' data sampled from $f(z|\theta)$. Suppose ν_i and ς_i are the mean and variance of z_i, then, following Sahu *et al.* (1997), one criterion for any $w > 0$ is

$$D = \sum_{i=1}^{n} \varsigma_i + [w/(w+1)] \sum_{i=1}^{n} (\nu_i - u_i)^2 \tag{9.1}$$

where $u_i = \max(\nu_i, s_i)$ if s_i is a censored time and $u_i = t_i$ if the time is uncensored.

9.2 CONTINUOUS TIME FUNCTIONS FOR SURVIVAL

Suppose event or survival times T are recorded in continuous time. Then the density $f(t)$ of these times defines the probability that an event occurs in the interval $(t, t + dt)$, namely

$$f(t) = \lim_{dt \to 0} \Pr(t \leq T \leq t + dt)/dt$$

with cumulative density

$$F(t) = \int_0^t f(u)du$$

From this density the information contained in duration times can be represented in two different ways. The first involves the chance of surviving until at least time t (or not undergoing the event before duration t), namely

$$S(t) = \Pr(T \geq t) = 1 - F(t)$$
$$= \int_t^\infty f(u)du$$

The other way of representing the information involves the hazard rate, measuring the intensity of the event as a function of time,

$$h(t) = f(t)/S(t)$$

and in probability terms, the chance of an event in the interval $(t, t + dt)$ given survival until t. From $h(t)$ is obtained the cumulative hazard $H(t) = \int_0^t h(u)du$, and one may also write the survivor function as $S(t) = \exp(-H(t))$.

As an example of a parameterised form of time dependence, we may consider the Weibull distribution for durations $W(\lambda, \gamma)$, where λ and γ are scale and shape parameters, respectively (Kim and Ibrahim, 2000). The Weibull hazard is defined as

$$h(t) = \lambda \gamma t^{\gamma - 1}$$

with survival function

$$S(t) = \exp(-\lambda t^\gamma)$$

and density

$$f(t) = \lambda \gamma t^{\gamma - 1} \exp(-\lambda t^\gamma)$$

The Weibull hazard is monotonically increasing or decreasing in time according to whether $\gamma > 1$ or $\gamma < 1$. The value $\gamma = 1$ leads to exponentially distributed durations with parameter λ.

To introduce stationary covariates x of dimension p, we may adopt a proportional form for their impact on the hazard. Then the Weibull hazard function in relation to time and the covariates is

$$h(t, x) = \lambda e^{\beta x} \gamma t^{\gamma - 1} \tag{9.2}$$

Under proportional hazards, the ratio of the hazard rate at a given time t for two individuals with different covariate profiles, x_1 and x_2 say, is

$$h(t, x_1)/h(t, x_2) = \exp(\beta(x_1 - x_2))$$

which is independent of time.

An equivalent form for the Weibull proportional hazards model in Equation (9.2) (Collett, 1994) involves a log-linear model for the durations t_i and assumes a specified error u_i, namely the extreme value (Gumbel) distribution. Then

$$\log(t_i) = \nu + \alpha x_i + \sigma u_i \tag{9.3}$$

where, in terms of the parameters in Equation (9.2), the scale is

$$\sigma = 1/\gamma$$

the intercept is

$$\nu = -\sigma \log(\lambda)$$

and the covariate effects are

$$\alpha_j = -\beta_j \sigma \quad j = 1, \ldots, p$$

Taking u_i as standard Normal leads to a log-Normal model for durations, while taking u_i as logistic leads to the log-logistic model for t (Lawless, 1982; Fahrmeir and Tutz, 2001).

In BUGS the Weibull density for durations is routinely implemented as

```
t[i] ~ dweib(lambda[i],gamma),
```

where the log of lambda[i] (or possibly some other link) is expressed as a function of an intercept and covariates, and gamma is the shape parameter. While the Weibull hazard is monotonic with regard to duration t, a non-monotonic alternative such as the log-logistic may be advantageous, and this may be achieved in BUGS by taking a logistic model for $y = \log(t)$. Here, t are observed durations, censored or complete.

Thus,

$$y_i \sim \text{Logistic}(\mu_i, \kappa) \tag{9.4}$$

where κ is a scale parameter and μ_i is the location of the ith subject. The location may be parameterised in terms of covariate impacts $\mu_i = \beta x_i$ on the mean length of log survival (rather than the hazard rate). The variance of y is obtained as $\pi^2/(3\kappa^2)$. The survivor function in the y scale is

$$S(y) = [1 + \exp(\{y - \mu\}/\sigma)]^{-1} \tag{9.5}$$

where $\sigma = 1/\kappa$. In the original scale, the survivor function is

$$S(t) = [1 + \{t/\theta\}^{\kappa}]^{-1} \tag{9.6}$$

where $\theta = e^{\mu}$.

Example 9.1 Reaction times An example of a parametric analysis of uncensored data is presented by Gelman *et al.* (1995, Chapter 16), and relates to response times on $i = 1, .. 30$ occasions for a set of $j = 1, .. 17$ subjects; 11 were not schizophrenic and six were diagnosed as schizophrenic. In Program 9.1, the first 11 cases are non-schizophrenic. As well as response times being higher for the latter, there is evidence of greater variability in reaction times for the schizophrenics.

For the non-schizophrenic group a Normal density for the log response times $y_{ij} = \log_e t_{ij}$ (i.e. a log-Normal density for response times) is proposed, with distinct means for each of the 11 subjects. We might alternatively adopt a heavier tailed density than the Normal for the schizophrenic group, but there are substantive grounds to expect distinct sub-types. Specifically, delayed reaction times for schizophrenics may be due to a general motor retardation common to all diagnosed patients, but attentional deficit may cause an additional delay on some occasions for some or all schizophrenics.

The observed times for non-schizophrenics are modelled as

$$y_{ij} \sim \mathrm{N}(\alpha_j, \omega)$$

with the means for subjects j drawn from a second stage prior $\alpha_j \sim N(\mu, \Phi)$. For the schizophrenics, the observed times are modelled as

$$y_{ij} \sim \mathrm{N}(\alpha_j + \tau G_{ij}, \omega)$$

$$\alpha_j \sim \mathrm{N}(\mu + \beta, \Phi)$$

where β and τ are expected to be positive. The G_{ij} are a latent binary classification of schizophrenic times, according to whether the Additional Attention Deficit (AD) impact was operative or not.

We accordingly assign $N(0, 1)$ priors for β and for τ, measuring the AD effect, with sampling confined to positive values. For the probabilities λ_1 and λ_2 of belonging to the AD group or not (among the schizophrenic patients) a Dirichlet prior is adopted, with weights of 1 on the two choices. Gelman *et al.* use the equivalent parameterisation $\lambda_1 = 1 - \lambda$ and $\lambda_2 = \lambda$ with the group indicators drawn from a Bernoulli with parameter λ. Note that the constraint $\tau > 0$ is already a precaution against 'label switching' in this discrete mixture problem. Convergence of parameters (over a three chain run to 20 000 iterations) is achieved by around iteration 8000 in terms of scale reduction factors between 0.95 and 1.05 on the unknowns, and summaries based on the subsequent 12 000.

We find an estimated median λ_2 of 0.12 (Table 9.1, Model A), which corresponds to that obtained by Gelman *et al.* (1995, Table 16.1). The excess of the average log response time for the non-delayed schizophrenic times over the same average for non-schizophrenics is estimated at $\beta = 0.32$, as also obtained by Gelman *et al.*

We follow Gelman *et al.* in then introducing a distinct variance parameter for those subject to attentional deficit, and also an additional indicator $F_j \sim \mathrm{Bern}(\omega)$ for schizophrenic subjects such that G_{ij} can only be 1 when $F_j = 1$. λ_2 is now the chance of an AD episode given that the subject is AD prone.

The second half of a three chain run of 20 000 iterations leads to posterior means $\lambda_2 = 0.66$, and $\omega = 0.50$ (Table 9.1, Model B). However, using the predictive loss criterion of Gelfand and Ghosh (1998) and Sahu *et al.* (1997), it appears that the more heavily parameterised model has a worse loss measure as in Equation (9.1), and so the simpler model is preferred. This conclusion is invariant to values of w between 1 and values of w so large that $w/(1 + w)$ is effectively 1.

Example 9.2 Motorettes Tanner (1996) reports on the analysis of repeated observations of failure times of ten motorettes tested at four temperatures. All observations are

Table 9.1 Response time models, parameter summary

Model A	Mean	St. devn.	2.5%	Median	97.5%
λ_1	0.877	0.029	0.814	0.879	0.929
λ_2	0.123	0.029	0.071	0.121	0.186
β	0.317	0.08	0.16	0.317	0.477
μ	5.72	0.05	5.63	5.72	5.81
τ	0.843	0.06	0.729	0.842	0.962
Model B					
λ_1	0.344	0.160	0.054	0.336	0.674
λ_2	0.656	0.160	0.326	0.664	0.946
β	0.261	0.100	0.082	0.253	0.472
μ	5.72	0.04	5.63	5.72	5.81
τ	0.552	0.245	0.228	0.468	1.099
ω	0.500	0.236	0.068	0.527	0.888

Table 9.2 Failure times of motorettes (* censored)

	Temperature (centigrade)			
Motorette	150	170	190	220
1	8064*	1764	408	408
2	8064*	2772	408	408
3	8064*	3444	1344	504
4	8064*	3542	1344	504
5	8064*	3780	1440	504
6	8064*	4860	1680*	528*
7	8064*	5196	1680*	528*
8	8064*	5448*	1680*	528*
9	8064*	5448*	1680*	528*
10	8064*	5448*	1680*	528*

right censored at the lowest temperature, and three motorettes are censored at all temperatures (Table 9.2).

The original times t are transformed via $W = \log_{10}(t)$, and a Normal density proposed for them with variance σ^2 and means modelled as

$$\mu_i = \beta_1 + \beta_2 V_i$$

where $V_i = 1000/(\text{temperature}+273.2)$. For censored times it is necessary to constrain sampling of possible values above the censored time; it is known only that the actual value must exceed the censored time. For uncensored cases, we follow the BUGS convention in including dummy zero values of the censoring time vector (W.cen[] in Program 9.2).

Tanner obtains $\sigma = 0.26$, $\beta_1 = -6.02$ and $\beta_2 = 4.31$. We try both linear and quadratic models in V_i and base model selection on the Schwarz criterion at the posterior mean. The posterior means on the censored failure times come into the calculations of the SBC's (Models A1 and B1 in Program 9.2). Less formal assessments might involve comparing (between linear and quadratic models) the average deviance or likelihood over iterations subsequent to convergence. It is important to centre the V_i (especially in the quadratic model) to guarantee early convergence of the β_j, which means that the intercept will differ from Tanner's.

Summaries in Table 9.3 are based on the last 4000 iterations of three chain runs to 5000 iterations. With N(0, 1000) priors on the β_j and G(1, 0.001) prior on $1/\sigma^2$, there is a slight gain in simple fit, as measured by the average likelihood, with the quadratic model. However, the SBC suggests that the extra parameter is of doubtful value, with the simpler model preferred. As an illustration of the predictions of the complete failure times for observations on incomplete or censored times, the times censored at 8064 for temperature 150°C are predicted to complete at 16 970 (median), with 95% interval (8470, 73 270) under the linear model. These appear as log10(t.comp) in Table 9.3.

Example 9.3 Log-logistic model As an example of log-logistic survival, we apply the logistic model (9.4)–(9.5) to the logs of the leukaemia remission times from the Gehan (1965) study. A G(1, 0.001) prior on $\kappa = 1/\sigma$ is taken in Equation (9.5) and flat priors on the treatment effect, which is the only covariate. Initially, κ is taken the same across all subjects (Model A).

A three chain run to 10 000 iterations (with 1000 burn-in) leads to an estimated mean treatment difference of 1.31 (i.e. longer remission times for patients on the treatment) and a median for $\sigma = 1/\kappa$ of 0.58. These values compare closely with those obtained by (Aitkin *et al.*, 1989, p. 297). Treatment and placebo group survival curves as in Equation (9.6) show the clear benefit of the treatment in terms of mean posterior probabilities up to $t = 50$ (Table 9.4).

In a second model (Model B), the scale parameter κ is allowed to differ between the treatment and placebo groups. The predictive loss criterion in Equation (9.1) suggests the simpler model to be preferable to this extension; the same conclusion follows from the pseudo Bayes factor based on Monte Carlo estimates of the CPO (Sahu *et al.*, 1997). It is, however, noteworthy that β_1 is enhanced in Model B, and that variability σ appears greater in the treatment group.

Table 9.3 Motorette analysis, parameter summary

Quadratic	Mean	St. devn.	2.5%	Median	97.5%
SBC at posterior mean	−22.1				
Log-likelihood	−3.3	6.2	−17.0	−2.7	7.1
β_1	3.35	0.07	3.21	3.34	3.50
β_2	5.43	0.85	4.00	5.34	7.32
β_3	14.01	6.33	2.91	13.52	27.75
σ	0.26	0.05	0.18	0.25	0.38
log10(t.comp)	4.63	0.38	4.01	4.58	5.48
Linear					
SBC at posterior mean	−19				
Log-likelihood	−4.6	6.0	−18.1	−4.0	5.5
β_1	3.48	0.06	3.38	3.48	3.62
β_2	4.34	0.46	3.49	4.32	5.31
σ	0.27	0.05	0.19	0.26	0.39
log10(t.comp)	4.26	0.24	3.93	4.23	4.82

Table 9.4 Treatment effect on remission times (Model A)

Parameter	Mean	St. devn.	2.5%	Median	97.5%
β_0	1.886	0.221	1.436	1.890	2.312
β_1 (Treatment effect on remission time)	1.312	0.354	0.652	1.296	2.046
σ	0.582	0.093	0.427	0.573	0.790

Group survival curves

	Placebo		Treated	
Time	Mean	St. devn.	Mean	St. devn.
1	0.958	0.026	0.995	0.005
2	0.882	0.053	0.985	0.011
3	0.791	0.072	0.971	0.019
4	0.701	0.083	0.954	0.026
5	0.616	0.090	0.934	0.034
6	0.541	0.092	0.913	0.041
7	0.475	0.091	0.890	0.048
8	0.419	0.089	0.866	0.055
9	0.371	0.087	0.842	0.062
10	0.331	0.083	0.817	0.068
11	0.296	0.080	0.791	0.073
12	0.266	0.076	0.766	0.079
13	0.240	0.073	0.741	0.083
14	0.218	0.070	0.716	0.088
15	0.199	0.067	0.692	0.092
16	0.182	0.064	0.668	0.095
17	0.168	0.061	0.645	0.098
18	0.155	0.058	0.623	0.101
19	0.143	0.056	0.601	0.103
20	0.133	0.053	0.580	0.105
21	0.124	0.051	0.560	0.106
22	0.116	0.049	0.541	0.108
23	0.109	0.047	0.522	0.109
24	0.102	0.045	0.504	0.109
25	0.096	0.043	0.487	0.110
26	0.090	0.042	0.471	0.110
27	0.085	0.040	0.455	0.111
28	0.081	0.039	0.440	0.111
29	0.077	0.038	0.426	0.111
30	0.073	0.036	0.412	0.110
31	0.069	0.035	0.399	0.110
32	0.066	0.034	0.386	0.109
33	0.063	0.033	0.374	0.109
34	0.060	0.032	0.363	0.108
35	0.057	0.031	0.352	0.108
36	0.055	0.030	0.341	0.107
37	0.053	0.029	0.331	0.106

(continues)

Table 9.4 (*continued*)

	Placebo		Treated	
Time	Mean	St. devn.	Mean	St. devn.
38	0.051	0.028	0.321	0.105
39	0.049	0.028	0.312	0.104
40	0.047	0.027	0.303	0.103
41	0.045	0.026	0.295	0.102
42	0.043	0.025	0.286	0.101
43	0.042	0.025	0.279	0.100
44	0.040	0.024	0.271	0.099
45	0.039	0.024	0.264	0.098
46	0.038	0.023	0.257	0.097
47	0.036	0.022	0.250	0.096
48	0.035	0.022	0.244	0.095
49	0.034	0.021	0.238	0.094
50	0.033	0.021	0.232	0.093

Example 9.4 Nursing home length of stay Morris *et al.* (1994) consider length of stay for a set of 1601 nursing home patients in terms of a treatment and other attributes (age, health status, marital status, sex) which might affect length of stay. Stay is terminated either by death or return home. We here estimate linear covariate effects in a proportional Weibull hazard

$$h(t, z) = \gamma t^{\gamma-1} \exp(\beta x)$$

This is equivalent to a regression of the logged length of stay on the regressors with a scaled error term u

$$\log(t) = \phi x + \sigma u$$

where $\sigma = 1/\gamma$ and $\phi = -\beta\sigma$. We obtain results on the treatment and attribute variables similar to those of Morris *et al.* These are the coefficients ϕ on the predictors of log length of stay, which is a close proxy for length of survival in the context. All covariates are categorical except age, which is converted to a spline form

$$\text{age.s} = \min(90, \text{age}) - 65.$$

Health status is based on numbers of activities of daily living (e.g. dressing, eating) where there is dependency in terms of assistance being required. Thus, health=2 if there are four or less ADLs with dependence, health=3 for five ADL dependencies, health=4 for six ADL dependencies, and health=5 if there were special medical conditions requiring extra care (e.g. tube feeding).

Convergence with a three chain run is achieved early and the summary in Table 9.5 is from iterations 500–2000. Personal attributes such as gender, health status, age and marital status all impact on length of stay. Married persons, younger persons and males have shorter lengths of stay, though the effect of age straddles zero. Married persons, often with a care-giver at home, tend to enter with poorer initial functional status, associated with earlier death. The experimental treatment applied in some nursing homes involved financial incentives to improve health status and (for the non-terminal patients) achieve discharge within 90 days; however, the effect is not towards lower

Table 9.5 Nursing home stays, parameter summary

	Mean	St. devn.	2.5%	Median	97.5%
Full model					
Intercept	5.739	0.165	5.437	5.735	6.070
Age	0.007	0.007	−0.007	0.007	0.020
Treatment	0.201	0.093	0.017	0.203	0.381
Male	−0.562	0.108	−0.772	−0.562	−0.342
Married	−0.262	0.128	−0.516	−0.259	−0.006
Health 3	0.045	0.131	−0.215	0.045	0.301
Health 4	−0.377	0.131	−0.636	−0.378	−0.117
Health 5	−0.872	0.166	−1.201	−0.873	−0.550
Scale	1.635	0.036	1.564	1.635	1.706
Reduced model					
Intercept	5.766	0.149	5.475	5.763	6.066
Age	0.007	0.007	−0.007	0.007	0.020
Treatment	0.198	0.089	0.022	0.197	0.373
Male	−0.565	0.112	−0.791	−0.566	−0.342
Married	−0.257	0.127	−0.499	−0.258	0.001
Health 4	−0.400	0.101	−0.594	−0.399	−0.205
Health 5	−0.897	0.149	−1.182	−0.897	−0.596
Scale	1.635	0.037	1.557	1.636	1.707

length of stay, possibly because patients in treatment homes were more likely to be Medicaid recipients (Morris *et al.*, 1994).

It would appear that the effect of health status level 3 is not clearly different from zero (i.e. from the null parameter of the reference health status), and so the groups 2 and 3 might be amalgamated. We therefore fit such a model (Model B in Program 9.4), and find its pseudo-marginal likelihood to be in fact higher than the model (−8959 vs. −9012) involving the full health status scale. The conventional log likelihood averages around −8548 for both models.

9.3 ACCELERATED HAZARDS

In an Accelerated Failure Time (AFT) model the explanatory variates act multiplicatively on time, and so affect the 'rate of passage' to the event; for example, in a clinical example, they might influence the speed of progression of a disease. Suppose

$$\nu_i = \beta_1 x_{1i} + \beta_2 x_{2i} + \ldots + \beta_p x_{pi} \tag{9.7a}$$

denotes a linear function of risk factors (without a constant). Then the AFT hazard function is

$$h(t, x) = e^{\nu_i} h_0(e^{\nu_i} t)$$

For example, if there is Weibull time dependence, the baseline hazard is

$$h_0(t) = \lambda\gamma t^{\gamma-1}$$

and under an AFT model, this becomes

$$\begin{aligned} h(t, x) &= e^{\nu_i}\lambda\gamma(te^{\nu_i})^{\gamma-1} \\ &= (e^{\nu_i})^{\gamma}\lambda\gamma t^{\gamma-1} \end{aligned} \tag{9.7b}$$

Hence the durations under an accelerated Weibull model have a density

$$W(\lambda e^{\gamma\nu_i}, \gamma)$$

whereas under proportional hazards the density is

$$W(\lambda e^{\nu_i}, \gamma)$$

If there is a single dummy covariate (e.g. $x_i = 1$ for treatment group, 0 otherwise), then $\nu_i = \beta x_i = \beta$ when $x_i = 1$. Setting $\phi = e^{\beta}$, the hazard for a treated patient is

$$\phi h_0(\phi t)$$

and the survivor function is $S_0(\phi t)$. The multiplier ϕ is often termed the acceleration factor.

The median survival time under a Weibull AFT model is

$$t.50 = [\log 2/\{\lambda e^{\gamma\nu_i}\}]/\gamma \tag{9.8}$$

In an example of a Bayesian perspective, Bedrick *et al.* (2000) consider priors for the regression parameters in Equation (9.7a) expressed in terms of their impact on median survival times in Equation (9.8) rather than as direct priors on the β_j.

Example 9.5 Breast cancer survival We consider the breast cancer survival times (in weeks) of 45 women, as presented by Collett (1994, p. 7). The risk factor is a classification of the tumour as positively or negatively stained in terms of a biochemical marker HPA, with $x_i = 1$ for positive staining, and $x_i = 0$ otherwise. We use a G(1, 0.001) prior on the Weibull shape parameter.

A three chain run of 5000 iterations shows early convergence of β and convergence at around iteration 750 for γ. The summary, based on the last 4000 iterations, shows the posterior mean of γ to be 0.935, but with the 95% interval straddling unity (Table 9.6). The posterior mean of the positive staining parameter β is estimated as around 1.1, and shows a clear early mortality effect for such staining. The CPOs show the lowest probability under the model for cases 8 and 9, where survival is relatively extended despite positive staining. The lowest scaled CPO (the original CPOs are scaled relative to their maximum) is 0.016 (Weiss, 1994).

Table 9.6 Breast cancer survival, parameter estimates

	Mean	St. devn.	2.5%	Median	97.5%
β	1.105	0.566	0.098	1.075	2.288
γ (Weibull shape)	0.935	0.154	0.651	0.931	1.266
t.50	92.0	23.8	54.7	89.1	147.8
Hazard ratio under proportional hazards	3.14	1.75	1.09	2.73	7.50

The posterior mean of the analytic median survival formula (9.8) for women with cancer classed as positively stained is around 92 weeks, a third of the survival time of women with negatively stained tumours. Under the proportional hazards model the hazard ratio would be $e^{\gamma\beta}$ which has a median of 2.7 similar to that cited by Collett (1994, p. 214), though is not precisely estimated both because of the small sample and because it involves a product of parameters.

9.4 DISCRETE TIME APPROXIMATIONS

Although events may actually occur in continuous time, event histories only record time in discrete units, generally called periods or intervals, during which an event may only occur once. The discrete time framework includes population life tables, clinical life table methods such as the Kaplan–Meier method, and discrete time survival regressions. Applications of the latter include times to degree attainment (Singer and Willett, 1993), and the chance of exit from unemployment (Fahrmeir and Knorr-Held, 1997). The discrete framework has been adapted to semi-parametric Bayesian models (Ibrahim *et al.*, 2001a) as considered below.

Consider a discrete partition of the positive real line,

$$0 < a_1 < a_2 < .. < a_L < \infty$$

and let A_j denote the interval $[a_{j-1}, a_j)$, with the first interval being $[0, a_1)$. The discrete distributions analogous to those above are

$$\begin{aligned} f_j &= \Pr(T \varepsilon A_j) \\ &= S_j - S_{j+1} \end{aligned} \quad (9.9)$$

where, following Aitkin *et al.* (1989),

$$\begin{aligned} S_j &= \Pr(T > a_{j-1}) \\ &= f_j + f_{j+1} + \ldots + f_L \end{aligned} \quad (9.10)$$

The survivor function at a_{j-1} is S_j and at a_j is S_{j+1} with the first survivor rate being $S_1 = 1$. The jth discrete interval hazard rate is then

$$h_j = \Pr(T \varepsilon A_j | T > a_{j-1}) = f_j / S_j$$

It follows that

$$h_j = (S_j - S_{j+1})/S_j$$

and so

$$S_{j+1}/S_j = 1 - h_j$$

So the chance of surviving through r successive intervals, which is algebraically

$$S_{r+1} = \prod_{j=1}^{r} S_{j+1}/S_j$$

can be estimated as a 'product limit'

$$S_{r+1} = \prod_{j=1}^{r} (1 - h_j)$$

The likelihood is defined over individuals i and periods j and a censoring variable w_{ij} is coded for the end point a_j of each interval $(a_{j-1}, a_j]$ and each subject, up until the final possible interval $(a_L, \infty]$. Suppose we have a non-repeatable event. Then if the observation on a subject ends with an event within the interval $(a_{j-1}, a_j]$, the censoring variable would be coded 0 for preceding periods, while $w_{ij} = 1$. A subject still censored at the end of the study would have indicators $w_{ij} = 0$ throughout.

In aggregate terms, the likelihood then becomes a product of L binomial probabilities, with the number at risk at the beginning of the jth interval being K_j. This total is composed of individuals still alive at a_{j-1} and still under observation (i.e. neither censored or failed in previous intervals). For individuals i still at risk in this sense (for whom $R_{ij} = 1$) the total deaths in the jth interval are

$$\sum_{R_{ij}=1} w_{ij} = d_j$$

The Kaplan–Meier estimate of the survival curve is based on the survival rates estimated from the binomial events with $K_j - d_j$ subjects surviving from n_j at risk. In practice, we may restrict the likelihood to times at which failures or deaths occur, i.e. when h_j is non-zero.

Some authors have taken the Kaplan–Meier approach as a baseline, but proposed non-parametric methods to smooth the original KM estimates. Thus, Leonard *et al.* (1994) suggest that the unknown density function of survival times $f(t)$ be obtained via an equally weighted mixture of hazard functions with m components

$$h_m(t, \xi, \eta) = m^{-1} \sum_{k=1}^{m} h(t, \xi_k, \eta) \tag{9.11}$$

where each $h(t, \xi_k, \eta)$ is a specific hazard function (e.g. exponential, Weibull), η denotes parameters of that function not varying over the mixture and ξ_k are components that do vary. The number of components may exceed the number of observations n, in which case some will be empty. The equally weighted mixture is analogous to kernel estimation and smooths f without assuming that f itself comes from a parametric family. This mixture has known component masses, and is easier to estimate and analyse than a discrete mixture model with unknown and unequal probabilities. The special case $\xi_1 = \xi_2 = \ldots = \xi_m$ means that f can be represented by a parametric density.

Example 9.6 Colorectal cancer: Kaplan–Meier method To illustrate the Kaplan–Meier procedure, we first consider data on survival in months in 49 colorectal cancer patients (McIllmurray and Turkie, 1987), as in Table 9.7.

We restrict the analysis to survival in the treatment group subject to linolenic acid, and in relation to five distinct times of death, namely 6, 10, 12, 24 and 32 months. Totals at risk K_j (for whom $R_{ij} = 1$) are defined according to survival or withdrawal prior to the start of the jth interval. Thus, two of the original 25 patients censored at one and five months are not at risk for the first interval where deaths occur, namely the interval (6, 9). For these two patients the survival rate is 1. At 6 months, two treated patients die in relation to a total at risk of $K_1 = 23$ patients, so the survival rate is $1 - 2/23 = 0.913$. The next patient, censored at nine months, is also subject to this survival rate. The survival rate changes only at the next death time, namely 10 months, when $K_2 = 20$ patients are at risk, and there are two deaths, so that the survival rate (moment estimate) is (0.9)(0.913)=0.822. This process is repeated at the next distinct death time of 12 months.

Table 9.7 Survival in patients with Dukes' C colorectal cancer and assigned to linolenic acid or control treatment. Survival in months (* = censored)

Linolenic acid (n=25)		Control (n = 24)	
1*	13*	3*	18*
5*	15*	6	18*
6	16*	6	20
6	20*	6	22*
9*	24	6	24
10	24*	8	28*
10	27*	8	28*
10*	32	12	28*
12	34*	12	30
12	36*	12*	30*
12	36*	15*	33*
12	44*	16*	42
12*			

A technique useful in many applications with survival times involves reformulating the likelihood to reveal a Poisson kernel (Fahrmeir and Tutz, 2001; Lindsey, 1995). Aitkin and Clayton (1980) show how for several survival densities, estimation is possible via a log-linear model for a Poisson mean that parallels a log-linear model for the hazard function. Here, although other approaches are possible, we use the equivalent Poisson likelihood for the outcome indicators w_{ij}. These have means

$$\theta_{ij} = R_{ij}h_j$$

where h_j is the hazard rate in the jth interval and has a prior proportional to the width of that interval, namely $h_j \sim G(c[a_j - a_{j-1}], c)$, where c represents strength of prior belief.

Taking $c = 0.001$, and three chains to 5000 iterations (with 1000 burn in) we derive the survival rates S_j at the five distinct times of death, as in Table 9.8, together with the hazard rates h_j.

Table 9.8 Colorectal cancer: survival and hazard rates at distinct death times

	Mean	St. devn.	2.5%	Median	97.5%
Survival probabilities					
S_1	0.913	0.062	0.759	0.926	0.990
S_2	0.821	0.086	0.621	0.834	0.951
S_3	0.627	0.116	0.374	0.637	0.824
S_4	0.548	0.130	0.270	0.556	0.773
S_5	0.437	0.153	0.111	0.447	0.703
Hazard					
h_1	0.087	0.062	0.010	0.074	0.241
h_2	0.100	0.071	0.013	0.085	0.286
h_3	0.236	0.118	0.067	0.216	0.513
h_4	0.127	0.127	0.004	0.087	0.464
h_5	0.202	0.199	0.005	0.141	0.733

Example 9.7 Colon cancer survival: non-parametric smooth via equally weighted mixture As an example of the equal weighted mixture approach, we consider data on colon cancer survival in weeks following an oral treatment (Ansfield *et al.*, 1977). Of 52 patients times, 45 were uncensored (ranging from 6–142 weeks) and seven patients had censored survival times. Leonard *et al.* (1994) investigated Weibull and exponential hazard mixtures in Equation (9.11), but were unable to find a stable estimate for the Weibull time parameter η in

$$h(t, \eta, \xi_k) = \eta \xi_k t^{\eta-1}$$

where ξ_k are location parameters which vary over components. They therefore used a set value of $\eta = 2$, and performed an exponential mixture analysis of the squared survival times ($s = t^\eta$), so that

$$h(s, \xi_k) = \xi_k$$

They assume the exponential means are drawn from gamma density $G(\alpha, \beta)$, where α is known but β is itself gamma with parameters κ and ζ. In the colon cancer example, they set $\alpha = \kappa = 0.5$ and $\zeta = 52^{-2}$; these prior assumption are consistent with average survival of a year in the original time scale.

Here the original Weibull mixture is retained in the untransformed time scale, and $m = 15$ components taken, with η a free parameter assigned an $E(1)$ prior. The equivalent assumption to Leonard *et al.* on the $G(\alpha, \beta)$ prior for the location parameters (in the original time scale) involves setting $\alpha = \kappa = 0.5$ and $\zeta = 1/52$.

Of interest are the smoothed survivor curve $S(t)$ and the density $f(t)$ itself. We obtain similar estimates of the former to those represented in Figures 1 and 2 of Leonard *et al.* (1994). The estimates of β and α/β (i.e. the parameters governing the Weibull means ξ_k) will be affected by using the full Weibull hazard. The last 4000 of a three chain run of 5000 iterations show β not precisely identified and skewed with median at around 115. The Weibull time parameter has a posterior mean estimated at 1.72, with 95% interval from 1.35 to 2.10. Figure 9.1 shows the resulting survivor curve up to 200 weeks.

The analysis is similar in terms of parameter estimates and survivor function whether constrained (monotonic) or unconstrained sampling of ξ_k is used, or whether the logs of ξ_k are modelled via Normal priors – instead of modelling the ξ_k as gamma variables (Model B in Program 9.7). In the latter case, the prior mean of $\log(\xi_k)$ is -4 with variance 1, so the ξ_k will have average 0.018.

However, the predictive loss criterion in Equation (9.1) is lower for constrained gamma sampling than unconstrained gamma sampling, because of lower variances of new times z_i when these times are censored. The log-Normal approach has a lower predictive loss than either gamma sampling option, and gives a slightly higher estimate of the Weibull time slope, namely 1.92 with 95% interval (1.45, 2.55).

9.4.1 Discrete time hazards regression

The usual methods for discrete time hazards assume an underlying continuous time model, but with survival times grouped into intervals, such that durations or failure times between a_{j-1} and a_j are recorded as a single value. Assume that the underlying continuous time model is of proportional hazard form

$$\lambda(t, z) = \lambda_0(t) \exp(\beta z) \tag{9.12}$$

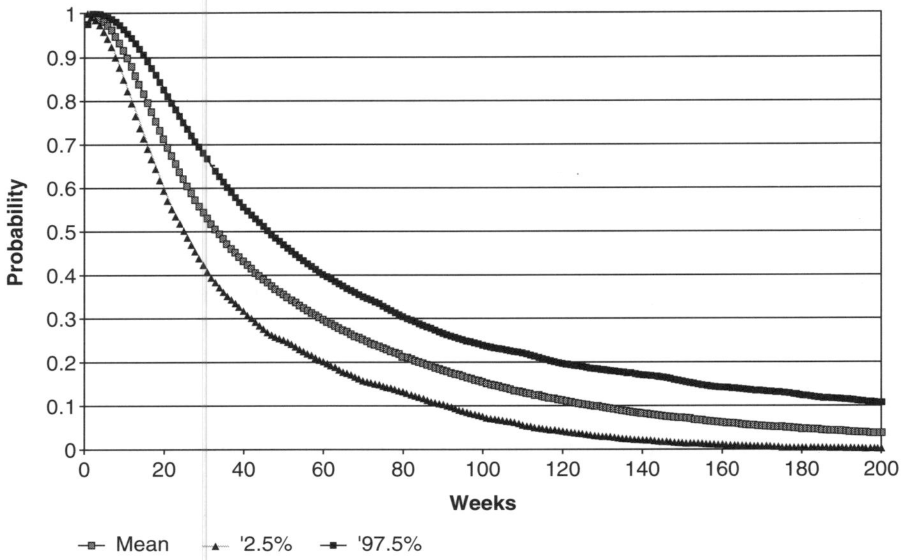

Figure 9.1 Weibull mixture analysis

with survivor function

$$S(t, x) = \exp(-\Lambda_0(t)e^{\beta z})$$

where the integrated hazard is denoted by

$$\Lambda_0(t) = \int_0^t \lambda_0(u)du \tag{9.13}$$

Then the conditional probability of surviving through the jth interval given that a subject has survived the previous $j-1$ intervals is

$$\begin{aligned} q_j &= \exp\left[-e^{\beta z}\int_{a_{j-1}}^{a_j} \lambda_0(u)du\right] \\ &= \exp\left[-e^{\beta z}\{\Lambda_0(a_j) - \Lambda_0(a_{j-1})\}\right] \end{aligned}$$

while

$$h_j = 1 - q_j$$

is the corresponding hazard rate in the jth interval $[a_{j-1}, a_j)$.

The total survivor function until the start of the jth interval is

$$S_j = \exp[-e^{\beta z}\Lambda_0(a_{j-1})]$$

Defining

$$\gamma_j = \ln[\Lambda_0(a_j) - \Lambda_0(a_{j-1})]$$

the likelihood of an event in interval $[a_{j-1}, a_j)$ given survival until then, can be written (Fahrmeir and Tutz, 2001; Kalbfleisch and Prentice, 1980) as

$$h_j S_j = \{1 - \exp(-e^{\beta z + \gamma_j})\} \prod_{k=1}^{j-1} \exp(-e^{\beta z + \gamma_k}) \tag{9.14}$$

Let $w_j = 1$ for an event in the jth interval and $w_j = 0$ otherwise. As to the regression term in Equation (9.14), we may allow $z[a_j]$ to be potentially time varying predictors. More generally, also let β_j denote a regression effect fixed within intervals, but that may vary between intervals. If the predictors themselves are time specific one may introduce lagged as well as contemperaneous effects (Fahrmeir and Tutz, 2001).

The typical log-likelihood contribution for an individual surviving $j-1$ intervals until either an event or censoring is then

$$w_j \log[1 - \exp(-\exp\{\gamma_j + \beta_j z[a_j]\})] - \sum_{k=1}^{j-1} \exp\{\gamma_k + \beta_k \mathrm{z}[\mathrm{a}_k]\}$$

This likelihood reduces to Bernoulli sampling over individuals and intervals with probabilities of the event π_{ij} modelled via a complementary log-log link, and with the censoring indicator forming the response. Thus, for a subject observed for r_i intervals until either an event or censoring

$$\begin{aligned} w_{ij} &\sim \text{Bernoulli}(\pi_{ij}) \quad i = 1, \ldots, n, j = 1, \ldots, r_i \\ \log\{-\log(1 - \pi_{ij})\} &= \gamma_j + \beta_j \mathrm{z}_i[\mathrm{a}_j] \end{aligned} \tag{9.15}$$

As well as fixed effect priors on γ_j and β_j, one can specify random walk priors, also called correlated prior processes (Gamerman, 1991; Sinha and Dey, 1997; Fahrmeir and Knorr-Held, 1997). For example, a first order random walk prior is

$$\gamma_{j+1} = \gamma_j + \varepsilon_j$$

where the ε_j are white noise with variance σ^2_γ. A variant of this is the local linear trend model

$$\begin{aligned} \gamma_{j+1} &= \gamma_j + \delta_j + \varepsilon_{1j} \\ \delta_{j+1} &= \delta_j + \varepsilon_{2j} \end{aligned}$$

where both ε_1 and ε_2 are white noise. Another option, again to avoid the parameterisation involved in assuming fixed effects, is for γ_j to be modelled as a polynomial in j (Mantel and Hankey, 1978). Smoothness priors may also be used for time varying regression coefficients β_j (Sargent, 1997). Thus, a first order random walk prior in a particular regression coefficient would be

$$\beta_{j+1} = \beta_j + \varepsilon_j$$

where the ε_j have variance σ^2_β.

Example 9.8 Longitudinal study of ageing Dunlop and Manheim (1993) consider changes in the functional status of elderly people (over age 70) using data from the US Longitudinal Study of Ageing, carried out in four interview waves in 1984, 1986, 1988 and 1990. This was a prospective study following an initial sample of around 5000, either through to death or the final wave. Dunlop and Manheim discuss the issues of modelling the probability of an initial disability in Activities of Daily Living (ADL) in terms of time-dependent covariates. A complete analysis would involve allowing for left censoring, since some people are disabled at study entry (in 1984).

They confine their analysis to people who were able in 1984 and consider transitions to disablement or loss of function on six ADLs: walking, dressing, bathing, toileting, feeding, and transferring (getting from chair to bed, and other types of short range mobility). They identify an empirical ordering for the average age at loss of function on these activities: the first disability is walking with an average age of 84 when disability commences, then bathing at age 87, transferring (age 90), dressing (age 92), toileting (age 93), and finally, feeding at age 100.

The analysis here follows Dunlop and Manheim in considering transitions to toileting disability. In terms of the empirical ordering of Dunlop and Manheim, it would be expected that people who are already limited in walking, bathing, transferring and dressing to have a higher chance of a move to toilet disability. In a subsidiary analysis, feeding status is also used as a predictor. Thus, time-dependent dummy covariates (present/absent) at waves 1, 2 and 3 on these disabilities are used to predict transitions to toileting disability in the intervals $(k, k+1)$, with $k = 1, 2, 3$. So $y_{ik} = 1$ if a transition to disability occurs for person i in the interval $(k, k+1)$.

A complementary log-log transform relates the probability that $y_{ik} = 1$ (given $y_{ik-1} = y_{ik-2} = ..0$) to the included covariates. As well as the ADL status variables, age and education (a continuous variable) are used to predict loss of functional status; these are divided by 100 for numerical reasons. Finally, a linear term in k itself (i.e. the study wave) is included; this is then a minimal form of the polynomial prior mentioned above.

Only persons observed through all four waves are included. For simplicity, observations where toilet status at k or $k+1$ is unknown are excluded, but cases with missing status on walking, bathing, transferring and dressing at waves 1 to 3 are included. This involves Bernoulli sampling according disability rates δ_{jk} specific to period k and activity j.

A three chain run to 2000 iterations is taken with initial value files including imputed values on the incomplete data (convergence is apparent by about iteration 300, and posterior summaries based on iterations 300–2000). We find, as do Dunlop and Manheim, that transition to toileting disability is positively related to age, and to preceding loss of status on walking, bathing and transferring (Table 9.9). A negative effect of education is obtained, as also reported by Dunlop and Manheim. However, in contrast to the results of Dunlop and Manheim, dressing status is not found to impact on this transition. Also, while Dunlop and Manheim obtain a non-significant impact of wave k itself, we obtain a clear gradient of increased chances of transition to disability with larger k.

If the analysis is extended to include preceding feeding status, essentially the same results are obtained (though convergence is delayed till around 1500 iterations). Thus, existing loss of walking, bathing or transferring status are positive predictors of loss of toileting status in the next interval. However, loss of feeding or dressing status, which tend to occur among only the very old, are not clear preceding predictors of loss of toileting status.

Program 9.8 also contains the code needed to make the coefficients on age, education and the ability variables time specific, as in Equation (9.15). Initial runs with this option suggested a slight lowering of the pseudo-marginal likelihood, and so results are not reported in full. There was a suggestion under this model that the education effect became less marked for waves 2 and 3 and enhancement of the linear time coefficient also occurred. Another modelling possibility is to use lagged ability variables.

Table 9.9 LSOA, onset of toileting disability

(a) Excluding Feeding	Mean	St. devn.	2.5%	Median	'97.5%
Intercept	−8.3	0.8	−9.9	−8.3	−6.68
Time (wave)	0.244	0.074	0.108	0.238	0.397
Age*	0.060	0.010	0.038	0.060	0.080
Education*	−0.040	0.014	−0.067	−0.040	−0.011
Walking	0.937	0.135	0.674	0.936	1.199
Bathing	0.567	0.154	0.264	0.567	0.871
Transferring	0.519	0.143	0.233	0.517	0.803
Dressing	−0.177	0.191	−0.553	−0.178	0.196
(b) Including feeding					
Intercept	−8.8	0.8	−10.2	−8.9	−7.4
Time (wave)	0.251	0.073	0.104	0.251	0.394
Age*	0.066	0.009	0.047	0.066	0.082
Education*	−0.038	0.014	−0.066	−0.038	−0.010
Walking	0.936	0.138	0.662	0.939	1.211
Bathing	0.564	0.157	0.249	0.566	0.857
Transferring	0.534	0.148	0.251	0.533	0.826
Dressing	−0.156	0.195	−0.547	−0.152	0.217
Feeding	−0.228	0.269	−0.781	−0.217	0.272

*Effects on original age and education scales

Example 9.9 Transitions in youth employment As an illustration of time varying predictor effects, this example follows Powers and Xie (2000) in an analysis of a random sample of young white males from the US National Longitudinal Survey on Youth relating to the transition from employment (at survey time) to inactivity (in the next survey). The sample contains 1077 subjects, and the analysis here relates to five transitions between surveys (namely changes between 1979 and 1980, between 1980–1981,.., up to 1983–1984).

Age effects on the outcome are defined by age bands 14–15, 16–17.. up to 22–23, and there are three time varying covariates: whether the subject graduated in the previous year; the local labour market unemployment rate; and whether the respondent had left home at survey time. Fixed covariates relate to father's education (none, high school or college), family structure not intact, family income in 1979, an aptitude score (Armed Services Vocational Aptitude Battery Test, ASVAB) and living in the Southern USA or not. There are fourteen coefficients for AIC and SBC calculations; here we the SBC adjusted for the parameter priors is considered (Klugman, 1992) – see Models A1 and B1 in Program 9.9.

A model with fixed (i.e. time stationary) effects on all coefficients shows peak rates of the outcome at ages 18–19 and 20–21, and a positive relation of the transition to inactivity with coming from a broken home (Table 9.10). (Parameter summaries are based on a three chain model to 2500 iterations, with early convergence at around 500 iterations). Though not significant the effects of high local unemployment and recent graduation are biased towards positive effects. This event is negatively related to aptitude, living in the South and having left home; the income effect is also predominantly negative, but the 95% credible interval just straddles zero.

Table 9.10 Employment transitions

	Mean	St. devn.	2.5%	Median	97.5%
Unemployment effect varying					
SBC	−1097				
Age Effects					
Age 14–15	−2.92	0.33	−3.58	−2.91	−2.30
Age 16–17	−2.85	0.25	−3.35	−2.85	−2.36
Age 18–19	−2.57	0.26	−3.08	−2.57	−2.05
Age 20–21	−2.57	0.29	−3.11	−2.57	−1.99
Age 22–23	−2.67	0.44	−3.57	−2.66	−1.83
Covariates					
FHS	0.226	0.156	−0.077	0.224	0.532
FCOL	0.072	0.178	−0.276	0.075	0.414
GRAD	0.190	0.176	−0.164	0.192	0.527
INCOME	−0.234	0.129	−0.492	−0.233	0.004
ASVAB	−0.440	0.074	−0.585	−0.440	−0.295
NONINT	0.379	0.153	0.080	0.380	0.677
SOUTH	−0.517	0.185	−0.891	−0.513	−0.163
SPLIT	−0.437	0.187	−0.817	−0.432	−0.080
Unemployment (by year)					
1979–80	0.257	0.210	−0.128	0.250	0.677
1980–81	0.243	0.200	−0.132	0.239	0.639
1981–82	0.236	0.192	−0.129	0.233	0.614
1982–83	0.225	0.189	−0.135	0.224	0.590
1983–84	0.210	0.195	−0.154	0.208	0.581
Stationary coefficient model					
SBC	−1056.8				
Age effects					
Age 14–15	−2.857	0.320	−3.538	−2.851	−2.232
Age 16–17	−2.787	0.237	−3.242	−2.784	−2.336
Age 18–19	−2.514	0.248	−3.005	−2.517	−2.033
Age 20–21	−2.517	0.277	−3.074	−2.499	−1.974
Age 22–23	−2.618	0.431	−3.487	−2.590	−1.846
Covariates					
FHS	0.226	0.149	−0.063	0.223	0.519
FCOL	0.063	0.181	−0.286	0.062	0.407
GRAD	0.177	0.178	−0.178	0.177	0.532
INCOME	−0.240	0.125	−0.499	−0.242	0.004
ASVAB	−0.439	0.074	−0.585	−0.438	−0.293
NONINT	0.382	0.150	0.089	0.387	0.677
UNEMP	0.189	0.167	−0.128	0.190	0.542
SOUTH	−0.509	0.180	−0.881	−0.504	−0.170
SPLIT	−0.444	0.180	−0.815	−0.439	−0.106

Allowing the unemployment coefficient to vary over time according to a random walk prior suggest a lessening effect of local labour market conditions, though no coefficient is significant in any year. This more heavily parameterised model leads to a clear worsening in the Schwarz criterion and also has a slightly worse predictive loss criterion (9.1) for values $w = 1, w = 100$ and $w = 10\,000$.

The previous analysis used a diffuse G(1, 0.001) prior on the precision $1/\sigma_\beta^2$. Sargent (1997) adopts an informative prior in the study he considers, namely a high precision consistent with small changes in β_j. For instance, one might say that shifts greater than ± 0.1 in the unemployment coefficient in adjacent periods are unlikely. If this is taken as one standard deviation (i.e. $\sigma_\beta = 0.1$), then the precision would have mean 100, and a prior such as G(1, 0.01) might be appropriate as one option (see exercises).

9.4.2 Gamma process priors

In the proportional hazards model

$$h(t_i, x_i) = h_0(t_i) \exp(\beta x_i)$$

a non-parametric approach to specifying the hazard h or cumulative hazard H is often preferable. Priors on the cumulative hazard which avoid specifying the time dependence parametrically have been proposed for counting process models, as considered below (Kalbfleisch, 1978). However, a prior may also be specified on the baseline hazard h_0 itself (e.g. Sinha and Dey, 1997).

Thus, consider a discrete partition of the time variable, based on the profile of observed times $\{t_1, \ldots t_N\}$ whether censored or not, but also possible referring to wider subject matter considerations. Thus, M intervals $(a_0, a_1], (a_1, a_2], \ldots (a_{M-1}, a_M]$ are defined by breakpoints at $a_0 \leq a_1 \leq \ldots \leq a_M$, where a_M exceeds the largest observed time, censored or uncensored, and $a_0 \geq 0$. Let

$$\delta_j = h_0(a_j) - h_0(a_{j-1}) \quad j = 1, \ldots M$$

denote the increment in the hazard for the jth interval. Under the approach taken by Chen *et al.* (2000), and earlier by workers such as Dykstra and Laud (1981), the δ_j are taken to be gamma variables with scale γ and shape

$$\alpha(a_j) - \alpha(a_{j-1})$$

where α is monotonic transform (e.g. square root, logarithm). Note that this prior strictly implies an increasing hazard, but Chen *et al.* cite evidence that this does not distort analysis in applications where a decreasing or flat hazard is more reasonable for the data at hand.

In practice, the intervals $a_j - a_{j-1}$ might be taken as equal length and α as the identity function. If the common interval length were L, then the prior on the δ_j would be set at $G(L, \gamma)$. Larger values of γ reflect more informative beliefs about the increments in the hazard (as might be appropriate in human mortality applications, for example).

The likelihood assumes a piecewise exponential form and so uses information only on the intervals in which a completed or censored duration occurred. Let the grouped times s_i be based on the observed times t_i after grouping into the M intervals. The cumulative distribution function (cdf) is

$$F(s) = 1 - \exp\left\{-e^{B_i} \int_0^s h_0(u)du\right\}$$

where B_i is a function of covariates x_i. Assuming $h_0(0) = 0$, the cdf for subject i is approximated as

$$F(s_i) = 1 - \exp\left\{-e^{B_i} \sum_{j=1}^{M} \delta_j (s_i - a_{j-1})^+\right\} \tag{9.16}$$

where $(u)^+ = u$ if $u > 0$, and is zero otherwise.

For a subject exiting or finally censored in the jth interval, the event is taken to occur just after a_{j-1}, so that $(s_i - a_{j-1})^+ = (s_i - a_{j-1})$. The likelihood for a completed duration s_i in the jth interval, i.e. $a_{j-1} < s_i < a_j$, is then

$$P_{ij} = F(x_i, a_j) - F(x_i, a_{j-1})$$

where the evaluation of F refers to individual specific covariates as in Equation (9.16), as well as the overall hazard profile. A censored subject with final known follow up time in interval j has likelihood

$$S_{ij} = 1 - F(x_i, a_j)$$

Example 9.10 Leukaemia remission To illustrate the application of this form of prior, consider the leukaemia remission data of Gehan (1965), with $N = 42$ subjects and observed t_i ranging from 1–35 weeks, and define $M = 18$ intervals which define the regrouped times s_i. The first interval $(a_0, a_1]$ includes the times $t = 1, 2$; the second including the times 3,4.. up to the 18th $(a_{17}, a_{18}]$ including the times 35,36. The mid intervals are taken as 1.5 (the average of 1 and 2), and then 3.5, 5.5,.. and so on up to 35.5. These points define the differences $a_1 - a_0, a_2 - a_1, ..$ as all equal to 2. The Gehan study concerned a treatment (6-mercaptopurine) designed to extend remission times; the covariate is coded 1 (for placebo) and 0 for treatment, so that end of remission should be positively related to being in the placebo group.

It may be of interest to assess whether a specific time dependence (e.g. exponential or Weibull across the range of all times, as opposed to piecewise versions) is appropriate if this involves fewer parameters; a non-parametric analysis is then a preliminary to choosing a parametric hazard. One way to gauge this is by a plot of $-\log\{S(u)\}$ against u, involving plots of posterior means against u, but also possibly upper and lower limits of $-\log(S)$ to reflect varying uncertainty about the function at various times. A linear plot would then support a single parameter exponential.

In WINBUGS it is necessary to invoke the 'ones trick' (with Bernoulli density for the likelihoods) or the 'zeroes trick' (with Poisson density for minus the log-likelihoods). With $B_i = \beta_0 + \beta_1$ Placebo, a diffuse prior is adopted on the intercept, and an N(0, 1) prior on the log of hazard ratio (untreated vs. treated). Following Chen *et al.* (2000, Chapter 10), a $G(a_j - a_{j-1}, 0.1)$ prior is adopted for $\delta_j, j = 1, .., M - 1$ and a $G(a_j - a_{j-1}, 10)$ prior for δ_M.

With a three chain run, convergence (in terms of scaled reduction factors between 0.95 and 1.05 on β_0, β_1 and the δ_j) is obtained at around 1300 iterations. Covariate effects in Table 9.11 are based on iterations 1500–5000, and we find a positive effect of placebo on end of remission as expected, with the remission rate about 3.3 times ($\approx$exp(1.2)) higher. The plot of $-\log S$ as in Figure 9.2 is basically supportive of single rate exponentials for both placebo and treatment groups, though there is a slight deceleration in the hazard at medium durations for the placebo group.

Table 9.11 Leukaemia regression parameters

	Mean	St. devn.	2.5%	Median	97.5%
Intercept	−6.45	0.35	−7.19	−6.44	−5.80
Treatment	1.22	0.36	0.53	1.24	1.95

Minus log survivor rates					Actual Survivorship rates		
	Untreated		Treated		Posterior means		
	Mean	St. devn.	Mean	St. devn.		Untreated	Treated
$-\log S_1$	0.06	0.04	0.02	0.01	S_1	0.947	0.986
$-\log S_2$	0.23	0.09	0.07	0.03	S_2	0.796	0.941
$-\log S_3$	0.39	0.12	0.12	0.05	S_3	0.676	0.902
$-\log S_4$	0.65	0.17	0.20	0.07	S_4	0.525	0.844
$-\log S_5$	0.94	0.23	0.28	0.09	S_5	0.394	0.783
$-\log S_6$	1.08	0.26	0.33	0.10	S_6	0.345	0.756
$-\log S_7$	1.40	0.33	0.42	0.13	S_7	0.255	0.697
$-\log S_8$	1.59	0.38	0.48	0.14	S_8	0.215	0.666
$-\log S_9$	1.85	0.44	0.55	0.16	S_9	0.169	0.624
$-\log S_{10}$	2.06	0.49	0.62	0.18	S_{10}	0.141	0.594
$-\log S_{11}$	2.20	0.53	0.66	0.19	S_{11}	0.124	0.573
$-\log S_{12}$	2.54	0.63	0.76	0.22	S_{12}	0.093	0.53
$-\log S_{13}$	2.91	0.74	0.87	0.25	S_{13}	0.069	0.485
$-\log S_{14}$	3.11	0.81	0.93	0.27	S_{14}	0.059	0.462
$-\log S_{15}$	3.32	0.88	0.99	0.29	S_{15}	0.05	0.441
$-\log S_{16}$	3.52	0.95	1.05	0.30	S_{16}	0.043	0.421
$-\log S_{17}$	3.72	1.01	1.11	0.32	S_{17}	0.037	0.402
$-\log S_{18}$	3.72	1.01	1.11	0.32	S_{18}	0.037	0.402

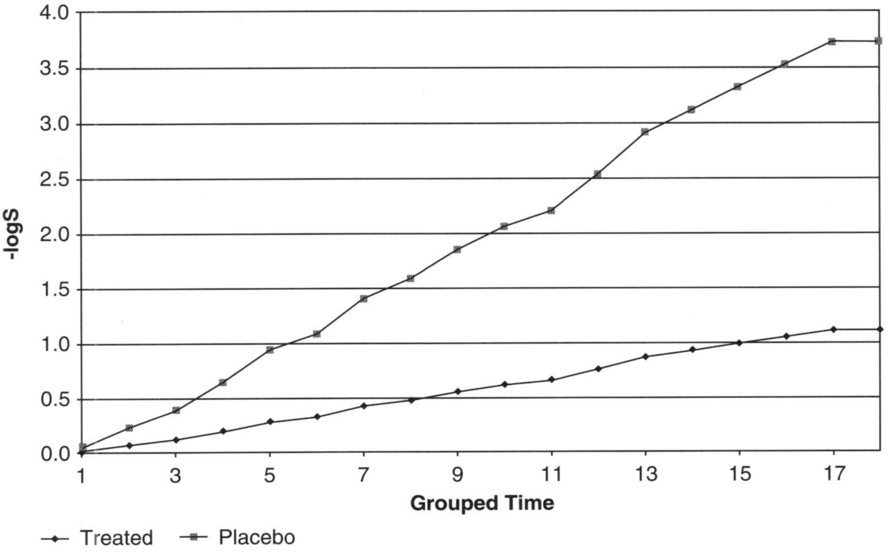

Figure 9.2 Plot of -log$S(u)$ versus u

9.5 ACCOUNTING FOR FRAILTY IN EVENT HISTORY AND SURVIVAL MODELS

Whether the event history or survival analysis is in discrete or continuous time, unobserved differences between subjects may be confounded with the estimated survival curve and the estimated impacts of observed covariates. While there is considerable debate regarding sensitivity of inferences to the specification of unobserved heterogeneity, this is an important aspect to consider, especially in complex models with time-varying effects of predictors or clustering of subjects.

Thus, frailty differences, whether modelled by parametric random effects or by non-parametric methods, provide a way to account for within-cluster correlations in event history outcomes (Guo and Rodriguez, 1992), or for multivariate survival times where an underlying common influence is present (Keiding *et al.*, 1997). Suppose subjects (patients, children) are arranged within aggregate units or clusters (hospitals, families) and event times are affected by cluster characteristics, known and unknown, as well as by the characteristics of individuals. Thus, for survival after surgery, patients are clustered within hospitals, while for age at pre-marital maternity, adolescent females are clustered according to family of origin (Powers and Xie, 2000). In these examples, random effects at cluster level are intended to account for unmeasured differences between clusters that may affect the outcome at the subject level.

Example 9.11 Unmeasured heterogeneity and proportional hazards in discrete time regression McCall (1994) discusses methods for assessing the proportional hazards assumption in discrete time regression in a single level example (with no clustering), in the context of data on joblessness durations in months. He considers tests for time varying coefficients β_j in Equation (9.15), which are equivalent to testing the proportional hazards assumption in the underlying continuous time model (e.g. Kay, 1977; Cox, 1972). In particular, he considers the need to allow for unmeasured heterogeneity when applying such tests.

McCall uses simulated data based on the real joblessness example, with a sample of $n = 500$ persons observed for a maximum of 60 intervals. An underlying continuous hazard is assumed with $\lambda_0(t) = 0.07$, and an individual level gamma frailty θ_i modifying the corresponding discrete time event probabilities, and distributed with mean and variance of 1, namely $\theta_i \sim G(1, 1)$. Thus, for subject i at interval j, the most general model involves time varying predictor x and z and time varying regression coefficients β_1 and β_2:

$$h_{ij} = [1 - \exp(-\theta_i \exp\{\gamma_j + \beta_{1j}x_{ij} + \beta_{2j}z_{ij}\})]$$

and

$$S_{ij} = \exp\left(-\theta_i \sum_{k=1}^{j-1} \exp(\gamma_k + \beta_{1k}x_{ik} + \beta_{2k}z_{ik})\right)$$

There is a concurrent 0.01 probability of being right-censored in any interval (e.g. via emigration or death in the joblessness example).

Here we consider only one of the scenarios of McCall, involving fixed predictors x and z, with fixed coefficients β_1 and β_2, and no trend in the γ_j. For $n = 100$, the covariates are generated via

$$x = 0.02 + e_1$$
$$z = 0.1x + e_2\sqrt{/}2$$

where e_1 and e_2 are standard Normal variables. This leads to a Bernoulli likelihood model as above generated by a complementary log-log link

$$w_{ij} \sim \text{Bernoulli}(\pi_{ij}) \quad i = 1, .. 100; j = 1, .. r_i$$
$$\log\{-\log(1 - \pi_{ij})\} = \gamma + \beta_1 x_i + \beta_2 z_i + \ln(\theta_i)$$

where $\beta_1 = -0.5$, $\beta_2 = 0.5$, $\gamma = \log(0.07) = -2.66$. The r_i are determined by the minimum duration at which an event is generated (at which point $w_{ij} = 1$ and no further observations are made), by censoring at 60 intervals, or by loss of follow up due to the concurrent exit rate of 0.01. The latter two categories are coded Fail=0, those undergoing an event as Fail=1 in Program 9.11. The data generation stage produces durations of stay, status at final observation (event or censored), and values for the two covariates. Values of θ_i are generated, but would be unobserved in a real application.

In the re-estimation stage, McCall allows (despite the generation mechanism) for Weibull baseline hazards with $\Lambda_{0j} = j^\rho$ and

$$\gamma_j = \gamma\,(j^\rho - (j-1)^\rho)$$

where $\rho = 1$ is the exponential case. Also, although the generation procedure assumes fixed regression coefficients, we may test for time varying coefficients using a linear dependence test, namely

$$\beta_{1j}{}^* = \beta_1 + \delta_1 j \tag{9.17a}$$

$$\beta_{2j}{}^* = \beta_2 + \delta_2 j \tag{9.17b}$$

with $\delta_1 = \delta_2 = 0$ if there is no time dependence. A quadratic form of this test may also be used.

The data are now analysed as if in ignorance of the generating meachanism. In applying a Bayesian analysis, certain issues regarding prior specification, especially around the δ parameters, may be noted. Since these parameters are multiplying large numbers j within the log-log link, there may be initial numerical problems (i.e. in the first 30 or 40 iterations) unless an informative prior is specified. Other stratagems such as rescaling time (e.g. by dividing by 100) may be adopted to reduce numerical problems, especially in early samples. Note that a logit rather than complementary log-log link is more robust to numerical problems (Thompson, 1977).

Rather than a multiplicative frailty, an additive frailty in $\lambda_i = \log(\theta_i)$ is taken in the re-estimation stage. Some degree of information in the prior for the variance of λ_i assists in identifiability. Thus, one might say, *a priori*, that the contrast between the 97.5th and 2.5th percentile of θ_i might be 1000 fold at one extreme and 10 fold at the other; corresponding percentiles of θ_i would be $30 = \exp(3.4)$ as against $0.03 = \exp(-3.5)$ and $3.2 = \exp(1.16)$ as against $0.32 = \exp(-1.14)$. The implied standard deviations, on the basis of approximate Normality of $\lambda_i = \log(\theta_i)$, are 1.7 and 0.6, respectively, with precisions $\phi = 1/\text{var}(\lambda)$ (respectively 3 and 0.33). A G(3, 2) prior on ϕ has 2.5th and 97.5th percentiles 0.33 and 3.6, and can be seen to permit a wide variation in beliefs about frailty.

Model A initially assumes no heterogeneity, and adopts a logit link with scaled time. Three chain runs to 2500 iterations showed early convergence (under 500 iterations) and

Table 9.12 Unemployment duration analysis

	Mean	St. devn.	2.5%	Median	97.5%
(A) No heterogeneity					
β_1	−0.40	0.18	−0.75	−0.40	−0.04
β_2	0.62	0.15	0.33	0.61	0.92
Avg. of β^*_{1j}	−0.29	0.19	−0.66	−0.29	0.07
Avg. of β^*_{2j}	0.26	0.13	0.02	0.26	0.52
$\delta_1(\times \text{time}/100)$	0.33	0.72	−1.09	0.34	1.72
$\delta_2(\times \text{time}/100)$	−1.19	0.55	−2.28	−1.18	−0.08
γ	−2.37	0.26	−2.92	−2.36	−1.90
ρ	1.10	0.03	1.04	1.10	1.16
(B) Log normal frailty					
β_1	−0.66	0.29	−1.31	−0.65	−0.06
β_2	0.93	0.20	0.57	0.90	1.34
Avg. of β^*_{1j}	−0.66	0.33	−1.38	−0.63	−0.10
Avg. of β^*_{2j}	0.67	0.27	0.21	0.64	1.26
$\delta_1(\times \text{time}/100)$	0.20	0.81	−1.41	0.21	1.79
$\delta_2(\times \text{time}/100)$	−0.97	0.68	−2.32	−0.97	0.40
γ	−3.66	0.81	−5.60	−3.48	−2.51
ρ	0.93	0.08	0.76	0.95	1.06
ϕ	0.73	0.51	0.17	0.59	2.15
(C) Discrete mixture					
γ_1	−3.61	0.88	−5.37	−3.70	−2.22
γ_2	−2.22	0.34	−2.87	−2.22	−1.56
β_1	−0.50	0.22	−0.91	−0.50	−0.07
β_2	0.72	0.17	0.41	0.72	1.07
Avg. of β^*_{1j}	−0.45	0.26	−0.99	−0.44	0.03
Avg. of β^*_{2j}	0.41	0.18	0.08	0.41	0.76
$\delta_1(\times \text{time}/100)$	0.21	0.80	−1.41	0.22	1.71
$\delta_2(\times \text{time}/100)$	−0.98	0.63	−2.17	−0.98	0.26
φ_1	0.36	0.16	0.11	0.33	0.73
φ_2	0.64	0.16	0.27	0.67	0.89
ρ	1.05	0.04	0.97	1.05	1.13

summaries are based on iterations 500–2500 (Table 9.12). This model shows a trend in the coefficient of z, and might be taken (erroneously) to imply non-proportionality. It also shows a significant Weibull shape parameter. Accuracy in reproducing the central covariate effects may need to take account of the trend estimated in the coefficient: thus averaging β^*_{1j} and β^*_{2j} (as in Equation (9.17)) over all intervals is a more accurate approach than considering the estimates of β and δ. The original values are reproduced in the sense that the 95% credible intervals for the average β^*_{1j} and β^*_{2j} contain the true values, but there appears to be understatement of the real effects.

Adopting a parametric frailty (log-Normal with $G(3, 2)$ prior on the precision of λ) in Model B again shows the estimated δ_1 in Equation (9.17) straddling zero, as it should do. (Summaries are based on a three chain run of 2500 iterations and 500 burn-in.)

There is still a tendency towards a declining effect for $z(\delta_2 < 0)$ to be identified, but the 95% interval clearly straddles zero. In contrast to the no frailty model, the posterior credible interval for ρ includes the null value for 1. The average β^*_{1j} and β^*_{2j} (denoted Beta.star[] in Program 9.11) in this model overstate (in absolute terms) the true effects, but are less aberrant than those of model A. In terms overall model assessment, the Pseudo-Marginal Likelihood (PML) based on CPO estimates shows model B to have a PML of −289.9, and so is preferred to Model A with PML of −294.3.

Analysis C uses non-parametric frailty assuming two components with masses φ_1 and φ_2 and differing intercepts on each component. Thus, if C_i is the latent group, we fit the model

$$\text{logit}(1 - \pi_{ij})\} = \gamma[\text{C}_\text{i}](j^\rho - (j - 1)^\rho) + (\beta_1 + \delta_1 j/100)x_i + (\beta_2 + \delta_2 j/100)z_i$$

A constraint on the terms γ_j is used for identifiability (preventing label switching). This produces similar results to Model B on the δ coefficients (i.e. both straddling zero) and ρ. The average β^*_{1j} and β^*_{2j} in this model slightly understate the true effects. The gain in the PML over Model A is relatively modest, namely −292.9 vs. −294.3, but the pattern of results is much more plausible.

In this connection, McCall argues that the simple linear trend tests are sensitive to whether unmeasured heterogeneity is allowed for, but less sensitive as to the form assumed for such heterogeneity (e.g. parametric or non-parametric). In the present analysis, a model without any form of heterogeneity might lead to incorrect inferences on proportionality, the size of regression coefficients and the shape of duration effects. This illustrates the potential inference pitfalls about duration or covariate effects if heterogeneity is present but not modelled.

Example 9.12 Pre-marital maternity We consider age at pre-marital maternity data analysed by Powers and Xie (2000), with the time variable being defined as age at maternity minus 11. These data relate to $n = 2290$ women, arranged in $m = 1935$ clusters (with the clustering arising from the fact that the survey includes at least two sisters from the same family). A birth outside marriage will typically be during the peak ages of maternity (15–34), and times for many women are right censored at their age of marriage. It is plausible to set an upper limit to censored times, such as 45 (age 56), and this considerably improves identifiability of the models described below.

As one of several possible approaches, we adopt a parametric mixture, at cluster level ($j = 1, .. m$), namely a $G(w, w)$ prior on unobserved factors θ_j with mean 1 and variance $1/w$. Thus

$$t_i \sim \text{Wei}(\nu_i, r)I(t^*_i, 45) \quad i = 1, \ldots n$$

$$\nu_i = \theta_j \mu_i$$

$$\log(\mu_i) = \alpha + \beta x_i$$

with t^*_i denoting censored times or zero for uncensored times, and $j = c_i$ denoting the cluster of the ith subject. Another option would be a discrete mixture of K intercepts with the group to which the jth cluster belongs being determined by sampling from a categorical density with K categories.

The first analysis (with which a mixture model can be compared) is a standard model with no frailty and a single intercept α. Starting values for three chains are based on null values and on the 2.5th and 97.5th points of a trial run; results are based on iterations

250–1000, given early covergence. The effects for binary indicators are reported as odds ratios, so values above 1 translate into 'positive' effects. The results are close to those reported by Powers and Xie, and show extra-marital maternity positively related to shorter education, and non-intact family of origin, and negatively related to weekly church attendance, Southern residence, and self-esteem (Table 9.13).

For the model including frailty, we consider a prior on w that reflects possible variations in the chance of this outcome after accounting for the influence of several social and religious factors. For example, a G(2, 1) prior on w is consistent with 2.5th and 97.5th percentiles on w of 0.3 and 5.5; these values are in turn consistent with ratios of the 97.5th to the 2.5th percentiles of θ_i of approximately 400 000 and five, respectively. This would seem a reasonable range of beliefs about the chances of such a behavioural outcome. By contrast, taking a relatively diffuse prior on w will generate implausible variations in the chance of the outcome – and possibly cause identifiability problems. One might experiment with other relatively informative priors on w to assess sensitivity. Mildly informative priors on r and α are also adopted, namely $r \sim G(3, 0.1)$ and $\alpha \sim N(-10, 100)$.

With a G(2, 1) prior on w, and applying the upper age limit to censored values, early convergence (at under 500 iterations) is obtained in the mixture analysis, and summaries are based on iterations 500–2000. The posterior mean of w is around 15, consistent with a relatively narrow range in the chance of the outcome (a ratio of the 97.5th to the 2.5th percentiles of θ_i of approximately 3). Compared to the analysis without frailty, Table 9.13 shows an enhanced impact of self-esteem. There is also an increase in the Weibull parameter. In fact, the pseudo-marginal likelihood and predictive loss criteria both show no gain in allowing frailty. Other model variations might be significant, though. Thus, a monotonic hazard could be improved on for these data: either by a more complex non-monotonic hazard (e.g. log-logistic) or by discretising the ages and taking a piecewise or polynomial function to model the changing impact of age.

9.6 COUNTING PROCESS MODELS

An alternative framework for hazard regression and frailty modelling is provided by counting process models (e.g. Aalen, 1976; Fleming and Harrington, 1991; Andersen *et al.*, 1993) which formulate the observations in a way enhancing the link to the broader class of generalized linear models. Consider a time W until the event of interest, and a time Z to another outcome (e.g. a competing risk) or to censoring. The observed duration is then $T = \min(W, Z)$, and an event indicator is defined, such that $E = 1$ if $T = W$ and $E = 0$ if $T = Z$.

The counting process $N(t)$ is then

$$N(t) = I(T \leq t, E = 1) \tag{9.18}$$

and the at risk function

$$Y(t) = I(T > t)$$

where $I(C)$ is the indicator function. If a subject exits at time T, his/her at risk function $Y(t) = 0$ for times exceeding T. So the observed event history for subject i is $N_i(t)$, denoting the number of events which have occurred up to continuous time t. Let $dN_i(t)$ be the increase in $N_i(t)$ over a very small interval $(t, t + dt)$, such that $dN_i(t)$ is (at most) 1 when an event occurs, and zero otherwise.

Table 9.13 Models for premarital birth; parameter summaries (parameters converted to odds ratios for binary predictors)

Model without frailty	Mean	St. devn.	2.5%	Median	97.5%
Intercept	−12.4	0.5	−13.2	−12.5	−11.2
Nonintact family (OR)	1.57	0.15	1.31	1.56	1.91
Mothers education under 12 years (OR)	1.70	0.17	1.38	1.69	2.03
Family Income*	−0.048	0.050	−0.148	−0.049	0.051
No of siblings	0.021	0.021	−0.018	0.021	0.06
South (OR)	0.82	0.07	0.70	0.82	0.96
Urban (OR)	1.03	0.09	0.87	1.02	1.21
Fundamental Protestant upbringing (OR)	1.55	0.16	1.25	1.56	1.85
Catholic (OR)	0.99	0.09	0.84	0.98	1.19
Weekly Church Attender (OR)	0.93	0.07	0.79	0.93	1.08
Traditional Sex Attitude Score	0.14	0.08	0.01	0.14	0.31
Self Esteem Score	−0.30	0.09	−0.47	−0.30	−0.13
Weibull Shape Parameter	3.67	0.13	3.35	3.71	3.90
Gamma mixture analysis					
Intercept	−13.7	0.3	−14.2	−13.7	−12.9
Gamma parameter (w)	14.7	3.1	9.2	14.5	21.5
Nonintact family (OR)	1.71	0.16	1.40	1.71	2.04
Mothers education under 12 years (OR)	1.92	0.21	1.54	1.93	2.36
Family Income*	−0.067	0.056	−0.177	−0.066	0.050
No of siblings	0.03	0.02	−0.02	0.03	0.07
South (OR)	0.77	0.07	0.65	0.77	0.93
Urban (OR)	1.04	0.08	0.86	1.03	1.20
Fundamental Protestant upbringing (OR)	1.65	0.18	1.30	1.64	1.99
Catholic (OR)	0.93	0.09	0.77	0.93	1.11
Weekly Church Attender (OR)	0.89	0.07	0.77	0.89	1.03
Traditional Sex Attitude Score	0.18	0.09	−0.01	0.19	0.36
Self Esteem Score	−0.39	0.10	−0.60	−0.39	−0.21
Weibull Shape Parameter	4.07	0.09	3.86	4.08	4.22

*Parameter converted to original scale

The expected increment in $N(t)$ is given by the intensity function

$$L(t)dt = Y(t)h(t)dt$$

with $h(t)$ the usual hazard function, namely

$$h(t)dt = \Pr(t \leq T \leq t + dt,\ E = 1|T \geq t)$$

In the counting process approach, the increment in the count (9.18) is modelled as a Poisson outcome, with mean given by the intensity function (e.g. in terms of time specific covariates). Under a proportional hazards assumption, the intensity is

$$L(t) = Y(t)h(t) = Y(t)\lambda_0(t)\exp(\beta x) \tag{9.19}$$

typically with $h(t) = \lambda_0(t)\exp(\beta x)$ as in the Cox proportional hazards model. The intensity may equivalently be written

$$L_i(t) = Y_i(t)d\Lambda_0(\mathrm{t})\exp(\beta x_i) \tag{9.20}$$

and so may be parameterised in terms of jumps in the integrated hazard Λ_0 and a regression parameter.

With observed data $D = (N_i(t),\ Y_i(t),\ x_i)$ the posterior for the parameters in (9.18) is

$$P(\beta,\ \Lambda_0|D) \propto P(D|\beta,\ \Lambda_0)P(\beta)P(\Lambda_0)$$

The conjugate prior for the Poisson mean is the gamma, so a natural prior for $d\Lambda_0$ has the form

$$d\Lambda_0 \sim G(cH(t),\ c) \tag{9.21}$$

where $H(t)$ expresses knowledge regarding the hazard rate per unit time (i.e. amounts to a guess at $d\Lambda_0(t)$), and $c > 0$ is higher for stronger beliefs (Sinha and Dey, 1997). The mean hazard is $cH(t)/c = H(t)$ and its variance is $H(t)/c$. Conditional on β, the posterior for Λ_0 takes an independent increments form on $d\Lambda_0$ rather than Λ_0 itself,

$$d\Lambda_0(t)|\beta,\ D \sim G(cH(t) + \Sigma_i dN_i(t),\ c + \Sigma_i Y_i(t)\exp(\beta x_i))$$

This model may be adapted to allow for unobserved covariates or other sources of heterogeneity ('frailty'). This frailty effect may be at the level of observations or for some form of grouping variable.

The above basis for counting processes is in terms of continuous time. In empirical survival analysis, the observations of duration will usually be effectively discrete, and made at specific intervals (e.g. observations on whether a subject in a trial has undergone an event will be made every 24 hours) with no indication how the intensity changes within intervals. So even for notionally continuous survival data, the likelihood is a step function at the observed event times.

If the observation intervals are defined so that at most one event per individual subject occurs in them, then we are approximating the underlying continuous model by a likelihood with mass points at every observed event time. Hence, the observation intervals will be defined by the distinct event times in the observed data.

The prior (9.21) on the increments in the hazard might then amount to a prior for a piecewise function defined by the observation intervals – this approach corresponds to a non-parametric estimate of the hazard as in the Cox regression (Cox, 1972). But the hazard might also be modelled parametrically (Lindsey, 1995).

One aspect of the counting process model is the ability to assess non-proportionality by defining time-dependent functions of regressors in hazard models of the form

$$h_i(t) = \lambda_0(t)\exp\{\beta x_i + \gamma w_i(t)\}$$

Thus $w_i(t) = x_i g(t)$ might be taken as the product of one or more covariates with a dummy time index $g(t)$ set to 1 up to time τ (itself a parameter), and to zero thereafter. This is consistent with proportional hazards if $\gamma = 0$.

Another possibility in counting process models applied to event histories is the modelling of the impact of previous events or durations in a subject's history. Thus, the intensity for the next event could be made dependent on the number of previous events, in what are termed birth models (Lindsey, 1995, 2001, Chapters 1 and 5; Lindsey, 1999).

Example 9.13 Leukaemia remissions We consider again the classical data from Gehan (1965) on completed or censored remission times for 42 leukaemia patients under a drug treatment and a placebo, 21 on each arm of the trial. A censored time means that the patient is still in remission. Here the observation interval is a week, and of the 42 observed times, 12 are censored (all in the drug group). There are 17 distinct complete remission times, denoted t.dist[ ] in Program 9.13. Termination of remission is more common in the placebo group, and the effect of placebo ($z_1 = 1$) vs. treatment ($z_1 = 0$) on exits from remission is expected to be positive.

The hazard is modelled parametrically, and for a Weibull hazard this may be achieved by including the $\log_e$ survival times, or logs of times since the last event, in the log-linear model for the Poisson mean (e.g. Lindsey, 1995; Aitkin and Clayton, 1980). Thus

$$L(t) = Y(t)\exp(\beta z + \kappa^* \log t)$$

where κ^* is the exponent in the Weibull distribution, κ, minus 1. We might also take a function in time itself:

$$L(t) = Y(t)\exp(\beta z + \zeta t)$$

and this corresponds to the extreme value distribution. For the Weibull, a prior for κ confined to positive values is appropriate (e.g. a G(1, 0.001) prior), while for ζ a prior allowing positive and negative values, e.g. an N(0, 1) density, may be adopted.

Three chain runs of 5000 iterations show early convergence on the three unknowns in each model. We find (excluding the first 500 iterations) a Weibull parameter clearly above 1, though some analyses of these data conclude that exponential survival is appropriate (Table 9.14). The 95% credible interval for the extreme value parameter is similarly confined to positive values. The extreme value model has a slightly lower pseudo-marginal likelihood than the Weibull model (−101.8 vs. 102.8); this is based on logged CPO estimates aggregated over cases with $Y(t) = 1$ (Y[i, j]=1 in Program 9.13). The exit rate from remission is clearly higher in the placebo group, with the coefficient on Z being entirely positive, and with average hazard ratio, for placebo vs. drug group, of $\exp(1.52) = 4.57$.

Example 9.14 Bladder cancer As an illustration of counting process models when there are repeated events for each subject, we consider the bladder cancer study conducted by the US Veterans Administrative Cooperative Urological Group. This involved 116 patients randomly allocated to one of three groups: a placebo group, a group receiving vitamin B6, and a group undergoing installation of thiotepa into the bladder. On follow up visits during the trial, incipient tumours were removed, so that an event

Table 9.14 Leukaemia treatment effect, Weibull and extreme value models

	Mean	St. devn.	2.5%	Median	97.5%
Weibull					
Intercept	−4.70	0.64	−6.06	−4.68	−3.52
Placebo	1.52	0.41	0.74	1.51	2.37
Shape	1.64	0.25	1.16	1.63	2.15
Extreme value					
Intercept	−4.31	0.49	−5.30	−4.30	−3.40
Placebo	1.56	0.42	0.76	1.55	2.39
Shape	0.090	0.030	0.029	0.091	0.147

history (with repeated observations on some patients) is obtained, with 292 events (or censorings) accumulated over the 116 patients. Times between recurrences are recorded in months, with many patients not experiencing recurrences (i.e. being censored).

A beneficial effect of thiotepa would be apparent in a more negative impact β_3 on the recurrence rate than the two other treatment options. We compare (a) Weibull vs. piecewise hazards, and (b) a subject level Normal frailty vs. a birth effect (modelling the impact ν of a count of previous recurrences), and follow Lindsey (2000) in using a criterion analogous to the AIC.

Summaries are based on three chain runs of 2500 iterations after early convergence (between 500–750 iterations in all model options discussed). The first two models use a Weibull parametric hazard with shape parameter κ. There is no apparent difference from the exponential null value ($\kappa = 1$) when frailty is included at the patient level (Table 9.15). However, omitting frailty and allowing for the influence of the number of previous events (also a proxy for frailty) shows the κ coefficient clearly below 1. There is a clear influence of previous events on the chance of a further one. However, neither model shows a clear treatment benefit.

The average deviance for the latter model is around 182, and must be used in a criterion that takes account of the extra parameters involved in the random frailty effects. The DIC criterion adds the parameter count to the average deviance and so is

Table 9.15 Models for bladder cancer; parameter summaries

	Mean	St. devn.	2.5%	Median	97.5%
Weibull hazard and patient frailty					
α	−3.22	0.31	−3.82	−3.22	−2.64
β_2	−0.008	0.316	−0.622	−0.010	0.598
β_3	−0.348	0.307	−0.972	−0.342	0.234
κ	1.028	0.101	0.824	1.027	1.223
σ^2	0.890	0.352	0.375	0.833	1.713
Weibull hazard and history effect					
α	−2.93	0.21	−3.36	−2.93	−2.54
β_2	0.093	0.169	−0.254	0.094	0.419
β_3	−0.137	0.183	−0.501	−0.135	0.210
κ	0.852	0.076	0.701	0.849	1.002
ν	0.572	0.105	0.370	0.572	0.781
Non-parametric hazard and history effect					
β_2	0.054	0.166	−0.274	0.055	0.382
β_3	−0.141	0.181	−0.509	−0.134	0.218
ν	0.503	0.107	0.296	0.499	0.718
Hazard profile (non-parametric hazard)					
$D\Lambda 0_1$	0.023	0.008	0.011	0.022	0.040
$D\Lambda 0_2$	0.057	0.014	0.033	0.056	0.088
$D\Lambda 0_3$	0.083	0.019	0.051	0.081	0.125
$D\Lambda 0_4$	0.075	0.018	0.044	0.074	0.121
$D\Lambda 0_5$	0.050	0.014	0.027	0.049	0.082

(*continues*)

Table 9.15 (*continued*)

	Mean	St. devn.	2.5%	Median	97.5%
Hazard profile (non-parametric hazard)					
$D\Lambda0_6$	0.088	0.023	0.049	0.085	0.139
$D\Lambda0_7$	0.030	0.013	0.011	0.028	0.060
$D\Lambda0_8$	0.033	0.014	0.012	0.031	0.066
$D\Lambda0_9$	0.054	0.020	0.022	0.052	0.098
$D\Lambda0_{10}$	0.020	0.012	0.005	0.018	0.048
$D\Lambda0_{11}$	0.015	0.010	0.002	0.012	0.039
$D\Lambda0_{12}$	0.061	0.023	0.026	0.058	0.118
$D\Lambda0_{13}$	0.025	0.014	0.005	0.023	0.057
$D\Lambda0_{14}$	0.028	0.017	0.005	0.024	0.072
$D\Lambda0_{15}$	0.020	0.014	0.003	0.017	0.052
$D\Lambda0_{16}$	0.022	0.016	0.003	0.018	0.061
$D\Lambda0_{17}$	0.037	0.023	0.008	0.032	0.094
$D\Lambda0_{18}$	0.027	0.021	0.003	0.023	0.082
$D\Lambda0_{19}$	0.014	0.016	0.000	0.009	0.056
$D\Lambda0_{20}$	0.014	0.014	0.000	0.009	0.052
$D\Lambda0_{21}$	0.015	0.014	0.000	0.011	0.050
$D\Lambda0_{22}$	0.031	0.023	0.004	0.026	0.089
$D\Lambda0_{23}$	0.017	0.016	0.000	0.012	0.057
$D\Lambda0_{24}$	0.038	0.026	0.005	0.032	0.103
$D\Lambda0_{25}$	0.022	0.022	0.001	0.014	0.082
$D\Lambda0_{26}$	0.022	0.022	0.000	0.015	0.080
$D\Lambda0_{27}$	0.026	0.026	0.001	0.019	0.100
$D\Lambda0_{28}$	0.028	0.028	0.001	0.019	0.101
$D\Lambda0_{29}$	0.040	0.041	0.001	0.027	0.152
$D\Lambda0_{30}$	0.065	0.065	0.002	0.049	0.258
$D\Lambda0_{31}$	0.071	0.073	0.001	0.047	0.250

187 for the history effect model. For the frailty model a subsidiary calculation gives an effective parameter count of 27.5, and deviance at the posterior mean of 146, so the DIC is 201. On this basis the 'history effect' model is preferred.

For a non-parametric hazard analysis one may either set one of the treatment effects to a null value or one of the piecewise coefficients. Here the first treatment effect (α) is set to zero. Applying the history effect model again shows κ below 1, but the average deviance is 176 and taking account of the 35 parameters, the DIC statistic (at 209) shows that there is no gain in fit from adopting this type of hazard estimation. The parametric hazard (combined with the history effect model) is therefore preferred. The rates $d\Lambda_0(t)$ for the first 18 months are precisely estimated and in overall terms suggest a decline, albeit irregular, in the exit rate over this period. Applying the remaining model option, namely non-parametric hazard with subject level frailty, and the assessment of its DIC, is left as an exercise.

9.7 REVIEW

While many of the earliest papers describing the application of MCMC methods to survival models (e.g. Kuo and Smith, 1992; Dellaportas and Smith,1993) are concerned with parametric survival functions, non-parametric applications are also developed in Hjort (1990) and Lo (1993). A major benefit of the Bayesian approach is in the analysis of censored data, treating the missing failure times as extra parameters, with the form of truncation depending on the nature of censoring (e.g. left vs. right censoring). Censoring is generally taken as non-informative but circumstances may often suggest an informative process.

Parametric survival models are often useful baselines for assessing the general nature of duration dependence and parametric frailty models have utility in contexts such as nested survival data; see Example 9.12 and Guo and Rodriguez (1992). However, much work since has focussed on Bayesian MCMC analysis of non-parametric survival curves, regression effects or frailty. For example, Laud *et al.* (1998) develop an MCMC algorithm for the proportional hazards model with beta process priors. Example 9.14 illustrates a comparison of a parametric hazard and non-parametric hazard based on the counting process model, while Example 9.11 considers a non-parametric frailty.

Recent reviews of Bayesian survival analysis include Ibrahim *et al.* (2001a), Kim and Lee (2002) and Rolin (1998).

REFERENCES

Aalen, O. (1976) Nonparametric inference in connection with multiple decrement models. *Scand. J. Stat., Theory Appl.* **3**, 15–27.

Aitkin, M. and Clayton, D. (1980) The fitting of exponential, Weibull and extreme value distributions to complex censored survival data using GLIM. *J. Roy. Stat. Soc., Ser. C* **29**, 156–163.

Aitkin, M., Anderson, D., Francis, B. and Hinde, J. (1989) *Statistical Modelling in GLIM.* Oxford: Oxford University Press.

Andersen, P., Borgan, Ø., Gill, R. and Keiding, N. (1993) *Statistical Models based on Counting* Processes. Berlin: Springer-Verlag.

Ansfield, F., Klotz, J., Nealon, T., Ramirez, G., Minton, J., Hill, G., Wilson, W., Davis, H. and Cornell, G. (1977) A phase III study comparing the clinical utility of four regimens of 5-fluorouracil: a preliminary report. *Cancer* **39**(1), 34–40.

Bedrick, E., Christensen, R. and Johnson, W. (2000) Bayesian accelerated failure time analysis with application to veterinary epidemiology. *Stat. in Med.* **19**(2), 221–237.

Chen, M., Ibrahim, J. and Shao, Q. (2000) *Monte Carlo Methods in Bayesian Computation.* New York: Springer-Verlag.

Collett, D. (1994) *Modelling Survival Data in Medical Research.* London: Chapman & Hall.

Cox, D. (1972) Regression models and life-tables (with discussion). *J. Roy. Stat. Soc., B*, **34**, 187–220.

Dellaportas, P. and Smith, A. (1993) Bayesian inference for generalized linear and proportional hazards model via Gibbs sampling. *Appl. Stat.* **42**, 443–460.

Dunlop, D. and Manheim, L. (1993) Modeling the order of disability events in activities of daily living. In: Robine, J., Mathers, C., Bone, M. and Romieu, I. (eds.), *Calculation of Health Expectancies: Harmonization, Censensus Achieved and Future Perspectives.* Colloque INSERM, Vol 226.

Dykstra, R. and Laud, P. (1981) A Bayesian nonparametric approach to reliability. *Ann. Stat.* **9**, 356–367.

Fahrmeir, L. and Knorr-Held, L. (1997) Dynamic discrete-time duration models: estimation via Markov Chain Monte Carlo. In: Raftery, A (ed.), *Sociological Methodology 1997.* American Sociological Association.

Fahrmeir, L. and Tutz, G. (2001) *Multivariate Statistical Modelling based on Generalized Linear Models, 2nd ed.* Berlin: Springer-Verlag.

Fleming, T. and Harrington, D. (1991) *Counting Processes and Survival Analysis.* New York: Wiley.

Gamerman, D. (1991) Dynamic Bayesian models for survival data. *J. Roy. Stat. Soc., Ser. C* **40**(1), 63–79.

Gehan, E. (1965) A generalized Wilcoxon test for comparing arbitrarily singly-censored samples. *Biometrika* **52**, 203–223.

Gelman, A., Carlin, J., Stern, H. and Rubin, D. (1995) *Bayesian Data Analysis*. CRC Press.

Gelfand, A. and Ghosh, S. (1998) Model choice: A minimum posterior predictive loss approach. *Biometrika* **85**, 1–11.

Gordon, I. R. and Molho, I. (1995) Duration dependence in migration behaviour: cumulative inertia versus stochastic change. *Environment and Planning A* **27**, 961–975.

Guo, G. and Rodriguez, G. (1992) Estimating a multivariate proportional hazards model for clustered data using the EM-algorithm, with an application to child survival in Guatemala. *J. Am. Stat. Assoc.* **87**, 969–976.

Hjort, N. (1990) Nonparametric Bayes estimators based on beta process in models for life history data. *Annals of Statistics* **18**, 1259–1294.

Ibrahim, J, Chen, M. and Sinha, D. (2001a) *Bayesian Survival Analysis*. Berlin: Springer-Verlag.

Ibrahim, J., Chen, M. and Sinha, D. (2001b) Criterion-based methods for Bayesian model assessment. *Statistica Sinica* **11**, 419–443.

Kalbfleisch, J. (1978) Non-parametric Bayesian analysis of survival time data. *J. Roy. Stat. Soc., Ser. B* **40**, 214–221.

Kalbfleisch, J. and Prentice, R. (1980) *The Statistical Analysis of Failure Time Data*. New York: Wiley.

Kay, R. (1977) Proportional hazard regression models and the analysis of censored survival data. *Appl. Stat.* **26**, 227–237.

Keiding, N., Anderson, P. and John, J. (1997) The role of frailty models and accelerated failure time models in describing heterogeneity due to omitted covariates. *Stat. in Med.* **16**, 215–225.

Kim, S. and Ibrahim, J. (2000) On Bayesian inference for proportional hazards models using noninformative priors. *Lifetime Data Anal*, **6**, 331–341.

Kim, Y. and Lee, J. (2002) Bayesian analysis of proportional hazard models. Manuscript, Penn State University.

Klugman, S. (1992). *Bayesian Statistics in Actuarial Sciences*. Kluwer.

Kuo, L. and Smith, A. (1992) Bayesian computations in survival models via the Gibbs sampler. In: Klein, J. and Goel, P. (eds.) *Survival Analysis: State of the Art*. Kluwer, pp. 11–24.

Kuo, L. and Peng, F. (2000) A mixture model approach to the analysis of survival data. In: Dey, D., Ghosh, S. and Mallick, B. (eds.), *Generalized Linear Models; a Bayesian Perspective*. Marcel Dekker.

Laud, P., Damien, P. and Smith, A. (1998) Bayesian nonparametric and semiparametric analysis of failure time data. In: Dey, D. *et al.*, (eds.) *Practical Nonparametric and Semiparametric Bayesian Statistics*. Lecture Notes in Statistics 133. New York: Springer-verlag.

Lawless, J. (1982) *Statistical Models and Methods for Lifetime Data*. New York: Wiley.

Leonard, T., Hsu, J., Tsui, K. and Murray, J. (1994) Bayesian and likelihood inference from equally weighted mixtures. *Ann. Inst. Stat. Math.* **46**, 203–220.

Lewis, S. and Raftery, A. (1995) Comparing explanations of fertility decline using event history models and unobserved heterogeneity. Technical Report no. 298, Department of Statistics, University of Washington.

Lindsey, J. (1995) Fitting parametric counting processes by using log-linear models. *J. Roy. Stat. Soc., Ser. C* **44**(2), 201–221.

Lindsey, J. (1999) *Models for Repeated Measurements, 2nd edn*. Oxford: Oxford University Press.

Lindsey, J. (2001) *Nonlinear Models for Medical Statistics*. Oxford: Oxford University Press.

Lo, A. (1993) A Bayesian Bootstrap for Censored Data. *Annals of Statistics*, **21**, 100–123.

McCall, B. (1994) Testing the proportional hazards assumption in the presence of unmeasured heterogeneity. *J. Appl. Econometrics* **9**, 321–334.

McIllmurray, M. and Turkie, W. (1987) Controlled trial of linolenic acid in Dukes' C colorectal cancer. *Br. Med. J.* **294**, 1260.

Mantel, N. and Hankey, B. (1978) A logistic regression analysis of response-time data where the hazard function is time dependent. *Comm. Stat.* **7A**, 333–348.

Morris, C., Norton, E. and Zhou, X. (1994) Parametric duration analysis of nursing home usage. In: Lange, N., Ryan, L., Billard, L., Brillinger, D., Conquest, L. and Greenhouse, J. (eds.), *Case Studies In Biometry*. New York: Wiley.

Powers, D. and Xie, Y. (1999) *Statistical Methods for Categorical Data Analysis*. Academic Press.

Rolin, J. (1993) Bayesian survival analysis. In: *Encyclopedia of Biostatistics*. New York: Wiley, pp. 271–286.

Sahu, S., Dey, D., Aslanidou, H. and Sinha, D. (1997) A Weibull regression model with gamma frailties for multivariate survival data. *Lifetime Data Anal* **3**(2), 123–137.

Sargent, D. (1997) A flexible approach to time-varying coefficients in the Cox regression setting. *Lifetime Data Anal.* **3**, 13–25.

Sastry, N. (1997) A nested frailty model for survival data, with an application to the study of child survival in Northeast Brazil. *J. Am. Stat. Assoc.* **92**, 426–435.

Singer, J. and Willett, J. (1993) It's about time: using discrete-time survival analysis to study duration and timing of events. *J. Educ. Stat.* **18**, 155–195.

Sinha, D. and Dey, D. (1997) Semiparametric Bayesian analysis of survival data. *J. Am. Stat. Assoc.* **92**(439), 1195–1121.

Spiegelhalter, D., Best, N., Carlin, B. and van der Linde, A. (2001) Bayesian measures of model complexity and fit. Research Report 2001–013, Division of Biostatistics, University of Minnesota.

Tanner, M. (1996) *Tools for Statistical Inference: Methods for the Exploration of Posterior Distributions and Likelihood Functions, 3rd ed.* Springer Series in Statistics. New York, NY: Springer-Verlag.

Thompson, R. (1977) On the treatment of grouped observations in survival analysis. *Biometrics* **33**, 463–470.

Volinsky, C. and Raftery, A. (2000) Bayesian Information Criterion for censored survival models. *Biometrics* **56**, 256–262.

Weiss, R. (1994) Pediatric pain, predictive inference and sensitivity analysis. *Evaluation Rev.* **18**, 651–678.

EXERCISES

1. In Example 9.2, apply multiple chains with diverse (i.e. overdispersed) starting points – which may be judged in relation to the estimates in Table 9.3. Additionally, assess via the DIC, cross-validation or AIC criteria whether the linear or quadratic model in temperature is preferable.

2. In Example 9.4, consider the impact on the covariate effects on length of stay and goodness of fit (e.g. in terms of penalised likelihoods or DIC) of simultaneously (a) amalgamating health states 3 and 4 so that the health (category) factor has only two levels, and (b) introducing frailty by adding a Normal error in the log(mu[i]) equation.

3. In Example 9.6, repeat the Kaplan–Meier analysis with the control group. Suggest how differences in the survival profile (e.g. probabilities of higher survival under treatment) might be assessed, e.g. at 2 and 4 years after the start of the trial.

4. In Program 9.8, try a logit rather than complementary log-log link (see Thompson, 1977) and assess fit using the pseudo Bayes factor or other method.

5. In Program 9.9 under the varying unemployment coefficient model, try a more informative Gamma prior (or set of priors) on $1/\sigma^2_\beta$ with mean 100. For instance try $G(1, 0.01)$, $G(10, 0.1)$ and $G(0.1, 0.001)$ priors and assess sensitivity of posterior inferences.

6. In Example 9.12, apply a discrete mixture frailty model at cluster level with two groups. How does this affect the regression parameters, and is there an improvement as against a single group model without frailty?

7. In Example 9.14, try a Normal frailty model in combination with the non-parametric hazard. Also, apply a two group discrete mixture model in combination with the non-parametric hazard; how does this compare in terms of the DIC with the Normal frailty model?

CHAPTER 10

Modelling and Establishing Causal Relations: Epidemiological Methods and Models

10.1 CAUSAL PROCESSES AND ESTABLISHING CAUSALITY

Epidemiology is founded in efforts to prevent illness by contributing to understanding the causal processes, or etiology, underlying disease. This includes establishing and quantifying the role of both risk factors and protective factors in the onset of ill-health. Risk factors include individual characteristics or behaviours, or external hazards that an individual is exposed to, that increase the chance that the individual, rather than someone selected randomly from the general population, will develop ill-health. External risk factors may relate to the environment or community, and include material factors (e.g. income levels), psychosocial and biological risk factors in human populations (e.g. Garssen and Goodkin, 1999). Epidemiological analysis extends to studies of disease in animal as well as human populations (Noordhuizen *et al.*, 1997).

Bayesian approaches to modelling in epidemiology, and in biostatistics more generally, have been the subject of a number of recent studies. Several benefits from a Bayesian approach, as opposed to frequentist procedures which are routinely used in many epidemiological studies, may be cited (Lilford and Braunholtz, 1996; Spiegelhalter *et al.*, 1999). These include model choice procedures that readily adapt to non-nested models; availability of densities for parameters without assuming asymptotic normality, and the formal emphasis on incorporating relevant historical knowledge or previous studies into the analysis of current information (Berry and Stangl, 1996). Also advantageous are Bayesian significance probabilities (Leonard and Hsu, 1999) which fully reflect all uncertainty in the derivation of parameters. Of particular interest is Bayesian model choice in situations where standard model assumptions (e.g. linear effects of risk factors in logit models for health responses) need to be critically evaluated. On the other hand, Bayesian sampling estimation may lead to relatively poor identifiability or slow converegence of certain types of models, including models popular in epidemiology such a spline regressions (Fahrmeir and Lang, 2001) and issues such as informativeness of priors and possible transformation of parameters become important in improving identifiability.

Most usually, causal processes in epidemiology involve multiple, possible interacting factors. Inferences about risk and cause are affected by the nature of the causal process,

Applied Bayesian Modelling P. Congdon
© 2003 John Wiley & Sons, Ltd ISBN: 0-471-48695-7

and by the setting and design of epidemiological studies, whether clinical or community based, and whether randomized trial as against observational study. The main types of observational study are case-control (retrospective) studies, cross-sectional prevalence studies, and prospective or cohort studies (Woodward, 1999). Measures of risk are governed by study design: for example, in a case-control study the focus may be on the odds ratio of being exposed given case as against control status (Breslow, 1996), whereas in a cohort study, the focus is on risk of disease given exposure status.

The major designs used have a wide statistical literature attached to them and statistical thinking has played a major role in the development of epidemiology as a science. Some authors, however, caution against routine application of concepts from multivariate analysis, such as using continuous independent variables to describe risk profiles, using product terms for evaluating interactions, or the notion of independent effects in the presence of confounding (Davey Smith and Phillips, 1992; Rothman, 1986, Chapter 1). Also, the goals of an epidemiological analysis may not coincide with a hypothesis testing approach.

Underlying generalisations from epidemiological studies, and guiding the application of statistical principles in them, are concepts of causality in the link between risk factor and disease outcome. Thus, a causal interpretation is supported by: (a) strength in associations, after controlling for confounding, evidenced by high risk ratios or clear dose-response relationships; (b) consistent associations across studies, different possible outcomes, and various sub-populations; (c) temporal precedence such that exposure predates outcome (subject to possible latency periods); (d) plausibility of associations in biological terms; and by (e) evidence of a specific effect following from a single exposure or change in a single exposure.

10.1.1 Specific methodological issues

Describing the relationship between an outcome and a given risk factor may be complicated by certain types of interaction between risk factors. Confounding, considered in Section 10.2, relates to the entangling or mixing of disease risks, especially when the confounder influences the disease outcome, and is also unequally distributed across categories of the exposure of interest. It may often be tackled by routine multivariate methods for correlated or collinear independent variables. Other major options are stratification and techniques based on matching, such as the matched pairs odds ratio (Rigby and Robinson, 2000).

Dose-response models (Section 10.3) aim to establish the chance of an adverse outcome occurring as a function of exposure level (Boucher *et al.*, 1998). Rothman (1986, Chapter 16) argues that the leading aim of epidemiological investigation is to estimate the magnitude of effect (e.g. relative risks) as a function of level of exposure. This may indicate categorisation of a continuous exposure variable and classical calculation of an effect according to a category of the exposure would require then sub-populations of sufficient size as a basis for precisely describing trend over categories. However, random effect models to describe trend (e.g. via state space techniques), especially in Bayesian implementations, may overcome such limitations (Fahrmeir and Knorr-Held, 2000). Establishing consistent relationships depends on the selected outcome and on the measurement of exposure: risk factors such as alcoholism or outcomes such as good health cannot be precisely operationalised and have to be proxied by a set of observable items, and so latent variable models come into play (Muthen, 1992).

Meta-analysis (see Section 10.4) refers to the combination of evidence over studies and hence plays a role in establishing consistency of associations: it provides a weighted average of the risk or treatment estimate that improves on rules-of-thumb, such as 'most studies show an excess risk or treatment benefit' (Weed, 2000). Findings of heterogeneity in risk parameters across studies need not preclude consistency. Meta-analysis may also play a role in more precisely establishing the strength of an association or dose-response relationship, e.g. in providing a summary estimate of relative risk with improved precision as compared to several separate studies.

10.2 CONFOUNDING BETWEEN DISEASE RISK FACTORS

As noted above confounding occurs when a certain risk factor – not the focus of interest or with an established influence in scientific terms – is unequally distributed among exposed and non-exposed subjects, or between treatment and comparison groups. Often, the interest is in a particular risk factor X and it is necessary to adjust for confounder variables Z. Mundt *et al.* (1998) cite the assessment of lung cancer risk due to occupational exposure when there is in practice mixing of risk due to smoking and occupational exposure. If smoking is more prevalent among occupations with a cancer risk then the relationship between cancer and the occupational exposure would be over-estimated. Complications arise in adjusting for confounders if they are not observed (i.e. not explicitly accounted for in the model) or if they are measured with error. A model erroneously omitting the confounder, or not allowing for measurement error if it is included, leads to under-estimation of the average impact of X on Y (Small and Fischbeck, 1999; Chen et al., 1999).

Whether Z is a confounder or not depends upon the nature of the causal pathways (Woodward, 1999). Z is not a confounder if Z causes X, and X is in turn a cause of Y (e.g. if Z is spouse smoking, X is exposure to tobacco smoke in a non-smoking partner, and Y is cancer in the non-smoking partner). If Z causes X and both X and Z were causal for Y, then Z is also not a confounder.

Generally, it is assumed that the true degree of association between the exposure and the disease is the same regardless of the level of the confounder. If, however, the strength of association between an exposure and disease does vary according to the level of a third variable then Z is known as an effect modifier rather than a confounder. This is essentially the same as the concept of interaction in log-linear and other models. It is possible that such a third variable Z is a confounder only, an effect modifier only, or both an effect modifier and confounder[1].

[1] Consider the example of Rigby and Robinson (2000) in terms of relative risk of an outcome (e.g. deaths from lung cancer) in relation to smoking (X) and tenure (Z), the latter specified as owner occupier ($Z = 1$), renter in subsidised housing ($Z = 2$), and private renter ($Z = 3$). Tenure would be a confounder, but not an effect modifier if the risk of cancer was higher among renter groups, but within each tenure category the relative risk of cancer for smokers as against non-smokers was constant at, say, 2. Suppose the mortality rate among non-smokers was 0.1 among owner occupiers, 0.15 among subsidised renters and 0.2 for private renters. Then the mortality rates among smokers in the 'tenure as confounder only' case would be 0.2 among owners, 0.3 among subsidised renters and 0.4 among private renters. The overall relative risk depends then upon the distribution of tenure between smokers and non-smokers. If smoking is less common among owner occupiers, then ignoring housing tenure in presentation or risk estimation would lead to overstating the overall relative risk of mortality for smokers (e.g. estimating it at 3 or 4 rather than 2).

Tenure would be an effect modifier in this example if the mortality rate among non-smokers was constant, at say 0.1, but the relative risk of cancer mortality for smokers as against non-smokers was higher among the renter subjects than the owner occupiers. There is then no association between tenure and mortality in the absence of smoking and differences in the relative risk between tenure categories reflect only effect modification. Tenure would be both a confounder and effect modifier when the global estimate of the smoking relative risk is influenced by the distribution of smoking across tenures, but the relative risk is concurrently different across tenures. If there is effect modification with genuine differences in relative risk (for smokers and non-smokers) according to the category of Z, then a global estimate of relative risk may make less substantive sense.

10.2.1 Stratification vs. multivariate methods

One method to reduce the effect of a confounder Z is to stratify according to its levels $(Z_1, .. Z_m\}$, and then combine effect measures such as odds ratios over strata, according to their precisions. Data from an Israeli cross-sectional prevalence study reported by Kahn and Sempos (1989) illustrate the basic questions (Table 10.1). Cases and non-cases (in terms of previous myocardial infarction) are classified by age (Z) and systolic blood pressure (X).

Age is related to the outcome because the odds ratio for MI among persons over 60 as against younger subjects is clearly above 1. The empirical estimate is

$$15 \times 1767/(188 \times 41) = 3.44$$

with $\log(\text{OR}) = 1.24$ having a standard deviation[2] of $(1/15 + 1/1767 + 1/188 + 1/41)^{0.5} = 0.31$. Moreover, age is related to SBP since with age over 60 as the 'outcome' and SBP over 140 as the 'exposure', the empirical odds ratio is

$$124 \times 1192/(616 \times 79) = 3.04$$

Providing there is no pronounced effect modification, it is legitimate to seek an overall odds ratio association controlling for the confounding effect of age. Suppose the cells in each age group sub-table are denoted $\{a, b, c, d\}$ and the total as $t = a + b + c + d$.

To combine odds ratios OR_i (or possibly log OR_i) over tables, the Mantel–Haenszel (MH) estimator sums $\nu_i = a_i d_i / t_i$ and $\delta_i = b_i c_i / t_i$ to give an overall odds ratio

$$\sum_i \nu_i / \sum_i \delta_i \qquad (10.1)$$

This is a weighted average of the stratum (i.e. age band) specific odds ratios, with weight for each stratum equal to $\delta_i = b_i c_i / t_i$, since

$$\{a_i d_i / (b_i c_i)\}\delta_i = a_i d_i / t_i = \nu_i$$

Table 10.1 Myocardial infarction by age and SBP

Age Over 60	MI Cases	No MI	All in SBP group
SBP > = 140	9	115	124
SBP < 140	6	73	79
All in Age Band	15	188	203
Age Under 60			
SBP > = 140	20	596	616
SBP < 140	21	1171	1192
All in Age Band	41	1767	1808

[2] The standard error estimate is provide by the Woolf method which relies on the Normality of log(OR).

The weights δ_i are proportional to the precision of the logarithm of the odds ratio under the null association hypothesis. For the data in Table 10.1 the estimator (10.1) for the overall odds ratio is 1.57, as compared to 0.95 and 1.87 in the two age groups.

A stratified analysis may become impractical if the data are dispersed over many subcategories or multiple confounders, or if the impact of the risk factor is distorted by categorisation, and multivariate methods such as logistic regression are the only practical approach. For example, logit regression methods are applicable to pooling odds ratios over a series of 2×2 tables from a case-control study, even if case/control status does not result from random sampling of a defined population (Selvin, 1998, Chapter 4). In the case of Bayesian estimation, credible intervals on the resulting effect estimates will be obtained without requiring Normality assumptions. This is especially important for small cell counts $\{a, b, c, d\}$, including the case when a cell count is zero such that the estimate $ad/(bc)$ is undefined.

Multivariate methods may be applied when there is intentional matching of a case with one or more controls on the confounders. The precision of a risk estimate (e.g. odds ratio or relative risk) will be increased if the control to case ratio M exceeds 1. As well as providing control for confounding *per se*, matching on risk factors with an established effect (e.g. age and cardiovascular outcomes, or smoking and lung cancer) may enhance the power of observational studies to detect impacts of risk factors of as yet uncertain effect (Sturmer and Brenner, 2000). Matched studies with a dichotomous outcome may be handled by conditional logistic regression, conditioning on the observed covariates in each matched set (each matched set of case and controls becomes a stratum with its own intercept). For 1:1 matching, this reduces to the standard logistic regression.

Example 10.1 Alcohol consumption and oesophageal cancer with age stratification One possible analysis of studies such as that in Table 10.1 is provided by a logit or log-linear model for the predictor combinations produced by the confounder and the exposure. For instance, in Table 10.1 the assumption of a common odds ratio over the two age groups corresponds to a model with main effects in age and blood pressure only, but without an interaction between them.

This example follows Breslow and Day (1980) and Zelterman (1999) in considering a case control study of oesophageal cancer (Y) in relation to alcohol consumption (X), where age (Z) is a confounding factor. Table 10.2 shows the age banded and all ages study data.

The unstandardised estimate of the overall odds ratio (from the all ages sub-table)

$$96 \times 666/(109 \times 104) = 5.64$$

However, the association appears to vary by age (the corresponding estimate for the 65–74 group being only 2.59). One might apply the Mantel–Haenszel procedure to obtain an aggregate effect pooling over the age specific odds ratios, though this is complicated by undefined odds ratios (from classical procedures) in the lowest and highest age groups.

An alternative is a log-linear model based on Poisson sampling for the frequencies f_{YXZ}, with case-control status being denoted Y (=1 for control and 2 for case). A log linear model (Model A) corresponding to a common odds ratio over the six age groups is then specified as

Table 10.2 Case control data on oesophageal cancer

Age group		Annual alcohol consumption Over 80 g	Under 80 g
25–34	Case	1	0
	Control	9	106
35–44	Case	4	5
	Control	26	164
45–54	Case	25	21
	Control	29	138
55–64	Case	42	34
	Control	27	139
65–74	Case	19	36
	Control	18	88
75+	Case	5	8
	Control	0	31
All ages	Case	96	104
	Control	109	666

$$f_{YXZ} \sim \mathrm{Poi}(\mu_{YXZ})$$
$$\log(\mu_{YXZ}) = \alpha + \beta_Y + \gamma_X + \delta_Z + \varepsilon_{YX} + \kappa_{YZ} + \eta_{XZ}$$

and the common odds ratio across sub-tables is estimated as $\phi = \exp(\varepsilon_{YX})$. Fairly diffuse N(0, 1000) priors are adopted for all the effects in this model. A two chain run with null starting values in one chain, and values based on a trial run in the other, shows convergence from 5000 iterations: the scale reduction factors for β_2 only settle down to within [0.9, 1.1] after then.

The Bayesian estimation has the benefit of providing a full distributional profile for ϕ; see Figure 10.1 with the positive skew in ϕ apparent. Tests on the coefficient (e.g. the

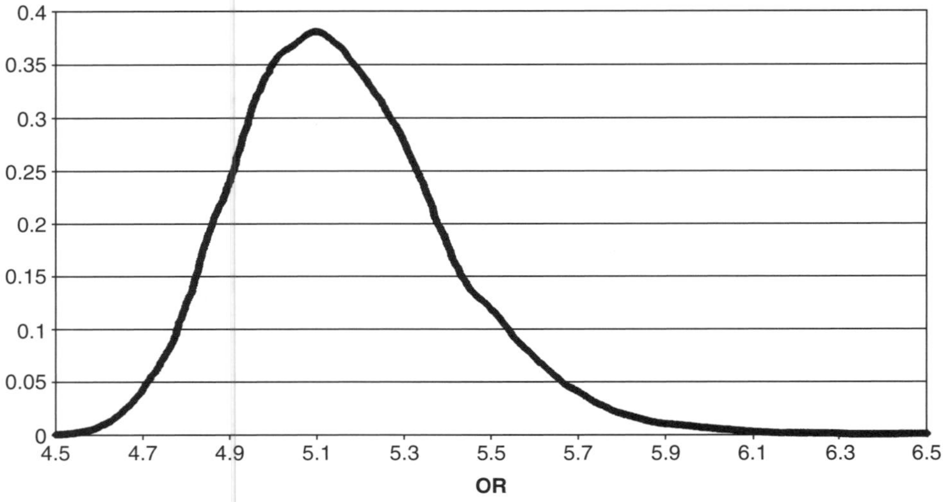

Figure 10.1 Posterior density of common odds ratio

probability that it exceeds 6) may be carried out by repeated sampling and accumulating over those iterations where the condition $\phi > 6$ is met. The posterior mean of ϕ (from iterations 5000–10 000) is 5.48 with 95% credible interval 3.68 to 7.83.

The pseudo (log) marginal likelihood (Gelfand and Dey, 1994) for Model A, obtained by accumulating over log CPOs, is −90.1. The worst fit (lowest CPO) is for the observation of 1 case in the 25–34 age band and high alcohol use. The predictive loss criterion[3] of Sahu *et al.* (1997), Ibrahim *et al.* (2001) and Gelfand and Ghosh (1998) is 1918, with $w = 1$. Classical criteria to assess whether this model is adequate include a χ^2 criterion comparing f_{22Z} with μ_{22Z} (see Breslow, 1996, Equation (7), namely

$$\chi^2 = \sum_Z (f_{22Z} - \mu_{22Z})^2 / \text{Var}(\mu_{22Z})$$

and this may be evaluated at the posterior mean of Model B. The relevant posterior means (with variances) are 0.35 (0.11), 4.1 (2.2), 24.5 (19.1), 40.3 (33.6), 23.7 (18.8) and 3.15 (2.2). This yields $\chi^2 = 6.7$ (5 d.f.), and so suggests Model A is adequate. Note that the posterior variances of μ_{22Z} are higher than maximum likelihood estimates (for instance, those provided by Zelterman (1999) stand at 0.22, 2.1, 7.8, 10.6, 6.3 and 1).

A model allowing for different odds ratios between caseness and high alcohol consumption according to the confounder level involves adding a three way interactions λ_{YXZ} to the above model. If these are taken to be fixed effects (Model B), then odds ratios for the lowest and highest age groups are still effectively undefined (though are no longer infinity providing the priors are proper). One may assess the fit of Model B as compared to Model A via the above general criteria; these are in fact in conflict, with the pseudo-marginal likelihood increasing to −86.5, but the predictive loss criterion worsening (as compared to Model A), namely to 1942. This worsening is in fact only apparent for small w, and for $w > 5$, the predictive loss criterion also favours allowing age specific odds ratios.

However, a random effects model (Model C) for the three way effects (and possibly other parameters) is also possible, and will result in some degree of pooling towards the mean effect, while maintaining age differentiation if the data require it (Albert, 1996). A random effects approach also facilitates a more complex modelling option, considered below, that involves choosing between a precision $1/\sigma^2_\lambda$ for λ_{YXZ} which is effectively equivalent to $\lambda_{YXZ} = 0$ (the common odds ratio model) and a precision which allows non-zero three way effects. Fitting Model C with a gamma prior G(1, 0.001) on $1/\sigma^2_\lambda$ gives a relatively small σ_λ for the λ_{YXZ} and age band odds ratios (phi.sr[] in Program 10.1) all between 5.45 and 5.6. This option has a better predictive loss criterion than Model B and broadly supports pooling over age bands.

Finally, we apply the specific model choice strategy, with $1/\sigma^2_\lambda$ set to 1 000 000 for effective equivalence to $\lambda_{YXZ} = 0$. A binary indicator chooses between this option and the prior $1/\sigma^2_\lambda \sim G(1, 0.001)$ with equal prior probability. The option $1/\sigma^2_\lambda = 1\,000\,000$ is chosen overwhelmingly, and so this procedure suggest age differentiation in the odds

[3] Let f_i be the observed frequencies, θ the parameters in the log-linear model, and z_i be 'new' data sampled from $f(z|\theta)$. Suppose ν_i and ς_i are the posterior mean and variance of z_i, then one possible criterion for any $w > 0$ is

$$D = \sum_{i=1}^{n} \varsigma_i + [w/(w+1)] \sum_{i=1}^{n} (\nu_i - f_i)^2$$

Typical values of w at which to compare models might be $w = 1$, $w = 10$ and $w = 100\,000$. Larger values of w put more stress on the match between ν_i and f_i and so downweight precision of predictions.

ratios is not required (and hence that age acts as a confounder rather than an effect modifier).

Example 10.2 Framingham follow up study for CHD An illustration of confounding influences in a cohort study is provided by data on development of coronary heart disease during an 18 year follow up period among 1363 respondents included in the Framingham study (Smith, 2000). At the start of follow up in 1948, the study participants were aged between 30 and 62, and the development of CHD among 268 participants is related to age, sex and Systolic Blood Pressure (SBP) at exam 1.

The aim is to control for confounding by age Z_1 and sex ($Z_2 = 1$ for males, 0 for females) in the relation between CHD onset and systolic blood pressure (X), which for these subjects ranges between 90 and 300. Specifically an odds ratio ϕ comparing CHD onset probability for participants with initial SBP above and below 165 mm Hg is sought. To investigate the impact of categorising continuous predictor variables, whether the risk factor itself or a confounder, different types of logit regression may be considered. The first is linear in the continuous predictors age and SBP, the second converts these predictors to categorical form, and the third considers non-linear functions of age and SBP.

The first logit regression takes both age and SBP as continuous with linear effects (Model A in Program 10.2). A three chain run[4] shows convergence at around 750 iterations and summaries are based on iterations 1000–5000. The resulting equation for CHD onset probability (with mean and posterior SD of coefficients) is

$$
\begin{aligned}
\text{logit}(\pi_i) &= \alpha + \beta_1 Z_1 + \beta_2 X + \beta_3 Z_2 \\
&= \underset{(0.9)}{-7.2} + \underset{(0.017)}{0.050 Z_1} + \underset{(0.002)}{0.0171 X} + \underset{(0.15)}{0.92 Z_2}
\end{aligned}
$$

Note that the maximum likelihood solution (from SPSS) is very similar, the only slight difference being that the ML estimation has a coefficient on age of 0.052 with SD of 0.015. Mildly informative N(0, 10) priors are used for the coefficients on the continuous predictors to avoid numerical overflow (which occurs if large sampled values for coefficients are applied to high values for Age or SBP). It may be noted that the logit link is more robust to extreme values in the regression term than alternatives such as the probit or complementary log-log links. Another option is to scale the age and SBP variables, for example to have a range entirely within 0 to 1 (e.g. dividing them by 100 and 300, respectively), or to apply standardisation.

In this first analysis, the original scales of age and SBP are retained and it is necessary obtain the average SBP in the group with SBP above 165 (namely 188.4) and the remainder (namely 136.5). The relevant odds ratio, under this continuous regressors model, is then the exponential of the coefficient for SBP times the difference $188.4 - 136.5 = 51.9$

$$\phi = \exp(\beta_2 \times 51.9)$$

As noted by Fahrmeir and Knorr-eld (2000), an advantage of MCMC sampling is that posterior densities of functionals of parameters (here of β_2) are readily obtained by

[4] Starting values are provided by null values, the posterior average from a trial run, and the 97.5th point from the trial run.

repeated sampling. Thus, the mean and 95% credible interval for ϕ are obtained as 2.44 (2.09, 2.89).

Standard fit measures (e.g. predictive loss criteria or pseudo marginal likelihood) may be applied. Thus the criterion in footnote 3 with $w = 1$ stands at 299.6 and the pseudo-marginal likelihood at -630.5. To further assess fit and predictive validity the risk probabilities π_i may be arranged in quintiles, and the cumulated risk within each quintile compared with the actual numbers in each quintile who developed the disease. Thus, among the sample members with the lowest 273 risk probabilities (approximately the lowest 20% of the 1363 subjects) we find the number actually developing CHD, then apply the same procedure among those ranked 274–546, and so on. There is some departure in the logit model prediction from the actual risk distribution, as in Table 10.3. Note that this is best done by monitoring the π_i (using the Inference/Summary procedure in BUGS) and then using other programs or spreadsheets to reorder the cases. Program 10.2 also contains the array rank[1363, 5] that monitors the quintile risk category of each subject.

We next consider a categorical regression analysis (Model B in Program 10.2), with the first continuous predictor SBP dichotomised at above and below 165, and the age predictor forming a four fold category, denoted AgeBand[] in Program 10.2: age under 50, between 50–54, between 55–59, and over 60. As usual a corner constrained prior is used with $\gamma_1 = 0$ and $\gamma_j \sim N(0, 1000)$ for $j = 2, 3, 4$. A three chain run[5] shows convergence at around 300 iterations and summaries are based on iterations 500–5000. Thus the model, with $I(s) = 1$ for s true, is

$$\text{logit}(\pi_i) = \alpha + \beta_1 Z_2 + \beta_2 I(X > 165) + \gamma[\text{AgeBand}]$$

Standard fit measures show a worse fit under this model. Thus, the predictive loss criterion stands at 303.2 and the pseudo-marginal likelihood at -641. The match of actual and predicted risk is assessed over the 16 possible risk probabilities, formed by the high and low categories of SBP and the four age groups (part b of Table 10.3). This shows an acceptable fit (cf. Kahn and Sempos, 1989). The posterior median of the odds ratio ϕ between high and low SBP subjects controlling for age confounding is estimated at around 2.74.

Specific types of nonlinear regression models have been proposed for representing risks (Greenland, 1998a). For example, a flexible set of curves is obtained using fractional polynomial models, involving the usual linear term, one or more conventional polynomial terms (squares, cubes, etc.), and one or more fractional or inverse powers (square root, inverse squared, etc.). A simple model of this kind in, say, SBP might be

$$\text{logit}(\pi_i) = \alpha + \beta_1 Z_1 + \beta_2 X + \beta_3 X^2 + \beta_4 X^{0.5}$$

For positive predictors X, $\log_e(X)$ can be used instead of $X^{0.5}$ to give a curve with a gradually declining slope as x increases (Greenland, 1995). In fact, inclusion of $\log_e(X)$ allows for the possibility of non-exponential growth in risk; for instance, $\exp(\beta \log_e(X)) = X^\beta$ can increase much slower than exponentially. Fractional polynomials and spline regression have been advocated as improving over simple categorical regression (Greenland, 1995); possible drawbacks are potentially greater difficulties in identifiability and convergence, and also the desirability of ensuring sensible dose-response patterns. For example, a polynomial model in SBP, while identifiable, might imply an implausibly declining risk at SBP above a certain point such as 275.

[5] Starting values are provided by null values, the posterior average from a trial run, and the 97.5th point from the trial run.

Table 10.3 Alternative logistic regression models to assess risk according to SBP (Framingham study)

(a) All Continuous Predictors treated as such

Quintile of risk Probability	Observed	Expected under logistic	Chi square
1st	16	22.3	1.80
2nd	30	35.6	0.88
3rd	64	48.5	4.95
4th	65	64.4	0.01
5th	93	98.7	0.33
Total	268	269.6	7.96

(b) Continuous predictors in category form

Summing over 16 Possible Risk Probabilities

Expected under logistic	Observed	Risk probability	Chi square
15.3	13	0.081	0.41
16.9	15	0.101	0.25
4.6	7	0.132	0.81
18.5	18	0.134	0.01
31.6	36	0.173	0.53
8.2	7	0.195	0.20
34.9	35	0.210	0.00
14.3	16	0.234	0.19
12.4	13	0.263	0.03
37.9	36	0.269	0.10
5.2	3	0.292	1.68
20.5	25	0.297	0.82
9.4	9	0.362	0.02
14.7	14	0.419	0.03
4.9	5	0.490	0.00
17.5	16	0.499	0.14
266.8	268		5.22

Here the coefficient selection procedure of Kuo and Mallick (1998) is applied to the specification

$$\begin{aligned}\text{logit}(\pi_i) &= \alpha + \beta_1 Z_1 + \beta_2 Z_1^2 + \beta_3 \log_e(Z_1)\\ &\quad + \beta_4 X + \beta_5 X^2 + \beta_6 \log_e(X) + \beta_7 Z_2\end{aligned}$$

Age and SBP are obtained by dividing the original values by 100 and 300, respectively. Binary selection indicators, with Bernoulli(0.5) priors, are applied to the coefficients $\beta_1 - \beta_6$. A single run of 10 000 iterations (see Model C in Program 10.2) shows β_2 and β_6 to have posterior selection probabilities exceeding 0.98, while the remaining coefficients have selection probabilities below 0.10. A third logit model is therefore estimated, namely

$$\text{logit}(\pi_i) = \alpha + \beta_1 Z_1^2 + \beta_3 \log(X) + \beta_4 Z_2 \tag{10.2}$$

This yields a slight improvement in pseudo-marginal likelihood over the linear continuous predictors model above (−628.5 vs. −630.5) and in the predictive loss criterion with w = 1 (namely 298.4 vs. 299.6). The parameter summaries for Model (10.2), from iterations 500–5000 of a three chain run, are in Table 10.4. The odds ϕ ratio is very similar to those previously obtained.

Example 10.3 Larynx cancer and matched case-control analysis The impact of matching to control for confounders and clarify the risk attached to an exposure of interest is illustrated by an example from Sturmer and Brenner (2000). They consider the utility of matching in case-control studies on risk factors whose effect is established and of no substantive interest. The interest is rather in the impact of a new suspected risk. They cite existing case-control findings on the link between larynx cancer and smoking (four categories, namely 0–7, 8–15, 16–25, over 25 cigarettes per day) and alcohol consumption (bands of 0–40, 40–80, 80–120, and over 120 grammes per day). Table 10.5 shows the relative distribution of cases and population between the 16 strata formed by crossing these two risk factors.

The impact of smoking and alcohol is established, and the interest is in the impact of case-control matching to assess the effect of a new putative risk X. We compare matched

Table 10.4 Nonlinear risk model, parameter summary

	Mean	St. devn.	2.5%	Median	97.5%
Odds Ratio	2.63	0.34	2.05	2.61	3.39
α	−1.17	0.51	−2.11	−1.20	−0.13
β_1	4.81	1.39	1.90	4.87	7.41
β_2	2.98	0.40	2.23	2.98	3.79
β_3	0.93	0.15	0.64	0.93	1.22

Table 10.5 Larynx cancer cases and controls across established risk factor combinations

Stratum identifier	Smoking rate (no. of cigarettes per day)	Alcohol consumption	Exposure risk to X (in population and controls) under moderate confounding	Proportion of cases belonging to stratum defined by known risk factors	Proportion of population belonging to stratum
1	0–7	0–40	0.01	0.010	0.168
2		41–80	0.02	0.024	0.140
3		81–120	0.03	0.017	0.053
4		Over 120	0.04	0.027	0.031
5	8–15	0–40	0.02	0.022	0.081
6		41–80	0.04	0.078	0.092
7		81–120	0.06	0.068	0.043
8		Over 120	0.08	0.095	0.023
9	16–25	0–40	0.03	0.066	0.081
10		41–80	0.06	0.103	0.09
11		81–120	0.09	0.127	0.045
12		Over 120	0.12	0.137	0.035
13	26+	0–40	0.04	0.012	0.043
14		41–80	0.08	0.037	0.034
15		81–120	0.12	0.054	0.025
16		Over 120	0.16	0.122	0.015

sampling, with controls sampled according to the case profile (i.e. the proportionate distribution among the 16 strata, as in the penultimate column in Table 10.5), with unmatched sampling. Under unmatched sampling, the sampling of controls is according to the population profile, given by the last column of Table 10.5.

For illustration, $M = 2$ controls are taken for each of 200 cases, and the exposure disease odds ratio in each stratum (the odds ratio of exposure to X given case-control status) is assumed to be 2. Sturmer and Brenner then generate samples with 200 cases and 400 controls to establish the power to detect this effect size under various assumptions about the confounding of the new risk factor X with the established risk factors Z_1 and Z_2 (smoking and alcohol consumption). In Program 10.3 it is necessary to generate both the stratum (defined by Z_1 and Z_2) from which an individual is sampled, and exposure status to X; for cases these are indexed by arrays Stratcase[] and Exp.case[].

Under the first assumption there is no confounding, with an exposure rate to the new factor X (proportion exposed to X in strata 1 to 16) set at 0.05 in all strata. Under an alternative moderate confounding assumption, the exposure rate rises in increments from 0.01 in the lowest smoking and alcohol intake group to 0.16 in the highest smoking and drinking group (see Table 10.5).

Sturmer and Brenner report higher powers to establish the assumed odds ratio of 2 under matched than unmatched sampling, and higher powers also under moderate confounding than no confounding. The analysis here confirms the ability of matched case-control sampling to obtain the correct odds ratio regardless of the confounding scenario, and the greater power to detect a positive odds ratio under moderate confounding rather than no confounding.

Under matched sampling both cases and controls are drawn to have the same distribution across the 16 strata, namely that in the penultimate column of Table 10.5. It is necessary to assess the power of the study to detect a positive relation between exposure and disease. The test used to establish the significance of the log of odds ratio (and hence power of the study) for each sample of 600 involves the empirical variance of the log of the odds ratio over all strata combined. It is preferable to use the log of the odds ratio to assess power as this is more likely to be approximately Normal, whereas the odds ratio itself is usually skewed.

Thus, let A, B, C and D be exposed cases, unexposed cases, exposed controls and unexposed controls respectively accumulated over all strata, with the stratum equivalents being a_j, b_j, c_j and d_j. So the variance of $\kappa = \log_e(\text{OR})$ is

$$1/A + 1/B + 1/C + 1/D$$

where $A = \Sigma_j a_j$, $B = \Sigma_j b_j$, $C = \Sigma_j c_j$ and $D = \Sigma_j d_j$. A refinement is to form the Mantel–Haenszel estimate OR_{MH} of the overall odds ratio, with weighting of the stratum odds ratios according to their precision.

A run of 5000 iterations with moderate confounding and matched case-control sampling with $M = 2$ leads to a power of 70.6% to detect a positive odds ratio at 2.5% significance (compared to 71.1% obtained by Sturmer and Brenner) and an estimated mean OR of 2.02. Estimates using the Mantel–Haenszel procedure are very similar, but involve slower sampling. When there is no confounding across the strata formed by Z_1 and Z_2, but still matched case-control sampling, the power is reduced to around 54% and the mean odds ratio is 2.12 (and median 2.01).

Under unmatched sampling with any degree of confounding the crude odds ratio is an overestimate. To allow for the fact that, under this type of sampling, controls are sampled disproportionately from strata with low exposure risk, one may adjust the crude odds ratio to take account of differential exposure to risk. One may obtain the ratio of average exposure to risk among cases as compared to average exposure among controls on the basis of a standard risk profile ρ_j (exposed to risk of X) over the strata. Thus, Table 10.5 shows the population (and control) risk profile under moderate confounding, and it can be seen that sampling from the case distribution $\pi_{\text{case}[j]}$ (penultimate column) leads to higher average exposure than sampling from the population distribution $\pi_{\text{pop}[j]}$ (last column). A run of 10 000 iterations estimates the median of the ratio $R_{\text{exp}} = \Sigma\rho_j\pi_{\text{case}[j]}/\Sigma\rho_j\pi_{\text{pop}[j]}$ at 1.93 on the basis of the actual sampling proportions over strata at each iteration. This is used to calculate adjusted totals $C' = C.R_{\text{exp}}$ of exposed controls, and $D' = 400 - C'$ of unexposed controls. The median crude OR is 3.98, and the median of the adjusted OR is then 1.98. The log of the adjusted OR is found to have a variance of 0.34^2 from a trial run, and from this a power of 52% (Sturmer and Brenner obtain 51%) to detect an association between disease and exposure is obtained, compared to 70.6% under matched sampling.

A wide range of alternative scenarios may be investigated; for example Sturmer and Brenner (2000) consider a strong confounding scenario with the exposure risk ranging from 0.005 in stratum 1 to 0.32 in stratum 16. Alternative numbers M of matched controls may also be taken (e.g. up to $M = 5$).

Example 10.4 Obesity and mental health Increasingly, health strategy and measures of health and clinical gain focus on improving quality of life, as well as extending life expectancy. These measures in turn depend upon valuations of health status, founded in utility theory, with different health states being assigned differing utilities ranging from 0 (death) to 1.0 (perfect health), or possibly scaled to run from 0 to 100. Following Doll, Petersen and Brown (2000), we analyse responses on an instrument used to assess health status and quality of life in both clinical and community settings, namely the Short Form 36 or SF36 questionnaire (Jenkinson *et al.*, 1993). This questionnaire has eight subscales, and includes items on particular aspects of physical and mental health and function. Here, observed subscale totals on the SF36 are used to measure the broader latent dimensions of physical and mental health. We then examine the associations between scores on these dimensions, actual reported illness status, and obesity, also observed directly.

Doll *et al.* report on studies finding an adverse impact of obesity on mental health, in addition to the established (and clinically plausible) impact of obesity on physical health. Other studies, however, have not found an association between emotional disturbance and obesity. Doll *et al.* suggest that some existing studies may not be controlling for confounding of the link between obesity (X) and mental health (F) by illness status (Z). Thus, obese people are more likely to have chronic illness, and once this is allowed for there may be no impact of obesity per se on emotional health. Specifically, Doll *et al.* combine obesity ($X_i = 1$ for yes,$= 0$ for no) and chronic illness ($Z_i = 1$ for yes, $= 0$ for no) into a composite indicator J_i. They find no difference in mental health between those with no illness and no obesity ($X_i = Z_i = 0$) and those obese only without being ill ($X_i = 1$, $Z_i = 0$).

The work of Doll *et al.* illustrates that a set of items may contain information on more than one latent dimension. Thus they use the eight items from the Short Form 36 Health Status Questionnaire to derive mental and physical health factor scores, though they assume these factors are uncorrelated (orthogonal) in line with the SF36 developers' recommendations (Ware *et al.*, 1994). In this connection, we consider six of the eight items of the SF36 recorded for 582 women aged 65–69 in the 1996 Health Survey for England. The selected items have values from 0 to 100, with the low score corresponding to most ill on all items and the high score to most well. Two items were excluded, because their distribution was highly spiked (concentrated on a few values) despite being nominally continuous variables. The density of the other scores is also skewed, with a bunching of values on all the items at 100 (the 'ceiling effect' in health status measurement). One might consider truncated sampling combined with a density allowing for skewness, and below a log-normal model is adopted – which reflects the minimum of the items being non-negative. Other options for sampling might be envisaged, such as a beta density or even a binomial, if we round non-integer values between 0 and 100.

In fact, the binomial provides a simple way of dealing with missing values in the health status outcomes, and is here used exclusively for that purpose – it provides an integer 'success' total V_{ij} between 0 and 100 in relation to a number at risk N_{ij} of 100 (for $i = 1, \ldots 582$ and $j = 1, 6$). It is necessary to impute missing values for the six SF36 items to be able to use the log-normal model (applied to the observed and imputed data combined as if it were all observed). The low rate of item missingness in these data is thus modelled according to

$$V_{ij} \sim \text{Bin}(\pi_{ij}, N_{ij})$$
$$\text{logit}(\pi_{ij}) = \gamma_{0j} + \gamma_{1j} \times V_i$$

where V_i is the total score on all six items, and is a (relatively crude) measure of overall health status. For illustration, a single imputation is used to 'fill out' the health status outcomes, though a full multiple imputation would use several imputations of the missing data, possibly generated under different non-response mechanisms.

We then relate the logged scores $v_1 - v_6$ on the six observed items, V_1 to V_6, (SF36 Physical Health, Pain, General Health, Vitality, Social Function, SF36 Mental Health) to the 2 hypothesised latent constructs, also denoted physical and mental health, with symbols F_1 and F_2. (Note that pain scores are higher for lower reported levels of pain.) Thus items 1–3 are assumed to be linked to the physical health factor, and items 4–6 to the mental health factor. For subject i

$$V_{1i} = \delta_1 + \beta_{11}F_{1i} + e_{1i}$$
$$V_{2i} = \delta_2 + \beta_{12}F_{1i} + e_{2i}$$
$$V_{3i} = \delta_3 + \beta_{13}F_{1i} + e_{3i}$$
$$V_{4i} = \delta_4 + \beta_{24}F_{2i} + e_{4i}$$
$$V_{5i} = \delta_5 + \beta_{25}F_{2i} + e_{5i}$$
$$V_{6i} = \delta_6 + \beta_{26}F_{2i} + e_{6i}$$

where the e_j are independent Normal errors with zero means (with $G(1, 1)$ priors on their precisions τ_j). For identifiability the constraint $\beta_{11} = \beta_{24} = 1$ is adopted (see Chapter 8). The factors are uncorrelated, and allowed to have free variances and means which differ by obesity status X, by illness type Z or by illness-obesity combined in analyses denoted (a), (b) and (c), respectively. Body mass X has categories below 20, 20–25, 25–30 and 30+, and illness Z has three categories (ill, slightly ill, well).

Thus in Model (a),

$$F_{ki} \sim N(\nu_{X_i k}, \phi_k)$$

with means ν_{jk} varying over obesity category j and the $k = 1, 2$ factors. The precisions $1/\phi_k$ are taken to be $G(1, 0.001)$. Since relativities between categories are the main interest, it may be assumed that $\nu_{1k} = 0$, with centred parameters then obtained as $\nu'_{jk} = \nu_{jk} - \bar{\nu}_k$. In Model (b) the means are defined over illness and factor:

$$F_{ki} \sim N(\nu_{Z_i k}, \phi_k)$$

and in Model (c) over eight joint obesity and illness categories, with well and slightly ill combined.

Convergence on all three models is apparent after 1000 iterations in Models (a) and (b) (5000 iterations in Model (c)) in a two chain run of 5000 iterations (10 000 in Model (c)), and applying the over-relaxation option. Starting values in one chain are null values, and for the other are based on trial preliminary runs. Fit is assessed via the predictive loss criterion of Ibrahim *et al.* (2001) and the pseudo-marginal likelihood of Gelfand (1995).

In Model (a) it appears that the obesity group means on the two factors show the worst health for the low BMI group; their physical health score of −0.32 is clearly worse than other levels of BMI and their emotional health is significantly negative (Table 10.6). It may be that low BMI is a proxy for certain types of emotional disturbance. The CPOs suggest potential outliers; for instance subject 447 has a low CPO on item 6, where the score is 0, despite having scores of 100 on social function. This model has pseudo marginal likelihood of −16 980 and predictive loss criterion (with $w = 1$) of 7747×10^3.

Table 10.6 Factor means by BMI and/or illness band

Model (a) Factor Means varying by BMI				
Physical Health Factor	Mean	St. devn.	2.5%	97.5%
Mean by BMI Band 1	−0.32	0.20	−0.68	0.03
Mean by BMI Band 2	0.30	0.07	0.15	0.44
Mean by BMI Band 3	0.13	0.07	−0.01	0.28
Mean by BMI Band 4	−0.11	0.10	−0.31	0.08
Mental Health Factor				
Mean by BMI Band 1	−0.10	0.04	−0.18	−0.01
Mean by BMI Band 2	0.08	0.02	0.04	0.12
Mean by BMI Band 3	0.03	0.02	−0.01	0.07
Mean by BMI Band 4	−0.01	0.03	−0.06	0.04
Factor Loadings				
β_{12}	0.87	0.06	0.77	0.99
β_{13}	0.49	0.04	0.43	0.57
β_{24}	2.03	0.17	1.72	2.41
β_{25}	1.93	0.19	1.58	2.33
Factor Variances				
Var(F1)	0.46	0.05	0.37	0.56
Var(F2)	0.05	0.01	0.04	0.07

(*continues*)

Table 10.6 (*continued*)

Model (b) Factor Means varying by Illness Type				
Physical Health Factor	Mean	St. devn.	2.5%	97.5%
Ill	−0.55	0.04	−0.63	−0.46
Slightly Ill	0.21	0.04	0.13	0.29
Well	0.34	0.04	0.26	0.42
Mental Health Factor				
Ill	−0.14	0.02	−0.17	−0.11
Slightly Ill	0.05	0.01	0.02	0.08
Well	0.09	0.01	0.06	0.11
Factor Loadings				
β_{12}	0.85	0.05	0.75	0.96
β_{13}	0.52	0.04	0.46	0.60
β_{24}	2.12	0.16	1.81	2.47
β_{25}	1.94	0.18	1.59	2.31
Factor Variances				
Var(F1)	0.31	0.04	0.25	0.39
Var(F2)	0.04	0.01	0.03	0.05
Model (c) Factor Means varying by Combined Illness and BMI Type				
Physical Health Factor	Mean	St. devn.	2.5%	97.5%
Ill and Low BMI	−0.48	0.33	−1.28	−0.04
Ill and Avg BMI	−0.04	0.10	−0.25	0.16
Ill and Above Avg BMI	−0.20	0.08	−0.36	−0.04
Ill and High BMI	−0.72	0.15	−0.95	−0.36
Well or Slight Ill, and Low BMI	0.35	0.18	0.01	0.70
Well or Slight Ill, and Avg BMI	0.43	0.08	0.29	0.59
Well or Slight Ill, & above avg BMI	0.36	0.08	0.21	0.51
Well or Slight Ill, and High BMI	0.30	0.09	0.13	0.48
Mental Health Factor				
Ill and Low BMI	−0.16	0.09	−0.35	0.01
Ill and Avg BMI	0.01	0.04	−0.06	0.08
Ill and Above Avg BMI	−0.03	0.03	−0.08	0.02
Ill and High BMI	−0.16	0.04	−0.23	−0.06
Well or Slight Ill, and Low BMI	0.08	0.06	−0.05	0.20
Well or Slight Ill, and Avg BMI	0.10	0.03	0.05	0.15
Well or Slight Ill, & above avg BMI	0.07	0.02	0.02	0.12
Well or Slight Ill, and High BMI	0.10	0.03	0.05	0.16
Factor Loadings				
β_{12}	0.80	0.05	0.70	0.92
β_{13}	0.47	0.03	0.40	0.54

Table 10.6 (*continued*)

β_{24}	2.04	0.14	1.78	2.33
β_{25}	1.94	0.17	1.63	2.28
Factor Variances				
Var(F1)	0.38	0.04	0.30	0.46
Var(F2)	0.04	0.01	0.03	0.05

A more convincing difference in mental health means is apparent for illness categories – analysis (b). The ill subjects have significantly worse mental health, though slightly ill as against well subjects do not differ in their mental health scores. This model has a higher pseudo marginal likelihood but worse predictive criterion than Model (a) – an example of conflict in model assessment criteria.

In a third analysis, analysis (c), the least two serious illness categories are combined and the resulting binary illness index crossed with the obesity categories. In terms of mental health a virtually flat profile in means over BMI can be seen for the less ill categories. Only when combined with more serious illness are both high BMI and low BMI associated with worse emotional health (though the interaction between low BMI and illness is not quite significant at the 5% level in terms of negative mental health). This model has a better predictive criterion than Models (a) or (b), but only improves in terms of pseudo-marginal likelihood over model (a). These findings replicate those of Doll *et al.* quite closely even though the analysis here is confined to one demographic group. Specifically, obesity does not have an independent effect on mental health and its impact is apparent only when combined with more serious illness.

10.3 DOSE-RESPONSE RELATIONS

Dose-response models typically aim to establish the probability of an adverse effect occurring as a function of exposure level (Boucher *et al.*, 1998), or of health gain from treatment inputs. They may derive from experiments involving human or animal subjects, or from observational and cohort studies. Evidence of a monotonic trend in the risk of disease over different exposure levels of a risk factor, lends support to a causal relationship, and provides a basis for public health interventions. A monotonic downward trend in disease risk with increased levels of a putative protective factor may also be relevant (e.g. cancer in relation to vegetable and fruit consumption).

The National Research Council (NRC, 1983) places dose-response assessment as one of series of stages in risk assessment, which includes hazard identification and hazard characterisation. Within the characterisation stage, risk assessment involves establishing a dose-response relationship and the site and mechanism of action. For example, in studies of developmental toxicology, hazard identification includes establishing whether new chemicals impair development before humans are exposed to them, and so the chemicals are evaluated in experimental animals to assess their effect on development. Hasselblad and Jarabek (1996) consider possible benefits of a Bayesian estimation approach in these situations, for example in obtaining the lower confidence point of the 'benchmark dose' that produces a 10% increase in the chance of a developmental abnormality.

Quantification of exposure and of the resulting risk are central in framing and assessing dose-response relations. In some circumstances, in designed trials or cohort studies, exposure to relevant risk factors may be intrinsically graded into a discrete number of levels, while in other instances an originally continuous exposure may be grouped into categories. Incidence rates may not be meaningful unless they are calculated for reasonably sized sub-populations, and if exposure is measured on a continuous scale then this is not possible (Rothman 1986, Chapter 16). One then typically compares estimates of effect for each category in comparison to a reference category (such as the lowest dosage exposure group). These may be obtained by regression methods, or by stratifying over a confounder at each level of the outcome, and forming a pooled estimate with weights based on a common standard for the effect at each level.

The latter method may be illustrated by case-control data from Doll and Hill (1950) on lung cancer in relation to daily smoking, with 60 matched female cases and controls and 649 male cases and controls (Table 10.7). The weights are based on the distribution of the two levels of the confounder (male, female) among the controls (Miettinen, 1972), so that male and female weights are respectively $w_1 = 0.915$ (=649/709) and $w_2 = 0.085$. An empirical estimate of the rate ratio of lung cancer for 1–4 cigarettes as compared to zero cigarettes is obtained by comparing the weighted total of the ratios of exposed cases to exposed controls with the weighted total of the ratios of unexposed cases to unexposed controls. These are 0.915(55/33)+0.085(12/7) and 0.915(2/27)+ 0.085(19/32), respectively, so that the estimated effect (here a rate ratio) is 5.07. For 5–14 and 15+ cigarettes the corresponding estimates are 7.98 and 12.09.

In a Bayes implementation, one would seek to allow for sampling uncertainty (e.g. illustrated by the small number of male cases at the lowest exposure level). Thus one might assume multinomial sampling conditional on the four totals (male controls, male cases, female controls, female cases). With a Dirichlet prior on the four sets of probabilities one obtains posterior mean rate ratio estimates for exposure levels $r = 2, 3, 4$ of 4.71 (s.d. 2.2), 7.24 (3.1) and 10.9 (4.6). The Bayes procedure[6] clarifies the uncertainty in the empirical estimates, and shows they overstate the risk relative to baseline exposure.

A possible drawback in using a categorisation with several (R) levels of an originally continuous risk factor means that confidence (credible) intervals in the resulting effect estimates do not reflect the relationship between possible patterns in the effect estimates and the continuity of the underlying variable. These considerations also apply if the

[6] The program and data (inits may be generated randomly) are:

```
model {# weights according to distribution of confounder among controls
M[1:2] ~ dmulti(w[1:2], TM)
w[1:2] ~ ddirch(alpha[1:2])
# distribution of male cases over exposure levels (level 1 is zero exposure with no cigarettes smoked)
a[1, 1:4] ~ dmulti(pi.case[1, 1:4], N[1])
# distribution of female cases over exposure levels
a[2, 1:4] ~ dmulti(pi.case[2, 1:4], N[2]);
# distribution of male controls over exposure levels
b[1, 1:4] ~ dmulti(pi.control[1, 1:4], M[1])
# distribution of female controls over exposure levels
b[2, 1:4] ~ dmulti(pi.control[2, 1:4], M[2]);
for (i in 1:2) {pi.case[i, 1:4] ~ ddirch(alpha[ ]);     pi.control[i, 1:4] ~ ddirch(alpha[ ])}
# rate (by sex i) among unexposed
for (i in 1:2) {SRR.div[i] <- w[i]*pi.case[i, 1]/pi.control[i, 1]
# rates by exposure j and sex
for (j in 2:4) {SRR.top[j, i] <- w[i]*pi.case[i, j]/pi.control[i, j]}}
for (j in 2:4) {SRR[j] <- sum(SRR.top[j, 1:2])/sum(SRR.div[1:2])}}
Data
list(alpha=c(1,1,1,1), # total male and female cases
N=c(649, 60), # total male and female controls
```

Table 10.7 Lung cancer cases and controls by exposure (Daily Smoking Levels)

	Cigarettes smoked daily	Cases	Controls
Males	0	2	27
	1–4	33	55
	5–14	250	293
	15+	364	274
	All levels	649	649
Females	0	19	32
	1–4	7	12
	5–14	19	10
	15+	15	6
	All levels	60	60

original variable is inherently ordinal, and not only if the categorisation derives from an originally continuous risk factor.

Suppose estimates have been produced by a procedure such as above or by a categorical regression (e.g. a logit model with disease outcome in relation to the risk variable in category form and with a corner constraint). Suppose the estimated odds or incidence ratios attached to category r derive from a population of size N_r, or person-years total T_r, then a weighted regression, linear or non-linear, involving the $R - 1$ effect estimates as dependent variable, is one way to model the trend (Rothman, 1986, p. 337). To model the trend, it is necessary to assign scores to each of the R categories. If the categorisation involved aggregating over originally continuous data, the average or median of a continuous exposure variable within each category might be used. For ordinal data it is common to assign uniformly incremented scores, though modelling of the cut-points on the underlying scale (see Chapter 3) might be a preliminary step.

It may be noted that a guide to the extent of non-linear impacts of the risk factor is provided by comparing fit measures between a categorical regression in that risk factor and a model with a linear trend in the corresponding category scores (Woodward, 1999). If the categorical regression does not improve markedly in fit over the linear scores model, this might be a preliminary to a linear effect model in the original (i.e. uncategorised) form of the exposure variable.

Greenland (1995) advocates spline regression to model dose-response effects. This still involves assigning breaks in the risk variable but allows more effectively for a changing gradient between disease rate and exposure, as the exposure varies over its range. This may be beneficial if there is non-linearity in the dose-response curve (Boucher *et al.*, 1998). There are, however, many other approaches to non-parametric regression which a Bayesian estimation approach may facilitate (Fahrmeir and Lang, 2001); efficient sampling for such models may be obtained using the BayesX software (http://www.stat.uni-muenchen.de/ lang/bayesx/bayesx.html).

```
M=c(649, 60), TM=709,
# male cases (2, 33, etc) then female cases by exposure (0 cigarettes, 1–4 cigarettes daily, 5–14, 15+)
a=structure(.Data=c(2, 33, 250, 364, 19, 7, 19, 15),.Dim=c(2, 4)),
# male controls (27, 55, etc) then female cases by exposure (0 cigarettes, 1–4 cigarettes daily, 5–14, 15+)
b=structure(.Data=c(27, 55, 293, 274, 32, 12, 10, 6),.Dim=c(2, 4))
```

10.3.1 Clustering effects and other methodological issues

Assuming that a categorical breakdown of the exposure variable or dose has been adopted, a binomial or Poisson analysis of responses Y_r from N_r units exposed is a standard one in assessing the strength of the risk as dosages X_r are varied. However, under certain types of sampling, especially if there is nesting of subjects, there may be excess variability relative to the binomial. (Similar considerations apply if there are multinomial outcomes at each dosage, such as say, no increase in morbidity, some increase, or mortality.)

An example is intra-litter correlation in experiments in developmental toxicology. Thus, let there be R levels of toxic exposure X_r and $j = 1, .. M_r$ litters of size N_{rj} at each exposure. Then a dose response analysis of a binary outcome (survival or death) may focuses on the outcomes

$$Y_{rj} \sim \text{Bin}(\pi_{rj}, N_{rj})$$

under a model for the proportions responding such as

$$\pi_{rj} = [1 + \exp(-\beta_0 - \beta_1 X_r)]^{-1} \tag{10.3}$$

Tests of dose effect involving the coefficient β_1 may, however, be affected by the level of intra-litter correlation, $\rho(X_r)$ as dosage changes. Instead suppose the response proportions are

$$\pi_{rj} = \tau_r/[\tau_r + \omega_r] \tag{10.4}$$

where

$$\tau_r = \tau[X_r] = \exp(\gamma_0 + \gamma_1 X_r) \tag{10.5}$$

$$\omega_r = \omega[X_r] = \exp(\delta_0 + \delta_1 X_r) \tag{10.6}$$

and the intra-litter correlation is

$$\rho[X_r] = [1 + \tau_r + \omega_r]^{-1} \tag{10.7}$$

This framework is consistent with a beta-binomial density with possibly overdispersed variance relative to the binomial, namely

$$\text{Var}(Y_{rj}|N_{rj}) = \pi_{rj}(1 - \pi_{rj})/N_{rj}[1 + (N_{rj} - 1)/(1 + \tau_r + \omega_r)]$$

Clustering reflects unobserved heterogeneity (frailty) for groups of related subjects. In dose-response models involving human subjects, especially in trials involving self-administration of treatment drug or placebo, outcomes may also be subject to another generally latent influence, namely subject compliance with treatment(s). Variations in compliance mean that actual exposure varies in an unknown way (Efron and Feldman, 1991; Zeger and Liang, 1991). Allowing for compliance amounts to modelling the true (and latent) exposure to a treatment on the basis of manifest variables (Dunn, 1999). The manifest variables would typically include both measures of compliance (e.g. bio-markers for drug absorption) and observed responses to treatment. This leads to structural equation models in which the latent exposure underlies both the observed outcomes and bio-markers. In clinical trials where there is a placebo group, certain additional assumptions may be reasonably made which facilitate analysis. For example, if treatment allocation is random, one may sometimes assume that the propensity to comply is similar between treated and control groups – even if the placebo group are not exposed to an active drug or treatment. On the other hand, there may be circumstances when compliance in the treatment group is related to prognosis (e.g. subjects with adverse symptoms may be more likely to comply) (Mark and Robins, 1993).

Dose-response relations may involve dependencies on both time and the dose itself, for example in bioassay where the analysis considers both the level of the mortality rate itself and the patterns to times to mortality. In animal experiments, a proportion of the animals may not be subject to experimental mortality at all, and go on to live their natural life. One might consider a mixture analysis of susceptibles who will die prematurely in the experiment and non-susceptibles (Pack and Morgan, 1990). But the analysis may be reduced to modelling the times to mortality of susceptible animals. Often a proportional hazard analysis (see Chapter 9) involving multiplicative functions of dose and time may be applied. Similar issues may occur in human disease onset or mortality, in terms of the joint impact of age at death or onset and a putative exposure such as smoking.

Dose-response relations may also be modified via multi-level approaches in order to take account of subject level covariates – and this might be one way to model concomitant time/age effects (Greenland, 1998b; Wijesinha and Piantadosi, 1995). Consider the logistic model in Equation (10.3). Both β_0 and β_1 may be made functions of subject covariates $v_{rj}(j = 1, .. M_r)$ for instance, via

$$\beta_{0r} = \exp(\lambda_0 v_{rj})$$
$$\beta_{1r} = \exp(\lambda_1 v_{rj}).$$

Example 10.5 Breast cancer and radiation Rothman (1986, Chapter 16) and Greenland (1995) are among those arguing against assessment of dose-response relations using categorical trend analysis. This might, for instance, involve selecting quintile or quartile breaks in a continuous risk factor X and then modelling trends in terms of category specific rates or odds ratios. The implicit assumption is that risk does not change within categories, whereas in fact there is often a trend within the category. One might ensure constancy of risk for all practical purposes by taking a larger number of categories than conventionally used (e.g. 10 or 20), but then risk estimates for each category tend to be unstable – though Bayesian smoothing methods may well attenuate this drawback. Alternatives are spline regression (see Example 10.6) which still involves selecting breaks in the continuous X, but allows for trends within categories, or regression with category scores. For instance, one might take as category scores the average or median exposure X_r within the rth category and then apply the usual regression methods – linear or power models, with a link function as appropriate.

To illustrate the role of scoring categories of a dose variable, we consider data from McGregor *et al.* (1977) on incidence of breast cancer among women from Nagasaki or Hiroshima, and aged 10 in 1945, according to four categories of radiation dose (Table 10.8). Following Rothman (1986) we can assign mid-category scores 0, 5, 55, and 150 and use as regression weights the total of person-years exposed T_r. If available, median or mean exposures within the three upper categories would be preferable options (in fact, a mean weighted for person years of exposure would be required).

The analysis is in terms of rates per 1000 Person-Years (PY). Following Rothman, this outcome may be modelled as a continuous score with Normal errors. Note that a more natural unit is per million person-years but this implies a very small precision and

Table 10.8 Radiation exposure and cancer cases

Exposure*	0	> 0– < 10	10– < 100	100 and over
Breast cancer cases	38	105	48	34
Peson-years exposed	208515	463086	164639	52185
Incidence rate per 1000 person-years	0.182	0.227	0.292	0.652

*> 0 – < 10 means over zero but under 10.

possible prior specification problems. A weighted regression is used with precisions for the rth category modelled as

$$P_r = T_r \tau$$

with τ an overall precision parameter and T_r the person year total for the rth category.

Following Rothman, a linear model is initially assumed. The linear slope of 0.00269 in the 1000 PY scale, with 95% interval from 0.0016 to 0.0038, compares to the confidence interval cited by Rothman (1986, p. 337) of 1.23 to 3.63 in the 1 000 000 PY scale. The predicted mean incidence for an exposure of 100 rads is 0.463 per 1000 women-years, with interval from 0.37 to 0.56.

An alternative model is quadratic in exposure as in Example 10.2, while still retaining the assumption of a continuous outcome. This yields an improved predictive loss criterion of 0.014 (with $w = 1$) as compared to one of 0.023 with the linear model. Note that there is a baseline or background mortality effect here (see Section 10.3.2), namely mortality at zero rads, so that prediction of mortality in the first category is not improved by power models.

Another sampling model might take the four observed counts of cases as Poisson with means $\nu_r = aT_r\mu_r$, with $\nu_r = 0.001T_r\mu_r$ for an analysis in incidence per 1000 person-years. This option in fact suggests the quadratic model is over-parameterised. This model yields improved predictions (*z*.scaled[] in Program 10.5, Model C) of the incidence rates as compared to the linear model in a continuous outcome, namely {0.204, 0.212, 0.308, 0.634} but has less precise predictions of the new data, and so has a worse predictive loss. Arguably, though, this approach more appropriately reflects the uncertainty in the observed data.

Example 10.6 Trend in CHD according to SBP, Framingham study The subject matter for this example is the CHD onset data for the Framingham cohort, but considering various options to improve dose-response modelling over the categorical predictor model in Example 10.2.

The first expedient is a finer subdivision of the SBP variable. This sub-division has six levels: under 140, 140–149, 150–159, 160–169, 170–179, and over 180. To assess trend, scores are then assigned using the observed mean exposure within categories $r = 1, .., 6$ and using as weights the observed numbers N_r in each category. The odds ratios for the six category breakdown are estimated as fixed effects via a logistic regression (Model A1 in Program 10.6). Relative to the SBP under 140 category, they range from 1.5 for SBP 140–149 (with 95% credible interval including values under 1) to 4.04 (Table 10.9).

Subsequently, fitting a weighted linear trend to the six posterior mean odds ratios (Model A2 in Program 10.6) gives a slope of 0.04 for the increase in the log OR for every

Table 10.9 Odds ratios for successive SBP bands

Odds Ratio for SBP	Mean	St. devn.	2.5%	97.5%
140–149	1.50	0.31	0.95	2.17
150–159	1.88	0.40	1.24	2.75
160–169	2.37	0.61	1.38	3.74
170–179	3.50	0.91	1.98	5.53
180+	4.04	0.88	2.64	6.01

unit increase in SBP. For the threshold of 165 mm Hg systolic blood pressure, the odds ratio estimated from this trend model is 2.65 with 95% interval from 2.3 to 2.8. The central estimate is in fact very similar to the estimate from the continuous predictor logistic model (see Example 10.2), but the credible interval is slightly narrower.

One might also compare the fit of a categorical model with a model linear in the scores attached to each category (Woodward, 1999). Model A3 is linear in the averages in each SBP band. Whereas the categorical regression in SBP has a DIC of 1277, this simpler model has a DIC of 1272 and suggests that non-linearity in the effect of SBP (on the logit of the incidence probability) may not be very marked. Note, though, that this analysis is conditioned on the break points chosen for the six levels.

A different perspective on possible non-linearity (Models B and C) is supplied by spline and state-space regression methods. As argued by Greenland (1995), spline regression may avoid some of the problems associated with redefining a continuous exposure into discrete categories. A simple categorical factor approach via logistic regression implies (for example) an implausible jump in risk moving from an SBP of 169 to one of 170. In a spline regression, by contrast, the within category lines have a non-zero slope, with smoothness also generated by using quadratic or cubic spline terms. Thus, a quadratic spline in SBP would involve linear and quadratic terms across the range of SBP, and spline terms operating only when category breaks (knot points) are exceeded.

If C_k denotes the kth break for SBP, then with notation for SBP, Age and sex as in Example 10.2, the spline terms D_k are zero if $X < C_k$ and have value $D_k = (X - C_k)^2$ if the threshold is exceeded. If there are K category breaks (e.g. $K = 5$ in the above analysis for SBP with cut-points C_k being 140, 150, 160, 170 and 180), then up to five extra terms are potentially added in the regression. Spline terms in age are denoted $E_k = (Z_1 - B_k)^2$ if the threshold B_k is exceeded and zero otherwise, so a full quadratic spline model in age with three cut-points (at ages 50, 55 and 60) would be represented as

$$\beta_1 Z_1 + \beta_2 Z_1^2 + \delta_1 E_1 + \delta_2 E_2 + \delta_3 E_3$$

A spline model has drawbacks regarding the number and location of knot points. It may be wise to start off with a minimal number of knot points, and then add additional ones if the data support them. For illustration and to improve identifiability, the quadratic terms in Age and SBP across the range (terms such as $\beta_2 Z_1^2$) are omitted, and single knots (i.e. $K_1 = K_2 = 1$) adopted in SBP and Age, respectively (with Age and SBP also divided by 10 for numeric stability). The knots are at SBP=180 and age 55. The choice of these points was based on earlier analyses with more than one knot in both age and SBP – though there remains a degree of arbitrariness unless the location of the knot point(s) is assumed unknown and itself assigned a prior.

With this framework, the odds ratio comparing subjects with SBP above and below 165 is based on the average SBP for those above and below the threshold. Here we consider males aged 50 in these SBP bands, and for males these averages are 185.3 and 135.4. The fitted logit model for the probability π of CHD onset is then

$$L = \text{logit}(\pi) = \alpha + \beta_1 Z_1 + \delta_1 E_1 + \beta_2 X + \gamma_1 D_1 + \beta_3 Z_2 \tag{10.8}$$

where Z_1 and X are scaled by 0.1. To obtain the relevant odds ratio, it is necessary to compare L for males aged 50 in the above and below 165 SBP groups. So Age is set to 50 in (10.8), and SBP at either 185.3 or 135.4. A profile of age effects (for the 18 ages 45, 46, up to 62) at the average SBP of 148 is obtained as

$$\text{Expit}(\alpha + \beta_1 Z_1 + \delta_1 E_1 + \beta_2 X + \gamma_1 D_1 + \beta_3 Z_2)$$

where $\text{Expit}(L) = \exp(L)/[1 + \exp(L)]$ is the inverse of the logit transform (Greenland, 1998a). With start points provided by the 2.5th and 97.5th percentiles of a trial run, convergence in a two chain run is obtained at around 1500 iterations. The spline term in age is not significant, in line with no clear quadratic effect in age (Table 10.10). There is a significant spline term in SBP, and additional terms might be experimented with. The posterior median for the odds ratio of 3.02 exceeds those obtained earlier. The pseudo marginal likelihood improves over Model A (−628.5 vs. −639).

The profile of CHD rates at various levels of SBP (for a male at average age 52.4) obtained from Model B shows a tailing off in the increased risks of CHD at very high SBP (Figure 10.2). This profile is stored in the vector SBP.eff[] in Program 10.6. Its shape may be an artifact of the data in that there happen to be no CHD cases among a small number (5) of subjects with SBP of 280 and over.

A final form of analysis of nonlinear risk effects (Model C) is relatively straightforward in terms of estimation through Bayesian sampling, and might serve as a basis for selecting knot points in a subsequent spline analysis. This model uses a form of state-space prior, and involves a large number of age and SBP categories, specifically 18 age categories and 43 SBP categories (in intervals of five on the original scale so that the groups are 90–95, 95–100, etc., up to 295–300). If β_j denotes the age parameters and γ_j the SBP group parameters, then a random walk prior is assumed, such that

$$\beta_j \sim N(\beta_{j-1}, \tau_\beta) \quad j = 2, .., 18$$
$$\gamma_j \sim N(\gamma_{j-1}, \tau_\gamma) \quad j = 2, \ldots 43$$

To assist identification, it is assumed that $\beta_1 = \gamma_1 = 0$. A two chain run shows convergence after 2000 iterations and the second half of a run of 5000 iterations produces gradients in the log-odds parameters for age and SBP groups (relative to the baselines,

Table 10.10 Quadratic spline in age and SBP, parameter summary

	Mean	St. devn.	2.5%	Median	97.5%
Odds ratio	3.03	0.47	2.23	3.02	4.09
α	−7.92	0.88	−9.60	−7.91	−6.13
β_1	0.51	0.16	0.21	0.51	0.83
β_2	0.22	0.03	0.16	0.22	0.28
β_3	0.92	0.15	0.63	0.92	1.22
δ_1	−0.021	0.525	−1.039	−0.022	1.019
γ_1	−0.020	0.009	−0.038	−0.020	−0.004

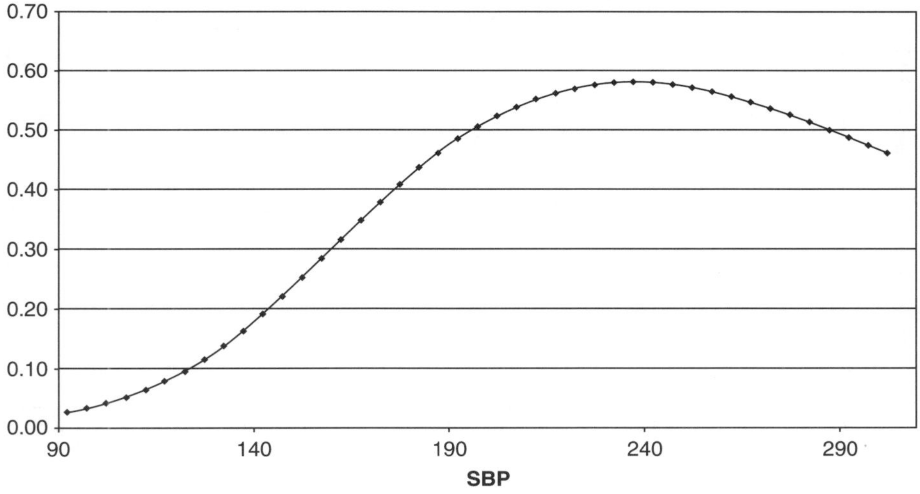

Figure 10.2 SBP profile of CHD rate (at average age)

age 45 and SBP of 90–95) as in Figures 10.3 and 10.4. This model improves over model A in terms of pseudo-marginal likelihood (−632.5 vs. −639), but has a broadly similar predictive loss measure (300.2 for $w = 1$ compared to 300.5 for Model A). Increasing w (e.g. to 1000) does, however, give a larger fit advantage to Model C – increasing w tends to downweight relatively imprecise predictions and emphasize the match between the actual data and the posterior means of the new data.

The results above use $G(1, 0.01)$ priors on $1/\tau_\beta$ and $1/\tau_\gamma$, and results of this type of analysis may well be sensitive to the priors on these parameters – as they govern the degree of smoothing. The posterior means for $\tau_\beta^{0.5}$ and $\tau_\gamma^{0.5}$ are 0.11 and 0.20. If, instead, G(5, 0.05) priors favouring relatively high precision are adopted, the posterior mean of

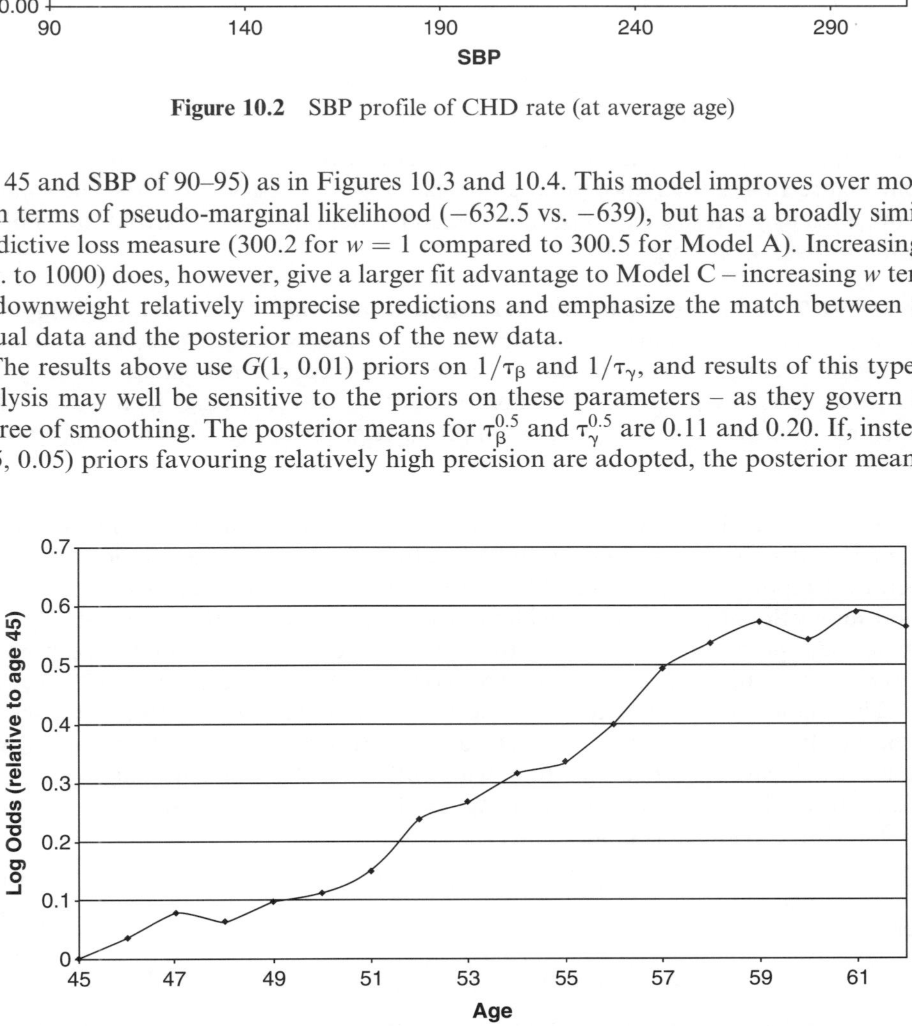

Figure 10.3 Age parameters

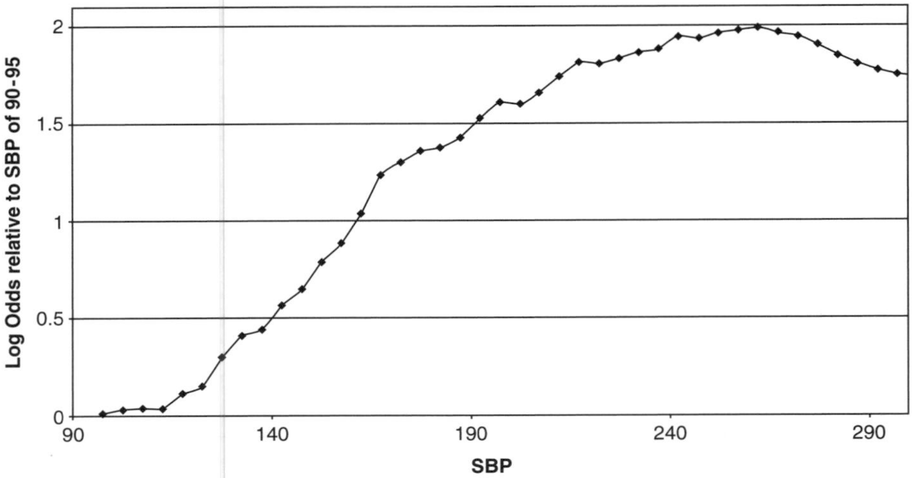

Figure 10.4 SBP parameters

$\tau_{\gamma}^{0.5}$ is reduced to 0.15, but the shape of the log-odds curve is very similar to Figure 10.4. From Figure 10.4, a final knot point at around 250 might be selected and the spline in the upper category restricted to be linear (see Greenland, 1998a) to avoid the implausible fall in risk at very high SBP.

Example 10.7 Cumulative mortality in relation to dose-time The previous example has illustrated how standard model assumptions (e.g. linearity in dose effects) may need to be critically examined. As mentioned above, dose-response modelling one may also need to consider the joint action of dose with other (confounding) factors, as well as departures from standard sampling assumptions (e.g. clustering as a source of binomial overdispersion).

To illustrate an alternative modelling structure, drawing on survival analysis concepts to represent the joint effects of dose and a (confounding) time index, consider an animal experiment reported by Pack and Morgan (1990). This involves deaths over a 13 day period among flour beetles sprayed with the insecticide pyrethrins B, where the focus is not only on endpoint or total mortality by the end of the period, but on cumulative mortality at days 1, 2, 3, . . , up to 13. Four dosage levels (mg/cm^2) were applied, 0.20, 0.32, 0.50 and 0.80. The relation of mortality to dose-time is expected to be differentiated by the sex k of the beetle. It may be noted that the structure of this example could be applied with suitable modifications in Example 10.6 (with SBP as the dose, and age parallel to time).

The probability of death at dose X_r in the jth time interval $\{t_{j-1}, t_j\}$, where $t_0 = 0$, $t_1 = 1, .. t_{13} = 13$ is modelled as

$$\pi_{rj} = C(t_j, X_r) - C(t_{j-1}, X_r) \quad (10.9a)$$

where

$$1/C(t, X) = [1 + \exp\{-\beta_1 - \beta_2 \log(X)\}]\ [1 + \mathrm{t}^{-\beta_3}\beta_4] \quad (10.9b)$$

Thus, separate dose-response and time-response models are present in Equation (10.9b). As an illustration of the potential benefits of parameter transformation, an alternative model expresses β_4 as exp (γ_4). Deaths are additionally classified by sex k, so this model involves parameters $\{\beta_{k1}, \beta_{k2}, \beta_{k3}, \beta_{k4}\}$, though selective equality of parameters might be investigated as one way to model simplification. As noted by Morgan (2000), the function in time in Equation (10.9b) is the cumulative density of a log-logistic, and for susceptible animals is the cumulative density of mortality over time, regardless of dose.

Prior substantive knowledge suggests the parameter β_4 may be larger than the others, and so the multivariate Normal precision matrix allows a wider range in its value. Its prior mean is set at 20 for both males and females. In the alternative parameterisation the γ_{k4} have prior means of 3, and variances 10. In the original parameterisation, convergence with a three chain model is not obtained after 150 000 iterations, with the parameter β_{24} failing to converge. By contrast, in the alternative parameterisation convergence occurs in all parameters by 20 000 iterations and the summary in Table 10.10 is based on the iterations 20 000–30 000.

The deaths data for higher day numbers are rather sparse, and asymptotic considerations applied in classical tests and often in simple Bayes assessments of fit (e.g. via AIC and BIC criteria) will be of some doubt (Pack and Morgan, 1990, p. 752). Therefore, bootstrap principles and/or predictive probability checks might be applied to assess fit. Accordingly we find the likelihoods of actual and replicate data to be closely comparable, and a predictive check to average about 0.34. A similar conclusion is reached by Pack and Morgan (1990), using Monte Carlo testing applied to the binomial deviance. They obtain a deviance of 92 for the model (10.9), and a range of deviances in 100 replicate data sets (sampled from the ML estimate and fitted with the same model) from 76 to 130, with a mean of 90 and variance of 150. They conclude that the fitted model is consistent with the simulated data.

Values of the dose and time profile parameters by sex of flour beetle, as in Table 10.11, are similar to those presented by Morgan (2000, p. 89). Note that the parameter β_{12} has an asymmetric posterior density, and with a flatter prior than the one used in Program 10.7 may become unstable. Morgan (2000) obtains a maximum likelihood estimate for β_{12} of 3.37 with standard error 0.33, and a 95% interval (assumed to be symmetric) from 2.72–4.02. It can be seen that a Bayesian analysis has an advantage in representing parameters that in fact have asymmetric densities.

Example 10.8 Boric exposure Examples 10.6 and 10.7 have considered Bernoulli/binomial sampling without allowing for overdispersion due to clustering or frailty

Table 10.11 Flour beetle mortality, parameter summary

Males	Mean	St. devn.	2.5%	Median	97.5%
β_{11}	4.99	1.36	3.26	4.82	8.86
β_{12}	3.59	0.88	2.44	3.48	6.04
β_{13}	2.71	0.20	2.34	2.71	3.09
γ_{14}	2.70	0.23	2.27	2.69	3.15
Females					
β_{21}	2.83	0.81	1.39	2.75	4.66
β_{22}	2.75	0.59	1.67	2.71	4.04
β_{23}	3.45	0.30	2.88	3.45	4.06
γ_{24}	4.00	0.38	3.30	4.00	4.77

Table 10.12 Boric acid exposure and in-utero damage

	Mean	St. devn.	2.5%	Median	97.5%
δ_0	4.12	0.73	2.77	4.09	5.58
δ_1	−7.67	1.92	−11.18	−7.56	−3.80
γ_0	1.34	0.72	−0.05	1.31	2.80
γ_1	−4.74	2.08	−8.93	−4.66	−0.70
π_1	0.058	0.010	0.040	0.058	0.080
π_2	0.074	0.009	0.058	0.074	0.095
π_3	0.094	0.011	0.075	0.093	0.116
π_4	0.130	0.026	0.083	0.128	0.186
ρ_1	0.019	0.014	0.004	0.016	0.056
ρ_2	0.036	0.020	0.010	0.032	0.085
ρ_3	0.068	0.028	0.028	0.063	0.136
ρ_4	0.231	0.065	0.121	0.227	0.374

effects. One might, for instance, in Example 10.6, introduce subject level random effects to represent unmeasured influences on CHD onset, though this would considerably slow estimation. To illustrate clustering as a source of binomial overdispersion and another source of departure from standard modelling assumptions, consider an experiment to assess in-utero damage to mice following exposure to boric acid (Slaton *et al.*, 2000). This is a constituent of many household products with suspected risks to humans. The exposures in terms of percent of acid in the mice feed were $X_1 = 0$ (control), $X_2 = 0.1$, $X_3 = 0.2$ and $X_4 = 0.4$, and the numbers of litters at each dose level were $M_1 = M_2 = M_3 = 27$ and $M_4 = 26$. The outcome is the number dead Y_{rj} among litters of size N_{rj}, with $j = 1, . . M_r$ and $r = 1, . . , 4$.

A three chain run with over-relaxation shows convergence at round 30 000 iterations in the four parameters $\{\beta_0, \beta_1, \gamma_0, \gamma_1\}$ in Equations (10.5)–(10.6). This gives estimates of these parameters, and of the intra-litter correlations, as in Equation (10.7), based on iterations 30 000–40 000 (Table 10.12).

These show the correlation as most pronounced at the highest dosage d_4, and cursory examination shows the high death rates within certain litters at this dose level. By comparison, the maximum likelihood and their standard errors as estimates obtained by Slaton *et al.* (2000) were

$$\gamma_0 = 1.54(1.10)$$
$$\gamma_1 = -5.19(3.07)$$
$$\delta_0 = 4.33(1.10)$$
$$\delta_1 = -8.26(2.99)$$

The beta-binomial mean proportion is

$$\pi_{rj}(X_r) = \tau_r/[\tau_r + \omega_r] = [(\delta_0 - \gamma_0) + \exp(\delta_1 - \gamma_1)X_r]^{-1}$$

so that the standard logit dose-response model is, from Equation (10.3), equivalent to letting $\gamma_0 - \delta_0 = \beta_0$ and $\gamma_1 - \delta_1 = \beta_1$. An absence of linear effect of dose in the mean proportions is then equivalent to $\delta_1 = \gamma_1$. One might therefore apply (in a sampling framework) a test of whether $\gamma_1 > \delta_1$, with a posterior probability over 0.95 or under

0.05 being broadly equivalent to rejecting $\gamma_1 = \delta_1$. A high probability that $\gamma_1 > \delta_1$ is consistent with an increasing dose-response.

Also, when $\tau(X_r)$ and $\omega(X_r)$ are exponential functions of dose, the intra-litter correlation is

$$\rho[X_r] = [1 + \exp(\gamma_0 + \gamma_1 X_r) + \exp(\delta_0 + \delta_1 X_r)]^{-1}$$

so an absence of dose-response effect in both correlation and mean only occurs if $\gamma_1 = \delta_1 = 0$. If this happens then $\rho[X_r] = \rho$ and the intra-litter correlation is constant. The hypothesis of constant correlation might be assessed by monitoring whether $\rho[X_r] > \rho[X_s]$ over pairs r,s. A rejection of $\rho[X_r] = \rho$ also amounts to rejecting $\rho[X_r] = 0$, which is the condition required for the standard binomial sampling model to be applicable.

Both inequalities $\gamma_1 > \delta_1$ and $\rho[X_r] > \rho[X_s]$ over pairs $r > s$ were confirmed with probability 1 (i.e. no samples were exceptions to these inequalities over iterations 30 000–40 000). Hence, there is both a dose effect and extra-binomial variation. One might assess these features via model fit criteria. Thus, the predictive loss criterion of footnote 3 is 267 under the binomial (with $w = 1$), but considerably lower at 174 under the clustered binomial. The latter model provides a much improved fit of deaths at the highest dose, for instance of $Y_{82} = 12$, with posterior mean $\nu_{82} = 9.8$ under the clustered model against $\nu_{82} = 2.2$ under the binomial.

Slaton *et al.* point out a high correlation between γ_1 and δ_1 and suggest an alternative parameterisation involving the parameters γ_j and $\beta_j = \gamma_j - \delta_j$. Adopting this here (Model C in Program 10.8) shows faster convergence (around 15 000 iterations with over-relaxation) and the correlation between γ_1 and β_1 is only -0.07.

Example 10.9 Compliance and response An illustration of dose-response modelling approaches where compliance is an issue is provided by simulated data from Dunn (1999). Here $n = 1000$ subjects are randomly divided in a 50:50 ratio between control and treated groups, with the outcome Y being a function of a latent true exposure F. The treated group has a higher coefficient on the true exposure than the control group in this simulation. Two fallible indicators C_1, C_2, of the compliance (e.g. bio-markers for active or placebo drugs) latent exposure are available. The first of these provides a scale for the unknown exposure F that is taken be centred at μ. The second has coefficient $\gamma = 1$ on F. The observed outcome Y is also related to the latent exposure, with the impact of F on Y allowed to differ according to assignment to treatment or otherwise.

Specifically, the simulated data is generated according to

$$\begin{aligned} Y_i &= \alpha + \beta_{G_i} F_i + \eta_i \\ C_{1i} &= F_i + e_{1i} \\ C_{2i} &= \gamma F_i + e_{2i} \\ F_i &= \mu + u_i \end{aligned} \qquad (10.10)$$

where $\mu = 70$, $\alpha = 50$, $\beta_2 = 4$ for treated subjects ($G_i = 2$) and $\beta_1 = 1$ for control subjects ($G_i = 1$). The variances of the normally distributed errors η_i, e_{1i}, e_{2i} and u_i are, respectively, $\tau_\eta = 225$, $\tau_1 = 144$, $\tau_2 = 225$ and $\tau_u = 225$ and their means are zero.

The model is re-estimated knowing only Y, C_1, C_2 and G. Both in setting priors on precisions and intercepts, and in sampling inverse likelihoods (to estimate CPOs), it is preferable to scale the data by dividing Y, C_1 and C_2 by 100. Otherwise, the variances are large and their estimation sensitive to prior assumptions. Initially, the same dose-response model as in Equation (10.10) is assumed, except that in contrast to the generating model, differential variances $\tau_{\eta 1}$ and $\tau_{\eta 2}$ of the errors η_i are adopted, according to patient treatment group, so that

$$Y_i = \alpha + \beta_{G_i} F_i + \eta_{i,\, G_i}$$

$G(1, 0.001)$ priors on $\phi_{\eta j} = 1/\tau_{\eta j}$, $\phi_j = 1/\tau_j$ and $\phi_u = 1/\tau_u$ are adopted. A three chain run with over-relaxation shows convergence at around 500 iterations, and the summary in Table 10.13 is based on iterations 1000–5000. The original parameters are reasonably accurately reproduced, when account is taken of the scaling.

In a second model, the C_j are taken as centred at ν_j and F to have zero mean and variance 1. This approach might be one among several options adopted in ignorance of the generating model. Because the variance of F is known, slopes of C_1 and C_2 on F may be estimated. Additionally, the intercept of Y is taken as differentiated by treatment. Thus

$$Y_i = \alpha_{G_i} + \beta_{G_i} F_i + \eta_{i,\, G_i}$$
$$C_{1i} = \nu_1 + \gamma_1 F_i + e_{1i}$$
$$C_{2i} = \nu_2 + \gamma_2 F_i + e_{2i}$$

This model has a considerably improved predictive loss criterion (0.32 vs. 0.54 for the first model, when $w = 1$), but slower convergence. The treatment differential on the effect of the latent exposure is still apparent, with β_2 over four times that of β_1.

Table 10.13 Compliance and latent exposure: parameter summary

	Mean	St. devn.	2.5%	Median	97.5%
1st model					
α	0.55	0.05	0.44	0.55	0.64
β_1	0.94	0.07	0.80	0.94	1.08
β_2	3.94	0.07	3.79	3.93	4.09
γ	0.995	0.009	0.978	0.995	1.012
μ	0.701	0.006	0.689	0.701	0.713
τ_1	0.015	0.001	0.013	0.015	0.017
τ_2	0.024	0.001	0.021	0.023	0.026
$\tau_{\eta 1}$	0.021	0.002	0.017	0.021	0.025
$\tau_{\eta 2}$	0.025	0.013	0.002	0.025	0.049
τ_u	0.022	0.001	0.019	0.022	0.025
2nd model					
α_1	1.21	0.02	1.19	1.21	1.22
α_2	3.30	0.04	3.26	3.30	3.35
β_1	0.134	0.015	0.116	0.134	0.152
β_2	0.603	0.027	0.559	0.604	0.644
γ_1	0.139	0.011	0.128	0.139	0.150
γ_2	0.155	0.011	0.142	0.154	0.168
ν_1	0.702	0.012	0.691	0.703	0.714
ν_2	0.696	0.013	0.682	0.696	0.710

10.3.2 Background mortality

A standard modelling assumption in controlled trials and laboratory experiments is that the responses of test subjects are due exclusively to the applied stimulus. In dose-response models this assumption means that the control probability (i.e. for subjects with no dose) of response is zero. However, multinomial or binary responses (e.g. for type of defect or for mortality) for such trials, where one or more intervention or treatment has been performed, may be subject to a background mortality effect. Such nonzero control response may need to be allowed for in dose-response studies.

At the simplest, consider a binary mortality or other response Y with the probability that $Y = 1$ modified to take account both of the chance of a background event and the chance of a dose-induced event. Thus, let α and $P(X)$ denote the respective chances of a background and treatment induced response, with corresponding random variables

$$Y_B \sim \text{Bern}(\alpha)$$

and

$$Y_M \sim \text{Bern}(P(X)\}$$

Then the overall probability that $Y = 1$ given a dosage X is a binary mixture

$$\Pr(Y = 1|X) = \alpha + (1 - \alpha)P(X)$$

This model has the effect of concentrating the dose-response curve modelled by $P(X)$ from (0, 1) into the range (α, 1). If Y is polytomous without ordering, or ordinal, and contains $S + 1$ categories, then Y_B and Y_M are multinomial, with

$$\Pr(Y_B = S) = \alpha_s(s = 0, 1, .. S)$$

where $\sum_0^S \alpha_s = 1$ and

$$\begin{aligned} \Pr(Y_M \geq s|X) &= H(\kappa_s + \beta X) \quad s = 1, .. S \\ &= 1 \qquad\qquad\qquad \text{s} = 0 \end{aligned} \tag{10.11}$$

where H is an inverse link and the dose effect is linear for the assumed link. This defines a proportional odds model for Y_M with cut points κ_s that are monotonically declining.

Example 10.10 Arbovirus injection This example involves the ordinal response data on deformity or mortality in chick embryos as a result of arbovirus injection (Xie and Simpson, 1999). Two viruses, Tinaroo and Facey's Paddock, were investigated, with 72 and 75 embryos, respectively, receiving these viruses. A further 18 embryos received no virus.

There are $S + 1 = 3$ outcomes: survival without deformity, survival with deformity, and death. There is one death (i.e. background mortality) among the controls. For the $g = 1, 2$ treatments (Tinaroo, Facey's Paddock), the probabilities of the responses may be expressed

$$\begin{aligned} \Pr(Y_M \geq s|X) &= H(\kappa_{gs} + \beta_g X) \quad s = 1, .. S \\ &= 1 \qquad s = 0 \end{aligned}$$

For the Tinaroo group, there were four dosage levels (in inoculum titre in terms of PFU/egg), namely 3, 20, 2400 and 88 000. For the Facey's Paddock group the doses were 3,

18, 30 and 90. These doses are subject to a log10 transform. We adopt the previous paramaterisation of the proportional odds model (see Chapter 3), with appropriate constraints on the κ_{gs}.

Follwing Xie and Simpson, the baseline mortality effect for the control group is taken to be binary rather than multinomial (excluding the option of survival without treatment induced deformity), and so only one parameter α is required. Note also that, to use the predictive loss criterion (footnote 3), it is preferable to use multinomial sampling where the data are dummy indicators $y_{ij} = 1$ if $Y_i = j$ and $y_{ik} = 0, k \neq j$ – as opposed to direct categorical sampling using Y as the data and the dcat( ) function. The two are equivalent ways of modelling the data.

N(0, 100) priors are adopted on the κ_{gs}, N(0, 10) priors on the β_g parameters[7], and a $B(1, 1)$ prior on α. A three chain run then shows convergence at 2500 iterations and the summary (Table 10.14) is based on iterations 2500–10 000. The mean posterior probability of background embryo mortality (from natural causes) stands at 0.13 compared to an estimate of 0.11 obtained by Xie and Simpson. The mortality rate is higher in the Tinaroo group as dosage increases and there are few surviving with deformity, whereas the Facey's Paddock group have relatively more embryos surviving, albeit with deformity. Accordingly, the β dose effect parameter is stronger for Tinaroo embryos and there is only a small difference in cut points κ_{21} and κ_{22} comparing the combined response of survival with deformity and death and the death response considered singly.

A second model introduces nonlinear effects in dose (adding a term in $1/X$), so that

$$\begin{aligned} \Pr(Y_M \geq s|d) &= H(\kappa_{gs} + \beta_g X + \gamma_g/X) \quad & s = 1, .. S \\ &= 1 & s = 0 \end{aligned}$$

Note that as X increases $1/X$ declines so a negative effect on $1/X$ is equivalent to X increasing risk. An N(0, 10) prior on the γ parameters is adopted. The analysis produces a negative effect on $1/X$ only for the first group, with mean (and standard deviations) on β_1 and γ_1 being 2.3 (0.8) and −2.3 (1.1), respectively. For the second group, the coefficient on $1/X$ is positive. This model produces no improvement in the predictive loss criterion (with $w = 1$) over the linear dose model, namely 85.4 as against 84.2, although both β_g and γ_g coefficients are significant.

Table 10.14 Arbovirus injection and chick embryo damage: parameter summary

	Mean	St. devn.	'2.5%	Median	'97.5%
α	0.13	0.05	0.05	0.13	0.23
β_1	2.27	0.66	1.28	2.14	3.88
β_2	3.48	1.19	1.69	3.26	6.04
κ_{11}	−6.21	1.92	−10.98	−5.85	−3.35
κ_{12}	−11.04	3.38	−19.31	−10.35	−6.11
κ_{21}	−5.08	1.83	−9.02	−4.74	−2.32
κ_{22}	−5.37	1.85	−9.35	−5.03	−2.56

[7] A more diffuse N(0, 100) prior on the β_g led to convergence problems. Moderately informative priors may be justified in terms of likely bounds on relative mortality between treatments or between treatment and baseline mortality.

10.4 META-ANALYSIS: ESTABLISHING CONSISTENT ASSOCIATIONS

Meta-analysis refers to methods for combining the results of independent studies into effectiveness of medical treatments, or into the impact of environmental or other health risks, and so form a prior evidence base for planning new studies or interventions (Hedges and Olkin, 1985). While randomised trials are the gold standard evidence for meta-analysis (e.g. on medical treatment effectiveness), meta-analysis may use other study designs, such as cohort and case control studies. The typical Bayesian approach aims at estimating underlying 'true' treatment or study effects, defined by random deviations from the average effect. If observations on each study include an outcome rate for a control and treatment group, then one may also model the average risk level or frailty of subjects in each trial.

Several possible outcomes may be considered as summarising study or trial results: examples are differences in proportions responding between treatment and control groups, the ratio of odds responding, or the ratio of proportions responding. With regard to previous sections, one might also pool the slopes of dose-response curves (DuMouchel and Harris, 1983) or odds ratios after allowing for confounders. DuMouchel (1996) presents an example of combining odds ratios from different studies, where studies differ in whether their odds ratio estimate controls for confounders. Whether or not the ith study did control for a given confounder defines a set of binary covariates that influence the estimates of underlying study effects in the meta-analysis over studies.

Bayesian methods may have advantages in handling issues which occur in meta-analysis, such as choice between fixed-effects vs. random-effects models, robust inference methods for assessing small studies or non Gaussian effects, and differences in underlying average patient risk between trials. Further questions which a Bayesian method may be relevant include adjusting a meta-analysis for publication bias, meta-analysis of multiple treatment studies, and inclusion of covariates (Smith *et al.*, 1995; Carlin, 1992; DuMouchel, 1990; Prevost *et al.*, 2000). Thus, whereas most medical meta-analyses involve two treatment groups (or treatment vs. control), Bayesian techniques can be used to compare either of the two main treatments with a common third treatment to improve estimation of the main treatment comparison (e.g. Hasselblad, 1998; Higgins and Whitehead, 1996). Publication bias occurs if studies or trials for meta-analysis are based solely on a published literature review, so that there may be a bias towards studies that fit existing knowledge, or are statistically significant.

The simplest meta-analysis model is when effect measures y_i, such as odds ratios for mortality or differences in survival rates for new as against old treatment, are available for a set of studies, together with estimated standard error s_i of the effect measure. For example, consider the log odds ratio as an effect measure. If deaths a_i and b_i are observed among sample numbers r_i and t_i under new and old treatments, then the odds ratio is

$$\{a_i/(r_i - a_i)\}/\{b_i/(t_i - b_i)\}$$

The log of this ratio may (for moderate sample sizes) be taken as approximately normal, with variance given by

$$s_i^2 = 1/a_i + 1/(r_i - a_i) + 1/b_i + 1/(t_i - b_i) \qquad (10.12)$$

Under a fixed effects model, data of this form may be modelled as

$$y_i \sim \mathrm{N}(\mu, s_i^2)$$

where μ might be estimated by a weighted average of the y_i and the inverses of the s_i^2 used as weights (since they are approximate precisions). Under a random effects model by contrast, the results of different trials are often still taken as approximately Normal, but the underlying mean may differ between trials, so that

$$y_i \sim \mathrm{N}(\nu_i, s_i^2) \tag{10.13}$$

where $\nu_i = \mu + \delta_i$ and the deviations δ_i from the overall mean μ, representing random variability between studies, have their own density. For example, if the y_i are empirical log odds, then μ is the underlying population log odds and the deviations around it might have prior density

$$\delta_i \sim \mathrm{N}(0, \tau^2)$$

The rationale for random effects approaches is that at least some of the variability in effects between studies is due to differences in study design, different measurement of exposures, or differences in the quality of the study (e.g. rates of attrition). These mean that the observed effects, or smoothed versions of them are randomly distributed around an underlying population mean.

We may make the underlying trial means functions of covariates such as design features, so that

$$\nu_i \sim \mathrm{N}(\mu_i, \tau^2)$$
$$\mu_i = \beta z_i$$

For instance, as mentioned above, DuMouchel (1996) considers odds ratios y_i from nine studies on the effects of indoor air pollution on child respiratory illness. These odds ratios were derived within each study from logistic regressions, either relating illness to thresholds of measured NO_2 concentration in the home, or relating illness to surrogates for high NO_2 (such as a gas stove). Thus, four of the nine studies actually measured NO_2 in the home as the basis for the odds ratio. In deriving the odds ratio, two of the nine studies adjusted for parental smoking, and five of the nine for the child's gender. Thus, in the subsequent meta-analysis, we can derive dummy indicators z_i for each study which describe the 'regression design', or confounders allowed for, in deriving the odds ratio.

10.4.1 Priors for study variability

Deriving an appropriate prior for the smoothing variance τ^2 may be problematic as flat priors may oversmooth – that is, the true means ν_i are smoothed towards the global average to such an extent that the model approximates the fixed effects model. While not truly Bayesian, there are arguments to consider the actual variability in study effects as the basis for a sensible prior. Thus DuMouchel (1996, p. 109, Equation (5)) proposes a Pareto or log-logistic density

$$\pi(\tau) = s_0/(s_0 + \tau)^2 \tag{10.14}$$

where $s_0^2 = n/\Sigma s_i^{-2}$ is the harmonic mean of the empirical estimates of variance in the n studies. This prior is proper but highly dispersed, since though the median of the density is s_0, its mean is infinity. The (1, 25, 75, 99) percentiles of τ are $s_0/99$, $s_0/3$, $3s_0$, $99s_0$. In BUGS the Pareto for a variable T is parameterised as

$$T \sim \alpha c^{\alpha} T^{-(\alpha+1)}$$

and to obtain the DuMouchel form involves setting $\alpha = 1$, $c = s_0$, and then $\tau = T - s_0$.

Other options focus on the ratio $B = \tau^2/(\tau^2 + s_0^2)$ with a uniform prior one possibility. The smaller is τ^2 (and hence B), the closer the model approximates complete shrinkage to a common effect as in the classical fixed effects model. (This is obtained when $\tau^2 = 0$.) Larger values of B (e.g. 0.8 or 0.9) might correspond to 'sceptical priors' in situations where exchangeability between studies, and hence the rationale for pooling under a meta-analysis, is in doubt. One might also set a prior directly on τ^2 directly without reference to the observed s_i^2. For instance, one may take the prior $\tau^{-2} \sim \chi^2(\nu)/\nu$, with the degrees of freedom parameter at values $\nu = 1$, 2 or 3 being typical choices. For a meta-analysis involving a relatively large number of studies, or studies with precise effects based on large samples, a vague prior might be appropriate, e.g.

$$\tau^{-2} \sim G(0.001, 0.001)$$

as in Smith *et al.*

Smith *et al.* (1995, p. 2689) describe how a particular view of likely variation in an outcome, say odds ratios, might translate into a prior for τ^2. If a ten-fold variation in odds ratios between studies is plausible, then the ratio of the 97.5th and 2.5th percentile of the odds ratios is 10, and the gap between the 97.5th and 2.5th percentiles for δ_i (underlying log odds) is then 2.3. The prior mean for τ^2 is then 0.34, namely $(0.5 \times 2.3/1.96)^2$, and the prior mean for $1/\tau^2$ is about 3. If a 20-fold variation in odds ratios is viewed as the upper possible variation in study results, then this is taken to define the 97.5th percentile of τ^2 itself, namely $0.58 = (0.5 \times 3/1.96)^2$. From this the expected variability in τ^2 or $1/\tau^2$ is obtained[8].

Example 10.11 Survival after CABG An example of the above random effects meta-analysis framework involves data from seven studies (Yusuf *et al.*, 1994) comparing Coronary Artery Bypass Graft (CABG) and conventional medical therapy in terms of follow-up mortality within five years. Patients are classified not only by study, but by a three-fold risk classification (low, middle, high). So potentially there are 21 categories for which mortality odds ratios can be derived; in practice, only three studies included significant numbers of low risk patients, and an aggregate was formed of the remaining studies.

Verdinelli *et al.* (1996) present odds ratios of mortality, and their confidence intervals for low risk patients in the four studies (where one is an aggregate of separate studies), namely[9] 2.92 (1.01, 8.45), 0.56 (0.21, 1.50), 1.64 (0.52, 5.14) and 0.54 (0.04, 7.09). The empirical log odds y_i and their associated s_i are then obtained by transforming the

[8] The upper percentile of τ^2 defines a 2.5th percentile for $1/\tau^2$ of 1/0.58=1.72. A $G(15, 5)$ prior for $1/\tau^2$ has 2.5th percentile of 1.68 and mean 3, and might be taken as a prior for $1/\tau^2$. If a hundredfold variation in odds ratios is viewed as the upper possible variation in study outcomes, a G(3, 1) prior is obtained similarly.

[9] The standard deviations of the odds ratios would usually have been derived by considering numbers (a_i, b_i, s_i, t_i) as in Equation (10.12) and exponentiating the 95% limits of the log-odds ratio. The original numbers are not, however, presented by Verdinelli *et al.* (1996).

above data on odds ratios and confidence limits. With a random effects model, a flat prior on the parameter τ^2 may lead to over-smoothing. To establish the appropriate degree of smoothing towards the overall effect μ, we first adopt the (weakly) data based prior (10.14) previously suggested by DuMouchel (1996).

A three-chain run for the low risk patient data shows early convergence. From iterations 5000–100 000 the estimated of the overall odds ratio in fact shows no clear benefit from CABG among the low risk patients (Table 10.15). The chance that the overall true effect is beneficial (i.e. that the pooled odds ratio μ exceeds 1) is 0.699. The deviance information criterion for this model, which partly measures the appropriateness of the prior assumptions, is 11.35.

A second analysis adopts a uniform prior on $\tau^2/(\tau^2 + s_0^2)$. This leads to a posterior mean for the overall odds ratio of 1.40 with 95% credible interval {0.25, 3.24}. The DIC is slightly improved to 10.9. Finally, as in DuMouchel (1990), the prior $\tau^{-2} \sim \chi^2(\nu)/\nu$ is taken with $\nu = 3$. This amounts to a 95% chance that τ^2 is between 0.32 and 13.3. This yields a lower probability that the overall odds ratio exceeds 1, namely 0.6, but the posterior mean for the overall effect is slightly higher at 1.52, with 95% interval {0.29, 4.74}. The DIC is again 10.9. The posterior median of τ^2 is 0.73.

Note that a relatively vague prior such as $\tau^{-2} \sim G(0.001, 0.001)$ or $\tau^{-2} \sim G(1, 0.001)$ leads to an overall odds ratio estimate with very large variance and essentially no pooling of strength: under the latter, the posterior 95% intervals for the odds ratios {0.9, 7.57}, {0.23, 1.65}, {0.52, 4.77} and {0.07, 6.06} are very similar to the original data. The DIC under this option worsens to 11.6.

Example 10.12 Thrombolytic agents after myocardial infarction An illustration of a meta-analysis where pooling of information is modified to take account of covariates is provided by mortality data from nine large placebo-control studies of thrombolytic agents after myocardial infarction, carried out between 1986 and 1993 (Schmid and Brown, 2000). Such covariates (if they have a clear effect on the trial outcome) mean the simple exchangeable model is no longer appropriate. In the thrombolytic studies, mortality rates were assessed at various times in hours t_i between chest pain onset and treatment, ranging from around 45 minutes to 18 hours.

The treatment effects y_i are provided as percent risk reductions,

$$100 - 100m_{1i}/m_{2i}$$

where m_{1i} is the treatment death rate and m_{2i} is the control death rate (Table 10.16). Hence, positive values of y show benefit for thrombolytics. Schmid and Brown provide

Table 10.15 CABG effects in lowest risk patient group

Study	Mean	St. devn.	2.5%	Median	97.5%
1. VA	1.98	1.16	0.75	1.67	5.07
2. EU	0.99	0.45	0.32	0.92	2.05
3. CASS	1.53	0.77	0.59	1.36	3.50
4. OTHERS	1.34	1.06	0.23	1.15	3.70
Meta Analysis (Overall Effect)	1.41	1.23	0.45	1.25	3.20

confidence intervals for these effect measures, so that sampling variances s_i^2 can be derived. In fact, they assume a model with constant observation variance,

$$\begin{aligned} y_i &\sim \mathrm{N}(\nu_i, \sigma^2) \\ \nu_i &\sim \mathrm{N}(\mu_i, \tau^2) \end{aligned} \tag{10.15}$$

Alternate models for μ_i are a constant regression ignoring the time covariate, $\mu_i = \gamma_0$ and a regression model

$$\mu_i = \gamma_0 + \gamma_1 t_i$$

where t_i is as in the third column of Table 10.16. We also consider a constant regression model $\mu_i = \gamma_0$, in which the sampling variances are taken equal to their observed values, so that

$$y_i \sim \mathrm{N}(\nu_i, s_i^2) \tag{10.16}$$

Consider first the model (10.15) with a common sampling variance. Here the observations y_i on the underlying μ_i are distorted by measurement error, and one may assume that $\tau^2 < \sigma^2$, or equivalently $1/\tau^2 > 1/\sigma^2$. This is achieved introducing a parameter $\pi \sim B(1, 1)$, and then dividing $1/\sigma^2$ by π, where $1/\sigma^2 \sim \mathrm{G}(1, 0.001)$. Under the empirical sampling variance model in Equation (10.16), a DuMouchel prior for τ is taken.

With a constant only regression, both models show early convergence in two chain runs, and inference is based on iterations 1000–20 000. The first option shows τ^2 around 85, the second has τ^2 around 45. The underlying treatment effects accordingly vary more widely under Equation (10.15), namely between 7.4 and 30.6, whereas under Equation (10.16) they are between 17.6 and 28.5. The mean percent risk reduction γ_0 is estimated as 19.5 under Equation (10.15) and 21.3 under Equation (10.16). The DIC is lower under model (10.16), namely 206.9 as against 209.7.

Introducing the time covariate, together with the common sampling variance assumption in Equation (10.15), shows that longer time gaps between onset and treatment reduce the mortality improvement. The mean for γ_1 is -1.2 with 95% interval $\{-2.3, -0.2\}$. Pooling towards the central effect is considerably lessened, and trial arms with longer time gaps (studies subsequent to ISIS-2 at 9.5 hours in Table 10.15) do not show a conclusive mortality benefit. Specifically, the 95% credible intervals for the corresponding ν_i include negative values, though the means are still positive. Adopting an alternative prior for the study effects

$$\nu_i \sim t_5(\mu_i, \tau^2)$$

slightly enhances the contrasts in posterior means ν_i, but still only four studies show no mortality reduction.

Example 10.13 Aspirin use: predictive cross-validation for meta analysis DuMouchel (1996) considers predictive cross-validation of meta-analysis to assess model adequacy (e.g. to test standard assumptions like Normal random effects). His meta-analysis examples include one involving six studies into aspirin use after heart attack, with the study effects y_i being differences in percent mortality between aspirin and placebo groups. The data (in the first two columns of Table 10.17) include standard errors s_i of the differences, and the pooled random effects model takes the precision of the ith study to be s_i^{-2}. A Pareto-type prior for τ, as in Equation (10.14), is based on the harmonic mean of the s_i^2. The model is then

Table 10.16 Studies of thrombolytics after myocardial infarction

			Treatment group		Control group		Death rate		% fall in death rate after treatment		
Study name	Year	Time (hours)	Deaths	Total	Deaths	Total	Treated	Control	Mean	LCL	UCL
GISSI-1	1986	0.75	52	635	99	642	0.082	0.154	46.9	27	61
ISIS-2	1988	1	29	357	48	357	0.081	0.134	39.6	6	61
USIM	1991	1.2	45	596	42	538	0.076	0.078	3.3	−45	35
ISAM	1986	1.8	25	477	30	463	0.052	0.065	19.1	−37	51
ISIS-2	1988	2	72	951	111	957	0.076	0.116	34.7	13	51
GISSI-1	1986	2	226	2381	270	2436	0.095	0.111	14.4	−1	28
ASSET	1988	2.1	81	992	107	979	0.082	0.109	25.3	2	43
AIMS	1988	2.7	18	334	30	326	0.054	0.092	41.4	−3	67
USIM	1991	3	48	532	47	535	0.090	0.088	−2.7	−51	30
ISIS-2	1988	3	106	1243	152	1243	0.085	0.122	30.3	12	45
EMERAS	1993	3.2	51	336	56	327	0.152	0.171	11.4	−25	37
ISIS-2	1988	4	100	1178	147	1181	0.085	0.124	31.8	13	46
ASSET	1988	4.1	99	1504	129	1488	0.066	0.087	24.1	2	41
GISSI-1	1986	4.5	217	1849	254	1800	0.117	0.141	16.8	2	30
ISAM	1986	4.5	25	365	31	405	0.068	0.077	10.5	−49	46
AIMS	1988	5	14	168	31	176	0.083	0.176	52.7	14	74
ISIS-2	1988	5.5	164	1621	190	1622	0.101	0.117	13.6	−5	29
GISSI-1	1986	7.5	87	693	93	659	0.126	0.141	11.0	−17	32
LATE	1993	9	93	1047	123	1028	0.089	0.120	25.8	4	42
EMERAS	1993	9.5	133	1046	152	1034	0.127	0.147	13.5	−7	30
ISIS-2	1988	9.5	214	2018	249	2008	0.106	0.124	14.5	−2	28
GISSI-1	1986	10.5	46	292	41	302	0.158	0.136	−16.0	−71	21
LATE	1993	18	154	1776	168	1835	0.087	0.092	5.3	−17	23
ISIS-2	1988	18.5	106	1224	132	1227	0.087	0.108	19.5	−3	37
EMERAS	1993	18.5	114	875	119	916	0.130	0.130	−0.3	−27	21

$$y_i \sim N(\nu_i, s_i^2)$$
$$\nu_i \sim N(\mu, \tau^2)$$

It can be seen from Table 10.17 that one study (AMIS) is somewhat out of line with the others, and its inclusion may be doubted on grounds of comparability or exchangeability; this study may also cast into doubt a standard Normal density random effects meta-analysis.

Such a standard meta-analysis using all six studies shows some degree of posterior uncertainty in τ. A two chain run to 10 000 iterations, with convergence by 1000 iterations, shows a 95% interval for τ ranging from 0.06–3.5. In five of the six studies the posterior standard deviation of ν_i is smaller than s_i, but for the doubtful AMIS study this is not true – compare $\text{sd}(\nu_6) = 0.95$ with the observed $s_6 = 0.90$ in Table 10.17. There is greater uncertainty about the true AMIS parameter than if it had not been pooled with the other studies. Despite the impact of this study the overall effect μ has posterior density concentrated on positive values, with the probability $\Pr(\mu > 0)$ being 0.944.

A cross-validatory approach to model assessment then involves study by study exclusion and considering criteria such as

$$U_k = \Pr(y_k^* < y_k | y[-k]) = \int \Pr(y_k^* < y_k | \theta, y[-k]) \pi(\theta | y[-k]) d\theta$$

where $y[-k]$ is the data set omitting study k, namely $\{y_1, y_2, .. y_{k-1}, y_{k+1}, .. y_n\}$. The quantity y_k^* is the sampled value for the kth study when the estimation of the model parameters $\theta = (\nu, \tau)$ is based on all studies but the kth. Thus new values for the first study percent mortality difference are sampled when the likelihood for the cross-validation excludes that study and is based on all the other studies 2, 3, .. n. If the model assumptions are adequate, then the U_k will be uniform over the interval (0, 1), and the quantities

Table 10.17 Aspirin use: cross-validation assessment of meta-analysis

	Observed Data		Cross validation			
	y_i	s_I	Predictive mean	Predictive SD	Predictive median	Predictive probability
UK1	2.77	1.65	1.03	1.97	0.96	0.782
CDPA	2.5	1.31	0.97	1.94	0.86	0.779
GAMS	1.84	2.34	1.24	1.96	1.17	0.590
UK2	2.56	1.67	1.09	1.96	1.02	0.740
PARIS	2.31	1.98	1.15	1.99	1.09	0.677
AMIS	−1.15	0.90	2.29	1.24	2.30	0.014

Standard meta-analysis

		Mean	St. devn.	2.5%	Median	97.5%
UK1	ν_1	1.76	1.17	−0.21	1.67	4.34
CDPA	ν_2	1.76	1.03	−0.07	1.68	3.97
GAMS	ν_3	1.40	1.29	−1.04	1.32	4.18
UK2	ν_4	1.69	1.17	−0.33	1.60	4.23
PARIS	ν_1	1.56	1.21	−0.59	1.46	4.24
AMIS	ν_2	−0.11	0.95	−2.02	−0.08	1.65

$$Z_k = \Phi^{-1}(U_k)$$

will be standard normal. A corresponding overall measure of adequacy is the Bonferroni statistic

$$Q = N \min_k (1 - |2U_k - 1|)$$

which is an upper limit to the probability that the most extreme U_k could be as large as was actually observed.

One may also sample the predicted true study mean ν_k^* from the posterior density $N(\mu_{[-k]}, \tau^2_{[-k]})$ based on excluding the kth study. This estimates the true mean for study k, had it not formed one of the pooled studies.

Applying the cross-validation procedure (Program 10.13) shows that the U_k for the AMIS study is in the lowest 2% tail of its predictive distribution (with predictive probability 1.4%). However, the Bonferroni statistic shows this may still be acceptable in terms of an extreme deviation among the studies, since $Q = 0.17$ (this is calculated from the posterior averages of the U_k). There is clear evidence that the AMIS study true mean is lower than the others, but according to this procedure, it is not an outlier to such an extent as to invalidate the entire hierarchical meta-analysis model or its random error assumptions. The posterior means ν_k^* (the column headed predictive means in Table 10.17) show what the pooled mean μ would look like in the absence of the kth study. The posterior mean ν_6^* for the AMIS study is about 2.29, with 95% interval 0.65 to 3.8, so that there is an unambiguous percent mortality reduction were this study not included in the pooling.

One may also assess the standard meta-analysis against a mixture of Normals

$$\nu_i \sim \mathrm{N}(\mu_{G_i}, \tau^2)$$

where the latent group G_i is sampled from a probability vector π of length 2, itself assigned a Dirichlet prior with elements 1. With the constraint that $\mu_2 > \mu_1$, this prior yields estimates $\pi_2 = 0.61$ and a credible interval for μ_2 that is entirely positive. The probability that G_i is 2 exceeds 0.6, except for the AMIS study where it is only 0.26. In fact, this model has a lower DIC than the standard meta-analysis (around 25 as compared to 25.8).

10.4.2 Heterogeneity in patient risk

Apparent treatment effects may occur because trials are not exchangeable in terms of the risk level of patients in them. Thus, treatment benefit may differ according to whether patients in a particular study are relatively low or high risk. Suppose outcomes of trials are summarised by a mortality log odds (x_i) for the control group in each trial, and by a similar log odds y_i for the treatment group. A measure such as $d_i = y_i - x_i$ is typically used to assess whether the treatment was beneficial. Sometimes the death rate in the control group of a trial, or some transformation of it such as x_i, is taken as a measure of the overall patient risk in that trial, and the benefits are regressed on x_i to control for heterogeneity in risk. Thompson *et al.* (1997) show that such procedures induce biases due to inbuilt dependencies between d_i and x_i.

Suppose instead the underlying patient risk in trial i is denoted ρ_i and the treatment benefits as ν_i, where these effects are independent. Assume also that the sampling errors s_i^2

are equal across studies and across treatment and control arms of trials, so that $\text{var}(x_i) = \text{var}(y_i) = \sigma^2$. Then, assuming normal errors, one may specify the model

$$y_i = \rho_i + \nu_i + u_{1i}$$
$$x_i = \rho_i + u_{2i}$$

where u_{1i} and u_{2i} are independent of one another, and of ρ_i and ν_i.

The risks ρ_i may be taken as random with mean R and variance σ^2_ρ. Alternatively, Thompson *et al.* take σ^2_ρ as known (e.g. $\sigma^2_\rho = 10$ in their analysis of sclerotherapy trials), so that the ρ_i are fixed effects. The ν_i may be distributed around an average treatment effect μ, with variance τ^2.

Another approach attempts to model interdependence between risk and effects. For example, a linear dependence might involve

$$\nu_i \sim \text{N}(\mu_i, \tau^2)$$
$$\mu_i = \alpha + \beta(\rho_i - R)$$

and this is equivalent to assuming the ν_i and ρ_i are bivariate Normal.

Example 10.14 AMI and magnesium trials These issues are illustrated in the analysis by McIntosh (1996) of trials into the use of magnesium for treating acute myocardial infarction. For the nine trials considered, numbers of patients in the trial and control arms N_{ti} and N_{ci} vary considerably, with one trial containing a combined sample ($N_i = N_{ti} + N_{ci}$) exceeding 50 000, another containing under 50 (Table 10.18).

It is necessary to allow for this wide variation in sampling precision for outcomes based on deaths r_{ti} and r_{ci} in each arm of each trial. McIntosh seeks to explain heterogeneity in treatment effects in terms of the control group mortality rates, $Y_{i2} = m_{ci} = r_{ci}/N_{ci}$. Treatment effects themselves are represented by the log odds ratio

$$Y_{i1} = \log(m_{ti}/m_{ci})$$

To reflect sampling variation, McIntosh models the outcomes Y_1 and Y_2 as bivariate normal with unknown means $\theta_{i,\,1:2}$ but known dispersion matrices Σ_i. The term σ_{11i} in Σ_i for the variance of Y_{i1} is provided by the estimate

$$1/\{N_{ti}m_{ti}(1 - m_{ti})\} + 1/\{N_{ci}m_{ci}(1 - m_{ci})\}$$

while the variance for Y_{i2} is just the usual binomial variance. The covariance σ_{12i} is approximated as $-1/N_{ci}$, and hence the 'slope' relating Y_{i1} to Y_{i2} in trial i is estimated as $\sigma_{12i}/\sigma_{22i}$. Table 10.18 presents the relevant inputs. Then the measurement model assumed by McIntosh is

$$Y_{i,\,1:2} \sim N_2(\theta_{i,\,1:2},\ \Sigma_i)$$

where $\theta_{i1} = \nu_i$, $\theta_{i2} = \rho_i$. One might consider a Multivariate t to assess sensitivity.

The true treatment effects ν_i, and true control group mortality rates, ρ_i, are then modelled as

$$\nu_i \sim \text{N}(\mu_i, \tau^2)$$
$$\rho_i \sim \text{N}(R, \sigma^2_\rho)$$

Table 10.18 Trial data summary: patients under magnesium treatment or control

	Magnesium		Control						
	Deaths	Sample size	Deaths	Sample size	Y_2	Y_1	Var(Y_2)	Var(Y_1)	Slope
Morton	1	40	2	36	0.056	−0.83	1.56	0.00146	−19.06
Abraham	1	48	1	46	0.022	−0.043	2.04	0.00046	−47.02
Feldsted	10	50	8	48	0.167	0.223	0.24	0.00035	−19.56
Rasmussen	9	35	23	135	0.170	−1.056	0.17	0.00105	−7.07
Ceremuzynski	1	25	3	23	0.130	−1.281	1.43	0.00493	−8.82
Schechter I	1	59	9	56	0.161	−2.408	1.15	0.00241	−7.41
LIMIT2	90	1150	118	1150	0.103	−0.298	0.021	0.00008	−10.86
ISIS 4	1997	27413	1897	27411	0.069	0.055	0.0011	2.35E-06	−15.52
Schechter II	4	92	17	98	0.173	−1.53	0.33	0.00146	−6.97

with $\mu_i = \alpha + \beta(\rho_i - R)$. If β is negative, this means that treatment effectiveness increases with the risk in the control group, whereas $\beta = 0$ means the treatment effect is not associated with the risk in the control group. The average underlying odds ratio ϕ for the treatment effect (controlling for the effect of risk) is obtained by exponentiating μ_1.

Inferences about β and $\phi = \exp(\alpha)$ may be sensitive to the priors assumed for the variances τ^2 and σ^2_ρ. We consider three options for the inverse variances $1/\tau^2$, namely a G(3, 1) prior (see above) and a more diffuse G(1, 0.001) option. The prior on $1/\sigma^2_\rho$ is kept at G(1, 0.001) throughout. The posterior estimate[10] of β declines as the informativeness of the prior on $1/\tau^2$ increases, with the probability that β is positive being highest (around 29%) under the G(3, 1) prior, and lowest (3%) under G(1, 0.001) priors. Hence, only under diffuse priors on $1/\tau^2$ is the treatment effect associated with the risk in the control group. The treatment odds ratio has a mean of around 0.62 with 95% interval {0.30, 1.13} under the G(3, 1) prior on $1/\tau^2$ and 0.74 {0.44, 1.10} under the G(1, 0.001) priors. Taking a multivariate Student t for $Y_{i,\,1:2}$ affects inferences relatively little, tending to reduce the chance of β being positive slightly; the degrees of freedom (with a uniform prior between 1 and 100) has a posterior mean of 51.

An alternative analysis follows Thompson *et al.* in taking the observed r_{ti} and r_{ci} as binomial with rates π_{ti} and π_{ci} in relation to trial populations N_{ti} and N_{ci}. Thus

$$r_{ti} \sim \text{Bin}(\pi_{ti}, N_{ti})$$
$$r_{ci} \sim \text{Bin}(\pi_{ci}, N_{ci})$$

The models for $y_i = \text{logit}(\pi_{ti})$ and $x_i = \text{logit}(\pi_{ci})$ are then

$$y_i = \rho_i + \nu_i$$
$$x_i = \rho_i$$

where the average trial risks ρ_i may be taken as either fixed effects or random. Under the fixed effects model we take $\sigma^2_\rho = 1$, while under the random effects model it is assumed that $1/\sigma^2_\rho \sim \text{G}(1, 0.001)$. The gain effects are modelled as above,

$$\nu_i \sim \text{N}(\mu_i, \tau^2)$$
$$\mu_i = \alpha + \beta(\rho_i - R)$$

Under the fixed effects option for ρ_i and $1/\tau^2 \sim \text{G}(1, 0.001)$ we obtain a probability of around 17% that β exceeds zero. Under random effects for ρ_i, inferences about β are sensitive to the prior assumed for $1/\tau^2$, as under the McIntosh model. Even for the more diffuse option, $1/\tau^2 \sim \text{G}(1, 0.001)$ there is a 12% chance that $\beta > 0$, while for $1/\tau^2 \sim \text{G}(3, 1)$ the chance that $\beta > 0$ is 36%. It may be noted that the more informative prior is associated with a lower DIC. As above, neither prior gives an overall treatment odds ratio ϕ with 95% interval entirely below 1.

10.4.3 Multiple treatments

The usual assumption in carrying out a meta-analysis is that a single intervention or treatment is being evaluated. The studies are then all estimating the same parameter, comparing the intervention with its absence, such as an effect size (standardised difference in means), relative risk or odds ratio.

[10] Two chain runs showed convergence at around 10 000 iterations and summaries are based on iterations 10 000–20 000.

However, in some contexts there may be a range of r treatment options, some studies comparing Treatment 1 to a control group, some studies comparing Treatment 2 to a control group, and some studies involving multi-treatment comparisons (control group, Treatment 1, Treatment 2, etc.). One may wish to combine evidence over $i = 1, ..n$ such studies, to assess the effectiveness of treatments $j = 1, ..r$ against no treatment (the placebo or control group is not considered a treatment), and to derive measures such as odds ratios comparing treatments j and k in terms of effectiveness.

Example 10.15 MI prevention and smoking cessation Hasselblad (1998) considers an example of three studies for short-term prevention of heart attack (myocardial infarction) using aspirin and heparin as possible alternative treatments. The outcome rates in the studies were five day MI rates, with only one study comparing $r = 2$ options with the placebo. Thus, the Theroux *et al.* (1988) study included no treatment (118 patients, of whom 14 had attacks):

- treatment 1: aspirin (four out of 121 patients having an MI);
- treatment 2: heparin (one from 121 patients had an MI).

The second study compared only aspirin with a placebo group, and the third study compared only heparin with a placebo group. There are then a total of $A = 7$ treatment or placebo arms over the three studies.

Because of the small number of studies, a random effects model is not practical. Following Hasselblad, the observations from each arm of each study are modelled in terms of study risk effects ρ_i, $i = 1, ..n$ (the log odds of MI in the control group), and treatment effects β_j, $j = 1, ..r$. The odds ratios $\text{OR}_{jk} = \exp(\beta_j - \beta_k)$ then compare heparin and aspirin, while the odds ratios $\phi_j = \exp(\beta_j)$, $j = 1, ..r$ compare the treatments with the placebo.

With N(0, 100) priors on all parameters, we find that the ϕ_j are unambiguously below unity, so both heparin and aspirin can be taken to reduce short term MI mortality. The 95% interval for odds ratio OR_{21} just straddles unity (Table 10.19), and as Hasselblad (1998) says, is 'more suggestive of a beneficial effect of heparin over aspirin than that from the Theroux study alone.'

A larger comparison by Hasselblad on similar principles involves 24 studies, evaluating smoking cessation programmes (and treatment success measured by odds ratio over 1). The control consisted of no contact, and there were three treatments: self-help programs, individual counselling, and group counselling. The majority of studies compare only one treatment with the placebo (e.g. individual counselling vs no contact) or two treatments (e.g. group vs. individual counselling), but two studies have three arms. One (Mothersill *et al.*, 1988) compares the two counselling options with self-help. Hence, there are $A = 50$ binomial observations.

Table 10.19 Multiple treatment comparison

	Mean	St. devn.	2.5%	Median	97.50%
OR_{21}	0.40	0.25	0.09	0.34	1.05
ϕ_1	0.36	0.12	0.17	0.34	0.63
ϕ_2	0.13	0.07	0.03	0.12	0.30

We assume a random study effect modelling cessation over all options including no contact,

$$\rho_i \sim N(R, \sigma^2)$$

where R is the grand cessation mean. This random variation is clearly present as shown by the credible interval (0.21, 0.78) on σ^2. The odds ratios ϕ_2 and ϕ_3 on the counselling options are clearly significant (Table 10.20), and that on self-help ϕ_1 suggests a benefit over no contact. The counselling options are in turn more effective than self-help.

To assess prior sensitivity, a two group mixture on ρ_i is adopted with

$$\rho_i \sim N(R_{G_i}, \sigma^2_{G_i})$$

and G_i denoting the latent group. This leads (Model C in Program 10.15) to low and high cessation rate studies being identified with posterior mean for R_2 (the high cessation group) being -1.15 against $R_1 = -3.35$. The treatment odds ratios are little changed however: ϕ_1 and ϕ_3 are slightly raised, OR_{21} is slightly reduced.

10.4.4 Publication bias

The validity of meta-analysis rests on encompassing all existing studies to form an overall estimate of a treatment or exposure effect. A well known problem in this connection is publication bias, generally assumed to take the form of more significant findings being more likely to be published. Insignificant findings are, in this view, more likely to be relegated to the 'file drawer'. One may attempt to model this selection process, and so give an indication of the bias in a standard meta-analysis based only on published studies. There is no best way to do this, and it may be advisable to average over various plausible models for publication bias.

A common approach to this problem is to assume differential bias according to the significance of studies. Assume each study has an effect size Y_j (e.g. log odds ratios or log relative risks) and known standard error s_j, from which significance may be assessed. Then, studies in the most significant category using simple p tests (e.g. p between 0.0001 and 0.025) have highest publication chances, those with slightly less significance (0.025 to 0.10) have more moderate publication chances, and the lowest publication rates are for studies which are 'insignificant' (with $p > 0.10$). Hence, if a set of observed (i.e. published) studies has N_1, N_2, and N_3 studies in these three categories, and there are M_1, M_2 and M_3 missing (unpublished) studies in these categories, then the true number of studies is $\{T_1, T_2, T_3\}$ where $T_1 = M_1 + N_1$, $T_2 = M_2 + N_2$, and $T_3 = M_3 + N_3$ and $T_3/N_3 > T_2/N_2 > T_1/N_1$.

Table 10.20 Smoking cessation: parameter summary

	Mean	St. devn.	2.50%	Median	97.50%
OR_{21}	1.69	0.22	1.30	1.68	2.16
OR_{31}	1.92	0.36	1.31	1.89	2.73
OR_{32}	1.14	0.19	0.81	1.13	1.55
R	−2.42	0.14	−2.69	−2.42	−2.15
ϕ_1	1.29	0.16	1.01	1.28	1.62
ϕ_2	2.16	0.12	1.92	2.15	2.41
ϕ_3	2.46	0.42	1.73	2.43	3.36
σ^2	0.42	0.15	0.21	0.39	0.78

The objective is to estimate an overall effect μ of exposure or treatment from the observed (i.e. published) effects. Suppose the Y_j are log relative risks, so that $\mu = 0$ corresponds to zero overall effect. For a Normal random effects model this is defined by

$$Y_j \sim \mathrm{N}(\nu_j, s_j^2)$$
$$\nu_j \sim \mathrm{N}(\mu, \tau^2)$$

where the ν_j model heterogeneity between studies. Given uninformative priors on τ^2 and μ, the posterior density of μ is essentially a normal density, with mean given by a weighted average of observed relative risks Y_j and the prior relative risk of zero, with respective weights

$$w_j = 1/s_j^2/[1/s_j^2 + 1/\tau^2]$$

on Y_j and $1 - w_j$ on zero. To allow for publication bias, one may modify this scheme so that weights also depend upon significance ratios $|Y_j/s_j|$.

Silliman (1997) proposes one scheme which in effect weights up the less significant studies so that they have a disproportionate influence on the final estimate of the treatment or exposure effect – one more in line with the distribution of the true number of studies. Suppose there are only two categories of study, those with higher significance probabilities, and those with lower significance. Then introduce two random numbers u_1 and u_2, and assign weights $W_1 = \max(u_1, u_2)$ and $W_2 = \min(u_1, u_2)$. This broadly corresponds to modelling the overall publication chance and the lesser chance attached to a less significant study. An alternative is to take $W_1 \sim U(0, 1)$ and $W_2 \sim U(0, W_1)$. Let $G_j = 1$ or 2 denote the significance category of study j. Then an additional stage is included in the above model, such that

$$\begin{aligned} Y_j &\sim \mathrm{N}(\nu_j, s_j^2) \\ \nu_j &= \gamma_j / W_{G_j} \\ \gamma_j &\sim \mathrm{N}(\mu, \tau^2) \end{aligned} \tag{10.17}$$

Givens *et al.* (1997) propose a scheme which models the number of missing studies in each significance category, i.e. M_1 and M_2 in the above two category example. They then introduce data augmentation to reflect the missing effects comparable to Y_j, namely $Z_{1j}, j = 1, .., M_1$ in the high significance category, and $Z_{2j}, j = 1, .., M_2$ in the lower significance category. The missing study numbers M_1 and M_2 are taken as negative binomial

$$M_j \sim NB(g_j, N_j)$$

where the prior for g_1 would reflect the higher publication chances in the high significance category; the Givens *et al.* simulation in Example 10.16 described below took $g_1 \sim U(0.5, 1)$ and $g_2 \sim U(0.2, 1)$. The missing effects are generated in a way consistent with their sampled category[11].

Example 10.16 Simulated publication bias We follow Givens *et al.* in generating a set of studies, only some of which are observed (published) subject to a known bias mechanism. Thus, the original known variances s_j^2 of 50 studies are generated according

[11] Implementing the Givens *et al.* approach in WINBUGS is limited by M_1 and M_2 being stochastic indices, and the fact that for loops cannot be defined with stochastic quantities.

to $s_j^2 \sim G(3, 9)$. The variance of the underlying study effects is $\tau^2 = 0.03$. These variance parameters are in fact close to those of a set of observed studies on excess lung cancer rates associated with passive smoking. A standard meta-analysis of all 50 studies (taking the s_j^2 as known) then gives an estimated overall relative risk, $RR = \exp(\mu)$ of 1.01 with 95% interval from 0.86 to 1.18. The priors used on μ and τ^2 are as in Givens *et al.* (1997, p. 229).

To reflect the operation of publication bias, a selective suppression is then applied. The 26 'positive' studies with

$$|Y_j/\hat{s}_j| \geq 0$$

and significance rates p therefore between 0 and 0.5 are retained in their entirety. The 24 negative studies with $|Y_j/\hat{s}_j| < 0$ are subjected to a 70% non-publication rate. In practice, this led here to retaining seven of the 24 negative studies and all the 26 positive studies (the seven retained had uniformly generated numbers exceeding 0.7, while non-publication applies to those with numbers under 0.7). So $M_1 = 0$ and $M_2 = 17$. A standard meta-analysis of this set of 33 studies gives an underlying central relative risk $\exp(\mu)$ of 1.26 with 95% interval from 1.17–1.51. This is considerably in excess of the 'true' RR in the data set of all 50 studies.

We then use the comparison of uniforms method of Silliman to compensate for bias, and the two mechanisms described above (Model A). Using the first mechanism gives an estimated mean *RR* of 1.18, and 95% interval from 1.005–1.39. This approach gives a clearer basis for doubting that the underlying *RR* over the studies exceeds unity. With the second mechanism, one obtains a posterior mean *RR* of 1.12 with 95% interval from 1.02–1.23.

To assess sensitivity to priors on the underlying study effects (Smith *et al.*, 1995) an alternative Student t prior is taken, in combination with the second mechanism, such that $\nu_j \sim t_5(\mu, \tau^2)$. This gives a mean *RR* of 1.11 with 95% interval from 1.006–1.23, so that conclusions are unaltered. Similar models might be envisaged, for example regression models for the publication probability in Equation (10.17) with a coefficient on the study significance ratio constrained to be positive:

$$\begin{aligned}
&Y_j \sim \mathrm{N}(\nu_j, s_j^2)\\
&\nu_j = \gamma_j / W_j\\
&\mathrm{logit}(W_j) = \beta Y_j / s_j\\
&\beta \sim \mathrm{N}(0, 1)
\end{aligned}$$

Applying this model here (Model B) gives an interval on *RR* of {0.99, 1.26}.

10.5 REVIEW

Bayesian epidemiology has drawn on wider ideas and developments in Bayesian statistics, but is oriented to specific concerns such as arise in the analysis of disease risk and causation. These include control for confounding influences on disease outcome where the confounder affects disease risk, and is also unequally distributed across categories of the main exposure; allowing for measurement errors in disease risk or outcome, perhaps drawing on information from calibration studies (Stephens and Dellaportas, 1992); the delineation of disease and risk factor distributions over time, attributes of individuals (e.g. age) and place (Ashby and Hutton, 1996); and the tailoring of hierarchical methods for combining information to the meta-analysis of medical intervention or risk factor studies. Because epidemiological applications often focus on the impact of well documented risk factors, framing of priors often involves elicitation of informative priors; this is so especially in clinical epidemiology in the sense of models for randomised trials, diagnostic tests, etc, as illustrated by ranges of priors (sceptical, neutral enthusiastic, etc.) possible in clinical trial assessment (Spiegelhalter *et al.*, 1999) or the use of informative priors in gauging diagnostic accuracy (Joseph *et al.*, 1995).

The above chapter has been inevitably selective in coverage of these areas. Thus while state space random walk models have been illustrated in Example 10.6 (and the modelling of place effects in disease outcomes in Chapter 7) more complex examples, in terms of identifiability issues, occur in disease models with age, period and cohort effects all present; methodological issues in this topic are discussed by Knorr-Held (2000). Recent work on measurement error modelling in epidemiology includes Richardson *et al.* (2001)

REFERENCES

Albert, J. (1996) Bayesian selection of log-linear models. *Can. J. Stat.* **24**(3), 327–347.

Ashby, D. and Hutton, J. (1996) Bayesian epidemiology. In: Berry, D. and Stangl, D. (eds.), *Bayesian Biostatistics*. New York: Dekker.

Baker, G., Hesdon, B. and Marson, A. (2000) Quality-of-life and behavioral outcome measures in randomized controlled trials of antiepileptic drugs: a systematic review of methodology and reporting standards. *Epilepsia* **41**(11), 1357–1363.

Berry, D. and Stangl, D. (1996). Bayesian methods in health related research. In: Berry, D. and Stangl, D. (eds.), *Bayesian Biostatistics*. New York: Dekker.

Boucher, K., Slattery, M., Berry, T. *et al.* (1998) Statistical methods in epidemiology: A comparison of statistical methods to analyze dose-response and trend analysis in epidemiologic studies. *J. Clin. Epidemiol.* **51**(12), 1223–1233.

Breslow, N. and Day, N. (1980) *Statistical Methods in Cancer Research: Vol I: The Analysis of Case-Control Studies*. Lyon: International Agency for Research of Cancer.

Breslow, N. (1996) Statistics in epidemiology: The case-control study. *J. Am. Stat. Assoc.* **91**(433), 14–28.

Carlin, J. (1992). Meta-analysis for 2×2 tables: a Bayesian approach. *Stat. in Med.* **11**, 141–159.

Chen, C., Chock, D. and Winkler, S. (1999) A simulation study of confounding in generalized linear models for air pollution epidemiology. *Environ. Health Perspectives* **107**, 217–222.

Davey Smith, G. and Phillips, A. (1992) Confounding in epidemiological studies: why 'independent' effects may not be all they seem. *Br. Med. J.* **305**, 757–759.

Doll, R. and Hill, A. (1950) Smoking and carcinoma of the lung. *Br. Med. J.* **ii**, 739–748.

Doll, H., Petersen, S. and Stewart-Brown, S. (2000) Obesity and physical and emotional well-being: associations between body mass index, chronic illness, and the physical and mental components of the SF-36 questionnaire. *Obesity Res.* **8**(2), 160–170.

DuMouchel, W. and Harris, J. (1983) Bayes methods for combining the results of cancer studies. *J. Am. Stat. Assoc.* **78**, 293–315.

DuMouchel, W. (1990) Bayesian meta-analysis. In: Berry, D. (ed.), *Statistical Methodology in the Pharmaceutical Sciences*. New York: Dekker.

DuMouchel, W. (1996) Predictive cross-validation of Bayesian meta-analyses (with discussion). In: Bernardo, J. *et al.* (eds.), *Bayesian Statistics V*. Oxford: Oxford University Press, pp. 105–126.

Dunn, G. (1999) *Statistics in Psychiatry*. London: Arnold.

Efron, B. and Feldman, D. (1991) Compliance as an explanatory variable in clinical trials. *J. Am. Stat. Assoc.* **86**, 9–17.

Fahrmeir, L. and Knorr-Held, L. (2000). Dynamic and semiparametric models. In: Schimek, M. (ed.), *Smoothing and Regression: Approaches, Computation and Application*. New York: Wiley, pp. 513–544.

Fahrmeir, L. and Lang, S. (2001) Bayesian inference for generalized additive mixed models based on Markov random field priors. *J. Roy. Stat. Soc., Ser. C (Appl. Stat.)* **50**, 201–220.

Garssen, B. and Goodkin, K. (1999) On the role of immunological factors as mediators between psychosocial factors and cancer progression. *Psychiatry Res.* **85**(1): 51–61.

Gelfand, A. (1995) Model determination using sampling-based methods, In: Gilks, W., Richardson, S. and Spiegelhalter, D. (eds.), *Markov Chain Monte Carlo in Practice*. London: Chapman & Hall, pp. 145–161.

Gelfand, A. and Dey, D. (1994) Bayesian model choice: asymptotics and exact calculations. *J. Roy Stat. Soc. B* **56**, 501–514.

Gelfand, A. and Ghosh, S. (1998) Model choice: A minimum posterior predictive loss approach. *Biometrika* **85**, 1–11.

Givens, G., Smith, D. and Tweedie, R. (1997) Bayesian data-augmented meta-analysis that account for publication bias issues exemplified in the passive smoking debate. *Stat. Sci.* **12**, 221–250.

Greenland, S. (1995) Dose-response and trend analysis in epidemiology: alternatives to categorical analysis. *Epidemiology* **6**(4), 356–365.

Greenland, S. (1998a) Introduction to regression models. In: Rothman, K. and Greenland, S. (eds.), *Modern Epidemiology, 2nd edition*. Lippincott; Williams and Wilkins.

Greenland, S. (1998b) Introduction to regression modeling. In: Rothman, K. and Greenland, S. (eds.), *Modern Epidemiology, 2nd edition*. Lippincott, Williams and Wilkins.

Hasselblad, V. and Jarabek, A. (1996) Dose-response analysis of toxic chemicals. In: Berry, D. and Stangl, D. (eds.), *Bayesian Biostatistics*. New York: Dekker.

Hasselblad, V. (1998) Meta-analysis of multitreatment studies. *Med. Decis. Making* **18**(1), 37–43.

Hedges, L. and Olkin, I. (1985). *Statistical Methods for Meta-analysis*. New York: Academic Press.

Higgins, J. and Whitehead, A. (1996) Borrowing strength from external trials in a meta-analysis. *Stat. Med.* **15**, 2733–2749.

Ibrahim, J., Chen, M. and Sinha, D. (2001) *Bayesian Survival Analysis*. Springer Series in Statistics. New York, NY: Springer.

Jenkinson, C., Coulter, A., and Wright, L. (1993) Short form 36 (SF36) health survey questionnaire: normative data for adults of working age. *Br. Med. J.* **306**, 1437–1440.

Kahn, H. and Sempos, C. (1989) *Statistical Methods in Epidemiology*. Oxford: Oxford University Press.

Knorr-Held, L. (2000). Bayesian modelling of inseparable space-time variation in disease risk. *Stat. in Med.* **19**, 2555–2567.

Kuo, L. and Mallick, B. (1998) Variable selection for regression models. *Sankhya* **60B**, 65–81.

Leonard, T. and Hsu, J. (1999) Bayesian Methods: an Analysis for Statisticians and Researchers. Cambridge: Cambridge University Press.

Lilford, R. and Braunholtz, D. (1996) The statistical basis of public policy: a paradigm shift is overdue. *Br. Med. J.* **313**, 603–607.

McGregor, H., Land, C., Choi, K., Tokuoka, S., Liu, P., Wakabayashi, T., and Beebe, G. (1977) Breast cancer incidence among atomic bomb survivors, Hiroshima and Nagasaki, 1950–69. *J. Nat. Cancer Inst.* **59**(3), 799–811.

McIntosh, M. (1996) The population risk as an explanatory variable in research synthesis of clinical trials. *Stat. in Med.* **15**(16), 1713–1728.

Mark, S. and Robins, J. (1993) A method for the analysis of randomized trials with compliance information. *Controlled Clinical Trials* **14**, 79–97.

Miettinen, O. (1972) Components of the Crude Risk Ratio. *Am. J. Epidemiology* **96**, 168–172.

Morgan, B. (2000) *Applied Stochastic Modelling*. Arnold Texts in Statistics. London: Arnold.

Mothersill, K., McDowell, I. and Rosser, W. (1988) Subject characteristics and long term post-program smoking cessation. *Addict Behav.* **13**(1), 29–36.

Mundt, K., Tritschler, J. and Dell, L. (1998) Validity of epidemiological data in risk assessment applications. *Human and Ecological Risk Assess.* **4**, 675–683.

Muthen, B. (1992) Latent variable modelling in epidemiology. *Alcohol Health and Res.* **16**, 286–292.

National Research Council (1983) *Risk Assessment in the Federal Government*. Washington, DC: National Academy Press.

Noordhuizen, J., Frankena, K., van der Hoofd, C. and Graat, E. (1997) *Application of Quantitative Methods in Veterinary Epidemiology*. Wageningen Press.

Pack, S. and Morgan, B. (1990) A mixture model for interval-censored time-to-response quantal assay data. *Biometrics* **46**, 749–757.

Prevost, T., Abrams. K. and Jones, D. (2000). Hierarchical models in generalized synthesis of evidence: an example based on studies of breast cancer screening. *Stat in Med.* **19**, 3359–3376.

Richardson, S., Leblond, L., Jaussent, I. and Green, P. J. (2001) Mixture models in Measurement error problems, with reference to epidemiological studies. Technical Report INSERM, France.

Rigby, A. and Robinson, M. (2000) Statistical methods in epidemiology. IV. confounding and the matched pairs odds ratio. *Disabil. Rehabil.* **22**(6), 259–265.

Rothman, K. (1986) *Modern Epidemiology*. New York: Little, Brown.

Sahu, S., Dey, D., Aslanidou, H. and Sinha, D. (1997) A Weibull regression model with gamma frailties for multivariate survival data. *Lifetime Data Anal.* **3**, 123–137.

Schmid, C. and Brown, E. (2000) Bayesian hierarchical models. *Meth. in Enzymology* **321**, 305–330.

Silliman, N. (1997) Hierarchical selection models with applications in meta-analysis. *J. Am. Stat. Assoc.* **92**(439), 926–936.

Slaton, T., Piegorsch, W. W. and Durham, S. (2000). Estimation and testing with overdispersed proportions using the beta-logistic regression model of Heckman and Willis. *Biometrics* **56**, 125–132.

Small, M. and Fishbeck, P. (1999) False precision in Bayesian updating with incomplete models. *Human and Ecological Risk Assess.* **5**(2), 291–304.

Smith, D. (2000) Cardiovascular disease: a historic perspective. *Japan J. Vet. Res.* **48**(2–3), 147–166.

Smith, T., Spiegelhalter, D. and Thomas, A. (1995) Bayesian approaches to random-effects meta-analysis: a comparative study. *Stat. in Med.* **14**, 2685–2699.

Smith, T., Spiegelhalter, D. and Parmar, M. (1996) Bayesian meta-analysis of randomized trials using graphical models and BUGS. In: Berry, D. and Stangl, D. (eds.) *Bayesian Biostatistics*. New York: Dekker.

Smith, D., Givens, G. and Tweedie, R. (2000) Adjustment for publication and quality bias in Bayesian meta-analysis. In: Stangl, D. and Berry, D. (eds) *Meta-Analysis in Medicine and Health Policy*. New York: Dekker.

Spiegelhalter, D., Myles, J., Jones, D. and Abrams, K. (1999) An introduction to Bayesian methods in health technology assessment. *Br. Med. J.* **319**, 508–512.

Stephens, D. and Dellaportas, P. (1992) Bayesian analysis of generalised linear models with covariate measurement error. In: Bernardo, J., Berger, J., Dawid, A. and Smith, A. (eds.). *Bayesian Statistics 4*. Oxford: Oxford University Press, pp. 813–820.

Sturmer, T. and Brenner, H. (2001) Degree of matching and gain in power and efficiency in case-control studies. *Epidemiology* **12**(1), 101–108.

Thompson, S., Smith, T. and Sharp, S. (1997) Investigating underlying risk as a source of heterogeneity in meta-analysis. *Stat. in Med.* **16**, 2741–2758.

Verdinelli, I., Andrews, K., Detre, K. and Peduzzi, P. (1996) The Bayesian approach to meta-analysis: a case study. Carnegie Mellon, Department of Statistics, Technical Report 641.

Weed, D. (2000) Interpreting epidemiological evidence: how meta-analysis and causal inference methods are related. *Int. J. Epidemiology* **29**, 387–390.

Wijesinha, M. and Piantadosi, S. (1995) Dose-response models with covariates. *Biometrics* **51**(3), 977–987.

Woodward, M. (1999) *Epidemiology*. London: Chapman & Hall.

Yusuf, S., Zucker, D., Peduzzi, P., Fisher, L., Takaro, T., Kennedy, J., Davis, K., Killip, T., Passamani, E. and Norris, R. (1994) Effect of coronary artery bypass graft surgery on survival: overview of 10-year results from randomised trials by the Coronary Artery Bypass Graft Surgery Trialists Collaboration. *Lancet* **344**, 563–570.

Zeger, S. and Liang, K. (1991) Comments on 'Compliance As An Explanatory Variable in Clinical Trials'. *J. Am. Stat. Assoc.* **86**, 18–19.

Zelterman, D. (1999) *Models for Discrete Data*. Oxford Science Publications. Oxford: Clarendon Press.

EXERCISES

1. In Example 10.2, try the square root transforms instead of the $\log_e$ transforms of age and SBP in the coefficient selection procedure (or include both square root and $\log_e$ transforms). Then fit a risk model using the most frequently selected terms and assess its fit.

2. In Example 10.4, replicate and assess the sensitivity of the analysis into illness and obesity by using (a) correlated physical and mental health status factors, and (b) using robust (heavy tailed) alternative for the density of the factors.

3. In Example 10.5 try adding a square root term rather than a square in rads to the power model with continuous outcome. Does this improve fit?

4. In Model A in Example 10.6, fit a logistic regression with categories under 130, 130–149, 150–169, 170–189, 190–209 and 210 and over, and then fit a weighted quadratic regression to the odds ratios (as in Model A2 in Program 10.6). How does this illustrate the nonlinear impact of SBP?

5. In the flour beetle mortality Example 10.7, consider the generalisation for dosage effects of the model

$$R_{rj} = C(t_j, X_r) - C(t_{j-1}, X_r)$$

where

$$C(t, X) = [1 + \lambda F(X)]^{-1/\lambda}\ [1 + t^{-\beta_3}\beta_4]^{-1}$$

and where $F(X) = \exp(\beta_1 + \beta_2 \log X) = \exp(\beta_1)X^{\beta_2}$. For identiability a prior $\lambda \sim U(0, 1)$ may be used with $\lambda = 1$ corresponding to the proportional odds model actually used in Example 10.7, and $\lambda = 0$ giving a Weibull model, when

$$C(t, X) = \exp(-e^{\beta_1}X^{\beta_2})\ [1 + t^{-\beta_3}\beta_4]^{-1}$$

6. Following Example 10.11, carry out a meta-analysis of the mortality outcomes following CABG for *high* risk patients, where the odds ratios and their confidence limits for seven studies are

$$\text{OR} = 0.58,\ 0.37,\ 0.43,\ 0.56,\ 0.27,\ 1.89,\ 0.95$$
$$\text{LCL} = 0.33,\ 0.15,\ 0.15,\ 0.19,\ 0.05,\ 0.31,\ 0.23$$
$$\text{UCL} = 1.01,\ 0.89,\ 1.26,\ 1.63,\ 1.45,\ 11.64,\ 3.83$$

7. In Example 10.12, apply a Bayesian significance test to obtain the probabilities that the ν_i are positive.

8. In Example 10.14, assess sensitivity of inferences to adopting a Student t prior rather than Normal prior for the study treatment effects, δ_i, with degrees of freedom an extra parameter (in both the McIntosh and Thompson *et al.* models).

Index

WILEY SERIES IN PROBABILITY AND STATISTICS
ESTABLISHED BY WALTER A. SHEWHART AND SAMUEL S. WILKS

The ***Wiley Series in Probability and Statistics*** is well established and authoritative. It covers many topics of current research interest in both pure and applied statistics and probability theory. Written by leading statisticians and institutions, the titles span both state-of-the-art developments in the field and classical methods.

Reflecting the wide range of current research in statistics, the series encompasses applied, methodological and theoretical statistics, ranging from applications and new techniques made possible by advances in computerized practice to rigorous treatment of theoretical approaches.

This series provides essential and invaluable reading for all statisticians, whether in academia, industry, government, or research.

ABRAHAM and LEDOLTER · Statistical Methods for Forecasting
AGRESTI · Analysis of Ordinal Categorical Data
AGRESTI · An Introduction to Categorical Data Analysis
AGRESTI · Categorical Data Analysis
ANDĚL · Mathematics of Chance
ANDERSON · An Introduction to Multivariate Statistical Analysis, *Second Edition*
*ANDERSON · The Statistical Analysis of Time Series
ANDERSON, AUQUIER, HAUCK, OAKES, VANDAELE, and WEISBERG · Statistical Methods for Comparative Studies
ANDERSON and LOYNES · The Teaching of Practical Statistics
ARMITAGE and DAVID (editors) · Advances in Biometry
ARNOLD, BALAKRISHNAN, and NAGARAJA · Records
*ARTHANARI and DODGE · Mathematical Programming in Statistics
*BAILEY · The Elements of Stochastic Processes with Applications to the Natural Sciences
BALAKRISHNAN and KOUTRAS · Runs and Scans with Applications
BARNETT · Comparative Statistical Inference, *Third Edition*
BARNETT and LEWIS · Outliers in Statistical Data, *Third Edition*
BARTOSZYNSKI and NIEWIADOMSKA-BUGAJ · Probability and Statistical Inference
BASILEVSKY · Statistical Factor Analysis and Related Methods: Theory and Applications
BASU and RIGDON · Statistical Methods for the Reliability of Repairable Systems
BATES and WATTS · Nonlinear Regression Analysis and Its Applications
BECHHOFER, SANTNER, and GOLDSMAN · Design and Analysis of Experiments for Statistical Selection, Screening, and Multiple Comparisons
BELSLEY · Conditioning Diagnostics: Collinearity and Weak Data in Regression
BELSLEY, KUH, and WELSCH · Regression Diagnostics: Identifying Influential Data and Sources of Collinearity
BENDAT and PIERSOL · Random Data: Analysis and Measurement Procedures, *Third Edition*
BERRY, CHALONER, and GEWEKE · Bayesian Analysis in Statistics and Econometrics: Essays in Honor of Arnold Zellner
BERNARDO and SMITH · Bayesian Theory
BHAT · Elements of Applied Stochastic Processes, *Second Edition*
BHATTACHARYA and JOHNSON · Statistical Concepts and Methods

*Now available in a lower priced paperback edition in the Wiley Classics Library.

BHATTACHARYA and WAYMIRE · Stochastic Processes with Applications
BILLINGSLEY · Convergence of Probability Measures, *Second Edition*
BILLINGSLEY · Probability and Measure, *Third Edition*
BIRKES and DODGE · Alternative Methods of Regression
BLISCHKE AND MURTHY · Reliability: Modeling, Prediction, and Optimization
BLOOMFIELD · Fourier Analysis of Time Series: An Introduction, *Second Edition*
BOLLEN · Structural Equations with Latent Variables
BOROVKOV · Ergodicity and Stability of Stochastic Processes
BOULEAU · Numerical Methods for Stochastic Processes
BOX · Bayesian Inference in Statistical Analysis
BOX · R. A. Fisher, the Life of a Scientist
BOX and DRAPER · Empirical Model-Building and Response Surfaces
*BOX and DRAPER · Evolutionary Operation: A Statistical Method for Process Improvement
BOX, HUNTER, and HUNTER · Statistics for Experimenters: An Introduction to Design, Data Analysis, and Model Building
BOX and LUCEÑO, · Statistical Control by Monitoring and Feedback Adjustment
BRANDIMARTE · Numerical Methods in Finance: A MATLAB-Based Introduction
BROWN and HOLLANDER · Statistics: A Biomedical Introduction
BRUNNER, DOMHOF, and LANGER · Nonparametric Analysis of Longitudinal Data in Factorial Experiments
BUCKLEW · Large Deviation Techniques in Decision, Simulation, and Estimation
CAIROLI and DALANG · Sequential Stochastic Optimization
CHAN · Time Series: Applications to Finance
CHATTERJEE and HADI · Sensitivity Analysis in Linear Regression
CHATTERJEE and PRICE · Regression Analysis by Example, *Third Edition*
CHERNICK · Bootstrap Methods: A Practitioner's Guide
CHILÈS and DELFINER · Geostatistics: Modeling Spatial Uncertainty
CHOW and LIU · Design and Analysis of Clinical Trials: Concepts and Methodologies
CLARKE and DISNEY · Probability and Random Processes: A First Course with Applications, *Second Edition*
*COCHRAN and COX · Experimental Designs, *Second Edition*
CONGDON · Bayesian Statistical Modelling
CONGDON · Applied Bayesian Modelling
CONOVER · Practical Nonparametric Statistics, *Second Edition*
COOK · Regression Graphics
COOK and WEISBERG · Applied Regression Including Computing and Graphics
COOK and WEISBERG · An Introduction to Regression Graphics
CORNELL · Experiments with Mixtures, Designs, Models, and the Analysis of Mixture Data, *Third Edition*
COVER and THOMAS · Elements of Information Theory
COX · A Handbook of Introductory Statistical Methods
*COX · Planning of Experiments
CRESSIE · Statistics for Spatial Data, *Revised Edition*
CSÖRGÖ and HORVATH · Limit Theorems in Change Point Analysis
DANIEL · Applications of Statistics to Industrial Experimentation
DANIEL · Biostatistics: A Foundation for Analysis in the Health Sciences, *Sixth Edition*
*DANIEL · Fitting Equations to Data: Computer Analysis of Multifactor Data, *Second Edition*
DAVID · Order Statistics, *Second Edition*
*DEGROOT, FIENBERG, and KADANE · Statistics and the Law
DENISON, HOLMES, MALLICK and SMITH · Bayesian Methods for Nonlinear Classification and Regression

*Now available in a lower priced paperback edition in the Wiley Classics Library.

DEL CASTILLO · Statistical Process Adjustment for Quality Control
DETTE and STUDDEN · The Theory of Canonical Moments with Applications in Statistics, Probability, and Analysis
DEY and MUKERJEE · Fractional Factorial Plans
DILLON and GOLDSTEIN · Multivariate Analysis: Methods and Applications
DODGE · Alternative Methods of Regression
*DODGE and ROMIG · Sampling Inspection Tables, *Second Edition*
*DOOB · Stochastic Processes
DOWDY and WEARDEN · Statistics for Research, *Second Edition*
DRAPER and SMITH · Applied Regression Analysis, *Third Edition*
DRYDEN and MARDIA · Statistical Shape Analysis
DUDEWICZ and MISHRA · Modern Mathematical Statistics
DUNN and CLARK · Applied Statistics: Analysis of Variance and Regression, *Second Edition*
DUNN and CLARK · Basic Statistics: A Primer for the Biomedical Sciences, *Third Edition*
DUPUIS and ELLIS · A Weak Convergence Approach to the Theory of Large Deviations
*ELANDT-JOHNSON and JOHNSON · Survival Models and Data Analysis
ETHIER and KURTZ · Markov Processes: Characterization and Convergence
EVANS, HASTINGS, and PEACOCK · Statistical Distributions, *Third Edition*
FELLER · An Introduction to Probability Theory and Its Applications, Volume I, *Third Edition*, Revised; Volume II, *Second Edition*
FISHER and VAN BELLE · Biostatistics: A Methodology for the Health Sciences
*FLEISS · The Design and Analysis of Clinical Experiments
FLEISS · Statistical Methods for Rates and Proportions, *Second Edition*
FLEMING and HARRINGTON · Counting Processes and Survival Analysis
FULLER · Introduction to Statistical Time Series, *Second Edition*
FULLER · Measurement Error Models
GALLANT · Nonlinear Statistical Models
GHOSH, MUKHOPADHYAY, and SEN · Sequential Estimation
GIFI · Nonlinear Multivariate Analysis
GLASSERMAN and YAO · Monotone Structure in Discrete-Event Systems
GNANADESIKAN · Methods for Statistical Data Analysis of Multivariate Observations, *Second Edition*
GOLDSTEIN and LEWIS · Assessment: Problems, Development, and Statistical Issues
GREENWOOD and NIKULIN · A Guide to Chi-Squared Testing
GROSS and HARRIS · Fundamentals of Queueing Theory, *Third Edition*
*HAHN · Statistical Models in Engineering
HAHN and MEEKER · Statistical Intervals: A Guide for Practitioners
HALD · A History of Probability and Statistics and their Applications Before 1750
HALD · A History of Mathematical Statistics from 1750 to 1930
HAMPEL · Robust Statistics: The Approach Based on Influence Functions
HANNAN and DEISTLER · The Statistical Theory of Linear Systems
HEIBERGER · Computation for the Analysis of Designed Experiments
HEDAYAT and SINHA · Design and Inference in Finite Population Sampling
HELLER · MACSYMA for Statisticians
HINKELMAN and KEMPTHORNE: · Design and Analysis of Experiments, Volume I: Introduction to Experimental Design
HOAGLIN, MOSTELLER, and TUKEY · Exploratory Approach to Analysis of Variance
HOAGLIN, MOSTELLER, and TUKEY · Exploring Data Tables, Trends and Shapes
*HOAGLIN, MOSTELLER, and TUKEY · Understanding Robust and Exploratory Data Analysis
HOCHBERG and TAMHANE · Multiple Comparison Procedures

*Now available in a lower priced paperback edition in the Wiley Classics Library.

HOCKING · Methods and Applications of Linear Models: Regression and the Analysis of Variables
HOEL · Introduction to Mathematical Statistics, *Fifth Edition*
HOGG and KLUGMAN · Loss Distributions
HOLLANDER and WOLFE · Nonparametric Statistical Methods, *Second Edition*
HOSMER and LEMESHOW · Applied Logistic Regression, *Second Edition*
HOSMER and LEMESHOW · Applied Survival Analysis: Regression Modeling of Time to Event Data
HØYLAND and RAUSAND · System Reliability Theory: Models and Statistical Methods
HUBER · Robust Statistics
HUBERTY · Applied Discriminant Analysis
HUNT and KENNEDY · Financial Derivatives in Theory and Practice
HUSKOVA, BERAN, and DUPAC · Collected Works of Jaroslav Hajek—with Commentary
IMAN and CONOVER · A Modern Approach to Statistics
JACKSON · A User's Guide to Principle Components
JOHN · Statistical Methods in Engineering and Quality Assurance
JOHNSON · Multivariate Statistical Simulation
JOHNSON and BALAKRISHNAN · Advances in the Theory and Practice of Statistics: A Volume in Honor of Samuel Kotz
JUDGE, GRIFFITHS, HILL, LÜTKEPOHL, and LEE · The Theory and Practice of Econometrics, *Second Edition*
JOHNSON and KOTZ · Distributions in Statistics
JOHNSON and KOTZ (editors) · Leading Personalities in Statistical Sciences: From the Seventeenth Century to the Present
JOHNSON, KOTZ, and BALAKRISHNAN · Continuous Univariate Distributions, Volume 1, *Second Edition*
JOHNSON, KOTZ, and BALAKRISHNAN · Continuous Univariate Distributions, Volume 2, *Second Edition*
JOHNSON, KOTZ, and BALAKRISHNAN · Discrete Multivariate Distributions
JOHNSON, KOTZ, and KEMP · Univariate Discrete Distributions, *Second Edition*
JUREČKOVÁ and SEN · Robust Statistical Procedures: Aymptotics and Interrelations
JUREK and MASON · Operator-Limit Distributions in Probability Theory
KADANE · Bayesian Methods and Ethics in a Clinical Trial Design
KADANE AND SCHUM · A Probabilistic Analysis of the Sacco and Vanzetti Evidence
KALBFLEISCH and PRENTICE · The Statistical Analysis of Failure Time Data
KASS and VOS · Geometrical Foundations of Asymptotic Inference
KAUFMAN and ROUSSEEUW · Finding Groups in Data: An Introduction to Cluster Analysis
KENDALL, BARDEN, CARNE, and LE · Shape and Shape Theory
KHURI · Advanced Calculus with Applications in Statistics
KHURI, MATHEW, and SINHA · Statistical Tests for Mixed Linear Models
KLUGMAN, PANJER, and WILLMOT · Loss Models: From Data to Decisions
KLUGMAN, PANJER, and WILLMOT · Solutions Manual to Accompany Loss Models: From Data to Decisions
KOTZ, BALAKRISHNAN, and JOHNSON · Continuous Multivariate Distributions, Volume 1, *Second Edition*
KOTZ and JOHNSON (editors) · Encyclopedia of Statistical Sciences: Volumes 1 to 9 with Index
KOTZ and JOHNSON (editors) · Encyclopedia of Statistical Sciences: Supplement Volume
KOTZ, READ, and BANKS (editors) · Encyclopedia of Statistical Sciences: Update Volume 1
KOTZ, READ, and BANKS (editors) · Encyclopedia of Statistical Sciences: Update Volume 2
KOVALENKO, KUZNETZOV, and PEGG · Mathematical Theory of Reliability of Time-Dependent Systems with Practical Applications
LACHIN · Biostatistical Methods: The Assessment of Relative Risks

NEWMAN · Biostatistical Methods in Epidemiology
OCHI · Applied Probability and Stochastic Processes in Engineering and Physical Sciences
OKABE, BOOTS, SUGIHARA, and CHIU · Spatial Tesselations: Concepts and Applications of Voronoi Diagrams, *Second Edition*
OLIVER and SMITH · Influence Diagrams, Belief Nets and Decision Analysis
PANKRATZ · Forecasting with Dynamic Regression Models
PANKRATZ · Forecasting with Univariate Box-Jenkins Models: Concepts and Cases
*PARZEN · Modern Probability Theory and Its Applications
PENA, TIAO, and TSAY · A Course in Time Series Analysis
PIANTADOSI · Clinical Trials: A Methodologic Perspective
PORT · Theoretical Probability for Applications
POURAHMADI · Foundations of Time Series Analysis and Prediction Theory
PRESS · Bayesian Statistics: Principles, Models, and Applications
PRESS and TANUR · The Subjectivity of Scientists and the Bayesian Approach
PUKELSHEIM · Optimal Experimental Design
PURI, VILAPLANA, and WERTZ · New Perspectives in Theoretical and Applied Statistics
PUTERMAN · Markov Decision Processes: Discrete Stochastic Dynamic Programming
*RAO · Linear Statistical Inference and Its Applications, *Second Edition*
RENCHER · Linear Models in Statistics
RENCHER · Methods of Multivariate Analysis, *Second Edition*
RENCHER · Multivariate Statistical Inference with Applications
RIPLEY · Spatial Statistics
RIPLEY · Stochastic Simulation
ROBINSON · Practical Strategies for Experimenting
ROHATGI and SALEH · An Introduction to Probability and Statistics, *Second Edition*
ROLSKI, SCHMIDLI, SCHMIDT, and TEUGELS · Stochastic Processes for Insurance and Finance
ROSENBERGER and LACHIN · Randomization in Clinical Trials: Theory and Practice
ROSS · Introduction to Probability and Statistics for Engineers and Scientists
ROUSSEEUW and LEROY · Robust Regression and Outlier Detection
RUBIN · Multiple Imputation for Nonresponse in Surveys
RUBINSTEIN · Simulation and the Monte Carlo Method
RUBINSTEIN and MELAMED · Modern Simulation and Modeling
RYAN · Modern Regression Methods
RYAN · Statistical Methods for Quality Improvement, *Second Edition*
SALTELLI, CHAN, and SCOTT (editors) · Sensitivity Analysis
*SCHEFFE · The Analysis of Variance
SCHIMEK · Smoothing and Regression: Approaches, Computation, and Application
SCHOTT · Matrix Analysis for Statistics
SCHUSS · Theory and Applications of Stochastic Differential Equations
SCOTT · Multivariate Density Estimation: Theory, Practice, and Visualization
*SEARLE · Linear Models
SEARLE · Linear Models for Unbalanced Data
SEARLE · Matrix Algebra Useful for Statistics
SEARLE, CASELLA, and McCULLOCH · Variance Components
SEARLE and WILLETT · Matrix Algebra for Applied Economics
SEBER · Linear Regression Analysis
SEBER · Multivariate Observations
SEBER and WILD · Nonlinear Regression
SENNOTT · Stochastic Dynamic Programming and the Control of Queueing Systems
*SERFLING · Approximation Theorems of Mathematical Statistics

*Now available in a lower priced paperback edition in the Wiley Classics Library.

LAD · Operational Subjective Statistical Methods: A Mathematical, Philosophical, and Historical Introduction
LAMPERTI · Probability: A Survey of the Mathematical Theory, *Second Edition*
LANGE, RYAN, BILLARD, BRILLINGER, CONQUEST, and GREENHOUSE · Case Studies in Biometry
LARSON · Introduction to Probability Theory and Statistical Inference. *Third Edition*
LAWLESS · Statistical Models and Methods for Lifetime Data
LAWSON · Statistical Methods in Spatial Epidemiology
LE · Applied Categorical Data Analysis
LE · Applied Survival Analysis
LEE · Statistical Methods for Survival Data Analysis, *Second Edition*
LePAGE and BILLARD · Exploring the Limits of Bootstrap
LEYLAND and GOLDSTEIN (editors) · Multilevel Modelling of Health Statistics
LIAO · Statistical Group Comparison
LINDVALL · Lectures on the Coupling Method
LINHART and ZUCCHINI · Model Selection
LITTLE and RUBIN · Statistical Analysis with Missing Data
LLOYD · The Statistical Analysis of Categorical Data
MAGNUS and NEUDECKER · Matrix Differential Calculus with Applications in Statistics and Econometrics, *Revised Edition*
MALLER and ZHOU · Survival Analysis with Long Term Survivors
MALLOWS · Design, Data, and Analysis by Some Friends of Cuthbert Daniel
MANN, SCHAFER, and SINGPURWALLA · Methods for Statistical Analysis of Reliability and Life Data
MANTON, WOODBURY, and TOLLEY · Statistical Applications Using Fuzzy Sets
MARDIA and JUPP · Directional Statistics
MASON, GUNST, and HESS · Statistical Design and Analysis of Experiments with Applications to Engineering and Science
McCULLOCH and SEARLE · Generalized, Linear, and Mixed Models
McFADDEN · Management of Data in Clinical Trials
McLACHLAN · Discriminant Analysis and Statistical Pattern Recognition
McLACHLAN and KRISHNAN · The EM Algorithm and Extensions
McLACHLAN and PEEL · Finite Mixture Models
McNEIL · Epidemiological Research Methods
MEEKER and ESCOBAR · Statistical Methods for Reliability Data
MEERSCHAERT and SCHEFFLER · Limit Distributions for Sums of Independent Random Vectors: Heavy Tails in Theory and Practice
*MILLER · Survival Analysis, *Second Edition*
MONTGOMERY, PECK, and VINING · Introduction to Linear Regression Analysis, *Third Edition*
MORGENTHALER and TUKEY · Configural Polysampling: A Route to Practical Robustness
MUIRHEAD · Aspects of Multivariate Statistical Theory
MÜLLER and STOYAN · Comparison Methods for Stochastic Models and Risks
MURRAY · X-STAT 2.0 Statistical Experimentation, Design Data Analysis, and Nonlinear Optimization
MYERS and MONTGOMERY · Response Surface Methodology: Process and Product Optimization Using Designed Experiments, *Second Edition*
MYERS, MONTGOMERY, and VINING · Generalized Linear Models. With Applications in Engineering and the Sciences
NELSON · Accelerated Testing, Statistical Models. Test Plans, and Data Analyses
NELSON · Applied Life Data Analysis

*Now available in a lower priced paperback edition in the Wiley Classics Library.